AF606396

Two w[illegible]k
loar[illegible]

Please return on or before the last
date stamped below.
Charges are made for late return.

MICROBIAL BIOPESTICIDES

MICROBIAL BIOPESTICIDES

Edited by

Opender Koul

Insect Biopesticide Research Centre
Jalandhar, India

and

G.S. Dhaliwal

Department of Entomology
Punjab Agricultural University
Ludhiana, India

London and New York

602.96
M

First published 2002 by Taylor & Francis
11 New Fetter Lane, London EC4P 4EE

Simultaneously published in the USA and Canada
by Taylor & Francis Inc,
29 West 35th Street, New York, NY 10001

Taylor & Francis is an imprint of the Taylor & Francis Group

© 2002 Taylor & Francis

Typeset in Arial Narrow by EXPO Holdings, Malaysia
Printed and bound in Great Britain by TJ International Ltd, Padstow, Cornwall

All rights reserved. No part of this book may be reprinted or reproduced or utilised in any form or by any electronic, mechanical, or other means, now known or hereafter invented, including photocopying and recording, or in any information storage or retrieval system, without permission in writing from the publishers.

Every effort has been made to ensure that the advice and information in this book is true and accurate at the time of going to press. However, neither the publisher nor the authors can accept any legal responsibility or liability for any errors or omissions that may be made. In the case of drug administration, any medical procedure or the use of technical equipment mentioned within this book, you are strongly advised to consult the manufacturer's guidelines.

British Library Cataloguing in Publication Data
A catalogue record for this book is available from the British Library

Library of Congress Cataloging in Publication Data
A catalog record for this book has been requested.

ISBN 0-415-27213-0

CONTENTS

PREFACE

In order to understand whether biotechnology is making a positive contribution to integrated pest management, it is useful to review the pest control biotechnologies presently available or in development. One focus of biotechnological research has been on improving natural enemies of pests as pest control agents. This has focused principally on pathogens of insect pests and their use as formulated biological pesticides. Emphasis has been placed on bacteria and viruses, largely because they are better understood and more easily manipulated, as opposed to fungi, protozoa, nematodes, etc. A key advantage of biological agents relative to chemical pesticides is their capacity both to kill pests and reproduce at the expense of pest thereby giving some control in the future pest generations.

The most common biological control has focused on insect pathogen *Bacillus thuringiensis* (Bt), which has been the principal target of product development, because it is less harmful to predators and parasitoids than broad-spectrum insecticides. Other insect pathogens are better adapted to having a continuous impact on pests in crops, such as viruses, fungi, nematodes and protozoa, which can cause continuing outbreaks and suppress pests under natural conditions. However, these organisms are as yet little developed as biopesticides. Insect viruses have a market in their natural form as biopesticides, mostly against caterpillar pests of forestry and field crops. Biotechnological research has focused on engineering of certain viruses to express genes whose toxins kill faster than the wild-type viruses.

This volume is designed to provide comprehensive treatment of microbials as biopesticides and various advances made in this direction in recent years. We have attempted to address the use of bacterial, fungal, viral and nematode based biopesticides in a balanced and complementary manner. We have also tried to focus on advantages and disadvantages of such materials along with their role in genetic engineering, because one has to ensure that environmental persistence of engineered organisms is not encouraged.

We are pleased to have contributors from various parts of the world, and who are very well known in their respective fields of biopesticides. We hope that this volume will provide a lot of material for further research exploration and discussion.

The volume has 9 chapters which deal with antiinsectan compounds from microorganisms, microbial biopesticides developed as inducible plant defensive systems, baculoviruses, their pathogenesis and role in IPM and also the impact of genetic engineering and tissue culture for their efficient utilization in insect control. Chapters on nematodes, fungi and bioherbicides have been included to consider specific problems in their utilization as biopesticides.

We wish to thank all the authors for their efforts and perseverance, as well as the many peer reviewers whose comments and suggestions invariably led to the improvement of chapters.

Opender Koul
G.S. Dhaliwal

Contributors

Dr Basil Arif
Great Lakes Forestry Center
Canadian Forest Service
Sault Ste. Marie
Ontario P6A 5M7
Canada

Dr Ramesh Arora
Department of Entomology
Punjab Agricultural University
Ludhiana 141 004
India

Dr Salvatore Arpaia
Metapontum Agrobios
S.S. 106 Jonica Km 448.2
1-75010, Metaponto (MT)
Italy

Dr John W. Barrett
Viral Immunology and Pathology
The John P. Robarts Research Institute
1400 Western Road
London, Ontario N6G 2V4
Canada

Dr G.S. Battu
Department of Entomology
Punjab Agricultural University
Ludhiana 141 004
India

Dr G.S. Dhaliwal
Department of Entomology
Punjab Agricultural University
Ludhiana 141 004
India

Dr Patrick F. Dowd
Bioactive Agents Research Unit
National Center for Agricultural Utilization Research
US Department of Agriculture
Agricultural Research Service
Peoria IL 61604
USA

Dr Harry K. Kaya
Department of Nematology
University of California
Davis CA 95616
USA

Dr G.G. Khachatourians
Bioinsecticide Research Laboratory
Department of Applied Microbiology and Food Science
College of Agriculture
University of Saskatchewan
Saskatoon S7N 5A8
Canada

Dr Albrecht M. Koppenhöfer
Department of Entomology
Rutgers University
New Brunswick NJ 08901
USA

Dr Opender Koul
Insect Biopesticide Research Centre
30 Parkash Nagar
Model Town
Jalandhar 144 003
India

Dr Robert J. Kremer
US Department of Agriculture
Agricultural Research Service
Cropping Systems and Water Quality Unit
and Department of Soil and Atmospheric Sciences
University of Missouri
Columbia MS 65211
USA

Dr Giuseppe Mennella
Istituto Sperimentale per l'Orticoltura
PO Box 48
I-84094 Pontecagnano (SA)
Italy

Dr G.S. Miranpuri
Department of Medical Microbiology and Immunology
436 Service Memorial Institute
1300 University Avenue
University of Wisconsin-Madison
Madison WI 53706
USA

Dr K. Narayanan
Project Directorate of Biological Control
PB No. 2491
Hebbal, Bangalore 560 024
India

Dr Subba Reddy Palli
Great Lakes Forestry Center
Canadian Forest Service
Sault Ste. Marie
Ontario P6A 5M7
Canada

Dr Mark Primavera
Great Lakes Forestry Center
Canadian Forest Service
Sault Ste. Marie
Ontario P6A 5M7
Canada

Dr Arthur Retnakaran
Great Lakes Forestry Center
Canadian Forest Service
Sault Ste. Marie
Ontario P6A 5M7
Canada

Dr Giuseppe L. Rotino
Istituto Sperimentale per l'Orticoltura
Via Paullese 28
I-26836 Montanaso Lombardo (LO)
Italy

Dr Francesco Sunseri
Dipartimento di Biologia
Difesa e Biotechnologie Agroforestali
Università degli Studi della Basilicata
Contrada Macchia Romana
I-85100 Potenza
Italy

Dr E.P. Valencia
Aventis Crop Science
Carrera 77A # 45-61
Bogota
Columbia

MICROBIAL BIOPESTICIDES: AN INTRODUCTION

1

Opender Koul[1] and G.S. Dhaliwal[2]
[1]Insect Biopesticide Research Centre, 30 Parkash Nagar, Model Town, Jalandhar 144 003, India, [2]Department of Entomology, Punjab Agricultural University, Ludhiana 141 004, India

Although microbial insecticides account for <1 per cent of global pesticide sales, microbial control of pests is gaining in importance. This is due to the improved performance and cost competitiveness of microbials in addition to increasing arthropod resistance to chemical insecticides. The use of microbial insecticides is growing at a rapid rate of 10–25 per cent per year. *Bacillus thuringiensis* has been the principle target of product development and accounts for most sales in US $75 million global market for biological control products (Stanes *et al.*, 1993).

In fact, microbial biopesticides provide a resource for discovering novel agents effective in controlling pests detrimental to agriculture, public health and forestry. Basic developments have been achieved in *B. thuringiensis*, *B. sphaericus*, entomopathogenic fungi and baculoviruses that may have a far reaching effect on growth of biopesticides.

As a product, *B. thuringiensis* (Bt) is valuable in IPM systems because it is much less harmful to predators and parasitoids than broad spectrum chemical pesticides, thus having potential to substitute for chemical products in "insecticide treadmill" situations and will allow the recovery of natural enemy populations. Like many biopesticides, it is often less effective on its own than a highly potent chemical product. However, in an IPM system, where it is used only when required thereby conserving natural enemies, its impact is augmented by the action of those natural enemies and can be economical and sustainable.

Bacillus thuringiensis (Bt)

Bacillus thuringiensis (Bt), a Gram-positive, motile, rod shaped bacterium produces a parasporal crystal composed of one or more proteins. When an insect ingests bacteria, the protoxin is first proteolyzed to an activated toxin protein that diffuses through the peritrophic membrane and binds to receptors on the midgut epithelium. The gut becomes paralyzed and the insect stops feeding. Once the activated toxin binds, the putative receptor-toxin interaction becomes irreversible, and the toxin is believed to insert into the membrane, causing a lesion or pore to form. The pore formation disrupts the potassium ion gradient, leading to microvilli swelling and destruction. The insect gut integrity is compromised, and the gut contents leak into the haemocoel. Spores germinate and the insect succumbs to a lethal septicemia (Starnes *et al.*, 1993), the process is comprehensively dealt with in chapter 3 of this volume.

Although Bt was first observed by Ishiwata in Japan in 1902, its scientific description was recorded in 1911 by Berliner in Germany. Steinhaus in USA established the commercial potential of Bt in 1950. In fact, from 1948 onwards nearly 9 Bt products have been registered with EPA alone (Table 1.1). The discovery of *Bt. tenebrionis* opened up new vistas for strains with novel activity. Strains can be separated based on flagellar, serotype, crystal morphology, SDS-PAGE and plasmid profiling. Several companies have claimed the collection of thousands of Bt isolates from every conceivable habitat. Many new crystal types have been discovered and characterization of genes and proteins from these isolates have resulted in several patents. With the discovery of numerous new toxins, a new classification system based entirely on the protein structure was proposed recently and is increasingly used for classification of new toxins (Table 1.2) (Lee *et al.*, 1998)

In addition to these proteins, Bt possesses other toxins such as α-, β- and γ-exotoxins and the vegetative insecticidal protein (Vip3A) (Estruch *et al.*, 1996).

Despite so many advantages, Bt has some constraints for commercial use. The limited host range is one of the constraints. In order to avoid this, trials have been

Table 1.1 EPA-registered Bt biopesticides

Microorganism	*Year of registration*	*Target pest*
1. *Bacillus popilliae*	1948	Japanese beetle larvae
2. *B. thuringiensis (Bt)*	1961	Lepidopteran larvae
3. *B.t. israelensis*	1981	Mosquito/black fly larvae
4. *B.t. aizawai*	1981	Lepidopteran larvae
5. *B.t. tenebrionis*	1988	Certain beetle larvae
6. *B.t. EG2348*	1989	Gypsy moth larvae
7. *B.t. EG2371*	1989	Lepidopteran larvae
8. *B.t. EG2424*	1990	Lepidopteran/Coleopteran larvae
9. *B.t.* + *Pseudomonas*	1991	Lepidopteran/Coleopteran larvae

Source: Starnes *et al.*, (1993)

Table 1.2 Classification of insecticidal proteins from Bt in respect of various hosts

Lepidopteran	*Dipteran*	*Coleopteran*
Cry1Aa (133.5)	Cry2Aa (70.9)	Cry3Aa (73.1)
Cry1Ab (131.0)	Cry4Aa (134.4)	Cry3Ba (74.2)
Cry1Ac (133.3)	Cry4Ba (127.8)	Cry7Aa (74.4)
Cry1Ba (138.0)	Cry10Aa (77.8)	Cry3Ca (73.8)
Cry1Ca (134.8)	Cry11Aa (72.4)	Cry1Ib (81.2)
Cry1Cb (134.0)	Cyt1Aa (27.4)	
Cry1Da (132.5)	Cyt2Aa (29.2)	
Cry1Fa (133.6)		
Cry9Aa (129.7)		
Cry2Aa (70.9)		
Cry2Ab (70.8)		
Cry1Ib (81.2)		

Source: Lee *et al.*, 1998; Figures in parentheses express mass in kDa.

carried out for controlling the European corn borer, a lepidopteran pest and Colorado potato beetle, through plasmid curing and transfer of a 150-Md plasmid from a *B.t. kurstaki* strain and an 88-Md plasmid from a *B.t. tenebrionis* strain, which were combined into one organism (Starnes *et al.*, 1993).

A second problem facing Bt is the risk of resistance. When Bt is still used as a single technology solution like its chemical predecessors, it is sprayed regularly and a range of insect pests are now developing resistance. In laboratory experiments, the Indian meal moth, *Plodia interpunctella* (Hubner) has developed 30 fold resistance in two generations and after 15 generations this resistance level against *B.t. kurstaki*-HD1 (Dipel) reached 100 times higher than in the untreated insects (McGaughey, 1985). Similarly, field populations of the diamondback moth, *Plutella xylostella* (Linnaeus) has shown 25–33 times less toxicity than susceptible insects (Tabashnik *et al.*, 1990) or *Heliothis virescens* (Fabricus) resistance to Cry1A has also exhibited cross resistance to other Bt toxins that differ significantly in structure and activity (Gould *et al.*, 1992).

The third problem related to Bt is that it lacks the most desirable biological property of a biological control agent, i.e. its ability to reproduce and perpetuate itself in crops. A key advantage of biological agents relative to chemical pesticides is their capacity to both kill pests, i.e. functional response and reproduction at the expense of pest, i.e. numerical response thereby giving some control in the future pest generation. Bt is not adapted to persist in the crops' environment and its commercial development has focused less on preserving its ability to reproduce and spread, but more on maximizing the effects of its insect-killing toxin. In other words, its commercial development has focused on using it like a chemical insecticide and not as a living biological control agent. This is true of most biopesticide development today. It also reflects the fact that the multi-national agrochemical industries, which have dominated biopesticide development, have traditional skills and interests which are limited to the production and marketing of pesticide-like products. However, it is beyond doubt that such commercial organisations have used various molecular techniques to develop products combining genes from various strains to increase the activity. In addition, fusion of genes from baculoviruses with Bt genes is being used to expand the host range of Bt These products are in various stages of development and commercialization.

Only few studies are available concerning the application of Bt to the foliage where the insect feeds. Bt is required to be ingested to act on the insect, therefore, methods of application of the microbe need greater attention. Use of Bt against forestry pests has been well studied (vanFrankenhuyzen, 1993; vanFrankenhuyzen *et al.*, 1997) where optimization of droplet size and distribution in a forest canopy has been recorded, which has improved their use and the control is equivalent to that of the most widely used chemicals.

There are a number of fruit feeding insects that are difficult to kill under field conditions. Accordingly, methods to engineer Bt gene(s) directly into the plant or into plant-colonizing microorganisms to address this problem and such plant genetic engineering programs for insect control are gaining momentum day by day. Maize, cotton, potato and tobacco have been genetically engineered to express the crystal protein (Cry) genes from Bt.

Although there could be a number of disadvantages to engineering Bt genes into plants, such as resistance development and public acceptance of engineered foods, commercial institutions see several advantages in addition to increased efficacy. The current transgenic plants, containing single Bt genes are just first generation plants and will be followed by more sophisticated second and third generation plants with greater flexibility for use in IPM system. A second generation of insect-resistant transgenic plants under development includes both Bt and non-Bt proteins with novel modes-of-action and different spectra of activity against insect pests (Estruch *et al.*, 1997). Future IPM programs will have a combination of genetically engineered and modified Bt microbial products, several types of engineered plants, and traditional Bt products.

However, looking from commercial point of view, to date several million kilograms of *B.t. kurstaki* HD-1 based products are produced annually in the United States and other countries. In the US alone usage is registered for nearly 30 crops and against over 100 pest insect species world-wide (Lee *et al.*, 1998). Similarly, a number of Bt screening trials have identified strains with proteins having a wide host range among lepidopterans, coleopterans and dipteran insect pests. *B.t. israelensis* is now a major commercial product for the control of mosquitoes and black flies which are human disease vectors worldwide. Recent X-ray crystal structure of *B.t. tenebrionis* CryIIIA protein (Li *et al.*, 1991) has opened new vistas for protein engineering and design of Bt insecticides.

Bacillus sphaericus (Bs)

Bacillus sphaericus (Bs) is another promising microbial agent for the control of mosquitoes. During the course of sporulation, Bs produces a parasporal crystalline body that contains several proteins toxic to mosquito larvae (Yap, 1990). Larvae of genus *Culex* are particularly susceptible to these toxins, those of *Anopheles* spp. are moderately susceptible, and larvae of *Aedes* spp. are quite resistant.

Identification of toxic components and sequencing of the cloned DNA (Arapinis *et al.*, 1988; Berry *et al.*, 1991) have established two major toxic proteins of 51- and 42-kDa individually translated from a single bicistronic transcription unit. Upon ingestion of the crystal by mosquito larvae, the 51 kDa and 42 kDa proteins dissolve in the alkaline gut and are processed proteolytically to protein 43 kDa and 39 kDa, respectively. It has been shown that 42 kDa protein did not exert toxicity unless accompanied by the 51 kDa protein (Broadwell *et al.*, 1990). The mechanism of toxic action at molecular level is not understood, but electron microscopy studies indicate that the primary location of action occurs in the cells of the gastric caecum and the posterior midgut where vacuolation and mitochondrial swelling are observed. Using radio labeled 51- and 42 kDa proteins, Davidson *et al.* (1990) failed to rescue the toxicity induced by 42 kDa-protein fed after the 51 kDa protein, if the order of treatment was reversed. An active form of 51 kDa protein remained associated with mid gut tissue for up to 24 hours and was available for interaction with the 42 kDa protein. However, in the absence of 51 kDa protein, the 42 kDa protein may bind to another site on the membrane.

Bs crystal toxin is lethal in low concentrations, persists in the aquatic environment for a long time, and is larvicidal in polluted water. It poses no hazard to non-target organisms, fish, wild life, or humans and provides a new and safe way to control pest vectors breeding in polluted wastewater. The greater residual larvicidal activity of Bs is an apparent advantage over *B.t. israelensis*. However, they are useful complements to each other in the overall strategy for mosquito control.

A limitation with Bs is that it has a limited host range. Even the sensitivity among mosquito species varies markedly which has been attributed to differences in receptor binding affinity and/or the site-of-action (Davidson, 1989). Further field resistance to Bs has been observed both in Brazil (Silva-Filha *et al.*, 1995) and in India (Rao *et al.*, 1995). The mechanism of this resistance is not known. However, high level resistance observed in laboratory experiments is shown to be due to decreased receptor binding in mid gut (Nielson-Leroux *et al.*, 1995).

Entomopathogenic fungi

Association of fungi with insects is well known and some of them do cause serious diseases in the hosts. About 750 fungal species belonging to 56 genera attack terrestrial and aquatic arthropods. The most commonly encountered fungi pathogenic to insects belong to three sub-divisions of the Eumycotina. Entomopathogenic fungi as commercial insect control agents have not made any impact so far. However, the product "Bio-1020" developed by Bayer is based on *Metarrhizium anisopliae* for the control of coleopteran pests on various ornamental crops (McDonald, 1991). Mycar is the product produced by Abbot Laboratories, USA, using *Hirsutella thompsonii* for the control of citrus mite, *Phyllocoptruta oleivora* (Ashmead). There are many other fungal species that are in commercial or experimental production stages in USA, Brazil, UK, India and some other countries. The most common species used are *Beauveria bassiana, Culicinomyces clavisporus, H. thompsonii, M. anisopliae, Nomuraea rileyi, Verticillum lecanii, Colletotrichum gloeosporioides, Cercospora rodmanii, Peniophora gigantea* and *Trichoderma viride* (McDonald, 1991) which have worldwide distribution. These fungi have specific insect targets like Colorado potato beetle, codling moth, European corn borer, pine caterpillar, mosquito larvae, citrus mites, spittle bugs, aphids, green bugs, white fly, thrips, grasshoppers, cockroaches and many other lepidopteran and coleopteran larvae (Zimmerman, 1993). Some recent studies have indicated the potential of *B. bassiana* and *V. lecanii* against English green aphid, *Sitobion avenae* (Fabricius) (Miranpuri and Khachatourians, 1995) and woolly elm aphid, *Eriosoma americanum* (Riley) (Miranpuri and Khachatourians, 1996).

Conidium is the infective unit in entomogenous fungi, which may penetrate the insect cuticle from a combination of mechanical pressure by the germ tube and enzymatic degradation of the cuticle. Strains of these fungi produce proteases and chitinases in liquid culture. It is interesting that proteases are produced first in cultures, which is consistent with the requirement for degradation of the protein matrix before the fungus can degrade chitin. Once through the cuticle, and having overcome host defense mechanisms, the fungus proliferates as hyphal bodies, which multiply by budding. The fungus spreads through the haemocoel and the insect normally dies 3–14 days after spore application. Toxins produced by these fungi also play a significant role in their pathogenicities and a number of such allelochemicals have been isolated and identified (Chapter 2, this volume).

As safety is one of the major factors in the development and practical use of a microbial pesticide, various pathological studies seem to show no adverse effects on man, mammals or plants. Various studies particularly with *M. anisopliae* via injection, inhalation, feeding, eye and skin applications against rats, rabbits, guinea pigs, mice, fish, quails, etc. have shown no pathological effects (Table 1.3).

The effectiveness of fungi in controlling insect pests is dependent on the environmental conditions prevailing after application, particularly with respect to relative humidity. In order to allow fungal growth at sub-optimal relative humidity, it is necessary to develop moisture-retaining formulations. In fact, soil is most suitable environment for biocontrol by fungi because it contains the right moisture content, normally suitable for fungal growth.

Sunlight and temperature are other constraints for use of fungal biopesticides as often-simulated sunlight and temperature inactivate them. Therefore, it is important to select isolates that grow rapidly at ambient temperature after spore application.

In order to improve the effectiveness of fungal pathogens, innovative biotechnological methods could be useful in manipulating desirable traits. Transformation systems and recombinant DNA techniques can give molecular insight of the system to improve genetics of these organisms in order to use them as insect control agents in the future.

Table 1.3 Symptoms of *M. anisopliae* against various animals during safety studies

Test animal	*Mode of action*	*Inference*
Rats	Injection, inhalation, feeding	No toxic effects, acute oral and dermal LD50 > 2000 mg/kg
Mice	Feeding, injection	No toxic or other pathological effects
Guinea pigs	Feeding, inhalation	No toxic effects observed
Fish	Conidia applied to water	No difference in mortality as compared to untreated animals
Quails	Feeding toxicity tests	No adverse effects
Rabbits	Skin and eye applications	No eye irritation or irritation to skin

Source: Zimmermann (1993)

Baculoviruses

Viruses are highly selective biocontrol agents. Baculoviruses are double-stranded DNA viruses very specific to insect hosts. They are rod shaped often embedded within a protective protein coat known as an occlusion body that contributes to their stability in the environment. Baculoviruses are categorized as: (i) nuclear polyhedrosis viruses (NPV) in which many virus particles are present in each occlusion body, composed of a polyhedrin protein, (ii) granulosis virus (GV) in which only one virus particle is present in each occlusion body, composed of granulin protein, and (iii) non-occluded viruses, which are transmitted as free particles. Out of the three categories, first two are ideal for integrated pest management, being safe to humans, the environment and non-target beneficial insects.

Although baculoviruses have been mainly isolated from Hymenoptera and Lepidoptera, few isolates from Diptera, Coleoptera, Neuroptera, Trichoptera and Crustacea are also known. Most of these viruses have been developed for forestry and agriculture purposes only.

Ingestion is the main route of their mode-of-infection. Once the virus is ingested by a lepidopteran larva, the occlusion body protein dissolves in the alkaline (pH 9.0–10.5) gut juice, releasing the virus particles. These particles infect gut cells, replicate and then spread to other organs as non-occluded virus. However, these virus particles become occluded late in the life cycle. When the larvae die, they release into the environment massive amounts of occlusion bodies that may infect more larvae. The time required to kill an insect depends on the species, but this time generally ranges from 3 to 7 days. It can be 3–4 weeks depending on the interactions between the virus, insect, and plant. Insect infection process is described in detail in chapter 6 of this volume.

The first baculovirus to be registered for commercial use was *Helicoverpa zea* (Boddie) NPV (HzNPV). The virus was produced and marketed by Sandoz in 1975 under the trade name Elcar (Ignoffo and Couch, 1981). HzNPV was successfully mass produced in China using semi-automated production techniques and used on cotton, soybean, tobacco, corn, sorghum, and tomato to control *H. armigera*

(Hubner). Hardee and Bell (1997) used HzNPV produced from 8 million *H. zea* larvae to treat 81,000 ha of wild early season host plants. Recently, a novel non-occluded virus was discovered from *H. zea* (Raina and Adams, 1995). The virus named Hz reproductive virus (HzRV) infects only the reproductive tissues of both males and females making them completely sterile. The virus is transmitted through eggs and sperm. Currently, baculoviruses are used on a large scale in Brazil where velvet bean caterpillar NPV protects 5900 ha of soybeans against this pest. Baculoviruses have also been used in Russia and China for many years. While in Russia they have been used for forest pest control, in China they are used for agricultural pests. In Europe, a number of companies have either developed or are developing baculoviruses based products like codling moth GV, European pine saw fly NPV, beet armyworm NPV and alfalfa looper NPV. Due to abundant and inexpensive labour, baculoviruses have been introduced into developing countries, and are thus ideally suited for production. According to Starnes *et al.* (1993) there are 4 strategies for using viral insecticides.

I. The virus spreads from limited applications and permanently regulates the insect population through a classical biological control.
II. An epizootic is established through vertical and horizontal transmission, but reapplication may be necessary because control is not permanent.
III. A vertical inoculum in the environment is conserved and reactivated through environmental manipulation.
IV. Repeated applications are used to control an insect population because there is no horizontal transmission of the virus - a strategy, which is widely used because of its effectiveness.

Baculoviruses are very susceptible to UV rays from sunlight. Several chemical agents, including optical brightners, have been tried as protectants for NPVs (Shapiro, 1992). Nearly 23 optical brightners have been tested and found effective against gypsy moths.

Baculoviruses are being utilized as expression vectors for human recombinant molecular biology. This has helped scientists to work on insertion of specific genes into the baculovirus-genome so that ability of baculovirus to kill early could be achieved. Because of the use of the baculoviruses both as expression system and insecticide, improvement in production *in vivo* and *in vitro* has been an effort of research. However, *in vitro* systems have not proved to be very fruitful because of lower yield, high costs, and sensitivity of cells to growth conditions, mutations and maintaining sterility.

Slow rate of kill remains a limitation with baculoviruses. The aim of genetic engineering of baculoviruses for use as insecticides is to combine the pathogenicity of the virus with the insecticidal action of a toxin, hormone or enzyme (Bonning and Hammock, 1996). In such cases, the polyhedrin gene sequences from the viral genome can be replaced with coding region of the foreign gene which will be under the strong polyhedrin gene promoter. In several cases this approach appears to be highly feasible (Maeda *et al.*, 1991; Hammock *et al.*, 1990; Stewart *et al.*, 1991; Tomalski and Miller, 1991). Among many recent developments is the expression of a scorpion toxin (AaIT) under the control of the promoter from an immediate early gene (iel) of AcMNPV causing earlier expression of the toxin (Jarvis *et al.*, 1996).

Other ways to reduce kill time could be improvement in formulations and application systems. Further reduction in production costs, optimization of field performance and overcoming of regulatory obstacles is necessary to enable the development of more economical and efficacious products. Environmental concerns associated with commercialization of recombinant viruses such as, interaction with vertebrates, beneficial and non-target species, displacement of natural viruses, and genetic stability of recombinant organisms still exists. Several modifications like polyhedrin-minus, co-occluded and pre-occluded viruses have been developed to avoid persistence of recombinant viruses in the nature (Wood, 1995).

Alongside efficacy, the production by fermentation is the other major technical challenge. Productivity will need to be improved substantially over current technology. In order to compete in major markets, large-scale production employing about 100,000 litre fermenters is the need of the day (Rice *et al.*, 1998). Culturing insect host cells on this scale is new territory with no guarantee of success. Also, whether a baculovirus product, which has been genetically modified will obtain a regulatory approval, remains a question to be answered.

Protozoa

Entomopathogenic protozoans occur in all the major groups of protozoa, but microsporidia is the major class of such pathogens. Protozoa based biopesticides have not made much impact so far and not many studies and products are available. Only registered material is *Nosema locustae* for grasshoppers, which was registered by EPA in 1980 (Colon, 1980). It is applied in a bait formulation and is used primarily for treatment of areas sensitive to chemical insecticides, for residual control in rangelands, and for home garden use (Henery, 1971). However, research is required to determine the earliest detectable stage, the detection threshold in grasshoppers, and a suitable method of sampling for expedient detection of the infection. Such research should facilitate the evaluation of *N. locustae* infection following field applications of the protozoan (Knoblett and Youssef, 1996).

For effective utilization of protozoan pathogens, molecular level characterization of species and intra-generic relativeness is required, which will help in proper identification of species, their mass production, genetic manipulation and subsequently proper utilization as microbial control agents.

Future of microbials

Although microbials are useful organisms for pest management, there are key constraints, which are a matter of continued focus in future. Most important amongst these are the regulatory constraints, particularly for engineered microbials where data requirements show redundancy. Adequate basic research is required at industrial level rather than at academic level, because there is currently a gap in cost-effective production technologies for fungi and viruses, economic incentives for development of scale up technologies and application technologies also need to be critically investigated.

Use of microbials also requires considerable attention towards pest behaviour, biology and population dynamics. One has to have a clear concept of differences between chemical and biological products and their usage methods. Therefore, a realistic picture of technologies needs to have complete transparency in usage. Many aspects of the candidate subject have been comprehensively dealt with in subsequent chapters of this book.

References

Arapinis, C., De La Torre, F. and Szulmajster, J. (1988) Nucleotide and deduced amino acid sequence of the *Bacillus sphaericus* 1593M gene encoding a 51.4 kDa polypeptide which acts synergistically with the 42 kDa protein for expression of the larvicidal toxin. *Nucleic Acids Res.*, **16**, 7731.

Berry, C., Hindley, J. and Oei, C. (1991) The *Bacillus sphaericus* toxins and their potential for biotechnological development. In K. Maramorosch (ed.), *Biotechnology for Biological Control of Pests and Vectors*, CRC Press, Boca Raton, Florida, pp. 35–51.

Bonning, B.C. and Hammock, B.D. (1996) Development of recombinant baculoviruses for insect control. *Ann. Rev. Entomol.*, **41**, 191–210.

Broadwell, A.H., Baumann, L. and Baumann, P. (1990) The 42- and 51-kilodalton mosquitocidal proteins from *Bacillus sphaericus* 2362: construction of recombinants with enhanced expression and *in vivo* studies of processing and toxicity. *J. Bacteriol.*, **172**, 2217–2223.

Colon, J.M. (1980) *Nosema locustae* exemption from requirement of tolerance. *Fed. Reg.*, **45**, 31312–31313.

Davidson, E.W. (1989) Variation in binding of *Bacillus sphaericus* toxin and wheat germ agglutinin to larval midgut cells of six species of mosquitoes. *J. Invertebr. Pathol.*, **53**, 251–259.

Davidson, E.W., Oei, C., Meyer, M., Bieber, A.L., Hindley, J. and Berry, C. (1990) Interaction of the *Bacillus sphaericus* mosquito larvicidal proteins. *Can. J. Microbiol.*, **36**, 870–878.

Estruch, J.J., Warren, G.W., Mullins, M.A., Nye, G.J., Craig, J.A. and Koziel, M.G. (1996) Vip3A, a novel *Bacillus thuringiensis* vegetative insecticidal protein with a wide spectrum of activities against lepidopteran insects. *Proc. Natl. Acad. Sci., USA,* **88**, 5389–5394.

Estruch, J.J., Carozzi, N.B., Desai, N., Duck, N.B., Warren, G.W. and Koziel, M.G. (1997) Transgenic plants: an emerging approach to pest control. *Nature Biotech*, **15**, 137–141.

Gould, F., Martinez-Ramirez, A., Anderson, A., Ferre, J., Silva, F.J. and Moar, W.J. (1992) Broad spectrum resistance to *Bacillus thuringiensis* toxins in *Heliothis virescens. Proc. Natl. Acad. Sci., USA*, **89**, 7986–7990.

Hammock, B.D., Bonning, B.C., Possee, R.D., Hanzlik, T.N. and Maeda, S. (1990) Expression and effects of the juvenile hormone esterase in a baculovirus vector. *Nature,* **344**, 458–461.

Hardee, D.D. and Bell, M.R. (1997) Area-wide management of bollworm/budworm with pathogens: results of six-year project and projections for the future. *Recent Res. Devel. Ent.*, **1**, 105–114.

Henery, J.E. (1971) Experimental application of *Nosema locustae* for control of grasshoppers. *J. Invertebr. Pathol.*, **18**, 389–394.

Ignoffo, C.M. and Couch, T.L. (1981) The nucleopolyhedrosis virus of *Heliothis* species as a microbial insecticide. In H.D. Burges (ed.), *Microbial Control of Pests and Plant Diseases 1970–1980*, Academic Press, New York, pp. 329–362.

Jarvis, D.L., Reilly, L.M., Hoover, K., Schultz, C., Hammock, B.D. and Guarino, L.A. (1996) Construction and characterization of immediate early baculovirus pesticides. *Biol. Cont.*, **7**, 228–235.

Knoblett, J.N. and Youssef, N.N. (1996) Detection of *Nosema locustae* (Microsporidia: Nosematidae) in frozen grasshoppers (Orthoptera: Acrididae) by using monoclonal antibodies. *J. Econ. Entomol.*, **89**, 841–847.

Lee, H., Cheong, H. and Gill, S.S. (1998) Microbial control of insects: use of bacterial insecticides. In G.S. Dhaliwal and E.A. Heinrichs (eds.), *Critical Issues in Insect Pest Management*, Commonwealth Publishers, New Delhi, pp. 87–117.

Li, J., Carrol, J. and Ellar, D.J. (1991) Crystal structure of insecticidal δ-endotoxin from *Bacillus thuringiensis* at 2.5 Å resolution. *Nature*, **353**, 815–821.

Maeda, S., Volrath, S.L., Hanzlik, T.N., Harper, S.A., Maddox, D.W., Hammock, B.D. *et al.* (1991) Insecticidal effects of an insect-specific neurotoxin expressed by a recombinant baculovirus. *Virology*, **184**, 777–780.

McDonald, D. (1991) Biopesticides- pesticides with a bright future. *Int. Pest Cont.*, **33**, 33–35.

McGaughey, W.H. (1985) Insect resistance to the biological insecticide *Bacillus thuringiensis. Science*, **229**, 193–195.

Miranpuri, G.S. and Khachatourians, G.G. (1995) Entomopathogenicity of *Beauveria bassiana* (Balsamo) Vuillemin and *Verticillium lecanii* (Zimmerman) toward English grain aphid, *Sitobion avenae* (Fab.) (Homoptera:Aphididae). *J. Insect Sci.*, **8**, 34–39.

Miranpuri, G.S. and Khachatourians, G.G. (1996) Bionomics and fungal control of woolly elm aphid, *Erisoma americanum* (Riley) (Eriosomatidae:Homoptera) on Saskatoon berry, *Amelanchier alinifolia. J. Insect Sci.*, **9**, 33–37.

Nielsen-Leroux, C., Charles, J.-F., Thiery, I. and Georghiou, G.P. (1995) Resistance in a laboratory population of *Culex quinquifasciatus* (Diptera: Culicidae) to *Bacillus sphaericus* binary toxin is due to a change in the receptor on midgut brush-border membrane. *Eur. J. Biochem.*, **228**, 206–210.

Raina, A.K. and Adams, J.R. (1995) Gonad specific virus of corn earworm. *Nature*, **374**, 770.

Rao, D.R., Mani, T.R., Rajendran, R. and Joseph, A.S. (1995) Development of high-level of resistance to *Bacillus sphaericus* in a field population of *Culex quinquefasciatus* from Kochi, India. *J. Am. Mosq. Cont. Assoc.*, **11**, 1–5.

Rice, M.J., Legg, M. and Powell, K.A. (1998) Natural products in agriculture- A view from the industry. *Pestic. Sci.*, **52**, 184–188.

Shapiro, M. (1992) Use of optical brightners as radiation protectants for gypsy moth (Lepidoptera: Lymantridae) nuclear polyhedrosis virus. *J. Econ. Entomol.*, **85**, 1682–1686.

Silva-Filha, M.-H., Regis, L., Nielsen-Leroux, C. and Charles, J.-F (1995) Low level resistance to *Bacillus sphaericus* in a field-treated population of *Culex quinquefasciatus* (Diptera: Culicidae). *J. Econ. Entomol.*, **88**, 525–530.

Starnes, R.L., Liu, C.L. and Marrone, P.G. (1993) History, use and future of microbial insecticides. *Am. Entomol.*, **39**, 83–91.

Stewart, L.M.D., Hirst, M., Ferber, M.L., Merryweather, A.T., Cayley, P.J. and Possee, R.D. (1991) Construction of an improved baculovirus insecticide containing an insect-specific toxin gene. *Nature*, **352**, 85–88.

Tabashnik, B.E., Cushing, N.L., Finson, N. and Johnson, M.W. (1990) Field development of resistance to *Bacillus thuringiensis* in diamondback moth (Lepidoptera: Plutellidae). *J. Econ. Entomol.*, **83**, 1671–1676.

Tomalski, M.D. and Miller, L.K. (1991) Insect paralysis by baculovirus-mediated expression of a mite neurotoxin gene. *Nature*, **352**, 82–85.

vanFrankenhuyzen, K. (1993) The challenge of *Bacillus thuringiensis*. In P.F. Entwistle, J.S. Cory, M.J. Bailey and S. Higgs (eds.), *Bacillus thuringiensis, an Environmental Biopesticide: Theory and Practice*, Wiley, New York, pp. 1–35.

vanFrankenhuyzen, K., Gringorten, L., Dedes, J. and Gauthier, D. (1997) Susceptibility of different instars of the spruce budworm (Lepidoptera: Tortricidae) to *Bacillus thuringiensis* var. *kurstaki* estimated with a droplet-feeding method. *J. Econ. Entomol.*, **90**, 560–565.

Wood, H.A. (1995) Genetically enhanced baculovirus insecticides. In M. Gunasekaran and D.J. Weber (eds.), *Molecular Biology of the Biological Control of Pests and Diseases of Plants*, CRC Press, Boca Raton, Florida, pp. 91–104.

Yap, H.H. (1990) Field trials of *Bacillus sphaericus* for mosquito control. In H.de Barjac and D.J. Sutherland (eds.), *Bacterial Control of Mosquitoes and Black Flies*, Rutgers University Press, New Brunswick, pp. 307–320.

Zimmermann, G. (1993) The entomopathogenic fungus *Metarhizium anisopliae* and its potential as a biocontrol agent. *Pestic. Sci.*, **37**, 375–379.

Antiinsectan Compounds Derived from Microorganisms

2

Patrick F. Dowd
Bioactive Agents Research Unit, National Center for Agricultural Utilization Research, US Department of Agriculture, Agricultural Research Service, Peoria IL 61604, USA

Introduction

Microorganisms comprise a diversity of organisms from radically different evolutionary background - Archaebacteria, Eubacteria and Eucaryota. The Archaebacteria include many organisms that live in extreme environments, such as very high temperatures. The Eubacteria include more "common" bacteria and filamentous bacteria such as actinomycetes. The eucaryotes include fungi and yeasts. These organisms can inhabit a diversity of habitats, including hot springs, subfreezing areas, deserts and oceans. As would be expected from the diverse organisms, a diversity of metabolites has been isolated, but relatively few groups have been studied in much detail for antiinsectan compounds. Of the groups studied, fungi have been the source of most of the secondary metabolites with activity against insects, followed by the actinomycetes. Bacteria, disproportionately represented by *Bacillus thuringiensis*, are probably the source of most antiinsectan proteins.

The sources of antiinsectan compounds reflect emphasis in past research areas by Government, university, and industry. For example, fermentation broths provide a readily screenable source of bioactivity for industry, especially when coupled with assays against other organisms or targets of agricultural or pharmaceutical interest. Although the ultimate goal may be to chemically synthesize the active compound (or compounds derived from an initial lead structure), derivation from a readily produced liquid fermentation potentially allows commercial scaleup if desired (e.g. avermectins, spinosyns, etc.). Combinatorial chemistry allows for rapid production of analogs of a lead structure, and multi-well plate assays used in conjunction with detectors such as microplate readers allow for rapid throughput (e.g. Borman, 1999). Due to the success of engineering plants to express modified forms of *B. thuringiensis* proteins, coupled with their somewhat limited efficacy range, bacterial fermentation screenings have been used to discover proteins with activity against insects (Purcell, 1997; Warren, 1997).

Because insecticides have been derived from compounds initially recognized as having activity against vertebrates (such as organophosphates and carbamates), compounds with vertebrate activity have been investigated against insects for both comparative purposes and to discover lead structures for activity against insects that can be made selective through derivatization. Primary among these compounds are the fungus derived "mycotoxins".

Ecological approaches have also been used to identify microbial compounds with activity against insects. The main areas of investigation have been insect pathogens, long term survival structures and plant endophytes. Some of this overlaps with the industry screening methods just described, since *B. thuringiensis* is an insect pathogen. Both secondary metabolites and proteins produced by insect pathogens have been intensively investigated (e.g. Claydon and Grove, 1982; Krasnoff *et al.*, 1996; Roberts, 1981; Warren, 1997). Long term survival structures, primarily mushrooms and sclerotia produced by fungi, have been investigated with some success for antiinsectan compounds (Daniewski *et al.*, 1995; Wicklow *et al.*, 1994). The logic used has been the same as was originally advanced for long term/relatively more important tissues from plants such as seeds (Janzen, 1977) with further refinement (Dowd, 1992b; Gloer, 1995a,b; Miller, 1986; Wicklow, 1984, 1988).

Because similar structures from different sources have been discovered through different approaches, and because previous discussions have involved an organismal and/or rationale-oriented presentation of antiinsectan compounds (e.g. Dowd, 1992b; Gloer, 1995a,b; Huang and Shapiro, 1971; Claydon and Grove, 1982; Miller, 1986; Roberts, 1981; Wicklow *et al.*, 1994), information in this review will be primarily grouped alphabetically by structural similarity, with division into bacterial, actinomycete and fungal compounds for secondary compounds. However, cyclic peptides will be treated separately. This organization should also help reduce repetitiveness when the same compound(s) are produced by diverse organisms. In addition, demonstrated/postulated modes of action will also be presented, along with QSAR information where available. Compounds that act in synergistic manner or occur in combinations from the same organisms will also be highlighted in a separate section. Due to the large volume of literature involved, only studies where apparently pure compounds have been used will typically be included, unless some novel source or mode of action is apparently involved. In cases where prior reviews of compounds have appeared, discussion will be more limited. Past reviews will also be utilized as sources of distribution of compounds (e.g. Cole and Cox, 1981; Turner, 1971; Turner and Aldridge, 1983). Types of abbreviations used include RGR for reduction in growth rate (based on weights of treated vs. control insects) and RFR for reduction in feeding rate (based on lower amounts of feeding by insects exposed to treated vs. control diets). In addition to reviews previously mentioned, other reviews have also been compiled and may present different perspectives (e.g. Ando, 1983; Ciegler, 1976; Cutler, 1987; Dev and Koul, 1997; Heisey *et al.*, 1988b; Mishra *et al.*, 1987; Wright *et al.*, 1981)

Antiinsectan allelochemicals

There are many allelochemicals with molecular weights less than 1000 that have demonstrated activity against insects. Many of these are fungus-derived and have been tested against the corn earworm, *Helicoverpa zea* (Boddie). Relative toxicity of these compounds run in the same oral assay (Dowd, 1988b) varies from significant mortality at less than 100 ppm to slight reductions in growth rates at several thousand ppm. However, activity at high concentrations may be biologically relevant to the concentration found in the source material. Extracts from producing organisms such as *Amanita muscarina* have been used as fly killers since the medieval times (Dickinson and Lucas, 1979).

Antiinsectan compounds derived from nonfilamentous bacteria

Alkyltetrahydrofuran acids

The compound 5-*n*-hexyl-tetrahydrofuran-2-acetic acid is a bioconversion product of 12-hydroxyoctadecanoic acid produced by *Bacillus lentus* (Huang *et al.*, 1996). This compound appeared to have juvenile hormone-like activity against pupae of *Carpophilus hemipterus* (Linnaeus), which show 15 per cent pupal-adult intermediates after being treated with 0.2 μl of a 1 mg/ml solution of methoprene (Huang *et al.*, 1996). The alkyltetrahydrofuran acid at 0.2 mg of pure material produced 80 per cent pupal-adult intermediates (Huang *et al.*, 1996).

Amino levulinic acid (5-amino-4-oxo-pentanoic acid)

5-(δ)-amino levulinic acid is produced by several bacteria that are phototrophic, such as *Anacystis nidulans, Anabaena variablis, Rhodobacter sphaeroides, R. palustris,*

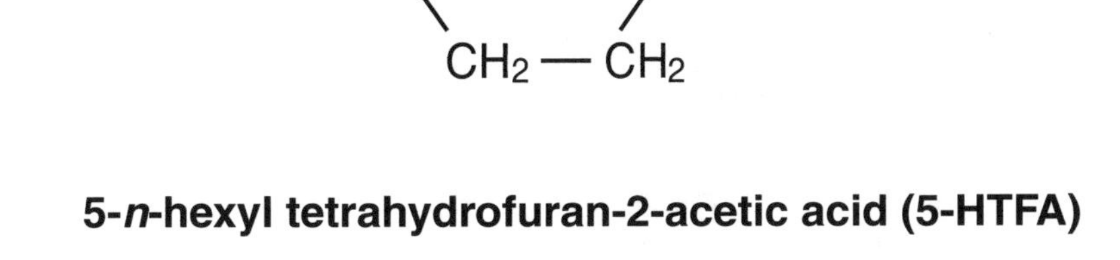

O O
H_3C—$(CH_2)_5$—CH CH—CH_2—C—OH
CH_2—CH_2

5-*n*-hexyl tetrahydrofuran-2-acetic acid (5-HTFA)

$H_2N—CH_2COCH_2CH_2COOH$

5-Amino Levulinic Acid

Chlorobium limicola, and *Chloroflexus aurantiacus* (Sasikala *et al.*, 1994). Chemotrophic bacteria, such as aerobes *Pseudomonas riboflavina* and *Propionibacterium shermanii*, and anaerobes such as *Methanosarcina barkeri*, *Methanobacterium thermoautotrophicum*, and especially *Clostridium thermoaceticum* (Sasikala *et al.*, 1994) also produce it. When administered orally to *H. zea* and *Spodoptera frugiperda* (J. E. Smith) at 250 ppm, 5-aminolevulinic acid produced RGRs of 23 and 14 per cent, respectively (Dowd, 1993). At 40 mM, 5-aminolevulinic acid produced up to 26 per cent mortality of *Trichoplusia ni* (Hubner) larvae after 3 days (Rebeiz *et al.*, 1988). It has been able to synergize the toxicity of porphyric insecticides such as dipyridyl (Rebus *et al.*, 1988) and 1,10-phenanthroline (Rebeiz *et al.*, 1990) as well as griseofulvin (Dowd, 1993), and as such has some activity on its own in its normal role as a protoporphyrin precursor.

Diabroticins

Diabroticins have been isolated from cultures of *Bacillus subtilis* (Stonard *et al.*, 1994). Diabroticins A and B had activity at 2–4 and 25–50 ppm orally, respectively against first instar larvae of *Diabrotica undecimpunctata howardi* Barber (Stonard *et al.*, 1994).

Thiolutin

The symbiotic bacteria *Xenorhabdus* sp. associated with the entomopathogenic nematode *Steinernema* (McInerney *et al.*, 1991) and some actinomycetes (see below)

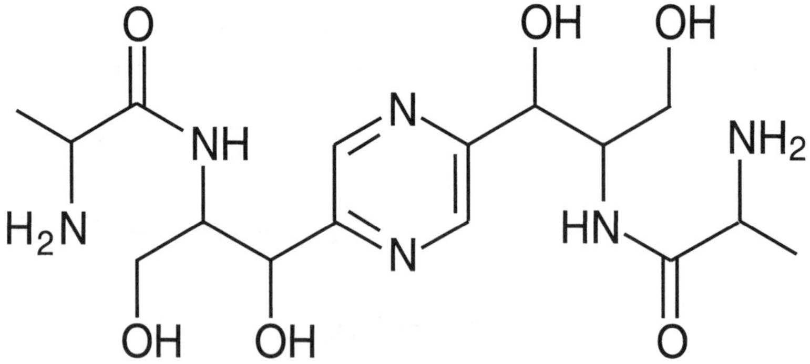

Diabroticin A

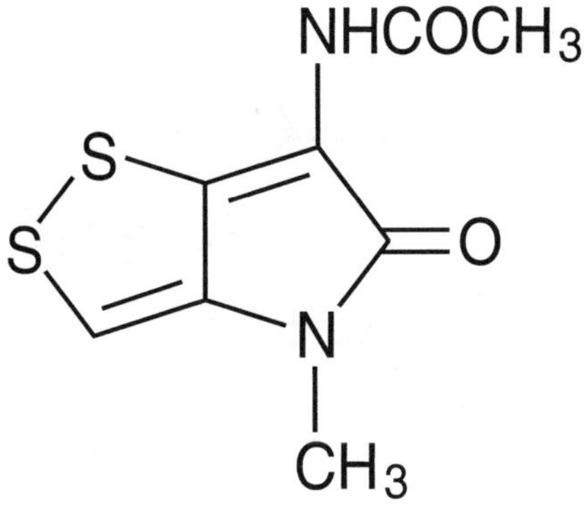

Thiolutin

produce thiolutin. It killed *Lucilia sericata* (Meigen) larvae at unspecified levels (McInerney *et al.*, 1991).

Thuringiensin (*β*-exotoxin)

A comprehensive review of thuringiensin, produced by *B. thuringiensis*, has been published (Sebesta *et al.*, 1981). It is effective against mites and insects in the orders Coleoptera, Diptera, Hymenoptera, Isoptera, Lepidoptera, and Orthoptera (Sebesta *et al.*, 1981). For *Galleria mellonella* (Linnaeus), the injected LD_{50} is 0.5 μg/g, but oral toxicity for most insect species is much lower compared to injection (Sebesta *et al.*, 1981).

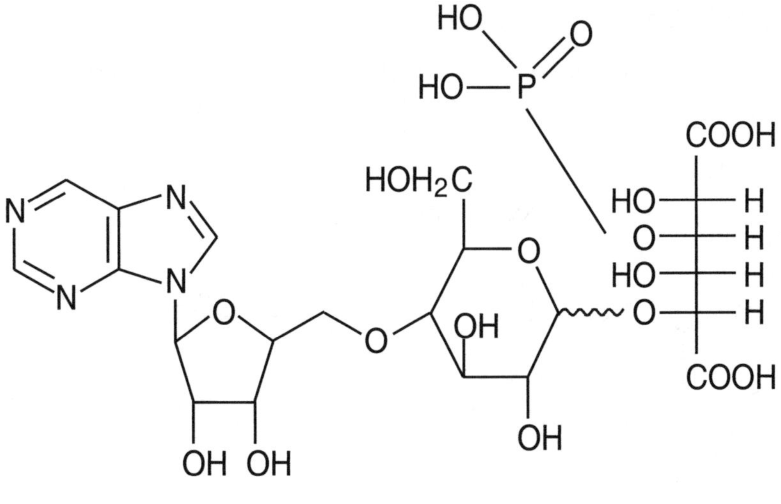

b-Exotoxin

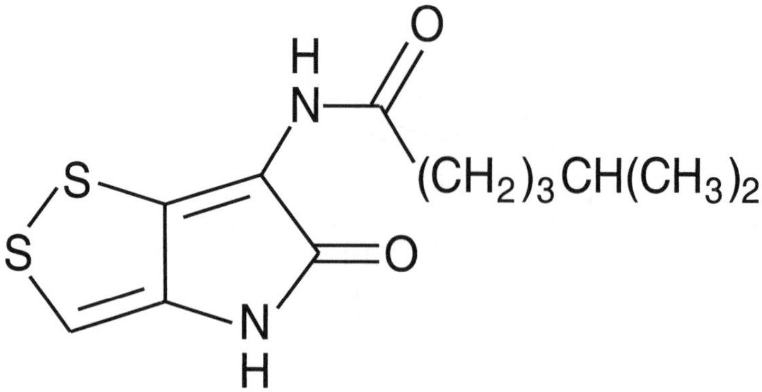

$$\text{(chemical structure of Xenorhabdin 1)}$$

Xenorhabdin 1

Thuringiensin affects the later stages of RNA synthesis, and the dephosphorlyated form is relatively nontoxic (Sebesta *et al.*, 1981).

Xenorhabdins

Xenorhabdins have been isolated from the symbiotic bacteria *Xenorhabdus* spp. that occur with the insect pathogenic nematodes *Steinernema* (McInerney *et al.*, 1991). When tested orally against *Helicoverpa punctigera* (Wallengren), the LD_{50} for xenorhabdin 2 (the only one tested) was 59.5 $\mu g/cm^2$ (McInerney *et al.*, 1991).

Antiinsectan compounds derived from actinomycetes

General antibiotics

An extensive review has appeared on the effects of "antibiotics" that are primarily produced by streptomycetes, on insects (Huang and Shapiro, 1971). Tests of antibi-

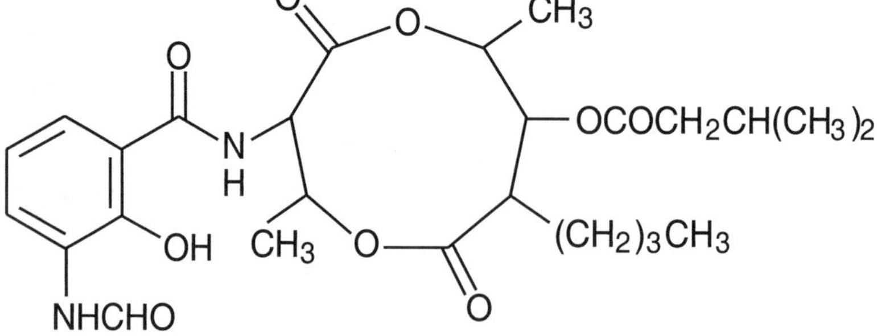

Actinomycin A

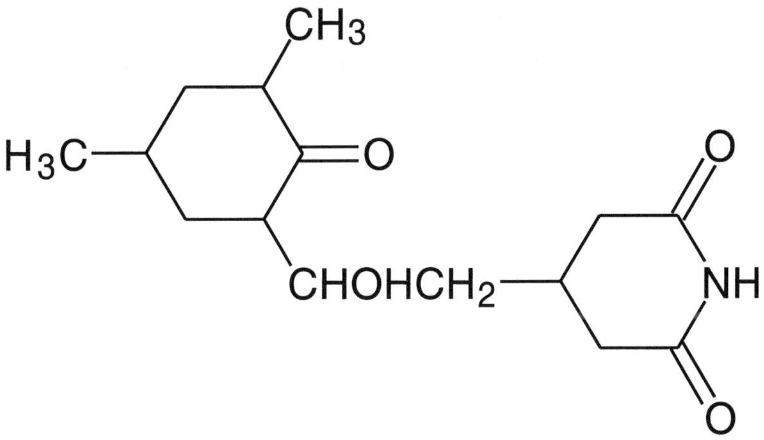

Cycloheximide

otics against insects have been reported since 1945 (Harries, 1967). Among several insect species and antibiotics reviewed (condensed in Table 2.1), the more active were actinomycin A, cycloheximide, and novobiocin although activity among insects was variable (Huang and Shapiro, 1971). The LC_{50} for cycloheximide was 5 ppm for *L. sericata* larvae (Cole and Rolinson, 1972). Novobiocin and actinomycin A inhibit RNA synthesis and cycloheximide inhibits protein synthesis (Jawetz *et al.*, 1982).

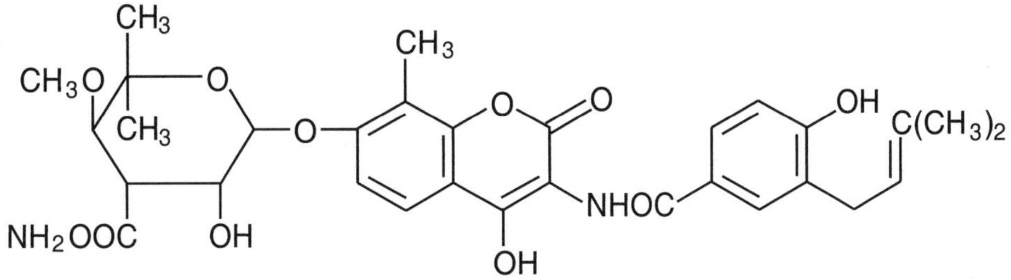

Novobiocin

A204A

A204A, a polyether related to monensin, is produced by *Streptomyces alvus* (Jones *et al.*, 1973). It produced 100 per cent mortality of *Aphis gossypii* Glover (>1000 ppm topical spray), *Epilachna varivestis* Mulsant adults (1000 ppm topical spray), *Oncopeltus fasciatus* (Dallas) (250 ppm topical spray), *Spodoptera eridania* (Cramer)

Table 2.1 Generalized toxicity of antibiotics to insects

Compound	*Lepidoptera*	*Homoptera*	*Coleoptera*	*Orthoptera*	*Mites*
amphotericin	L	L	–	–	M,H
ampicillin	–	L	–	–	M
anisomycin	–	–	–	L	–
anthelmycin	–	M	–	–	H
antimycin A	L	L	M	L,H	M
actinobolin	–	–	–	–	L
bacitracin	L	–	–	L	L
brevicidin	–	L	–	–	–
candidicin	– L	–	–	L	
chloramphenicol	M	–	–	–	–
chloromycetin	–	L	–	–	M
chlortetracycline	–	–	H	–	–
cytovirin	L	L,M,H	–	–	H
cycloheximide	L,H	L,M	–	–	H
erythromycin	L	–	–	–	M
flavensomycin	–	–	–	M	–
fumagillin	L	L	–	–	–
fungichromin	L	L	–	–	–
furacin	–	–	–	L	–
griseoviridin	L	–	–	–	L
hygromycin	L	L,H	–	–	–
kanamycin	L	L	–	–	M
mitomycin	–	M	–	–	M
neomycin	L	–	–	M,H	M
novobiocin	H	L,M,H	–	–	H
nybomycin	– L,M	–	–	–	
nystatin	–	–	–	L	M
oleandromycin	–	M	–	–	M
pactamycin	–	L	–	L	H
penicillin	L	L	–	–	–
oxytetracycline	–	L,M	L,M	–	L
phleomycin	L	–	–	–	–
phytoactin	M	L	–	–	M
polymixin	–	–	–	L,H	M
puromycin	–	L	–	–	H
ristocetin	L	–	–	–	–
sarkomycin	–	L	–	–	M
streptomycin	–	–	L,M	M	–
streptovitacin	L	L,M,H	–	–	M
terramycin	–	–	–	L,M	L
thipyrametin	–	–	–	L	–
trisulfamic	–	–	–	L	–
tylosin	L	–	–	–	L
tyrothricin	L	–	–	–	L
vancomycin	L	–	–	–	L
viomycin	–	–	–	L	–
viridogrisein	–	L	–	–	L
venturicidin	M	–	–	–	–
xanthomycin	–	–	–	H	–

L = low, M = moderate, H = high; relative to other compounds tested on the same species. Multiple entries indicate information on multiple species. Modified from Huang and Shapiro (1971). Most structures are available in Glasby (1976).

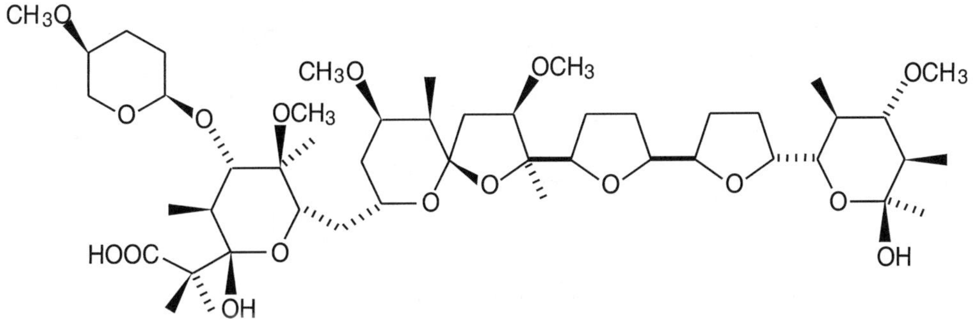

Antibiotic A204A

(100 ppm topical spray), and *Tetranychus urticae* Koch (25 ppm topical spray) (Bauer *et al.*, 1981).

L-alanosine

L-alanosine, L-2-amino-3-(hydroxynitrosamino)-propionic acid, has been isolated from different species of *Streptomyces* including *S. alanosinicus* (Matsumoto *et al.*, 1984). When administered in diets at 5 ppm, L-alanosine inhibited head capsule removal in 50 per cent of *Leucania separata* larvae. Similar effects were noted with *Mamestra brassicae* (Linnaeus) and *Bombyx mori* (Linnaeus) larvae at 10 ppm in diets (Matsumoto *et al.*, 1984).

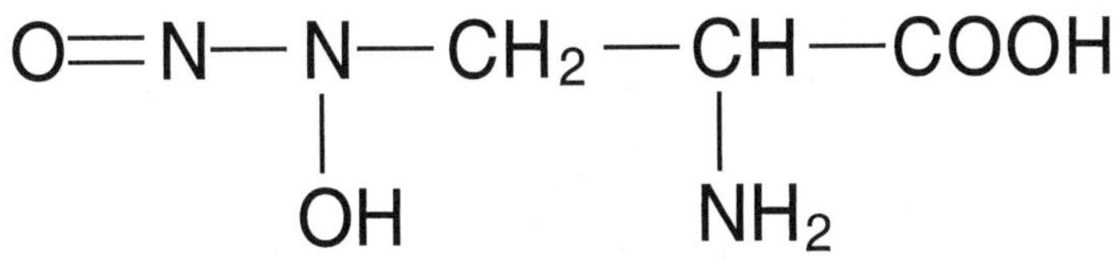

$$O{=}N{-}N(OH){-}CH_2{-}CH(NH_2){-}COOH$$

L-Alanosine

Allosamidin

Allosamidin has been isolated from a *Streptomyces* sp. (Sakuda *et al.*, 1986). When larvae were injected with allosamidin, the EI_{50}s were 2 and 4 μg for *B. mori* and 5th instar *L. separata*, respectively (Sakuda *et al.*, 1987). It inhibits chitinase from *B. mori*, with a Ki of ca. 0.1 μM (Koga *et al.*, 1986). The LC_{50} of allosamidin towards *Camptochironomus tentans* (Fabricius) cell chitinase was 0.46 μM (Spindler and Spindler-Barth, 1994). The diastereoisomer isoallosamidin was about 800 times less active (Spindler and Spindler-Barth, 1994).

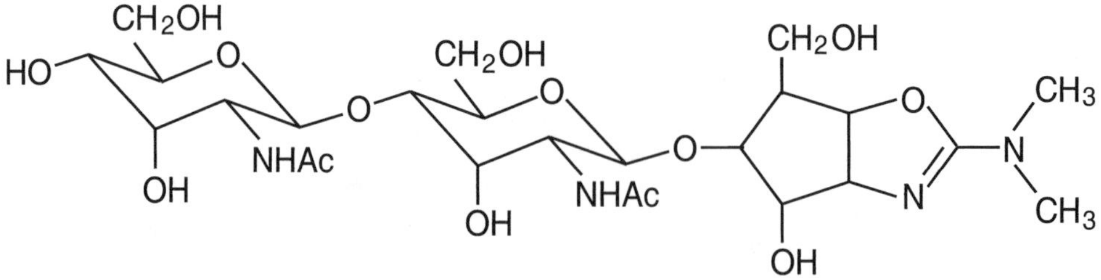

Allosamidin

Antibiotic B-41

Antibiotic B-41 is a macrocyclic lactone (U.S. patent 3,984,563) with unspecified insecticidal and acaricidal activity (Aoki *et al.*, 1976).

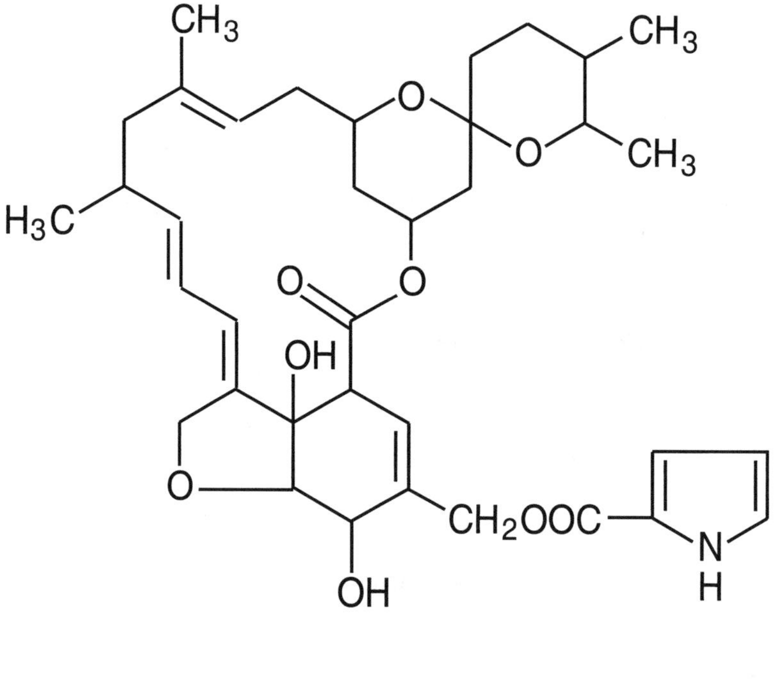

Antibiotic B41

Aplasmomycin

Aplasmomycin is produced by *Streptomyces griseus* (Hokko Chemical Co., 1982). At 100 ppm, it caused 100 per cent mortality of *T. urticae* after 13 days when mites were treated with the solution (Hokko Chemical Co., 1982).

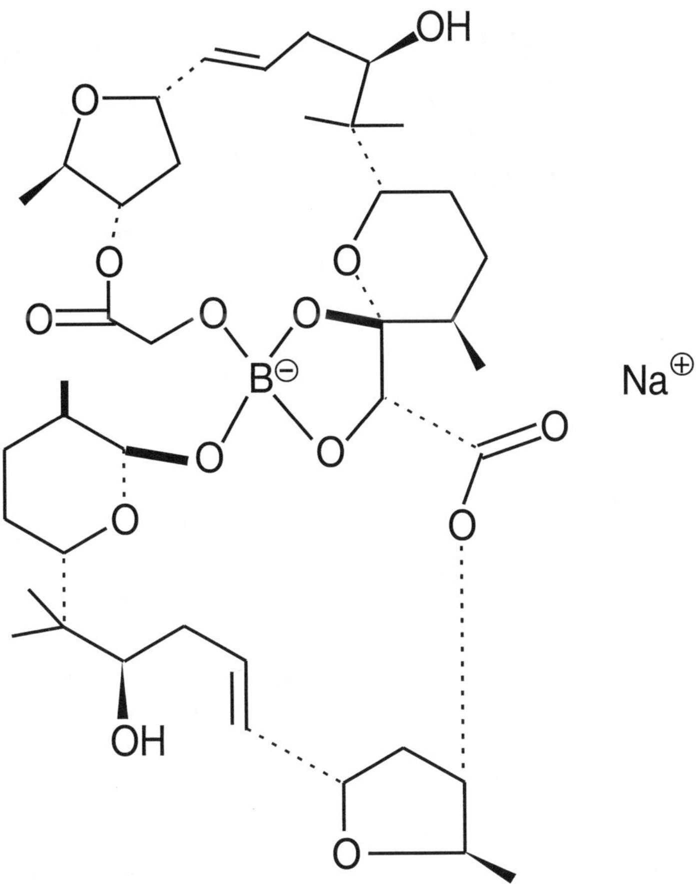

Aplasmomycin

Aromatic nitro compounds

Aromatic nitro compounds were isolated from *Streptomyces griseus* var. *autotrophicus* (Nair *et al.*, 1995). At 6.25 ppm all of the analogs killed 100 per cent of *Aedes aegypti* (Linnaeus) larvae after 24 h (Nair *et al.*, 1995).

Avermectins

Avermectins were originally isolated from the actinomycete *Streptomyces avermitilis* (Strong and Brown, 1987). The insecticidal activity of avermectins has been extensively reviewed (Strong and Brown, 1987; Fisher, 1997; Jansson and Dybas, 1998). Commercial forms are ivermectin (22,23-dihydroavermectin B_1) for livestock pests and abamectin (80% avermectin B_{1a}: 20% avermectin B_{1b}) for horticultural pests (Strong

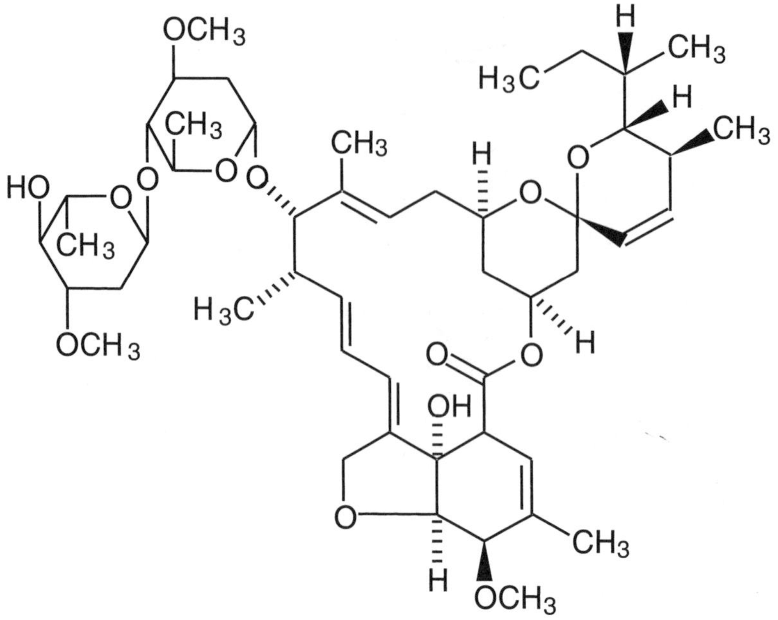

Avermectin B_{1a}

and Brown, 1987). Great differences in efficacy are reported for closely related insects. Although topical toxicity of abamectin to *H. zea* and *Heliothis virescens* (Fabricius) is very similar, oral toxicity is 30-fold lower for *H. zea* compared to *H. virescens* (Bull, 1986). The differences in toxicity for the two species were attributed to differences in GABA receptors, as other parameters such as penetration and metabolism were very similar (Bull, 1986). Reported sublethal effects have included distended cuticle, but antifeedant effects have not clearly been discriminated (Strong and Brown, 1987; Fisher, 1997; Jansson and Dybas, 1998). Structure-activity optimization has been reported (Mrozik, 1994). For example, derivatizing avermectin B_1 with a 4‴- epi-MeNH group decreased the LC_{90} towards *Spodoptera eridania* (Cramer) from 8 to 0.004 ppm (Mrozik, 1994). Avermectins are known to bind specifically to different chloride channel proteins (Mrozik, 1994). They appear to potentiate glutamate and/or GABA to produce chloride ion movement into nerve cells, resulting in irreversible paralysis (Jansson and Dybas, 1998). Abamectin resistance has been reported for *Leptinotarsa decemlineata* (Say) and both oxidative and hydrolytic enzymes appear to be involved (Clark *et al.*, 1992)

Aureothin

Aureothin is produced by *Streptomyces thiolutens* (Oishi *et al.*, 1969). When 1.2 μg was applied topically, 80 per cent of *Callosobruchus chinensis* (Linnaeus) were killed (Oishi *et al.*, 1969). In addition, 1.0 μg applied topically killed 75 per cent of *Pieris*

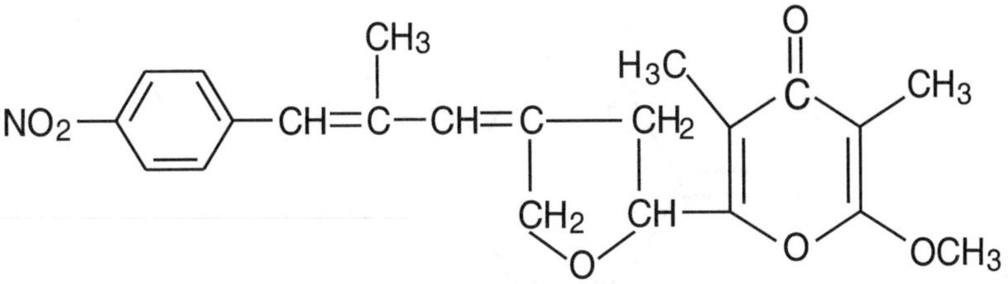

Aureothin

rapae (Linnaeus), while when *Tetranychus kanazawa* Kishida were sprayed with a 60 μg/ml solution, 95 per cent were killed (Oishi *et al.*, 1969).

Bafilomycins

Bafilomycins are produced by *Streptomyces griseus* (Kretschmer *et al.*, 1985). Dietary treatments with 1000 ppm solutions indicated activity against *Ceratitis capitata* (Wiedemann), *Dysdercus intermedius* Distant, *Phaedon cochleariae* (Fabricius), and *Plutella xylostella* (Linnaeus) (Kretschmer *et al.*, 1985). Bafilomycin A_1 apparently disrupts uric acid formation in *Drosophila hydei* by inhibiting a vacuolar ATPase involved in ion (and hence fluid) regulation when injected at 5 μM (Bertram *et al.*, 1991). This work was confirmed using X-ray microanalysis and microelectrodes on Malpighian tubules (Wessing *et al.*, 1993). Bafilomycin A_1 also inhibits vacuolar ATPase from tobacco cells (Matsuoka *et al.*, 1997).

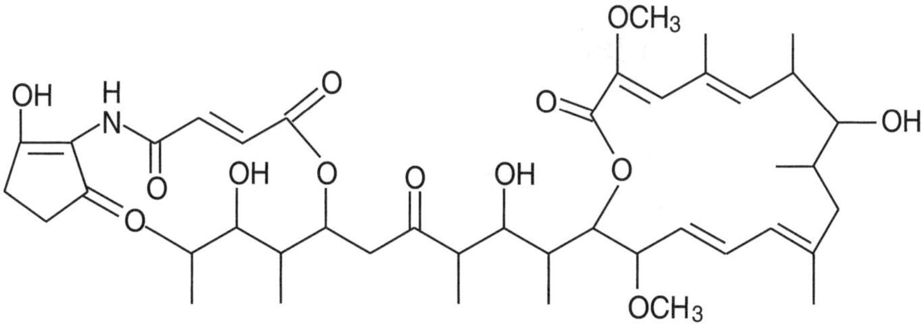

Bafilomycin

Citromycin

Citromycin is produced by *Streptomyces hygroscopicus*. When fed orally to *Musca domestica* Linnaeus and *Blatella germanica* (Linnaeus) at 500 ppm, citromycin caused 30 and 50 per cent mortality at the end of the assay period, respectively (Kubo *et al.*,

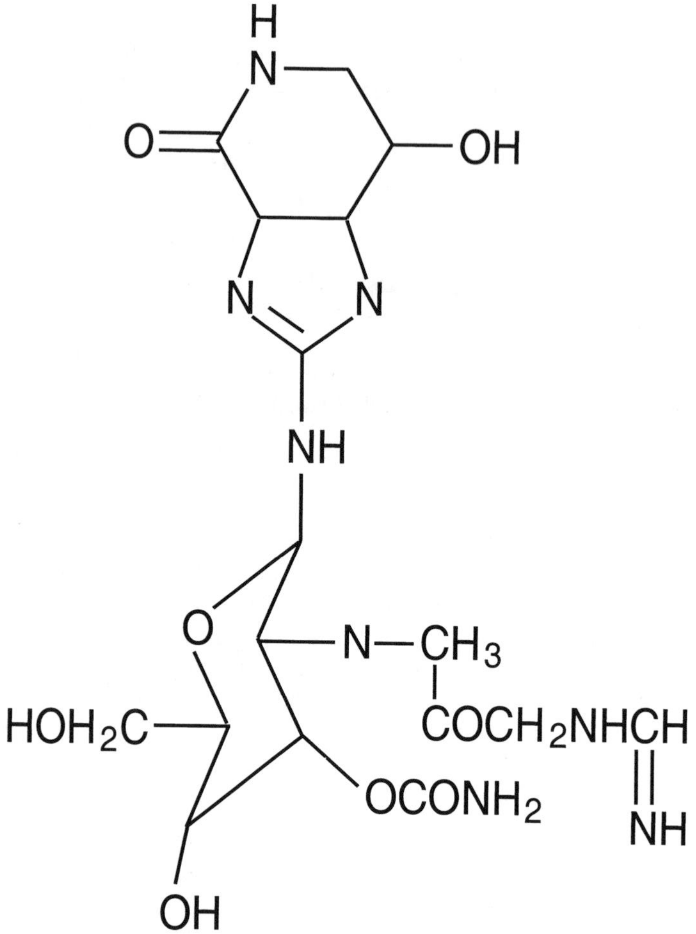

Citromycin

1981). Toxicity to *B. germanica* was slow to occur and may be related to toxicity or other effects on the Malpighian tubules (Kubo *et al.*, 1981).

Concanamycins

Concanamycins are produced by *Streptomyces diastatochromogenes* (Westley *et al.*, 1984). Concanamycin A is active against several insect species to an unspecified degree (Westley, 1978). Concanamycin A inhibits vacuolar ATPase from tobacco cells (Matsuoka *et al.*, 1997). Presumably it would also act similar to bafilomycins on ion regulation in Malpighian tubules in insects.

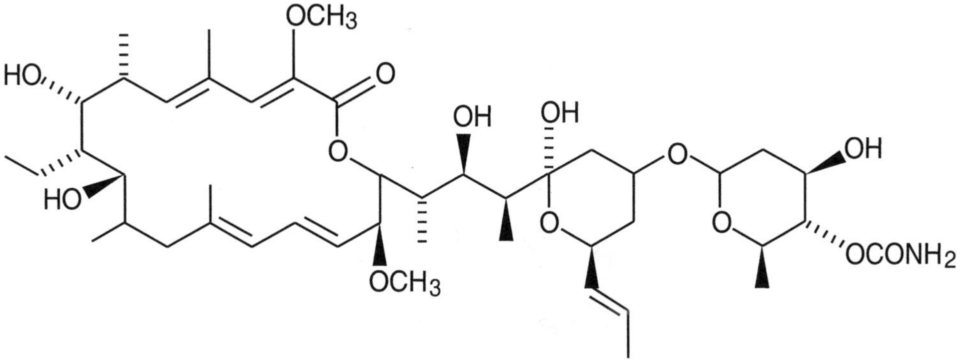

Concanamycin A

Dioxapyrrolomycin

Dioxapyrrolomycin was originally isolated from a culture of *Streptomyces fumanus* (Addor *et al.*, 1992; Hunt and Treacy 1998; Kuhn, 1997). It is moderately active towards different insects and mites (Hunt and Treacy, 1998), with LC_{50}s of 10, 32, 40 and > 100 ppm for *T. urticae, H. virescens, S. eridania* and *Empoasca abrupta*, respectively (Addor *et al.*, 1992). Although it had undesirable toxicity to vertebrates, it served as a lead structure for some effective analogs (Hunt and Treacy, 1998). The mode of action of this compound appears to be uncoupling of oxidative phosphorylation (Hunt and Treacy, 1998).

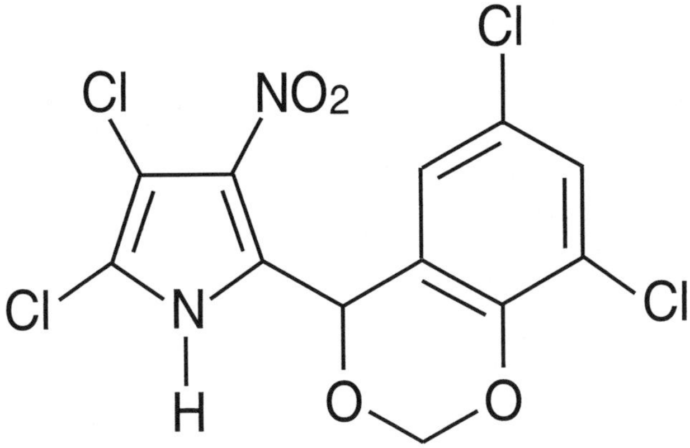

Dioxapyrrolomycin

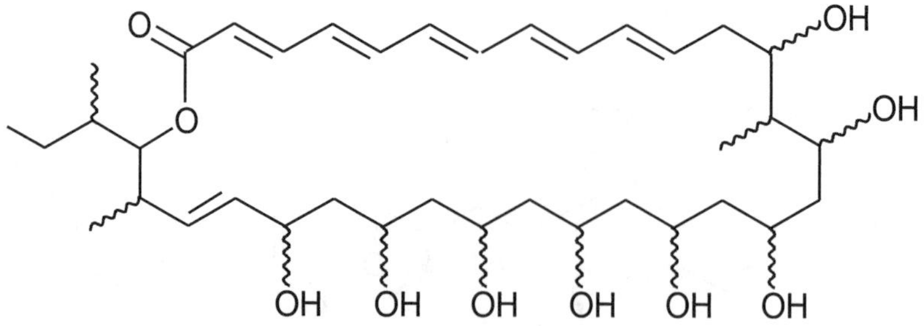

Faerifeungin A

Faeriefungins

Faeriefungin has been isolated from *Streptomyces griseus* var. *autotrophicus* (Nair *et al.*, 1989). It is actually a mixture of two compounds differing only in the presence or absence of a methyl group, faeriefungins A and B. At 100 ppm the faeriefungin mix caused 100 per cent mortality to *A. aegypti* larvae (Nair *et al.*, 1989).

Griseulin

Griseulin was produced by *Streptomyces griseus* var. *autotrophicus* (Nair *et al.*, 1993, 1995). At 6.25 ppm, all *A. aegypti* larvae were killed within 24 h (Nair *et al.*, 1995).

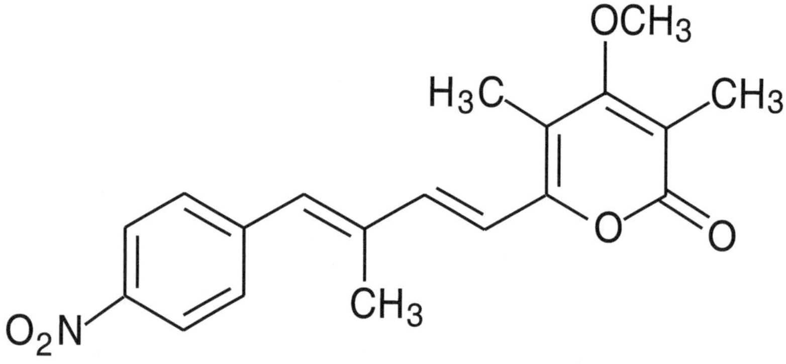

Griseulin

Leucanicidin

Leucanicidin is produced by *Streptomyces halstedii* (Isogai *et al.*, 1984). It killed 100 per cent of *L. separata* larvae after 4 days, when treated with 20 ppm of the compound (Isogai *et al.*, 1984).

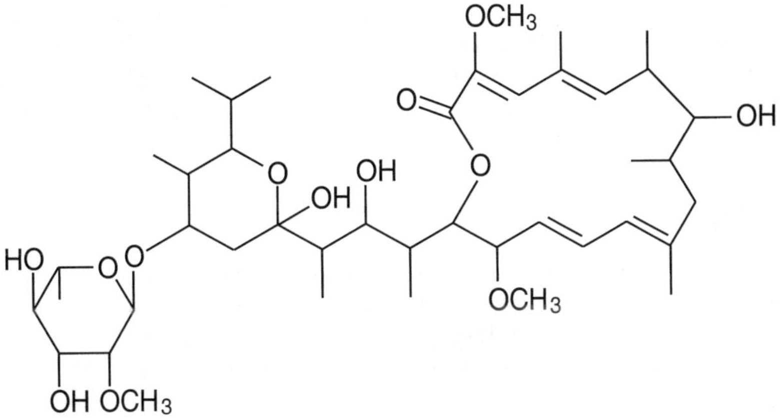

Leucanicidin

Milbemycins

Milbemycins are macrolide compounds isolated from *Streptomyces* such as *S. hygroscopicus* subsp. *aureolacrimosus* (Takiguchi *et al.*, 1980; Mishima, 1983). The α- series contain a tetrahydrofuran ring, but the β- series lack this ring, with 15 α- series and

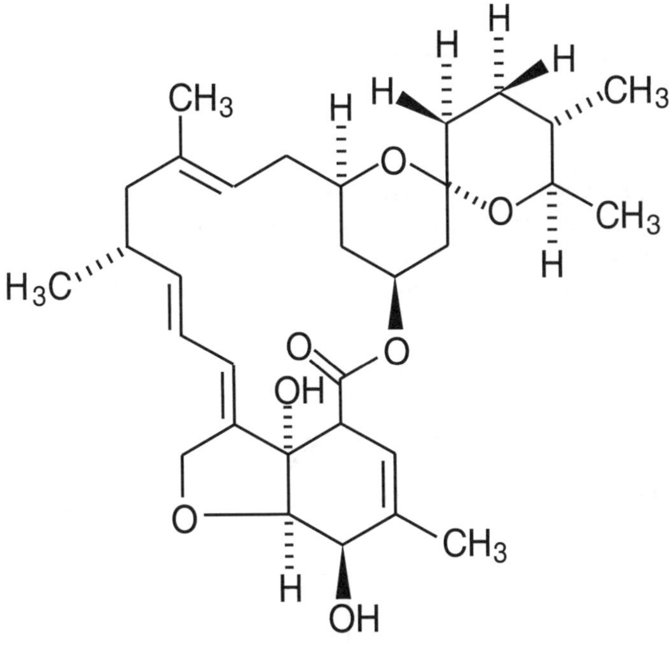

Milbemycin A

5 β- series reported in 1983 (Mishima, 1983). They are effective against mites, aphids and caterpillars (Mishima, 1983). For example, the LD_{50} for milbemycin D was 0.3 and 0.03 ppm topically against *T. urticae* and *Panonychus citri* (McGregor), respectively (Mishima, 1983). Milbemycins are believed to act like avermectins by opening chloride channels (Jansson and Dybas, 1998).

Nikkomycins

Nikkomycins have been isolated from *Streptomyces tendae* (Dähn *et al.*, 1976). They are known to be potent inhibitors of insect chitin synthase (Brillinger, 1979). However, the chitin synthase of *Hyalophora cecropia* (Linnaeus) (I_{50} of 1×10^{-3} M) was relatively unaffected compared to that of *T. ni* (I_{50} of 6×10^{-9} M) (Cohen and Casida, 1982).

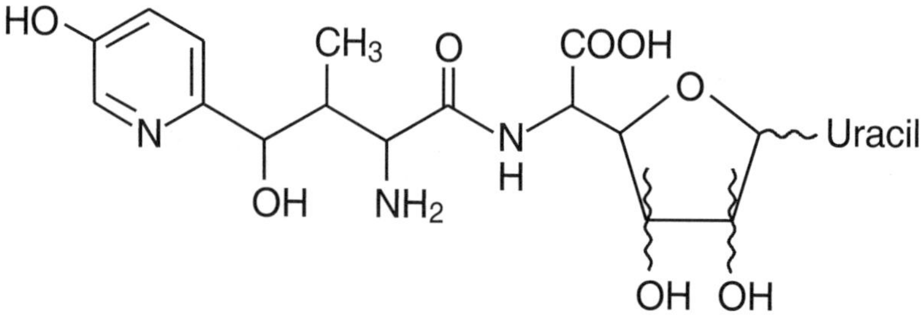

HO CH3 O COOH O N Uracil N H OH NH2 OH OH

Nikkomycin

Oxohygrolidin

Oxohygrolidin is produced by *Streptomyces hygroscopicus* (Kretschmer *et al.*, 1985). When 1000 ppm solutions were applied to diets, unspecified activity was noted against *Ceratitis capitata* (Wiedermann), *Dysdercus intermedius* Distant, *Phaedon cochleariae* Fabricius and *Plutella xylostella* (Linnaeus) (Kretschmer *et al.*, 1985).

O OH OH O OH O OCH3

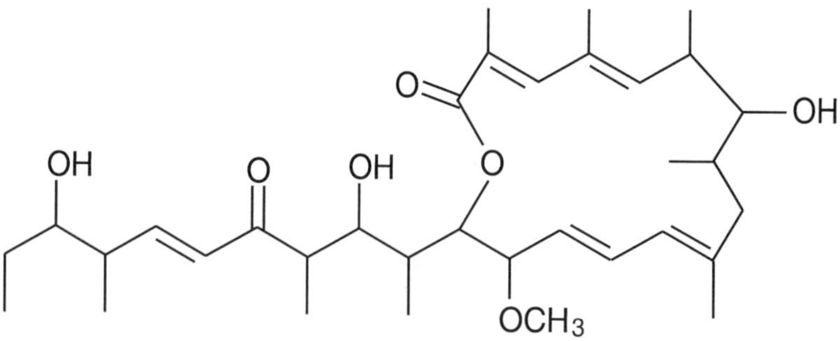

Oxohygrolidin

Table 2.2 Toxicity of Piericidin A and B against various insects

Insect	*Concentration*	*Mortality (%)* *Piericidin A*	*Piericidin B*
Pieris rapae	0.2 % dip	50	
	24 μg topical	56.7	48.3
Bombyx mori	0.05 μg topical	100	
	2.4 μg topical	90.0	88.8
Chilo simplex	64 μg topical	14.3	
	24 μg topical	56.7	60.0
Musca domestica	0.1 % contact	3.0 (40% knockdown)	
	1.0 μg topical	58.0	46.3
Blatella germanica	8.0 μg topical	80.0	30.0
Myzus persicae	20 ppm spray	75.8	74.9
Tetranychus urticae	5 ppm dip	90.2	26.0

Piericidins

Piericidins are produced by *Streptomyces mabaraensis* (Tamura *et al.*, 1963; Takahashi *et al.*, 1968). Piericidin A and B are known to be toxic to many insect species (Tamura *et al.*, 1963; Takahashi *et al.*, 1968) as shown in Table 2.2. The homology of piericidin A to coenzyme Q has been noted (Hall *et al.*, 1966). Piericidin A reacts with DPNH (NADPH) dehydrogenase at a site possibly the same as rotenone, and also inhibits succinic dehydrogenase by competing with the coenzyme Q binding site (Hall *et al.*, 1966). Various effects of piericidins have been reviewed (Yoshida and Takahashi, 1978). There are several examples where piericidins have inhibited the mitochondrial electron transport system, including that of insects. Piericidins inhibit electron transport by binding to the same site as rotenone, the ubiquinone site, where an unsaturated side chain and a free phenolic hydroxyl group in the pyridine have been found to be important, such as for cockroach mitochondrial preparations (Yoshida and Takahashi, 1978).

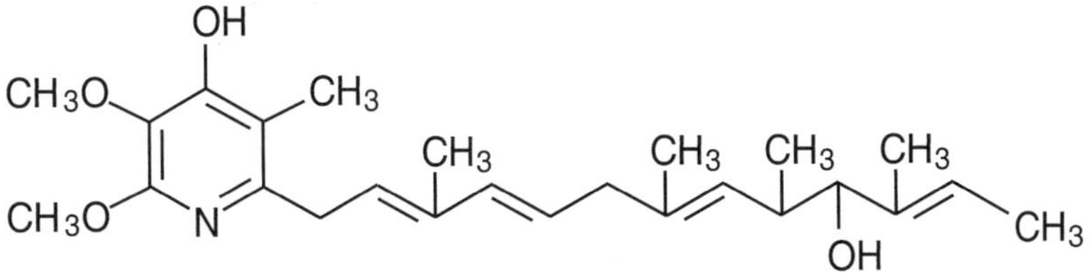

Piericidin A

Racemomycins

Racemomycins are produced by *Streptomyces lavendulae* (Kubo *et al.*, 1981). Activity against insects increases with increasing numbers of β-lysine residues, although absence of the streptolinidine ring greatly reduces insecticidal activity. When given orally to *M. domestica* at 1000 ppm, mortality after 48 h was 30, 50 and 80 per cent for racemomycin A, C and B, respectively. When given orally to *B. germanica* at 500 ppm, mortality after

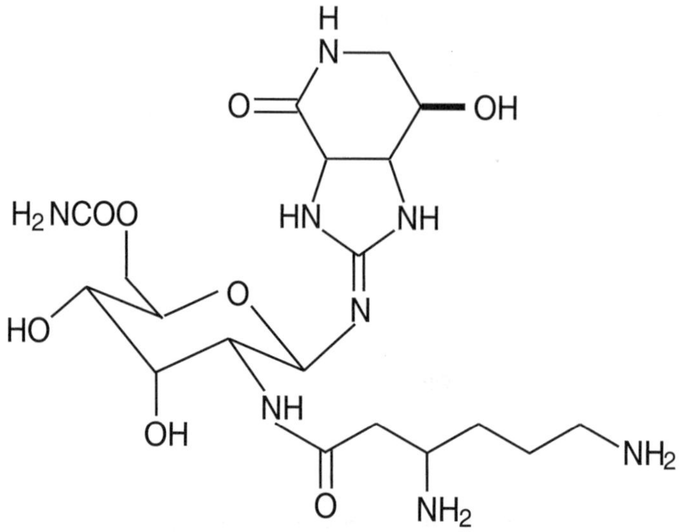

Racemomycin A

5 days was 30, 50 and 40 per cent for racemomycin A, C, and B, respectively (Kubo *et al.*, 1981). Toxic effects on *B. germanica* were slow to occur, and may be due to effects on Malpighian tubules. It was also observed that racemomycin-D accumulated after dosing of *Bombyx mori* (Linnaeus) (Kubo *et al.*, 1981).

Rhodaplutin

Rhodaplutin is produced by *Nocardioides albus* (Dellweg *et al.*, 1988). Applied to dietary materials at 400 ppm or less, unspecified activity was noted against *D. intermedius, M. persicae, P. cochlearia, P. xylostella*, and *T. urticae* (Dellweg *et al.*, 1988).

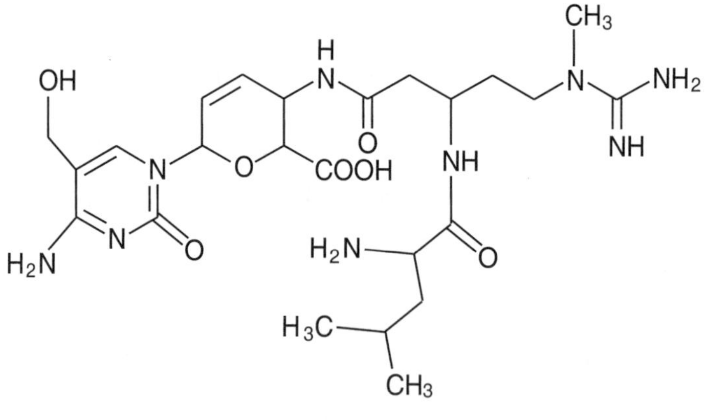

Rodaplutin

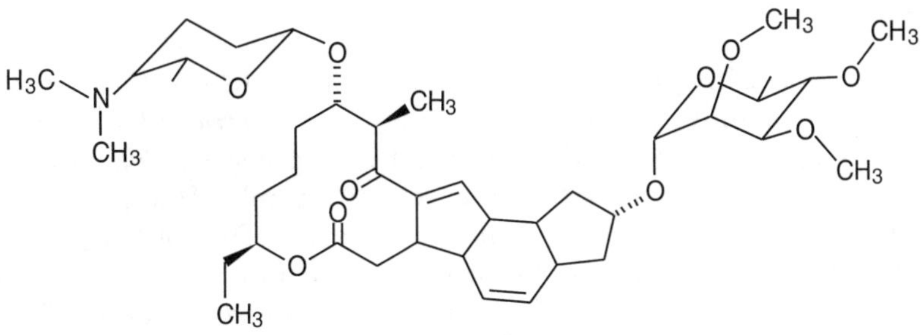

Spinosyn A

Spinosyns

Spinosyns are commercially available compounds that were originally isolated from the actinomycete *Saccharopolyspora spinosa* (DeAmicis *et al.*, 1997; Sparks *et al.*, 1999). Several forms are produced in the fermentation broth, but the most common are spinosyn A and D, which produce 0.3 and 0.8 ppm LC_{50}s, respectively for *H. virescens* larvae in drench assays (Sparks *et al.*, 1999). Good activity is generally noted against Diptera, Hymenoptera, Siphonaptera, and Thysanoptera but varies more with Coleoptera, and is generally poor against aphids and nematodes (Sparks *et al.*, 1999). Presence or absence of a single methyl group at any position can reduce activity against *H. virescens* by more that 200 times (DeAmicis *et al.*, 1997). Spinosyns cause persistent activation of nicotinic acetylcholine receptors and alter the function of GABA-gated chloride channels (Sparks *et al.*, 1999). Spinosyns can be as active as many synthetic pyrethroids against Lepidoptera (Sparks *et al.*, 1998).

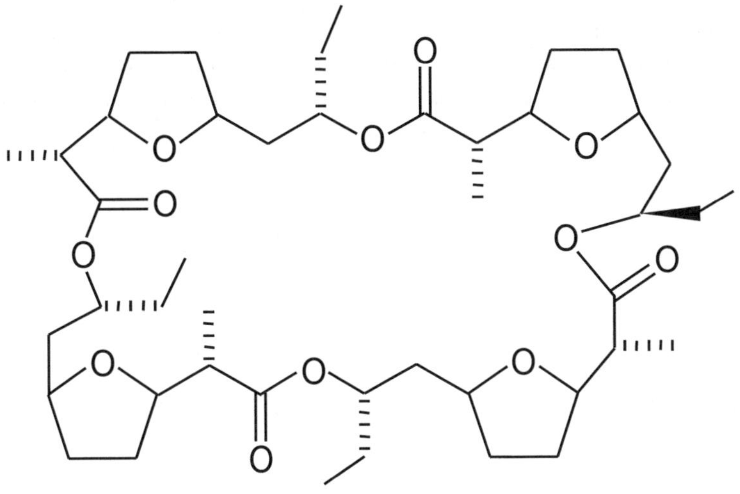

Tetranactin

Tetranactin

Tetranactin is produced by the actinomycete *Streptomyces aureus* (Oishi *et al.*, 1970; Ando, 1983). It is most active against mites, e.g. *Tetranychus cinnabarinus* (Boisduval) (LC_{50} 4.8 μg/ml) and lower in activity against *C. chinensis* (LD_{50} 0.8 μg/weevil) and *Culex pipiens molestus* larvae (LC_{50} 7μg/ml) (Ando, 1983). It is also less active against *B. germanica* and *M. domestica*, with LD_{50}s of > 4 μg/insect for both (Ando, 1983). A combination of dinactin, trinactin and tetranactin produced 84.8 per cent mortality at 1 μg/adult *C. chinensis* (Oishi *et al.*, 1970). Tetranactin acts by uncoupling oxidative phosphoralization, as demonstrated in cockroach mitochondria (Ando, 1983).

Thiolutin

Thiolutin is produced by *Streptomyces albus, S. celloflavus* (Cole and Rolinson, 1972) and the bacterium *Xenorhabdus* (McInerney *et al.*, 1991). Thiolutin produced an LC_{50} of 32 ppm for larvae of *L. sericata* (Cole and Rolinson, 1972).

Valinomycin

Streptomyces fulvissimus, S. roseochromogenes and *S. griseus* var. *flexipertum* (Heisey *et al.*, 1988a, b) produce valinomycin. When tested against *M. domestica*, there was no activity when topically applied (Pansa *et al.*, 1983). However, when injected the LD_{50} values for males and females were 0.02 and 0.03 μg/fly, respectively, and when administered orally, 100 per cent mortality occurred after 2.6 μg/fly was consumed (Pansa *et al.*, 1973). When injected into *Periplaneta americana* (Linnaeus) the LD_{50} value was 0.15 for males and 0.50 for females. The LD_{50} for *A. aegypti* larvae was 8 ppm (Pansa *et al.*, 1973). The LC_{50} values for *Epilachna varivestis* Mulsant, *T. urticae*, and an unspecified mosquito were 35, 3, and 2–3, respectively (Heisey *et al.*, 1988a,b). At sublethal concentrations, valinomycin caused an irregular heartbeat in *Periplaneta americana* (Linnaeus) (Pansa *et al.*, 1973). Valinomycin is an ionophore highly selective for K^+ ions, causing them to move across lipid membranes and as a result interferring with active transport (Harold *et al.*, 1974).

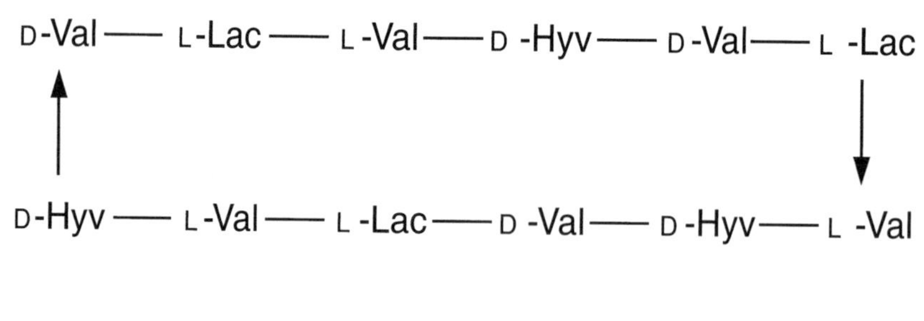

D-Val — L-Lac — L-Val — D-Hyv — D-Val — L-Lac

↑ ↓

D-Hyv — L-Val — L-Lac — D-Val — D-Hyv — L-Val

Valinomycin

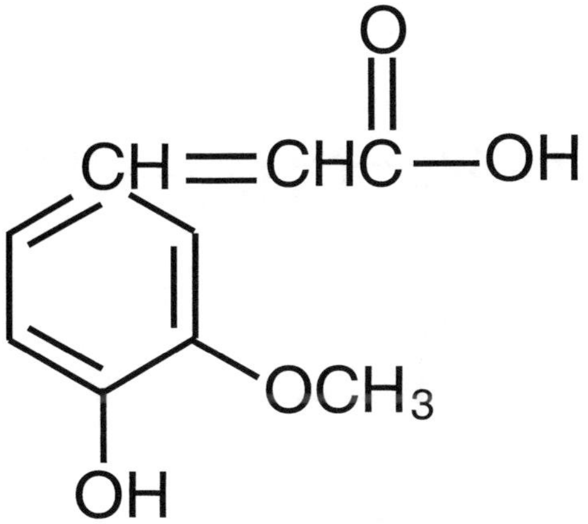

Ferulic Acid

Antiinsectan compounds derived from fungi

There are several instances where compounds originally isolated from higher plants and with activity reported against insects have also been isolated from fungi. Some of the more common ones are described below.

Phenolics

Several species of fungi produce phenolic compounds that are the same or similar to those that occur in higher plants. These compounds include ferulic acid (produced by *Rhizoctonia*), cinnamic acid and methyl cinnamate (produced by *Fusarium, Penicillium, Rhizoctonia, Lentinus, Inocybe, Stereum*, and *Ceratostomella*), coumaric acid (produced by *Rhizoctonia*, and *Eurotium*), benzoic acid and derivatives (produced by*Aspergillus, Eurotium, Hansenula, Lambertella, Mycotorula, Penicillium,*

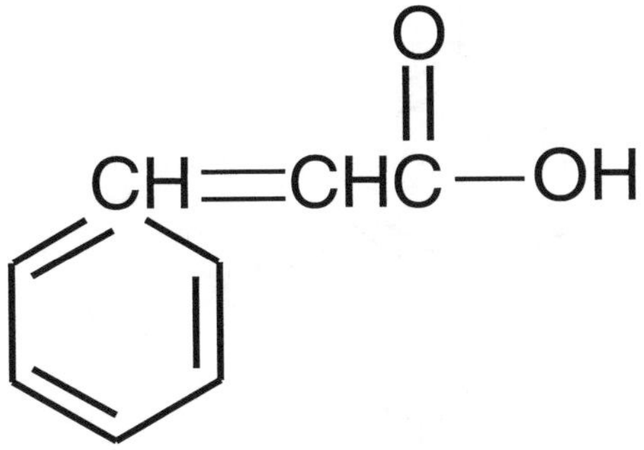

Cinnamic Acid

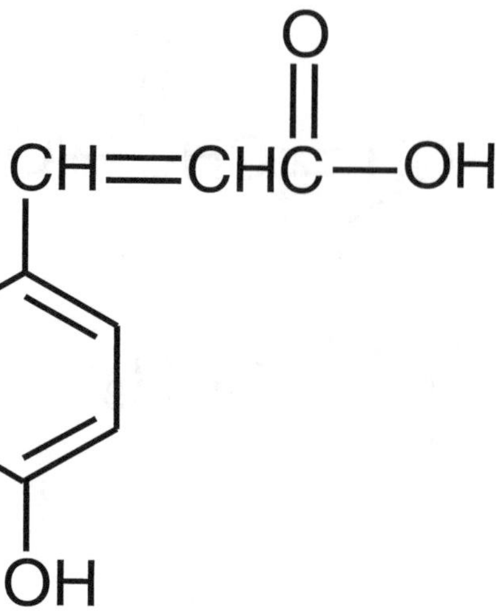

Coumaric Acid

Polyporus, Lentinus), and vanillic acid (from *Rhizoctonia leguminicola* and *R. solani*) (Turner, 1971; Turner and Aldridge, 1983). All of these compounds, generally considered derived from higher plants, have had antiinsectan activity described (Reese, 1977; Dowd and Vega, 1996).

Poly(acetyl)enes

Polyacetylenes are produced by several species of fungi, many of which are mushroom formers (Turner, 1971; Turner and Aldridge, 1983). The fungi that produce poly-

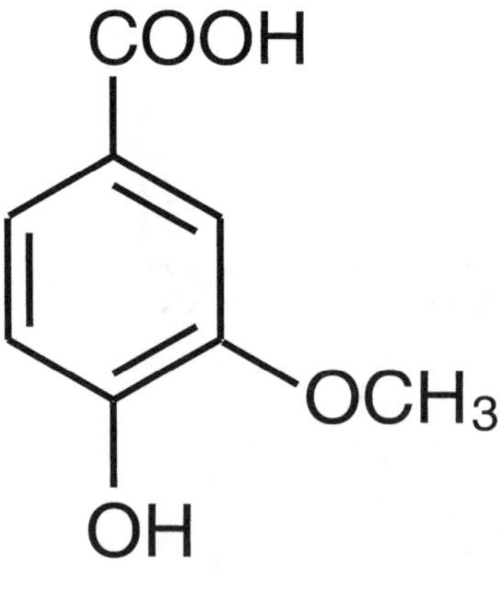

Vanillic Acid

$$CH_3CH{=}CH[C{\equiv}C]_2CH{=}CHCO_2H$$

Polyacetylene

acetylenes include *Agrocybe, Camorophyllus, Clitocybe, Collybia, Coprinus, Daedalea, Drosophila, Fistulina, Fomes, Gymnophilus, Hydnum, Kuehneromyces, Peniophora, Pleurotus, Polyporus, Psilocybe, Ramaria, Resinicium, Russula, Serpula, Stereum*, and *Trichoderma* (Turner, 1971; Turner and Aldridge, 1983). Most of the polyacetylenes produced by fungi are C-9 or C-10 (vs. C-13 for plants) and may be identical to, or cis-isomers of plant derived forms. Polyacetylenes can be photoactivated and plant-derived ones are especially toxic to insects (Arnason *et al.*, 1983).

Terpenoids

Several species of fungi produce terpenoids that are also the same or similar to those produced by higher plants. Geraniol and derivatives are produced by *Ceratocystis* spp. and *Trichothecium* spp., linalool is produced by *Agaricus, Boletus, Ceratocystis,*

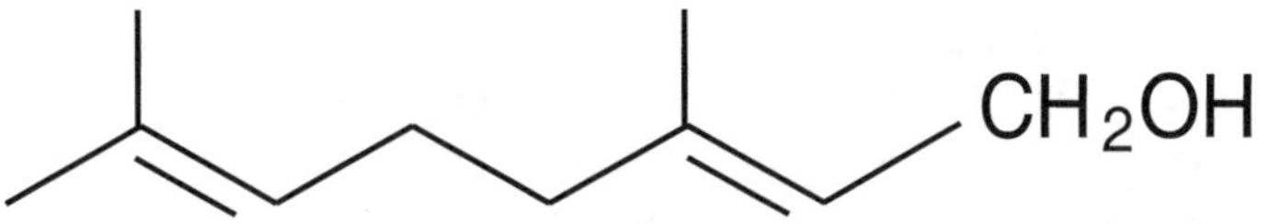

CH_2OH

Geraniol

Lactarius, Phellinus and *Trichothecium* spp.; α- and β-pinene are produced by *Cronartium fusiforme*, and limonene has been produced by *Cronartium fusiforme* and *Gyromitra esculenta* (Turner, 1971; Turner and Aldridge, 1983). Activity of these compounds against insects has been described in several reviews (e.g. Gershenzon and Croteau, 1991).

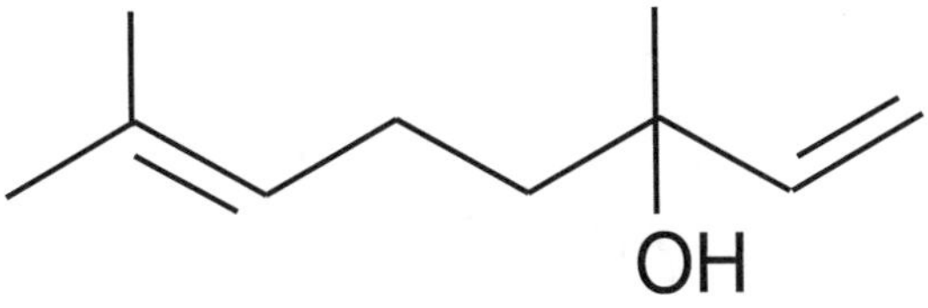

OH

Linalool

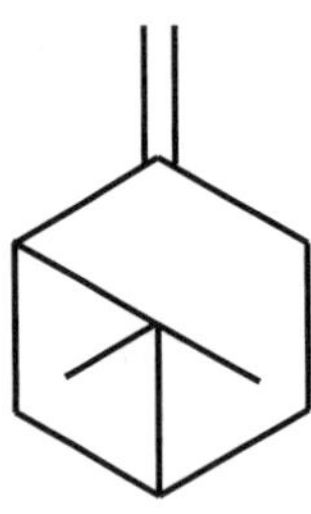

β-Pinene

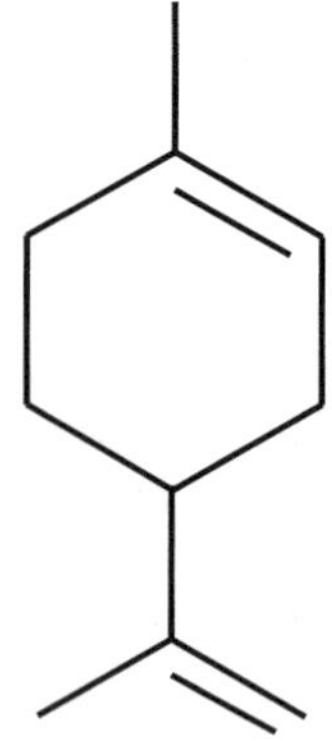

Limonene

Aflatoxins

Primarily *Aspergillus flavus* and *A. parasiticus* (Cole and Cox, 1981) produce aflatoxins. The effects of aflatoxins have been extensively studied on insects, and much earlier work has been comprehensively reviewed (Wright *et al.*, 1981). Earlier metabolites obtained in the aflatoxin biosynthetic pathway were shown to be less toxic to first instar *Ostrinia nubilalis* (Hubner) (Jarvis *et al.*, 1984). For instance aflatoxin at 1000 ppm caused 100 per cent mortality and sterigmatocystin at 60,000 ppm caused 45 per cent mortality, but versicolorin A, averufin and norsorlinic acid had no effect at 60,000 ppm (Jarvis *et al.*, 1984). Sterigmatocystin, an aflatoxin precursor, is also ca. 5–10× less toxic to *H. zea* and *S. frugiperda* (Dowd, 1992b). However, sterigmatocystin at 1000 ppm was more toxic to *Tyrophagus putrescentiae* (Schrank) where no adults could be produced compared to aflatoxin where 249 adults were produced at a

O O
O
O O OCH3

Aflatoxin B$_1$

similar level of treatment (Rodriguez *et al.*, 1980). Reproductive anomalies are frequently noted in insects administered sublethal levels (Wright *et al.*, 1981). Aflatoxins undergo enzymatic activation to an epoxide, which reacts with guanidine residues in DNA that convert to formamido pyrimide (Lillehoj, 1992). The presence of this anomaly in DNA causes errors during transcription (Lillehoj, 1992). When administered orally, some insects, such as flies and caterpillars (Wright *et al.*, 1981) are highly susceptible (mainly oral assays) while others, such as cockroaches (Scherertz *et al.*, 1978) and sap beetles (Dowd, 1992a) are relatively resistant. Studies with housefly ovaries indicated interference with DNA and rRNA synthesis (Al-adil *et al.*, 1973). Metabolic studies suggest that in *H. zea* and *S. frugiperda* midguts, aflatoxin B_1 is primarily converted to other B or G forms (Dowd, 1992a), which have similar toxicity (Cole and Cox, 1981). However, sap beetles appear to convert aflatoxin B_1 to aflatoxin Q (Dowd, 1992a; Dowd, unpublished), which is 18 times less toxic in vertebrate assays (Cole and Cox, 1981). Multiple genes appear to be involved in the resistance of different *Drosophila* strains to aflatoxins, but the mechanism(s) involved have not been clearly defined (see Dowd, 1992a).

Aflatrems

Aflatrem has been isolated from *Aspergillus flavus* (Gallagher *et al.*, 1980a; TePaske *et al.*, 1992), *Aspergillus parasiticus*, and *Aspergillus subolivaceus* sclerotia (TePaske *et al.*, 1992). It caused no significant activity at 100 ppm against caterpillars and beetles examined (Dowd *et al.*, unpublished data). The related compound *β*-aflatrem was also isolated from *A. flavus* sclerotia. This compound caused a 57 per cent RGR against *H. zea* at 100 ppm orally, but had no effect on *Carpophilus hemipterus* (Linnaeus) at this concentration (TePaske *et al.*, 1992).

Aflavanines

Aflavanine and dihydroxyaflavanine have been isolated from *Aspergillus flavus* (Gallagher *et al.*, 1980c) including the sclerotia of *Aspergillus flavus* (Wicklow and Cole, 1982). At 25 ppm orally, dihydroxyaflavanine produced 33.0 and 35.5 per cent

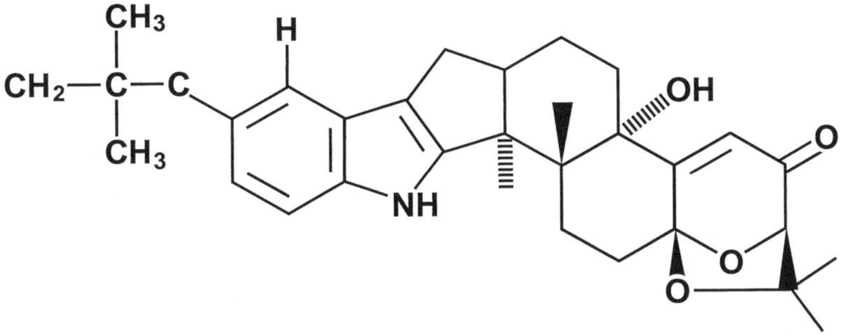

β-aflatrem

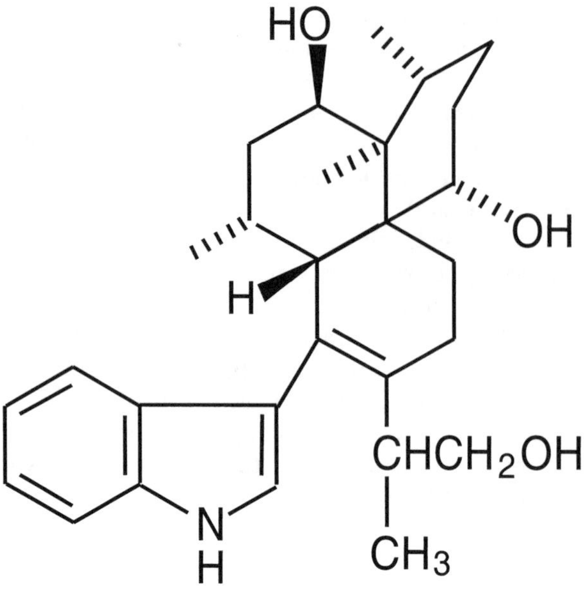

Dihydroxyaflavanine

RGRs for *H. zea* and *S. frugiperda*, respectively (Dowd *et al.*, 1988). Also, at 25 ppm oral treatment, dihydroxyaflavanine produced a 75 and 37 per cent RFR for *C. hemipterus* adults and larvae, respectively, while other analogs were inactive (Wicklow *et al.*, 1988). Additional aflavinines were isolated from the sclerotia of *A. flavus* (Gloer *et al.*, 1988). These analogs were inactive against *C. hemipterus* at 100 ppm but produced RFR at 400–1100 ppm, which are naturally occurring levels (Gloer *et al.*, 1988). Three new aflavanines were isolated from the sclerotia of *A. tubingensis* (TePaske *et al.*, 1989a). The keto form was the only active one of the three, and caused a 68 per cent RGR and 38 per cent RFR in *H. zea* larvae and *C. hemipterus* larvae at 125 ppm. Some of these were found in lower amounts from the sclerotia of *A. flavus, A. parasiticus* (TePaske *et al.*, 1989a), and *A. sulphureus* (Laakso *et al.*, 1993). Small amounts of aflavanine derivatives described from *A. tubingensis* (TePaske *et al.*, 1989) were also isolated from *A. leporis* (TePaske *et al.*, 1991). Isolated from both *A. tubingensis* (TePaske *et al.*, 1991) and ascostromata of *Eupenicillium crustaceum, E. molle*, and *E. reticulisporum* (Wang *et al.*, 1995) was 10,23-dihydro-24,25dehydroaflavinine. In addition, 10,23,24,25- tetrahydro-24-hydroxyaflavanine was isolated from ascostromata of *E. crustaceum* (Wang *et al.*, 1995). Limited acti vity of these two aflavanines was noted at 100 ppm against *H. zea* and *C. hemipterus*, but when the 10,23- compound was tested a concentration near the natural level (ca. 3000 ppm) in *E. crustaceum*, a 79 per cent RGR for *H.zea* and a 42 per cent RFR for *C. hemipterus* larvae was noted (Wang *et al.*, 1995).

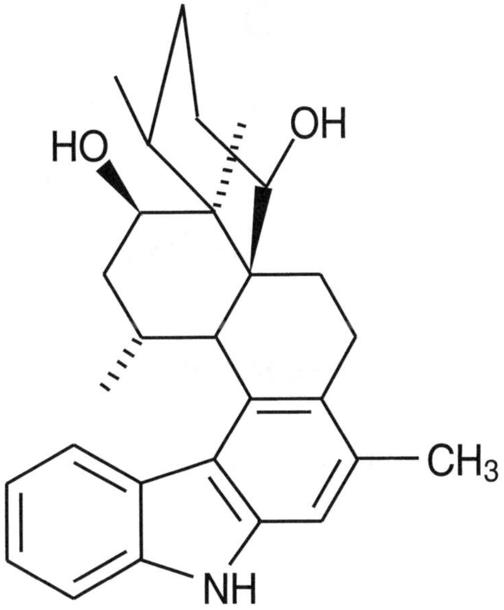

Aflavazole

Aflavazole

Aflavazole was isolated from the sclerotia of *Aspergillus flavus* (TePaske *et al.*, 1990). It produced a 75 and 62.5 per cent RFR for *C. hemipterus* larvae and adults, respectively at 200–600 ppm (TePaske *et al.*, 1990; Wicklow *et al.*, 1988).

Arenarins

Arenarins were isolated from the sclerotia of *Aspergillus arenarius* (Oh *et al.*, 1998a). Arenarin A produced a 13 per cent RFR for *C. hemipterus* adults and larvae, while

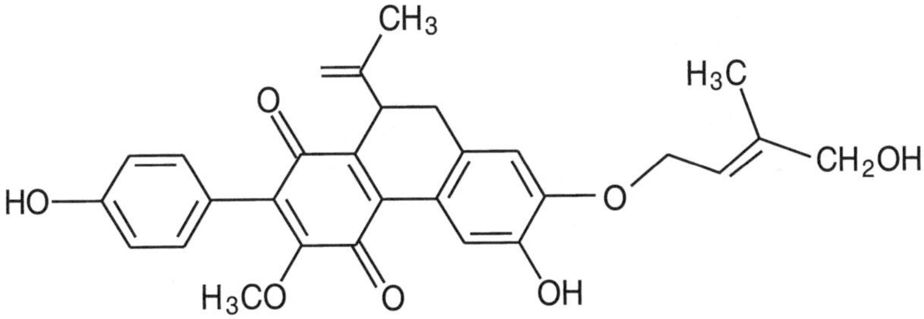

Arenarin A

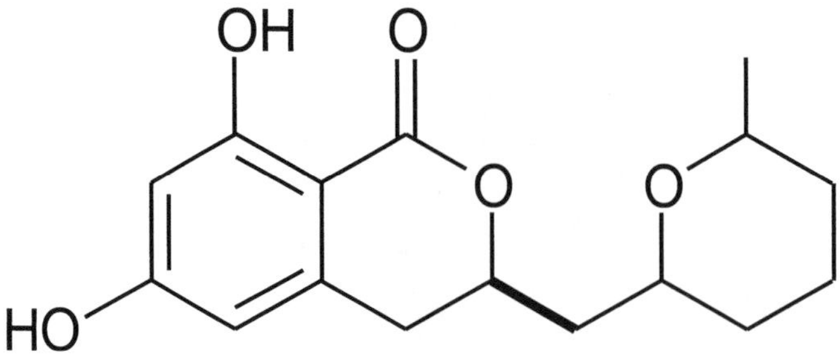

Asperentin

arenarins B and C produced a 20 and 13 per cent RFR for *C. hemipterus* adults at 100 ppm orally, respectively (Oh *et al.*, 1998a).

Asperentin

Aspergillus, Cladosporium, and *Eurotium* fungi (Grove and Pople, 1981) produce asperentin and derivatives. Asperentin produced an LD_{50} of 12.5 μg/adult *Calliphora vicina* Robineau-Desvoidy in 24 h (Grove and Pople, 1981). The 8- methyl ether and 5′-hydroxy- derivatives were essentially inactive at 50 μg/fly, but the 4″-hydroxy-derivative did produce 50 per cent knockdown within 1 h of treatment (Grove and Pople, 1981).

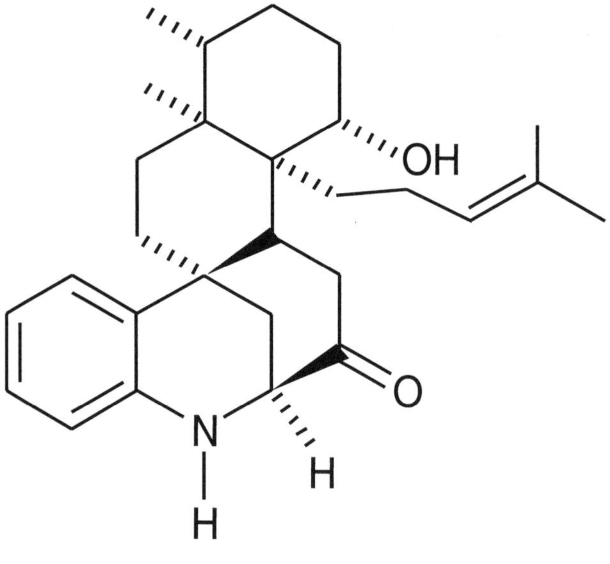

Aspernomine

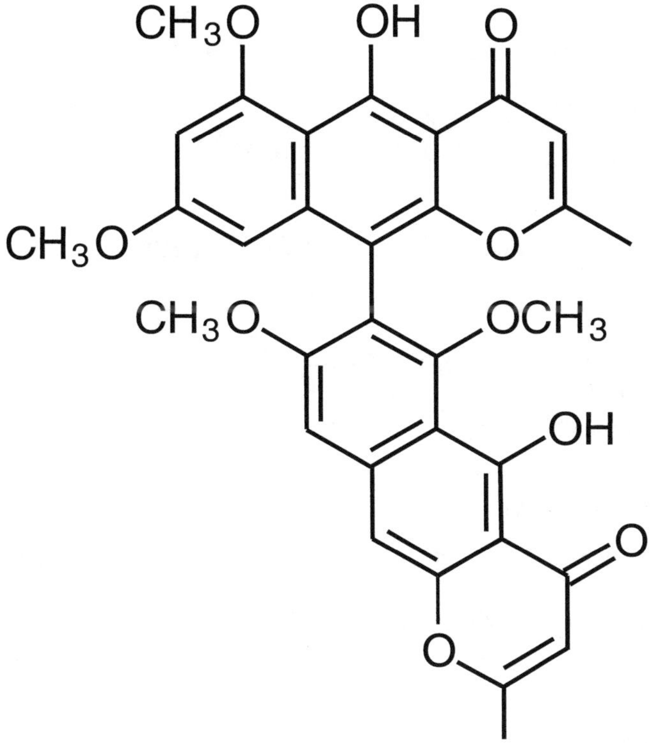

Aurasperone A

Aspernomine

Aspernomine was isolated from the sclerotia of *Aspergillus nomius* (Staub *et al.*, 1992). It produced a 35 per cent RGR in *H. zea* larvae at 25 ppm orally (Staub *et al.*, 1992).

Aurasperones and fonsecinones

Aurasperones and fonsecinones are produced by different species of *Aspergillus* (Turner and Aldridge, 1981; Gloer *et al.*, unpublished). At 200 ppm most tested compounds produced a 40–60 per cent RGR for *H. zea* and *C. hemipterus* adults and larvae (Dowd *et al.*, unpublished). At 100 ppm *in vitro*, most caused 40–60 per cent inhibition of NADH oxidase activity from *H. zea* and *C. hemipterus* larvae (Dowd *et al.*, unpublished).

Azoxybenzenes

The azoxybenzenes 4′-hydroxymethylazoxybenzene-4-carboxylic acid, and azoxybenzene-4,4″-dicarboxylic acid have been isolated from an entomopathogenic strain of

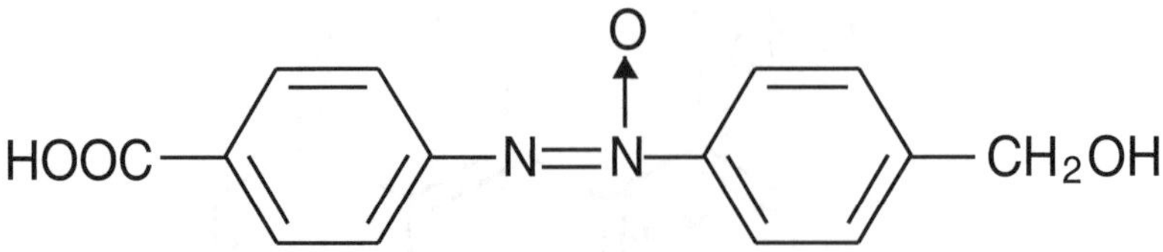

Azoxybenzene Derivative

Entomophthora virulenta (Claydon and Grove, 1978). When 2.5 μg of the hydroxy acid was injected into *C. vicina* adults, 77.5 per cent mortality resulted in 72 h; at 2.7 μg per fly dose, the diacid was inactive (Claydon, 1978).

Binaphthalene derivative

A binaphthylene derivative was isolated from an unidentified endophytic fungus from eastern larch (*Larix laricina*) (Findlay *et al.*, 1997b). It was toxic to the spruce budworm, *Choristoneura fumiferana* (Clemens) (96% mortality) at 400 ppm (Calhoun *et al.*, 1992).

Brevianamides

Penicillium brevicompactum and *Penicillium expansum* (Birch and Wright, 1969; Patterson *et al.*, 1987) produce brevianamides. Brevianamide A reduced *Drosophila*

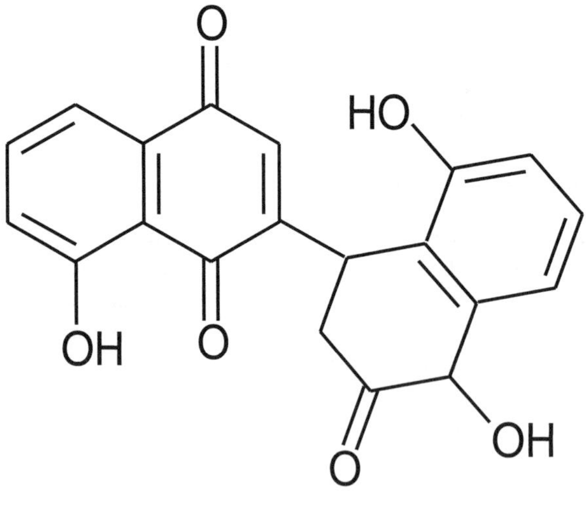

Binaphthalenyl Derivative

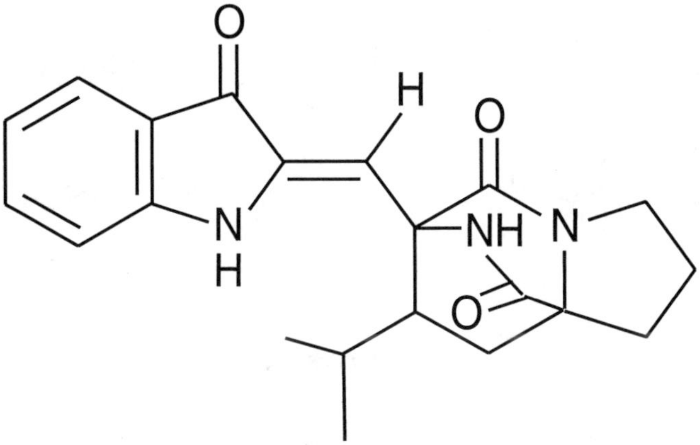

Brevianamide

melanogaster Meigen feeding by 83 per cent and caused 78 per cent mortality of *Spodoptera littoralis* (Boisduval) at 10 ppm (Patterson *et al.*, 1987).

Carbonarins

Several carbonarins have been isolated from sclerotia of *Aspergillus carbonarum* (Gloer *et al.*, unpublished). Several of them are active against insects. Carbonarin G

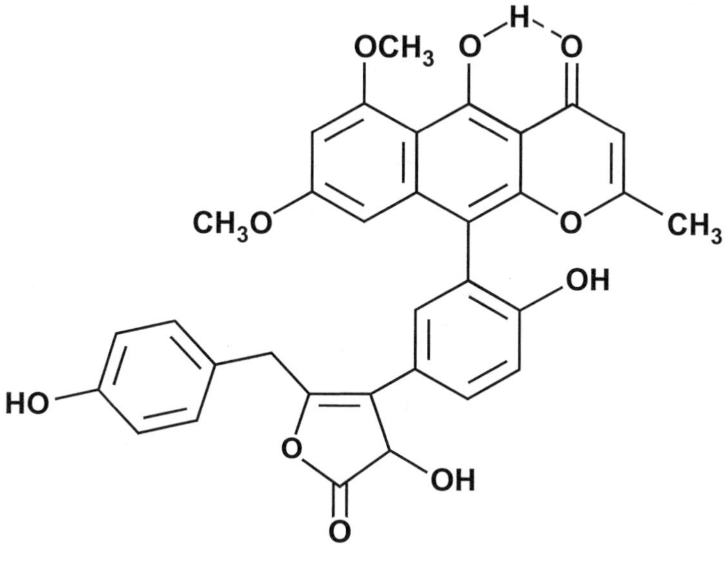

Carbonarin

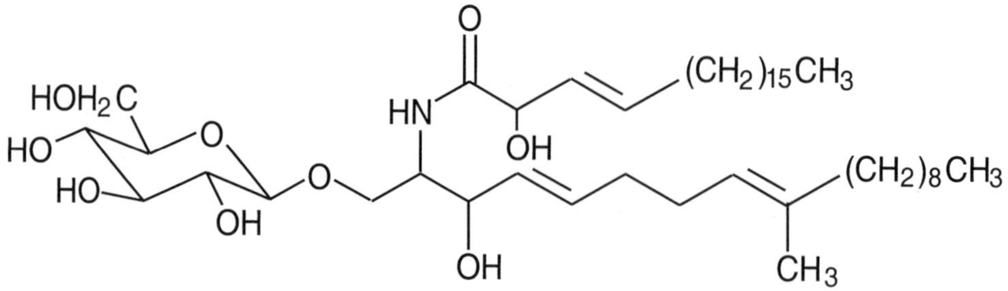

$$\text{Cerebroside D structure}$$

Cerebroside D

and H produced a 33.3 and 20.0 per cent RFR for *C. hemipterus* adults and a 31.2 and 18.8 per cent RFR for *C. hemipterus* larvae, respectively (Dowd *et al.*, unpublished data). Carbonarins are capable of inhibiting NADH oxidase (Dowd *et al.*, unpublished data).

Cerebrosides

Cerebrosides have been isolated from *Schizophyllum commune* (Mizuno, 1995), *Pachybasium* (= *Trichoderma*) sp. and the mycoparasite *Humicola fuscoatra* (Wicklow *et al.*, 1998). Cerebroside D produced a 47 per cent RGR for *H. zea* at 100 ppm orally (Wicklow *et al.*, 1998).

Chaetoglobosins

Chaetoglobosins have been isolated from *Chaetomium* spp., *Penicillium aurantio-virens* and *Diplodia macrospora* (Turner and Aldridge, 1983). Chaetoglobosin C produced

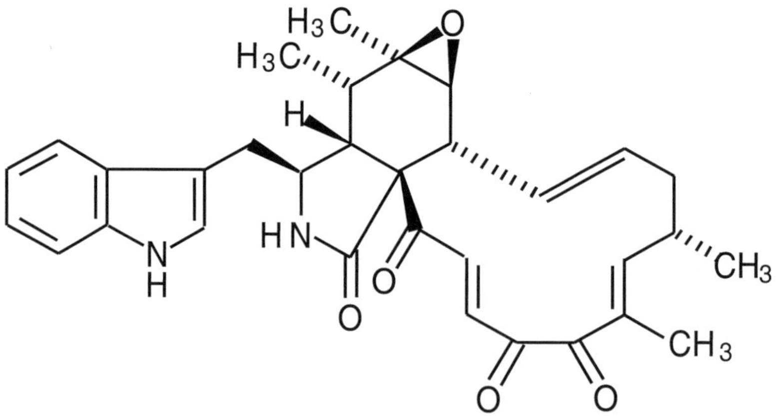

Chaetoglobosin C

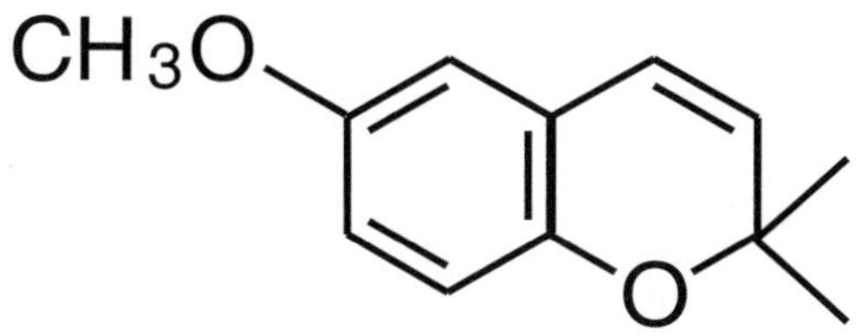

Lactarius Chromene

18.1 and 10.1 per cent RGRs for *H. zea* and *S. frugiperda*, respectively, at 25 ppm orally (Dowd, 1992b). Chaetoglobosin D, F_{ex}, and 19-O-acetyl chaetoglobosins A and D were isolated from *Chaetomium brasiliense* (Oh *et al.*, 1998b). All but chaetoglobosin F_{ex} were responsible for most of the antiinsectan activity in the ethyl acetate extract (Oh *et al.*, 1998b), where for example chaetoglobosin A at 100 ppm caused a 98.6 per cent RGR for *H. zea* (Dowd *et al.*, unpublished data). Chochliodinol produced no mortality of adult *Tribolium confusum* Duval at 800 ppm, but did totally inhibit reproduction (Rao and De las Casas, 1974).

Chromenes

Chromenes similar to precocenes have been reported from *Lactarius fuliginosus* and *L. picinus* (Conca *et al.*, 1981) and may occur in *L. fumosus* as well (Dowd and Miller, 1990). Extracts from *L. fuliginosus* and *L. fumosus* v. *fumosus* produced some precocious-like effects on *Oncopeltus fasciatus* (Dallas) nymphs exposed to filter paper with ca. 0.6 mg/cm^2 extract (Dowd and Miller, 1990). These extracts also produced significant mortality of *O. fasciatus* nymphs (0.6 mg/cm^2), as well as *H. zea* larvae (at 250 ppm orally). Chromene-positive reactions were noted with compounds separated by thin layer chromatography, but pure compounds were not isolated (Dowd and Miller, 1990).

Citrinin

Citrinin has been isolated from *Crotalaria crispata* and several species of *Penicillium* (Turner and Aldridge, 1983). Citrinin at 1000 ppm reduced growth rates of *T. confusum, Lasioderma serricorne* (Fabricius) and *Attagenus unicolor* (Brahm) by 6, 37 and 40 per cent, respectively (Wright *et al.*, 1980). Citrinin reduced numbers of *Tyrophagus putrescentiae* (Schrank) developing to adults by ca. 4.5 fold at 100 ppm (Rodriguez et al., 1980). Citrinin produced up to 78 per cent mortality of *Ephestia kuehniella* Zeller at 100 ppm orally (Wright and Harein, 1982). Citrinin reduced feeding of *D. melanogaster* by 50 per cent and produced 48 per cent mortality of *S. littoralis* at 10 ppm orally (Patterson *et al.*, 1987). In another study, the ED_{50} for *D. melanogaster* treated with citrinin was 140 μg/ml (Dobias *et al.*, 1980). Citrinin produced 22.5 per cent mortality and a 48.7 per cent RGR for *H. zea* and 79.7 per cent mortality and a 62.2 per cent RGR for *S. frugiperda* at 25 ppm orally (Dowd, 1989a). The structure of Malpighian tubules, which are analogous to the vertebrate kidney (a

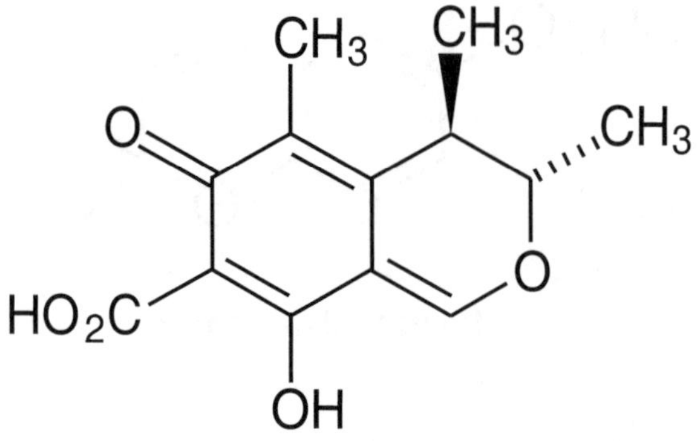

Citrinin

target site for citrinin) also, showed deterioration at these levels (Dowd, 1989a). Citrinin is reported to inhibit cholesterol synthesis in rats (Endo and Kuroda, 1976), so the same mechanism is also possible in insects.

Clitocine

Clitocine (6-amino, 5-nitro-4-imino (ribofuranosyl) pyramidine) is produced by the mushroom *Clitocybe inversa* (Kubo *et al.*, 1986). It had unspecified biological activity against *Pectinophora gossypiella* (Saunders).

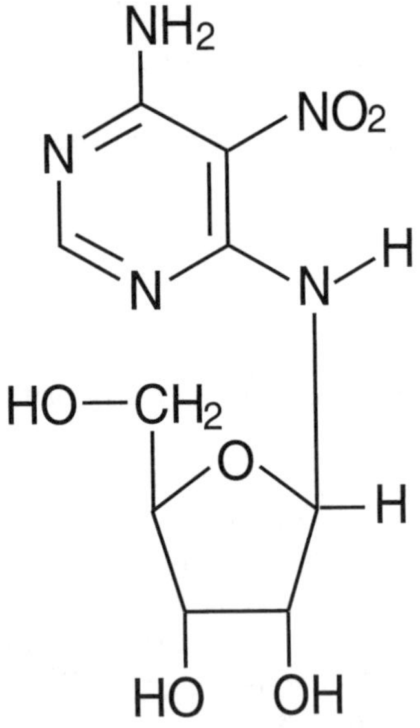

Clitocine

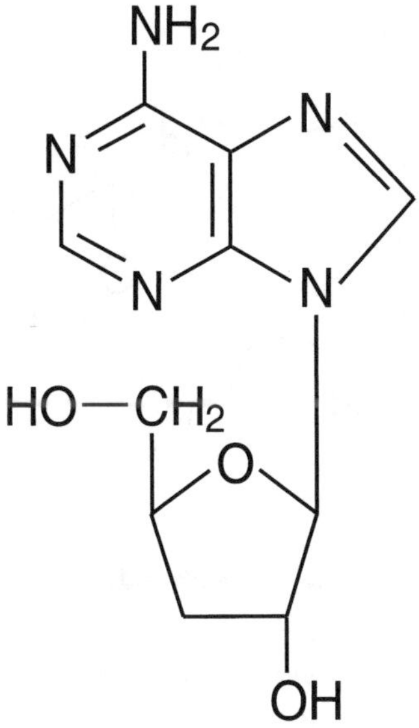

Cordycepin

Cordycepin

Cordycepin (3′-deoxyadenosine) is produced by the entomopathic fungus *Cordyceps militaris*, and *Aspergillus nitulans* (Cunningham *et al.*, 1951; Bentley *et al.*, 1951; Hanessian *et al.*, 1966). When injected into *Galleria mellonella* (Linnaeus) larvae, the LD_{50} was 30 $\mu g/g$ (Roberts, 1981).

Culmorin

Culmorin is produced by *Fusarium culmorum, F. nivale* (Turner and Aldridge, 1983) and *F. graminearum (F. roseum)* (Greenhalgh *et al.*, 1984a,b). Culmorin produced a

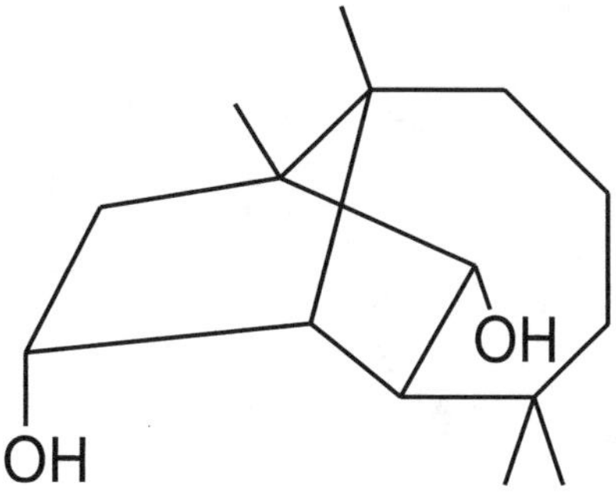

Culmorin

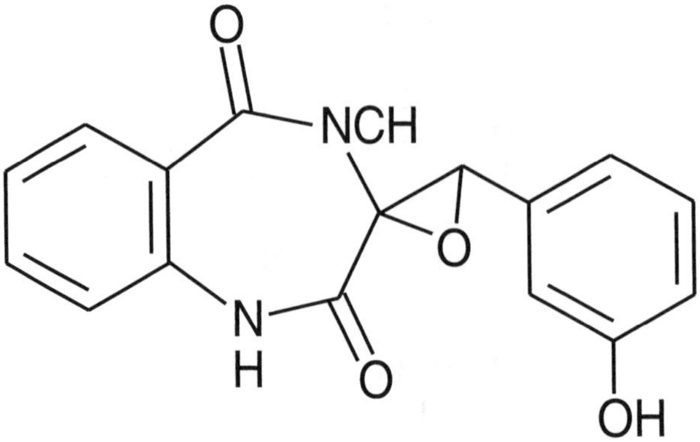

Cyclopenol

15.0 and 0.9 per cent RGR for *H. zea* and *S. frugiperda*, respectively, when tested at 25 ppm orally (Dowd *et al.*, 1989).

Cyclopenol

Cyclopenol is produced by several species of *Penicillium* (Turner and Aldridge, 1983). Cyclopenol caused a 52 per cent reduction in feeding by *D. melanogaster* and 26 per cent mortality of *S. littoralis* at 302×10^{-5}M (Patterson *et al.*, 1987).

Cyclopiazonic acid

Cyclopiazonic acid is produced by different species of *Aspergillus* and *Penicillium* (Turner, 1971; Turner and Aldridge, 1983). Initial symptoms when 2.5 μg of

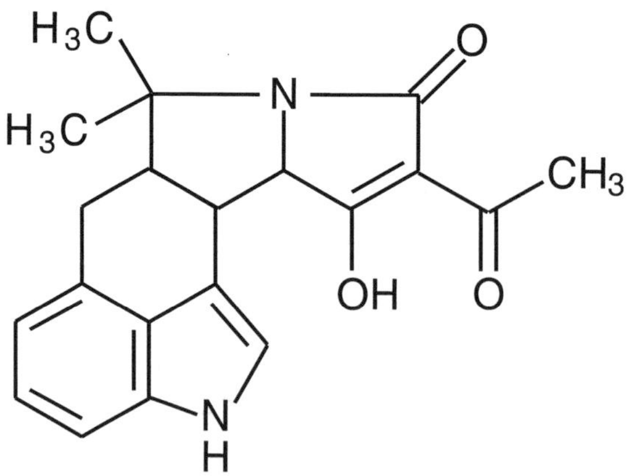

Cyclopiazonic Acid

cyclopiazonic acid was injected were stiffening of the body and swelling of the thorax of *B. mori*, which disappeared in a few hours (Yokota *et al.*, 1981). This type of activity appears to be due to interference in hydrostatic pressure regulation, similar to sclerotiamide (see below). A specific target at the cellular level is endoplasmic reticulum calcium transport ATPase (Riley and Showker, 1991). This information suggests that ATPase in the Malpighian tubules is being affected, altering water regulation, as has been reported for bafilomycin (see above).

Cytochalasins

Cytochalasins are produced by a diversity of fungi, including *Aspergillus clavatus, A. terreus, Chalara microspora, Helminthosporium dematioideum, Metarrhizium anisopliae, Phomopsis* spp., *Rosellinia necatrix*, and *Zygosporium masonii* (Turner and Aldridge, 1983). Cytochalasin H produced 13.0 and 57.6 per cent RGRs for *H. zea* and *S. frugiperda*, respectively, at 25 ppm orally (Dowd *et al.*, 1988). Chaetochalasin A was isolated from *Chaetomium brasiliense* (Oh *et al.*, 1998b). Chaetochalasin A produced a 25 per cent RFR for adult *C. hemipterus* at 150 ppm orally, but was inactive against *H. zea* at the same concentration (Oh *et al.*, 1998b). Cytochalasins can inhibit sugar transport across membranes (Kuo and Lampen, 1974).

Dendrodochin

Dendrodochin is produced by *Dendrodochium foxicum* (Lysenko and Vittinev, 1976). Dendrodochin caused 41 per cent mortality to *G. mellonella* larvae at 0.3 per cent oral administration (Lysenko and Vittinev, 1976). Dendrodochin increased the diameter of Malpighian tubules in the *G. mellonella* larvae.

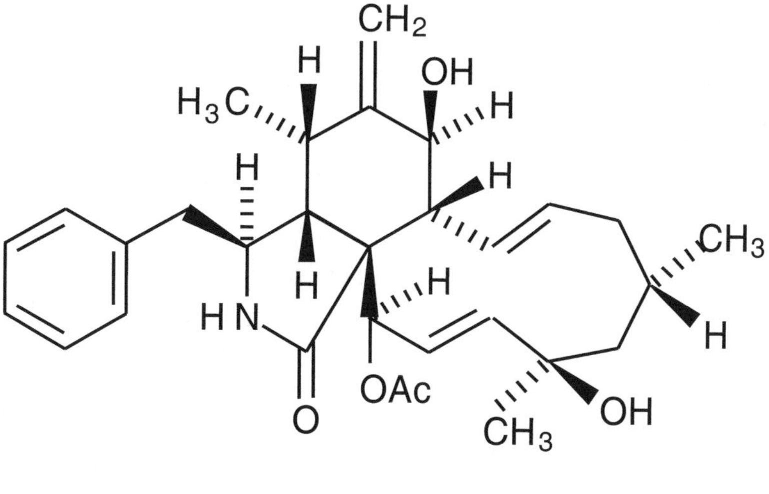

Cytochalasin H

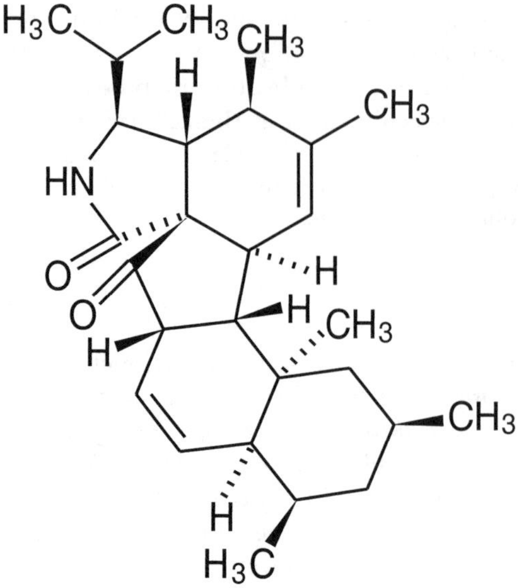

Chaetochalasin

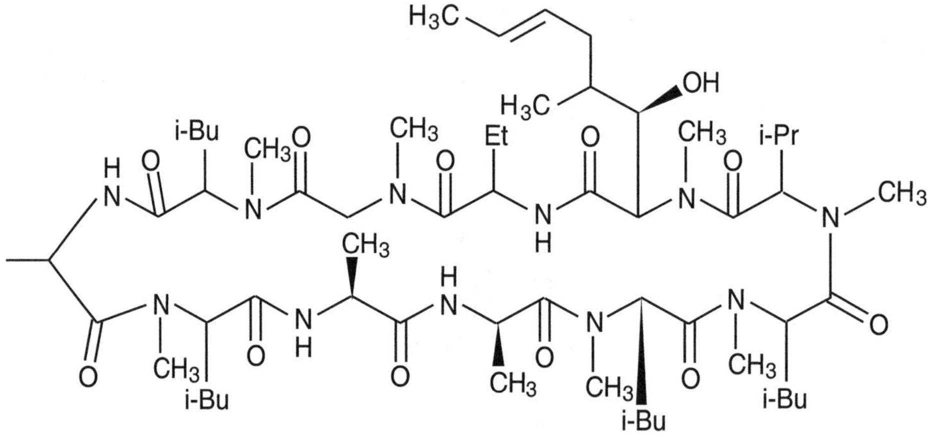

Dendrodochin A

Dermocybin

Dermocybin has been isolated from different species of *Dermocybe* (Besl and Blumreisinger, 1983). Dermocybin inhibited *D. melanogaster* survival to pupation by ca. 27 per cent at 350 ppm.

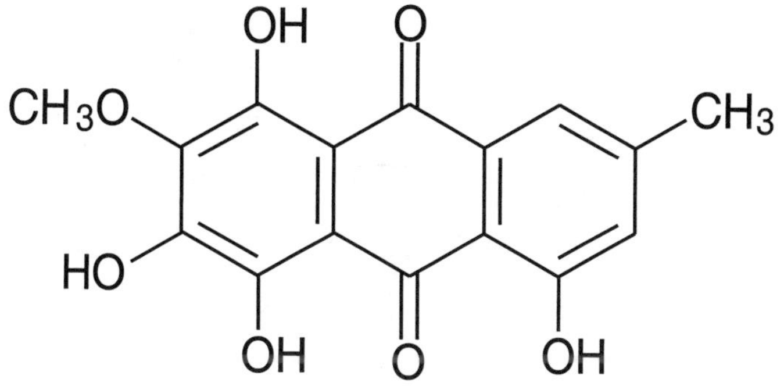

Dermocybin

E-64 and related thiol protease inhibitors

E-64 (L-*trans*-epoxysuccinyl-leucylamino-(4-guanidino)-butane) has been isolated from *Aspergillus japonicus* (Hanada *et al.*, 1978). Related compounds estatins A and B from *Myceliophthora thermophila* (Yaginuma *et al.*, 1989), cathestatins A and B from *Penicillium citrinum* (Woo *et al.*, 1995), AM4299 A and B from *Chromelosporium fulvum* (Morishita *et al.*, 1994) and kojistatin A from *Aspergillus oryzae* (Sato *et al.*, 1996) have also been reported. E-64 has been used frequently to characterize insect gut protease activity. E-64 also has been tested *in vivo* in some cases, mainly against beetles, which generally appear to have mostly thiol proteases (Murdock *et al.*, 1987). E-64 at 0.01 per cent in diets of *Callosobruchus maculatus* (Fabricius) produced mortality of ca. 43 per cent (Murdock *et al.*, 1988). At 0.35 mM in diets, E-64 produced 54 per cent mortality relative to controls on week old *Diabrotica undecimpunctata* Mannerheim larvae, and survivors gained almost no weight (Edmonds *et al.*, 1996). When administered in diets at 1 per cent to *T. confusum* larvae, 60 per cent mortality resulted and survivors had a 65 per cent RGR (Chen *et al.*, 1992). E-64 at 2000 ppm reduced survival of larvae to adults of *Sitophilus oryzae* (Linnaeus) by 75 per cent (Pittendrigh *et al.*, 1997). In one instance, E-64 produced RGRs of 64 per cent for larvae of *L. decemlineata* fed leaves treated with 200 ppm solutions (Wolfson and Murdock, 1987). In another study, E-64 at 1 $\mu g/cm^2$ reduced larval survival to adult of *L. decemlineata* by 53 per cent (Bolter and Latoszek-Green, 1997).

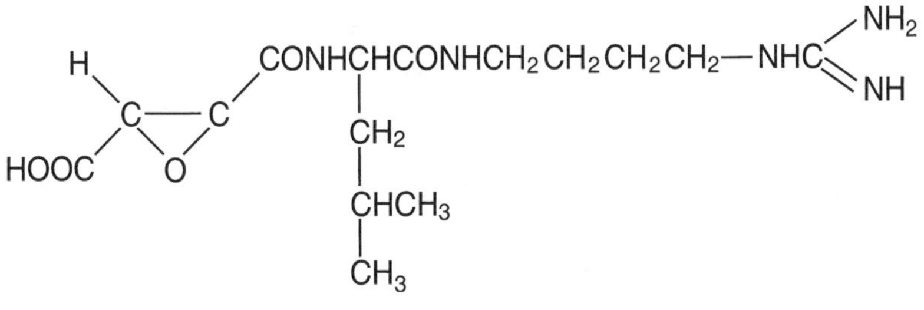

E-64

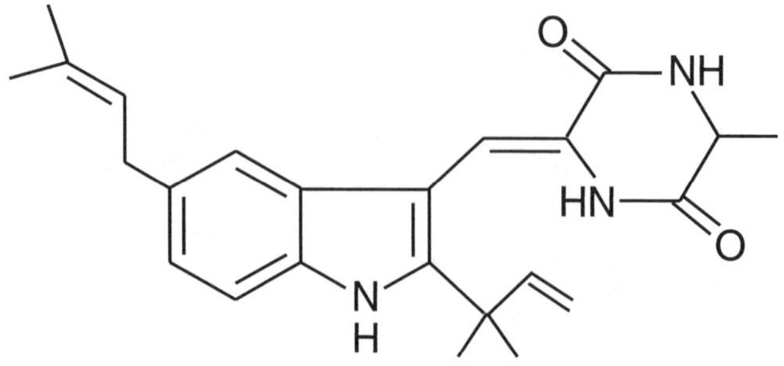

Isoechinulin A

Echinulins

Echinulins are derived from *Aspergillus* spp. (Nagasawa *et al.*, 1975) and additional *Aspergillus* (Turner and Aldridge, 1983), including *A. ochraceus* (DeGuzman *et al.*, 1992). Isoechinulin A (structure Nagasawa *et al.*, 1979) inhibited the growth of *B. mori* to an unspecified degree at 1000 ppm (Nagasawa, 1975). Cycloechinulin caused a 33 per cent RGR for *H. zea* at 100 ppm orally (DeGuzman *et al.*, 1992).

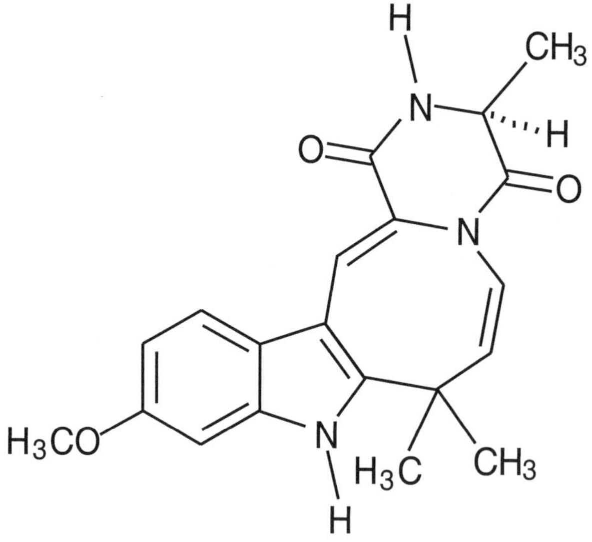

Cycloechinulin

Emodin

Emodin has been isolated from several species of fungi, including *Acroscyphus sphaerophoroides*, *A. aculeatus*, *Aspergillus wentii*, *Cetraria cucullata*, *Drechslera* sp.,

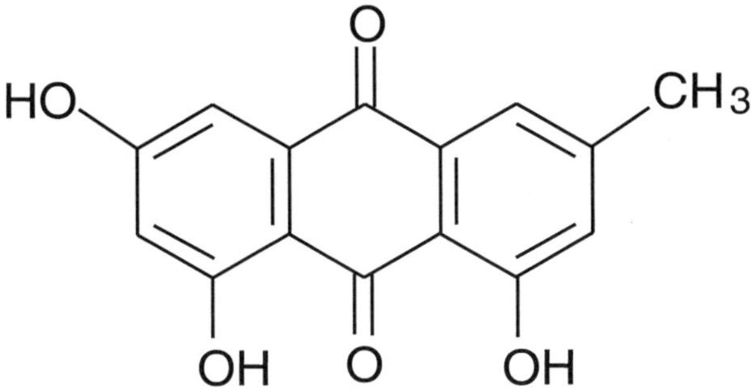

Emodin

Eurotium, Hypocrea, Nephoroma, Penicillium, Phoma, Pyrenochaeta, Talaromyces, and *Xanthoria* (Wells *et al.*, 1975; Turner and Aldridge, 1983). Emodin was a feeding deterrent to *Lymantria dispar* (Linnaeus) causing 46.7 per cent mortality at 45 ppm over the entire larval development period (Trial and Dimond, 1979). Emodin inhibited survival of *D. melanogaster* to pupation by ca. 32 per cent at 150 ppm and by ca. 66 per cent at 350 ppm (Besl and Blumreisinger, 1983).

Epiamauromines

Epiamauromines and derivatives are produced by *A. ochraceous* (DeGuzman *et al.*, 1992). Epiamauromine and *N*-methylepiamauromine caused a 30 and 17 per cent RGR of *H. zea* larvae, respectively, at 100 ppm oral administration (DeGuzman *et al.*, 1992). The structure is nearly the same as that of amauromine, isolated from the fungus *Amauroascus*, which is a vasodilator (Takase *et al.*, 1984). Presumably, the compounds could function similarly in insects.

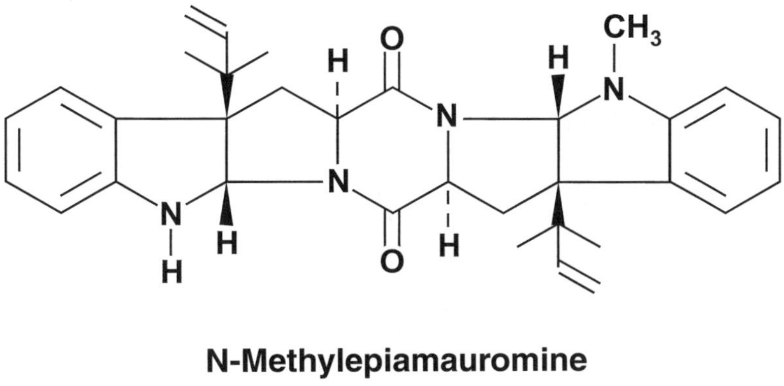

N-Methylepiamauromine

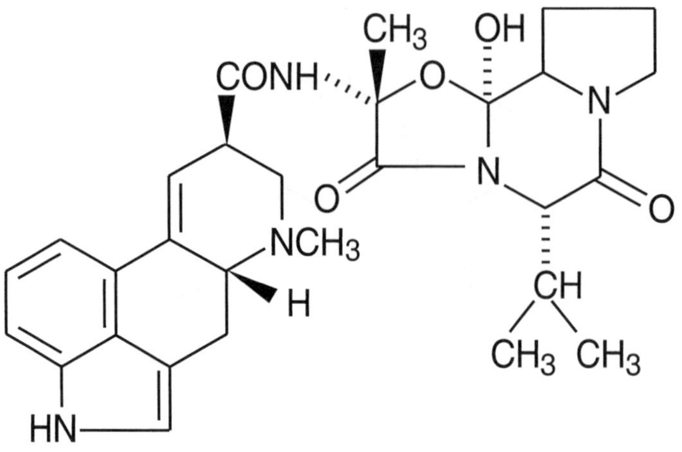

Ergovaline

Ergovaline

Ergovaline appears to be produced by the fungal endophyte of tall fescue, *Acremonium lolii/ arundinacea* and may be responsible for some activity against insects (Rowan and Gaynor, 1986; Siegel *et al.*, 1990; Roylance *et al.*, 1994). Ergolines display a high affinity for 5-hydroxy tryptamine binding sites, but many also have a high affinity for adrenergic and dopanergic binding sites (Glennon, 1987).

Eugenetin

Cylindrocarpon sp. (Turner and Aldridge, 1983) and *Aspergillus* (Gloer *et al.*, unpublished) produce Eugenetin. Eugenetin caused a 91 per cent RGR of *H. zea* larvae at 400 ppm (Dowd *et al.*, unpublished). Sublethal effects noted have included depigmented areas of the cuticle, which may indicate interfering with tanning (Dowd *et al.*, unpublished).

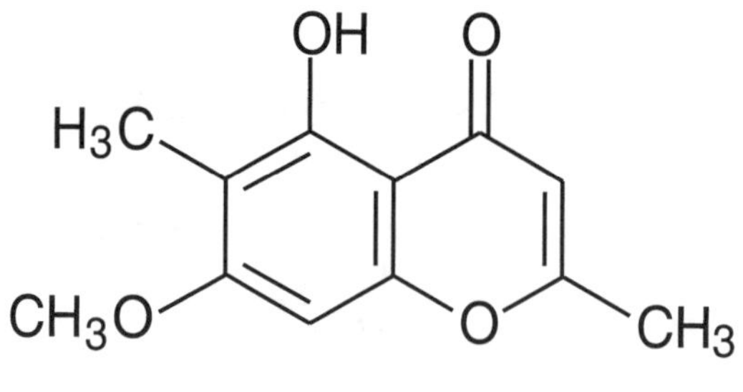

Eugenetin

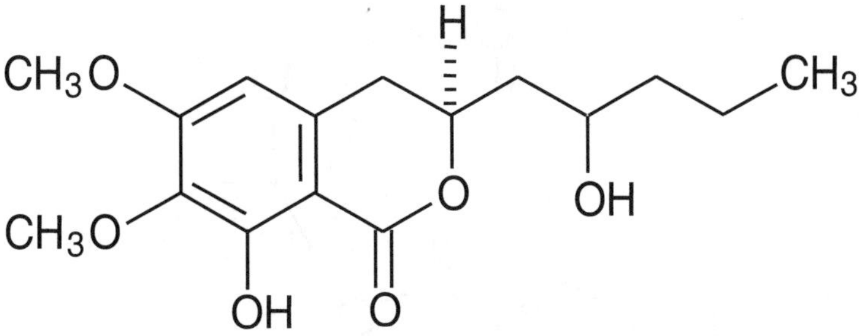

CH_3O ... CH_3 ... OH ... OH O ... H

Fusarentin Derivative

Fusarentins

Fusarentins were isolated from an entomopathogenic strain of *Fusarium larvarum* (Grove and Pople, 1979). Both fusarentin 6-methyl ether (10 μg) and fusarentin 6,7-dimethyl ether (7 μg) produced 30 per cent mortality after 3 days of treatment (Claydon *et al.*, 1979).

Fusaric acid

Fusaric acid is produced by several species of *Fusarium* (Turner and Aldridge, 1983). When 20 μg was injected into *C. vicina* adults, 70 per cent mortality resulted after 3 days (Claydon *et al.*, 1977). However, fusaric acid at 250 ppm reduced growth of *H. zea* to some extent but was less active against *S. frugiperda* (Dowd, 1988b). Fusaric acid appears to be more important as a synergist for cooccurring compounds (Dowd, 1988b, 1989b).

Griseofulvin

Griseofulvin has been isolated from several species of *Penicillium* (Turner and Aldridge, 1983) including the sclerotia of *P. raistrickii* (Belofsky *et al.*, 1998a). The related compound, 6-desmethylgriseofulvin, also has been isolated from the sclerotia of *P. raistrickii* (Belofsky *et al.*, 1998a). Griseofulvin interfered with cell division in *Antheraea polyphemus*

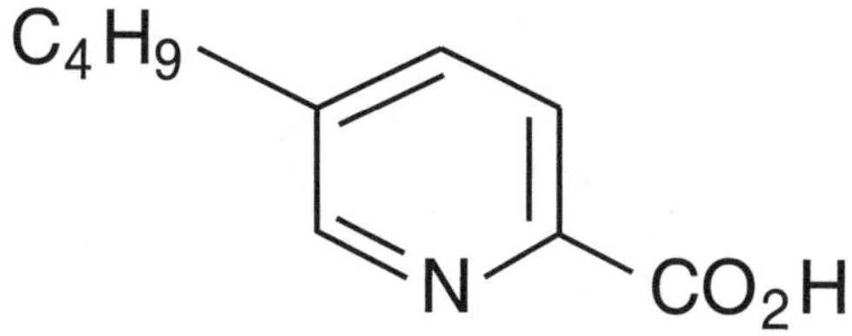

C_4H_9 ... N ... CO_2H

Fusaric Acid

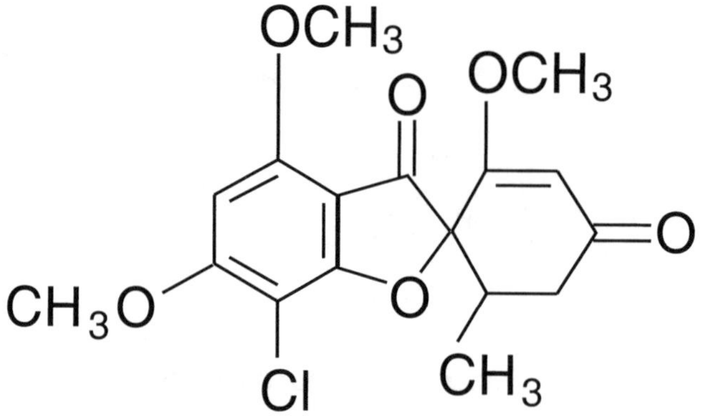

Griseofulvin

(Cramer). Griseofulvin at 18.5 ppm caused 71 per cent mortality of second instar *Aedes atropalpus* (Coquillett) larvae (Anderson, 1966). Griseofulvin at ca. 10 ppm caused cuticular abnormalities, including poor tanning and aberrant muscle attachment in larvae of *A. atropalpus* (Anderson, 1966). Griseofulvin appeared to be the main source of antiinsectan activity of *P. raistrickii* sclerotia. When griseofulvin was tested at natural concentration (400–500 ppm), it produced a 70 per cent RGR of *H. zea* and a 30 per cent RFR for *C. hemipterus* larvae (Belofsky *et al.*, 1998a). Griseofulvin was photo-activated when fed to *H. zea* and *S. frugiperda* (Dowd, 1993). There is also evidence that it interferes with porphyrin synthesis in these insects (Dowd, 1993). Griseofulvin appears to be metabolized primarily by oxidative enzymes in *S. frugiperda* and *H. zea* (Dowd, 1993).

Heptelidic acid

Heptelidic acid and analogs are produced by *Anthostoma avocetta, Chaetomium globosum, Glidocladium virens, Trichoderma viridae* (Turner and Aldridge, 1983) and the

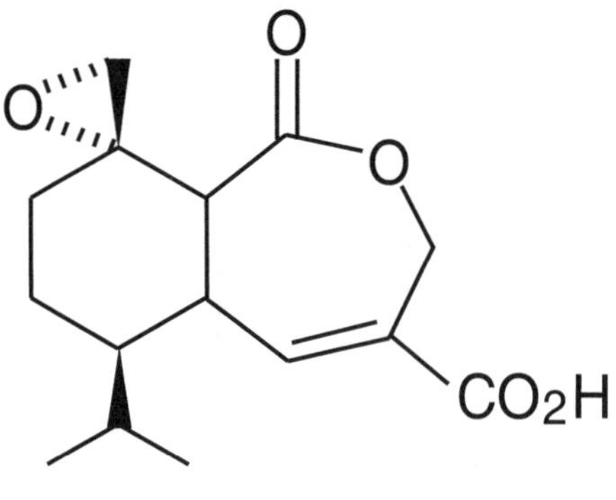

Heptelidic Acid

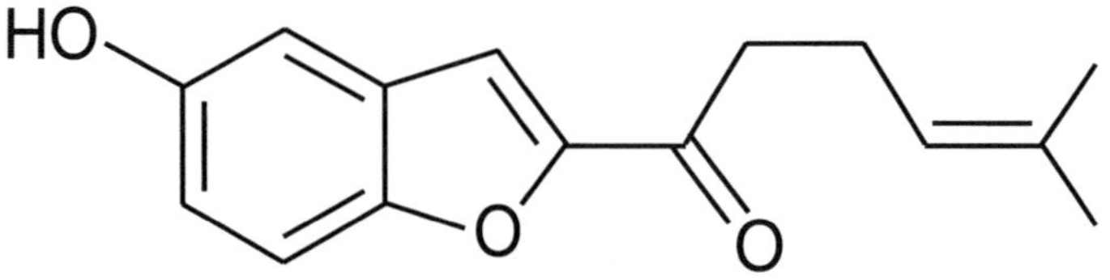

HO O O

5-Hydroxybenzofuran Derivative 1

spruce fungal endophyte *Phyllosticta* sp. (Calhoun *et al.*, 1992). Heptelidic acid (2.2 μmol), hydroheptelidic acid (16.8 μmol) and heptelidic acid chlorohydrin (4.7 μmol) killed 78, 88 and 100 percent of *C. fumiferana* after 2 weeks when incorporated into diets (Calhoun *et al.*, 1992).

Hexenylbenzofurans

Two hexenylbenzofurans [5-hydroxy-2-(1′-oxo-5′methyl-4′-hexenyl) benzofuran and 5-hydroxy-2-(1′-hydroxy-5′-methyl-4′-hexenyl)benzofuran] have been isolated from an unidentified endophytic fungus of wintergreen (*Gaultheria procumbens*) (Findlay *et al.*, 1997a). The first listed compound produced 36 per cent mortality of *Christoneura fumiferana* (Clemens) at 0.8 μmol per insect orally (Findlay *et al.*, 1997a).

Hydroquinone derivatives

The compound 2-methyl hydroquinone has been reported from *Penicillium urticae, Nectria erubescens, Phoma* spp. and *Scropulariopsis brumptii* (Heisey *et al.*, 1988a). Insecticidal properties had been reported previously (Bottger *et al.*, 1951; Questel and Gertler, 1952; Heisey *et al.*, 1988b). For example, 4lbs/100 gal of 2-methyl hydroquinone applied to maize leaves killed 37 per cent of *O. nubilalis* larvae (Questel and Gertler, 1952).

OH CH_3 OH

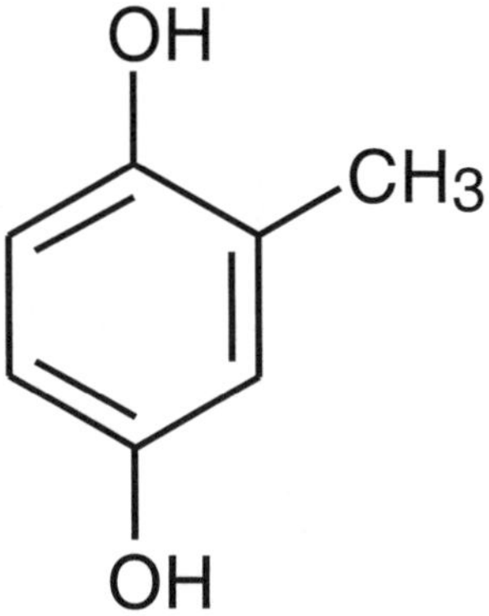

2-Methyl-hydroquinone

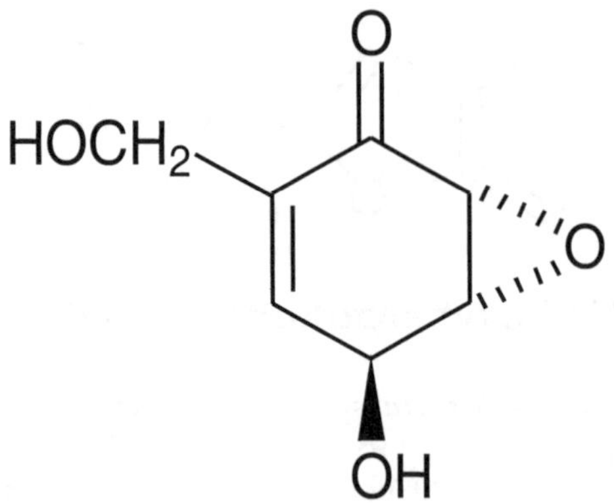

(+)-Isoepoxydon

Isoepoxydons

Isoepoxydon has been isolated from *Penicillium urticae* (Turner and Aldridge, 1983) and *Poronia punctata* (Gloer and Truckenbrod, 1988). It was active against *H. zea* in the 1000 ppm range (Dowd *et al.*, unpublished data).

Isopimaratrienes

Two diterpenoid isopimaratrienes have been isolated from an unidentified endophytic fungus of balsam fir (Findlay *et al.*, 1995b). Both compounds produced 44 per cent mortality of *C. fumiferana* at 6 μmol when incorporated into diets.

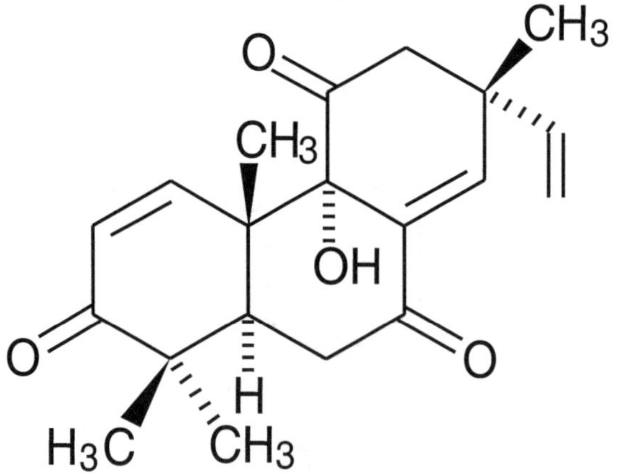

Isopimaratriene Derivative

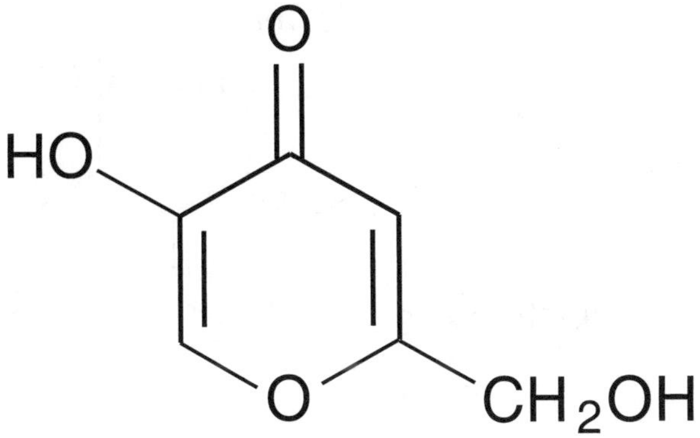

Kojic Acid

Kojic acid

Kojic acid is produced by several species of *Aspergillus* and *Penicillium* (Cole and Cox, 1981). At 15 per cent in a water source, kojic acid caused mortality and delayed development of surviving *Oncopeltus fasciatus* (Dallas) nymphs (Beard and Walton, 1969). At concentrations greater than 0.25 per cent, kojic acid killed *Aedes atropalpus* (Coquillett) larvae within a few days, while concentrations at 0.01 percent retarded development and prevented molting (Beard and Walton, 1969). The ED_{50} for kojic acid orally against *D. melanogaster* was 1562 ppm (Dobias *et al.*, 1977). At 250 ppm orally, kojic acid produced ca. 20 percent RGR for *H. zea*, but had no negative effect on *S. frugiperda* (Dowd, 1988b). However, kojic acid appears more important as an inhibitor of metalloenzymes, which allows it to interact synergistically with cooccurring metabolites and inhibit insect and plant defensive enzymes (Dowd, 1988a,b, 1994, 2000).

Kotanins and related compounds

Kotanins and isokotanins were isolated from the sclerotia of *Aspergillus alliaceus* (Laakso *et al.*, 1994). Kotanin and desmethylkotanin had been previously isolated from *A. glaucus* (Buchi *et al.*, 1971; Turner and Aldridge, 1983), and *A. flavus* sclerotia (TePaske *et al.*, 1992). Kotanin was inactive against *H. zea* and *C. hemipterus* at 100 ppm orally, but produced a 23 per cent RFR for *C. hemipterus* adults at 1000 ppm, a relevant natural concentration (Laakso *et al.*, 1994). Desmethylkotanin was inactive against *H. zea* and *C. hemipterus* (TePaske *et al.*, 1992). Isokotanin B produced a 21 per cent RFR for *C. hemipterus* larvae, and isokotanin C produced a 19 per cent RFR for *C. hemipterus* adults and a 11 per cent RGR for *H. zea* larvae at 100 ppm oral administration. Isokotanin A produced a 17 per cent RFR for *C. hemipterus* adults at 500 ppm orally, a near natural concentration (Laakso *et al.*, 1994). Kotanins are capable of inhibiting NADH oxidase (Dowd *et al.*, unpublished).

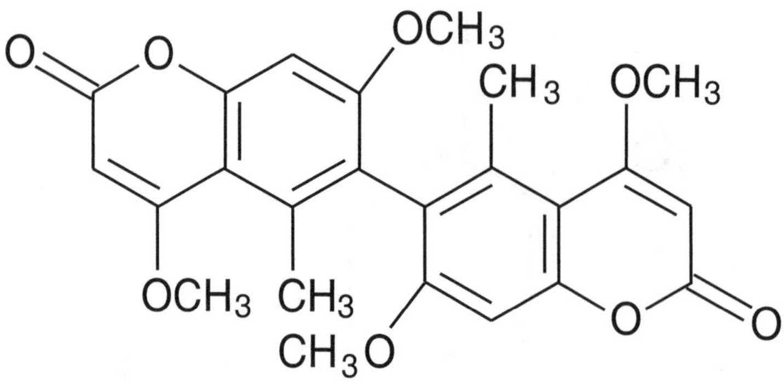

Isokotanin

A related compound, aflavarin was isolated from *A. flavus* sclerotia (TePaske *et al.*, 1992). It caused no effect on *H. zea* at 25 ppm orally, but caused a 66 and 53 per cent RFR for *C. hemipterus* adults and larvae, respectively, at the same concentration (TePaske *et al.*, 1992).

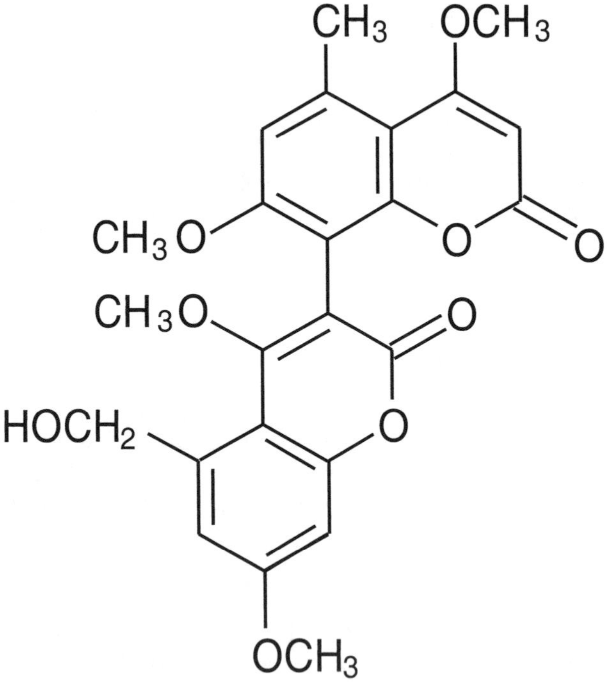

Aflavarin

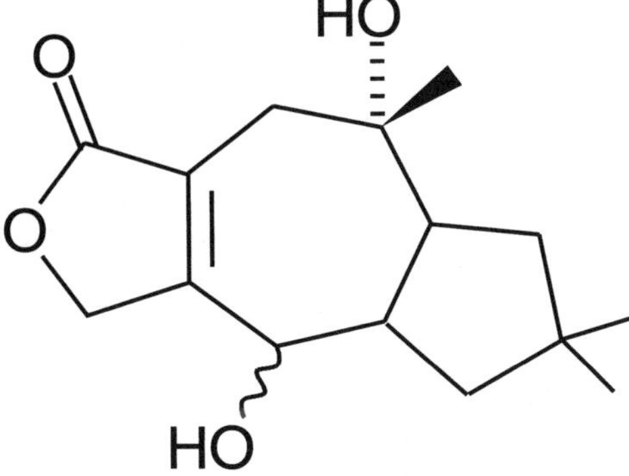

Lactarorufin A

Lactarans

Lactarorufins have been isolated from *Lactarius rufus* (Nawrot *et al.*, 1986). Activity range was lactarorufin A > B >isolactarorufin for *T. confusum* (Nawrot *et al.*, 1986). A mixture of monohydroxylactones from *Lactarius necator* also produced significant antifeedant activity towards *T. confusum* (Nawrot *et al.*, 1986). An extensive evaluation of those and related lactarane lactones towards *T. confusum, Trogoderma granarium* Everts, and *S. granarius* has been reported (Daniewski *et al.*, 1993a,b,1995). Among these most active compounds were lactarorufin A, 3-O-ethyl-8-dehydro-lactarorufin A, and 5-deoxy-8-dehydro-lactarolide B (Daniewski *et al.*, 1995). Combined results indicated that lactaranes and marasmanes were more active than isolactaranes (Daniewski *et al.*, 1995). Fewer hydroxyl groups produce greater activity of furans, but no consistent effect is noted with lactones (Daniewski *et al.*, 1995). Compounds with native C-8 moieties are more active.

Leporin A

Leporin A was isolated from *Aspergillus leporis* (TePaske *et al.*, 1991). Leporin A caused a 35.6 per cent RGR in *H. zea* at 25 ppm orally.

Macrophorins

Macrophorins are known from *Macrophora* sp. (Sassa and Yoshikoshi, 1983), and *Penicillium brevi-compactum* (Ayer *et al.*, 1990). A macrophorin analog was isolated from the ascostromata of *Eupenicillium crustaceum* and *E. molle* (Wang *et al.*, 1995). It occurred at 15,000 ppm in ascostromata of *E. crustaceum*, and at 2300 ppm caused a 63 per cent RGR for *H. zea* and a 40 and 69 per cent RFR for *C. hemipterus* adults and larvae, respectively (Wang *et al.*, 1995).

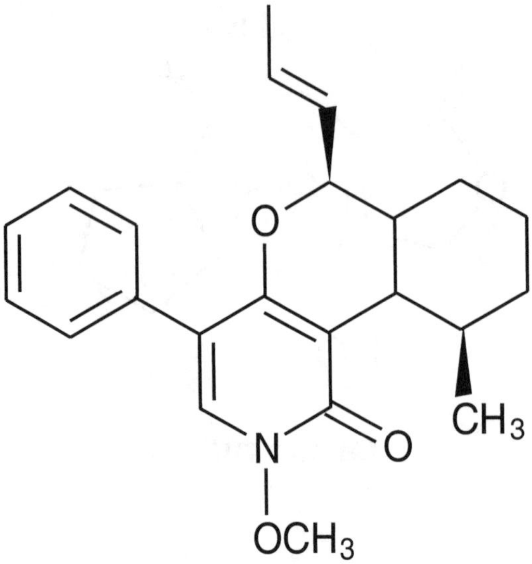

CH3

O

N

O

OCH3

Leporin A

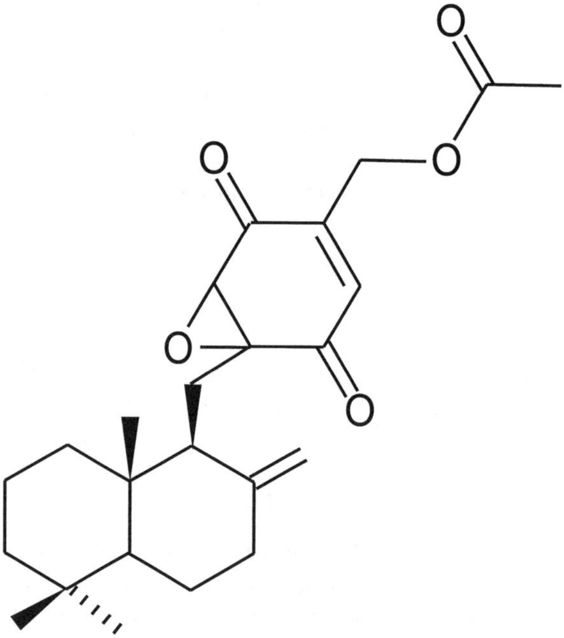

O

O

O

O

O

O

Macrophorin Derivative

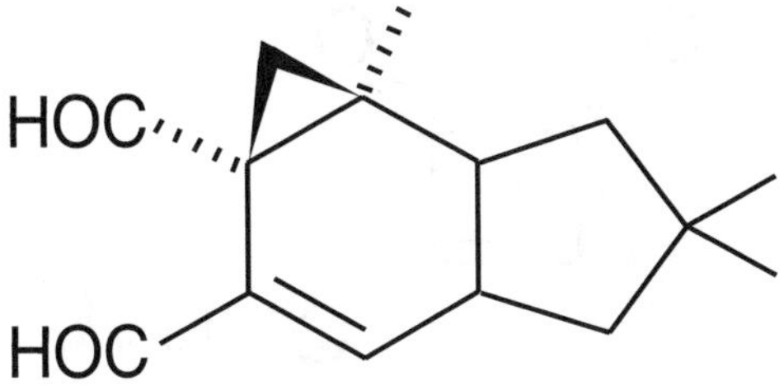

Isovelleral

Marasmanes

Marasmanes have been isolated from different species of *Lactarius* mushrooms, *Lentinellus ursinus*, and *Marasmius* spp. (Turner and Aldridge, 1983; Daniewski *et al.*, 1995; Camazine *et al.*, 1983). Compounds such as velleral and isovelleral are formed when mushrooms are broken (Magnusson *et al.*, 1972, 1973). Precursors are stored in the hyphae (Camazine and Lupo, 1984). Many of these compounds have antifeedant activity (Daniewski *et al.*, 1993a,b). In a structure-activity comparison (Daniewski *et al.*, 1995), 13-hydroxy- marasm-7(8)-3n-5-oic-acid γ-lactone and 5,8 α-dihydroxy-13-nor-marasm-7-one were the most active of the class against *T. confusum, T. granarium*, and *S. granarius* (Daniewski *et al.*, 1995). (see also lactarans).

Melleins

Melleins are reported from *Aspergillus ochraceus, Apiosphora camptospora, Cercospora taiwanensis* (Turner and Aldridge, 1983) and *Fusarium larvarum* (Grove and Pople, 1979). (+)-Mellein (5.5 μg) injected into adult *C. vicina* produced 13 per cent mortality after 3 days, although 87 per cent knockdown was noted after initial injection (Claydon *et al.*, 1979).

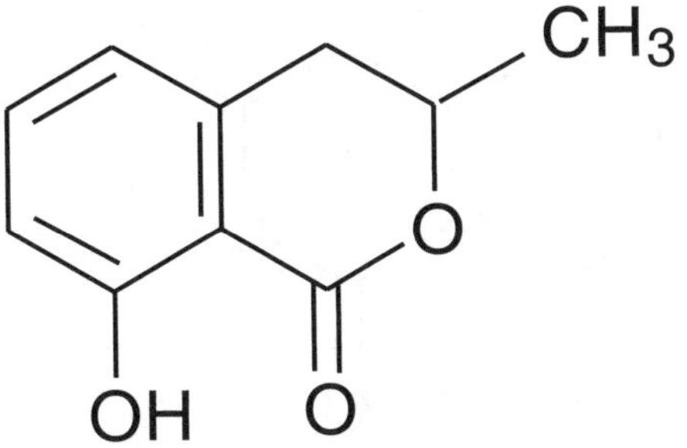

Mellein

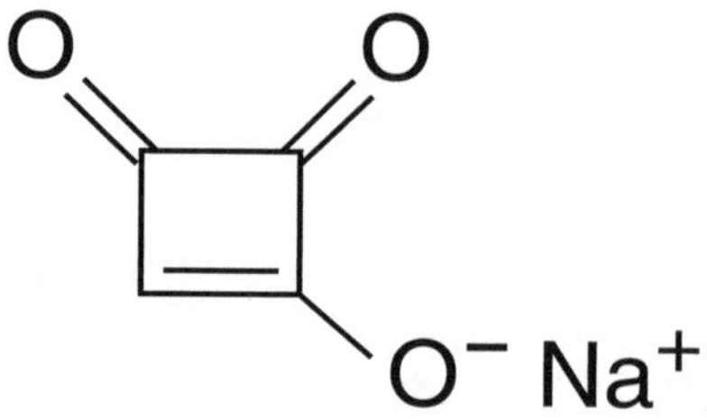

Moniliformin

Moniliformin

Moniliformin is produced by several species of *Fusarium* (Turner and Aldridge, 1983). Moniliformin produced a 22 per cent RGR for *S. frugiperda* at 250 ppm (Dowd, 1992b).

Monocerins

Monocerins are reported from *Drechslera monoceras* (= *Helminthosporium monoceras*) and *Fusarium larvarum* (Grove and Pople, 1979). Monocerin (17.5 μg) injected into adult *C. vicina* produced 42 per cent mortality after 3 days, but was inactive against *A. aegypti* (Claydon *et al.*, 1979).

Muscimol and related compounds

Muscimol and ibotenic acid have been isolated from *Amanita muscaria, A. pantherina* and related species of mushrooms (Turner and Aldridge, 1983; Beutler and der Darderosian, 1981; Konno, 1995; Mizuno, 1995). Muscimol killed 45 percent of

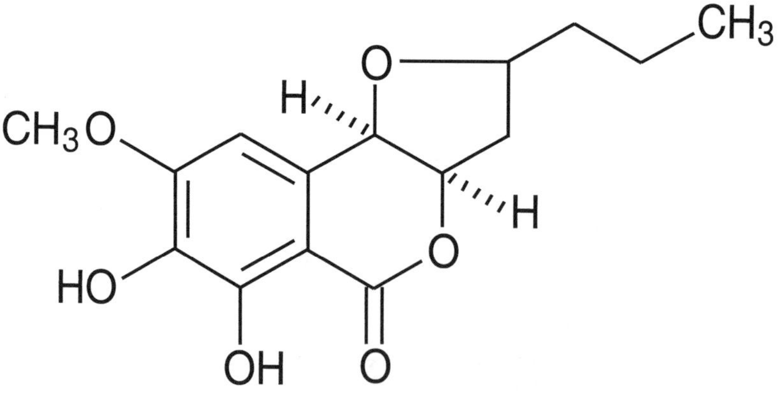

Monocerin Derivative

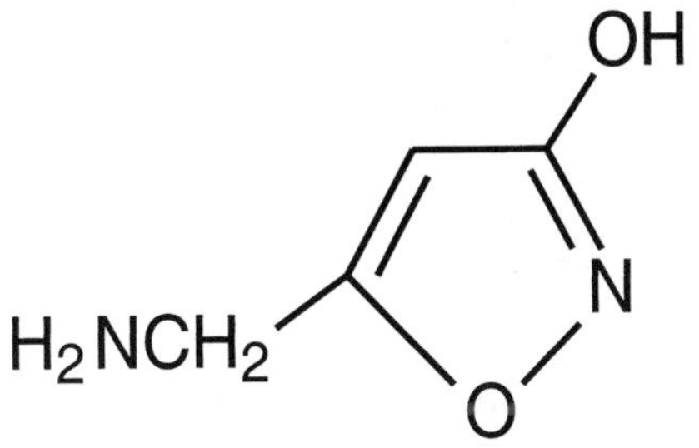

Muscimol

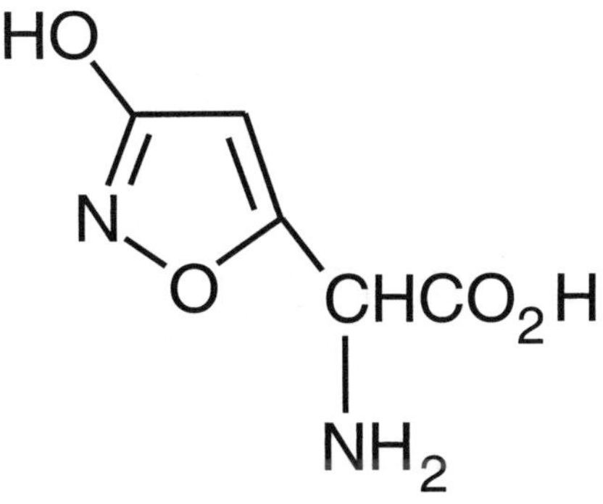

Ibotenic Acid

M. domestica at 40 mg/food quantity in 1 day, and killed 100 per cent of *A. aegypti* at 20 mg/larva in 25 min (Eugster, 1969). Muscimol is reported as a GABA- agonist for cockroach CNS extracts (Sattelle, 1990). Ibotenic acid killed 28 percent of adult *M. domestica* at 40mg/food quantity and killed at least 90 percent of *A. aegypti* larvae at 20 μg/insect (Eugster, 1969). Ibotenic acid has been reported as a glutamate agonist (Konno, 1995). Tricholomic acid has been isolated from *Tricholma muscaria*, and is similar to ibotenic acid in activity against flies (Konno, 1995). Muscarine, also from *Amanita* spp. interacts with specific receptors in both vertebrate and insect nervous systems (e.g. Shankland, 1976).

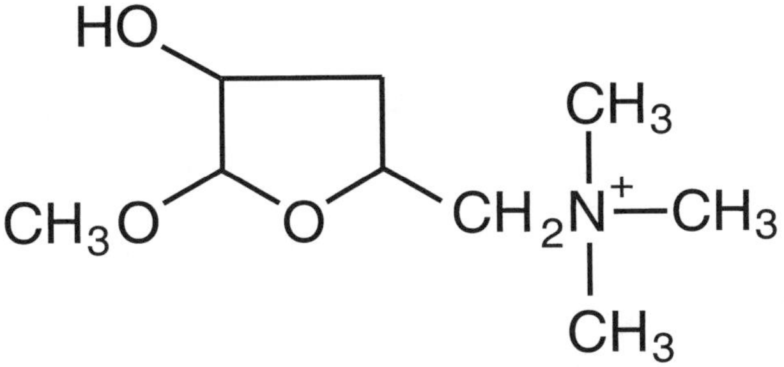

Muscarine

Mycophenolic acid

Mycophenolic acid has been isolated from *Penicillium* spp., *Verticicladella abientina*, and *Septoria notorum* (Turner and Aldridge, 1983). Mycophenolic acid is reported as an inhibitor of DNA synthesis and altered glial filament organization in cancer cell cultures (Lipsky and Silverman, 1987). It showed some activity against *H. zea* at 250 ppm (Dowd, unpublished).

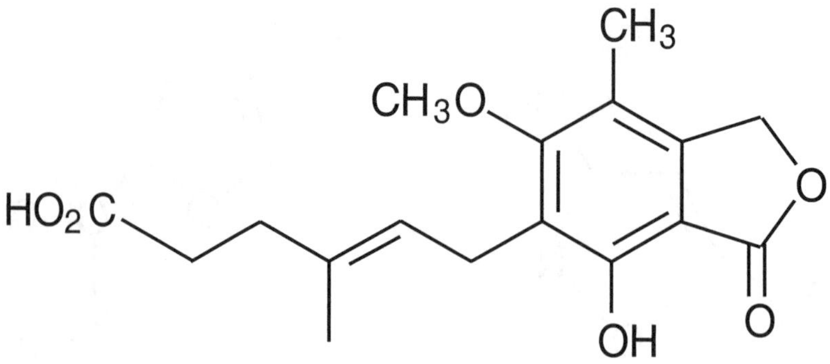

Mycophenolic Acid

Naphthazarin pigments

The naphthazarin pigments and relatives fusaruban, anhydrofusarubin, and javanicin were isolated from entomopathogenic strains of *Fusarium solani* and *F. javanicum* (Claydon *et al.*, 1977). When injected into *C. vicina* adults, mortalities after three days were 40, 55 and 50 per cent for anhydrofusarubin (0.5 μg), javanicin (0.5 μg) and fusarubin (7 μg), respectively (Claydon *et al.*, 1977). These compounds are known to chelate metal ions (Claydon *et al.*, 1977), so several possible targets exist for the site of action (see under synergism).

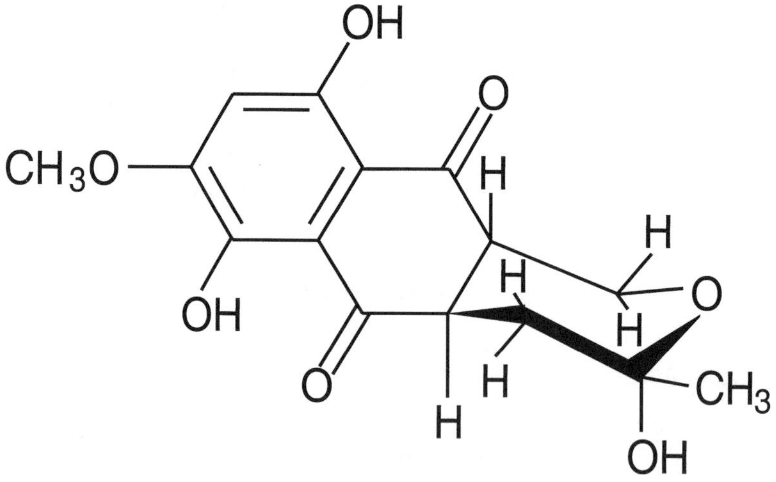

Anhydrofusarubin

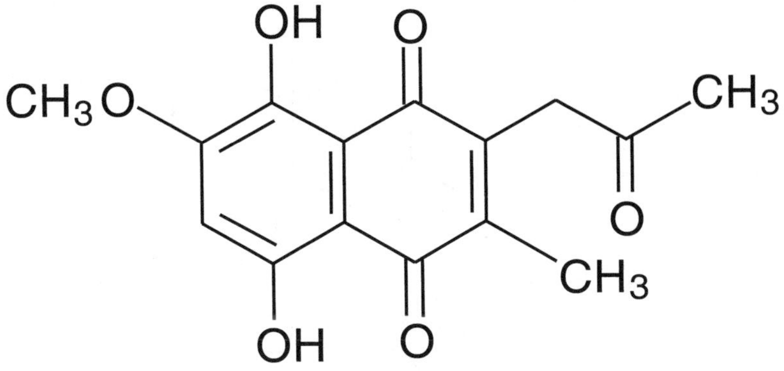

CH3O, OH, O, OH, O, CH3, O, CH3

Javanicin

Nigragillin

Aspergillus niger (Isogai *et al.*, 1975) produces nigragillin. Nigragillin caused immediate knockdown, occasionally followed by death when administered topically at 5 μg/g to *B. mori* larvae (Isogai *et al.*, 1975).

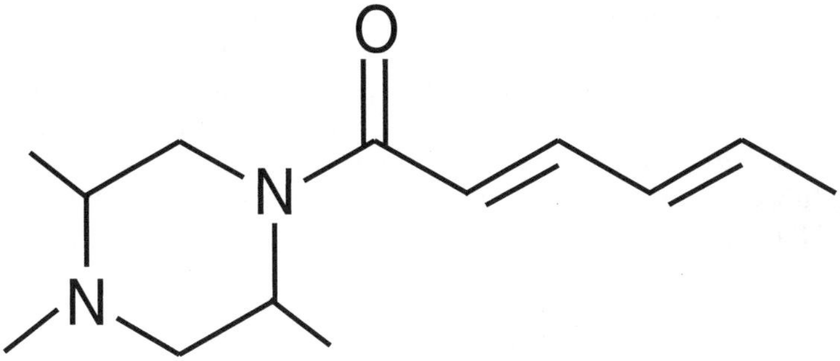

O, N, N

Nigragillin

Nominine

Nominine has been isolated from the sclerotia of *Aspergillus nomius* (Gloer *et al.*, 1989), and *A. alliaceus* (Laakso *et al.*, 1994). It was particularly active against *H. zea* at 25 ppm, since it produced 40 per cent mortality and 96.6 percent RGR. It was also active against *C. hemipterus* (Gloer *et al.*, 1989).

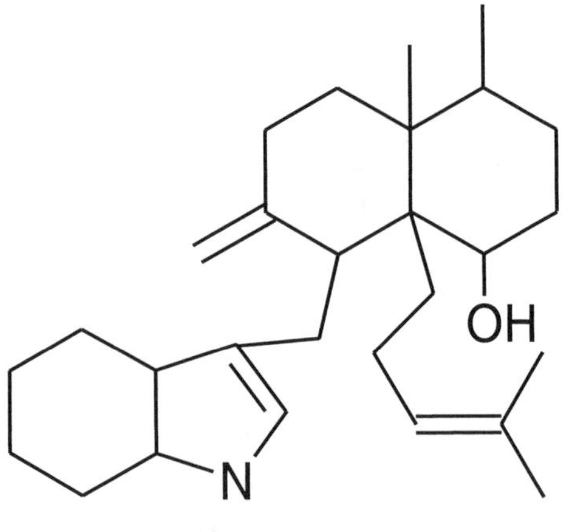

Nominine

Ochratoxins

Ochratoxin A has been isolated from *Penicillium* and *Aspergillus* spp. (Turner and Aldridge, 1983), including *A. carbonarius* sclerotia, where it occurs at ca. 50 ppm (Wicklow *et al.*, 1996). While ochratoxin A at 1000 ppm reduced the growth rate of *Attagenus unicolor* (Brahm) by 80 percent, it was almost inactive against *T. confusum* and *L. serricorne* (Wright *et al.*, 1980). Ochratoxin A produced up to 78 per cent mortality of *E. kuehniella* at 10 ppm orally (Wright and Harein, 1982). Ochratoxin A reduced the numbers of *Tyrophagus putrescentiae* (Schrank) larvae developing to adults by ca. 4.5 fold at 100 ppm (Rodriguez *et al.*, 1980).

This compound also produced a 50 per cent reduction in feeding by *D. melanogaster* and 40 percent mortality of *S. littoralis* at 10 ppm after oral treatment (Patterson *et al.*, 1987). Ochratoxin A at 10 ppm killed 100 per cent of *Plodia interpunctella* (Hubner) moths (Sargent, 1974) and induced 38.5 and 92.3 percent mortality and a 78.6 and 66.7 per cent RGR in *H. zea* and *S. frugiperda* larvae at 25 ppm

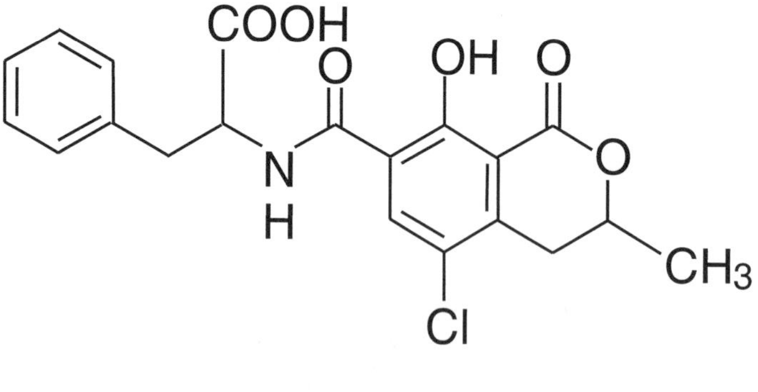

Ochratoxin A

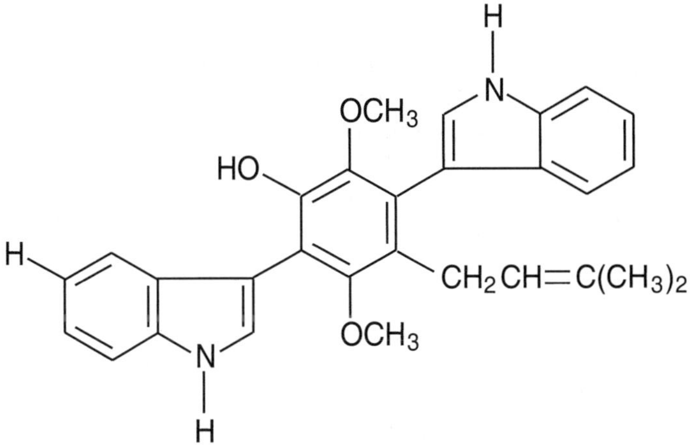

Ochrindole

oral administration respectively (Dowd, 1989a). The levels tested adversely affected the structure of the Malpighian tubules of these insects.

Ochrindoles

Ochrindoles were isolated from sclerotia of *Aspergillus ochraceus* (DeGuzman *et al.*, 1994). Ochrindole A produced a 30 percent RGR for *H. zea* larvae and a 20 RFR for *C. hemipterus* adults at 200 ppm orally (DeGuzman *et al.*, 1994).

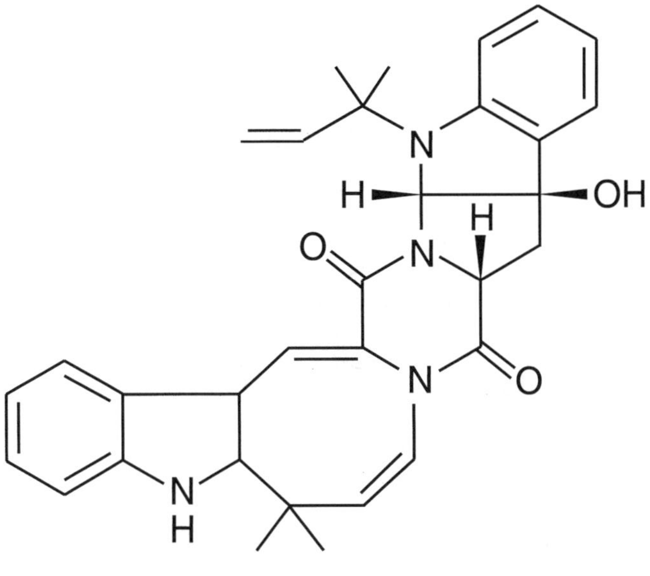

Okaramine A

Okaramines

Okaramines have been isolated from *Penicillium simplicissimum* (Hayashi *et al.*, 1989, 1995; Hayashi and Sakaguchi, 1998; Murao *et al.*, 1988). Relative activity against *B. mori* larvae for okaramines A-G (estimated from graphs) was 8, 0.2, 8, 20, >100, >100, and 40 ppm, respectively (Hayashi *et al.*, 1989, 1995; Hayashi and Sakaguchi, 1998). Okaramine B also had similar activity against 2nd instar *Spodoptera exigua* (Hübner) larvae (Hayashi *et al.*, 1989).

Paraherquamides and related compounds

Penicillium charlesii and *P. paraherquei* (Turner and Aldridge, 1983) produce paraherquamides. Published patents for these compounds have suggested insecticidal activity, but the most closely related data provided is for nematodes, where "stiffening" was noted (Goegelman *et al.*, 1988). A related compound, sclerotiamide, was isolated from the sclerotia of *Aspergillus sclerotiorum* (Whyte *et al.*, 1996). Sclerotiamide produced 45 percent mortality and a 98 percent RGR of survivors of *H. zea* when tested orally at 200 ppm (Whyte *et al.*, 1996). Symptomology was unusual, with the intersegmental membrane swollen beyond the cuticular plates in some individuals at sublethal doses (Whyte *et al.*, 1996; Dowd unpublished data). Dead insects often showed shriveling and blackening consistent with a ruptured cuticle and subsequent oxidation of DOPA compounds in the hemolymph by phenyloxidases (Whyte *et al.*, 1996; Dowd unpublished data). Effects were similar to those noted with sublethal concentrations

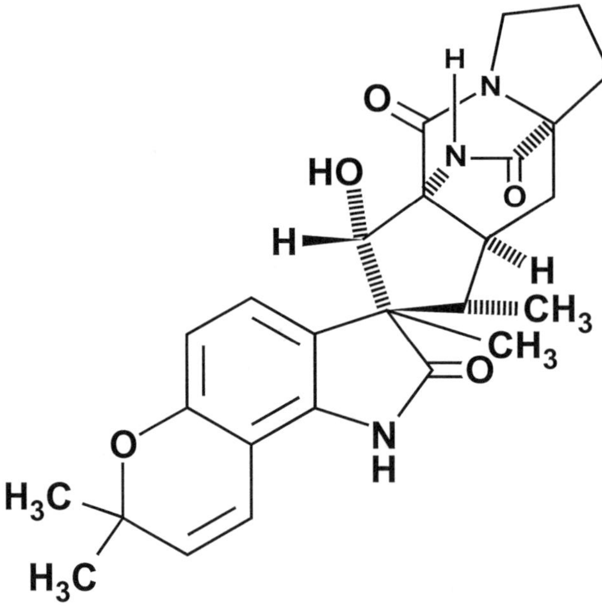

Sclerotiamide

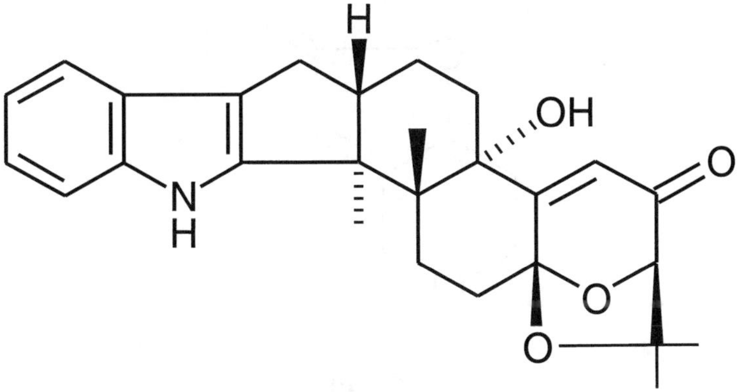

Paspalinine

of cyromazine on *H. zea* (Dowd, unpublished data), and *Manduca sexta* (Johannsen) (Hughes *et al.*, 1989). A possible explanation of the effects of sclerotiamide on caterpillars and nematodes at this point is that it interferes with regulation of hydrostatic pressure of the hemocoele, such that pressure is increased above normal levels. Sclerotiamide also caused 44 and 40 per cent RFR for *C. hemipterus* larvae and adults, respectively (Whyte *et al.*, 1996).

Paspalanines and derivatives

Paspalanine has been isolated from *Claviceps paspali* (Gallagher *et al.*, 1980b), *Aspergillus flavus* cultures (Turner and Aldridge, 1983), *A. flavus* sclerotia (TePaske *et al.*, 1992), the sclerotia of *A. nomius* (Staub *et al.*, 1993), and the ascostromata of *E. shearii* (Belofsky *et al.*, 1995). Paspalinine had no activity against *H. zea* or *C. hemipterus* at 25 ppm (Staub *et al.*, 1993) but produced a 75 per cent RGR of *H. zea* and a 62 per cent RFR for *C. hemipterus* larvae at 100 ppm orally (Belofsky *et al.*, 1995). Other paspalanine

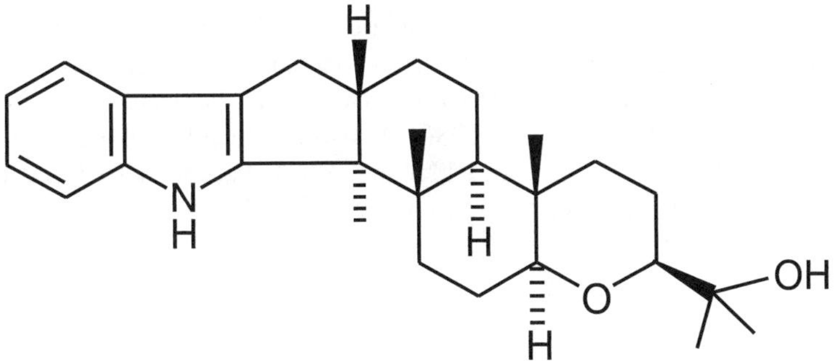

Paspaline

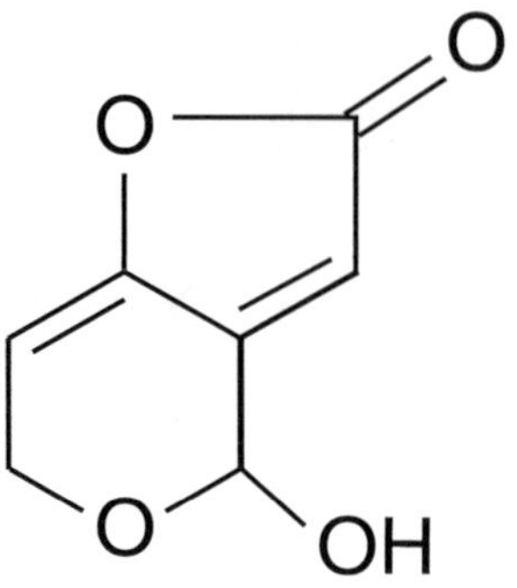

O, O, OH

Patulin

analogs also have significant activity against insects. For example, 14-hydroxypaspalanine and 14-(N,N-dimethyl-L-valyloxy) paspalinine have been isolated from the sclerotia of *A. nomius* (Staub *et al.*, 1993). The 14- hydroxy and 14-dimethylvalyl compounds produced a 91 and 88 percent RGR for *H. zea* larvae at 25 ppm oral administration (Staub *et al.*, 1993).

Paspaline has been isolated from *Claviceps paspali* (Turner and Aldridge, 1983), *Aspergillus flavus* sclerotia (TePaske *et al.*, 1992), and *A. alliaceus* sclerotia (Laakso *et al.*, 1994). Paspaline given orally to *H. zea* and *C. hemipterus* was inactive at 100 ppm (Laakso *et al.*, 1994).

Patulin

Patulin is produced by several species of *Penicillium* (Turner and Aldridge, 1983). Patulin had no oral activity against larvae of *Lucilia sericata* (Meigen) but did produce 95 percent knockdown of adult *D. melanogaster* at 10 $\mu g/cm^2$ in 24 h contact toxicity assays (Cole and Rolinson, 1972). Patulin had no effect on *D. melanogaster* feeding at 10 ppm, and only slightly reduced (8.2%) feeding by *S. littorallis* larvae (Patterson *et al.*, 1987). Other studies have also reported that patulin has little effect on *D. melanogaster* (Reiss, 1975). Patulin caused 38.4 per cent mortality of *G. mellonella* larvae at 1 per cent orally (Lysenko and Vititnev, 1976). Patulin decreased the diameter of Malpighian tubules in this species (Lysenko and Vititnev, 1976). Patulin inhibited growth of *T. confusum, L. serricorne* and *Attagenus megatoma* (Fabricius) by 23, 27 and 40 per cent, respectively (Wright *et al.*, 1980). It also produced 40 per cent mortality of *E. kuehniella* at 1000 ppm (Wright and Harein, 1982). Patulin has caused disruption of cell membranes in cell cultures (Riley and Showker, 1991), so a similar mode of action is likely to occur in insects.

Paxillines

Paxillines have been isolated from *Penicillium paxilli* (Springer *et al.*, 1975; Mantle *et al.*, 1990; Nozawa *et al.*, 1988; Gallagher *et al.*, 1980b; Mantle and Weedon, 1994) and

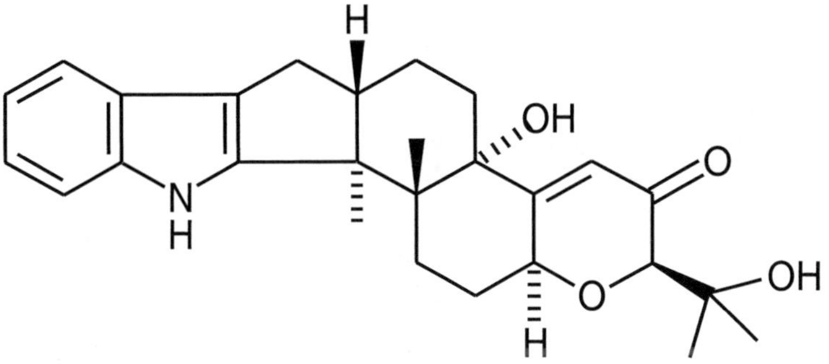

Paxilline

ascostromata of *Eupenicillium shearii* (Belofsky *et al.*, 1995). The paxillines isolated from *E. shearii* had varying activity against insects, with 21-isopentenylpaxilline, paxilline, 7-hydroxy-13-dehydroxypaxilline, 12-dehydroxypaxilline, and 2,18-dioxo-2,18-seco-paxilline producing a 55, 83, 65, 0 and 17 percent RGR, respectively for *H. zea*, and a 0, 38, 38, 12 and 19 percent RFR, respectively for *C. hemipterus* larvae at 100 ppm oral treatment (Belofsky *et al.*, 1995). When fed orally at 25 ppm, paxilline caused a 35.8 and 28.9 per cent RGR for *H. zea* and *S. frugiperda*, respectively (Dowd, *et al.*, 1988). Paxilline also caused 85 per cent reduction in feeding damage by *S. frugiperda* when added to cotton leaf disks (75 mg) at 50 μg level (Belofsky *et al.*, 1995). This compound inhibits GABA receptors in rat brain and *Torpedo* electric organ (Gant *et al.*, 1987), so it is likely that it is acting on GABA receptors in insects as well (Dowd *et al.*, 1988).

Penicillic acid

Penicillic acid has been isolated from several species of *Penicillium* and *Aspergillus* (Turner and Aldridge, 1983). Penicillic acid produced 94 percent reduction in feeding by *D. melanogaster* and 90 per cent mortality of *S. littoralis* at 10 ppm oral administration (Patterson *et al.*, 1987). It also produced 26.4 and 31.9 percent RGRs for *H. zea* and *S. frugiperda*, respectively at 25 ppm oral treatment (Dowd, 1989a).

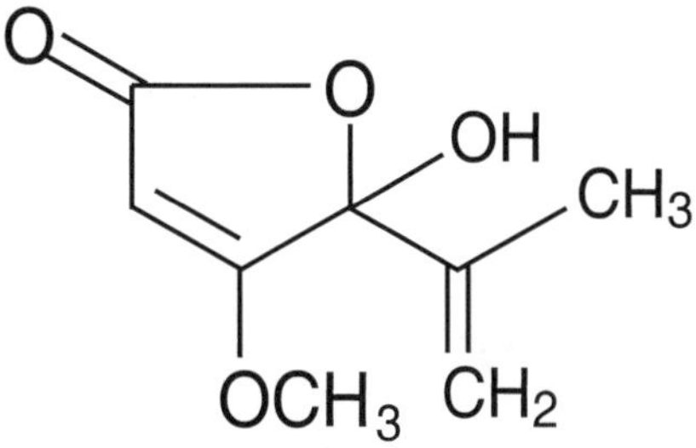

Penicillic Acid

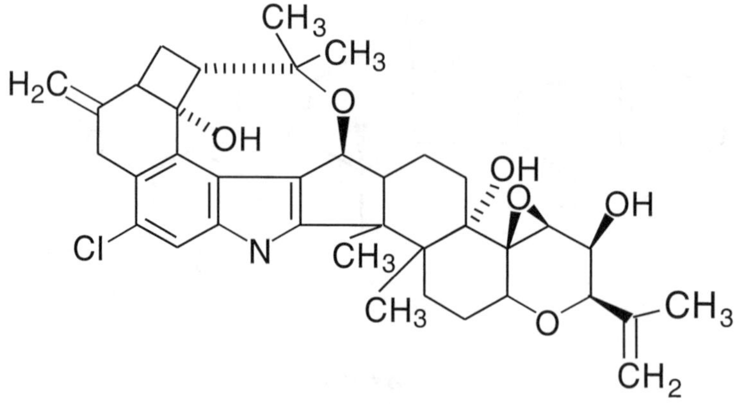

Penitrem A

Penicillic acid also appears to be important as a synergist for cooccurring compounds (Dowd, 1989a).

Penitrems and related compounds

Penitrem A has been isolated from *Penicillium crustosum* and *P. cyclopium* (Turner and Aldridge, 1983). At 25 ppm orally, penitrem A produced a 76.6 percent RGR for *S. frugiperda*, while at 0.25 ppm it produced a 81.9 percent RGR for *H. zea* (Dowd *et al.*, 1988). Penitrem B and secopenitrem B were isolated from sclerotia of *Aspergillus sulphureus* (Laakso *et al.*, 1992b). At 100 ppm, secopenitrem B caused 32 percent mortality and 87.0 percent RGR of *H. zea*, but had no effect on *C. hemipterus*. Topically, secopenitrem B at 2 μg per 2 mg larva caused a 66.2 percent RGR of *H. zea*. Penitrem B caused 87.4 percent RGR for *H. zea*, however, no effect on *C. hemipterus* larvae has been observed after 100 ppm oral administration of this compound (Laakso *et al.*, 1992b). Another penitrem analog, 10-oxo-11, 33-dihydropenitrem B, was isolated from the sclerotia of *A. sulphureus* (Laakso *et al.*, 1993). This penitrem analog caused a 95 percent RGR for *H. zea* and a 33 RFR for *C. hemipterus* adults at 25 ppm (Laakso *et al.*, 1993). Penitrem A and 6-bromo penitrem E have produced convulsions at 0.3 ppm in *B. mori*, when given orally to these larvae (Hayashi *et al.*, 1993a,b). Presumably, penitrems act at GABAnergic sites (Dowd *et al.*, 1988).

Peramine

Apparently peramine (Rowan *et al.*, 1986) is produced by the fungal endophyte of tall fescue, *Acremonium lolii arundinacea* (Rowan and Gaynor, 1986; Siegel *et al.*, 1990; Roylance *et al.*, 1994). Peramine is responsible for aphid, *Rhopalosiphum padi* (Linnaeus) resistance in tall fescue (determined indirectly, Siegel *et al.*, 1990) and Argentine stem weevil, *Listronotus bonariensis* (Kuschel) resistance in perennial rye grass (Rowan *et al.*, 1986). The *L. bonariensis* feeding was significantly deterred by

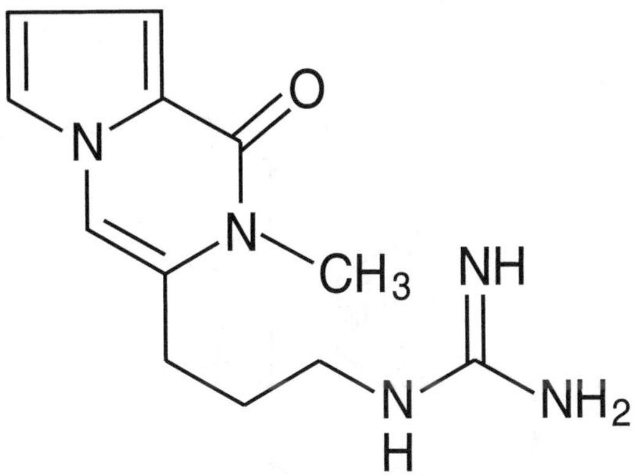

Peramine

10 ppm of peramine (Rowan and Gaynor, 1986), and peramine hydrobromide caused overall mortality of 20 percent from hatch to pupation (Rowan *et al.*, 1990). Peramine appears to inhibit activity of P-450 (unspecific monooxgenase) detoxifying activity in *Spodoptera eridania* (Cramer) (Dubis *et al.*, 1992).

Phomalactone

Phomalactone has been obtained from *Hypomyces rosellus, Nitrospora* sp. and *Phoma minispora* (Turner and Aldridge, 1983) and strains of the insect pathogen *Hirsutella thompsonii* var. *synnematosa* (Krasnoff and Gupta, 1994). At 1000 ppm orally, (+)-phomalactone produced 44.2 percent mortality after 5 days (Krasnoff and Gupta, 1994).

Pyripyropene

Pyripyropene A is reported from *Aspergillus fumigatus* (Kim *et al.*, 1994) and the ascostromata of *Eupenicillium reticulisporum* (Wang *et al.*, 1995). It produced a 62 percent RGR for *H. zea* at 50 ppm orally (Wang *et al.*, 1995). Pyripyropene A is

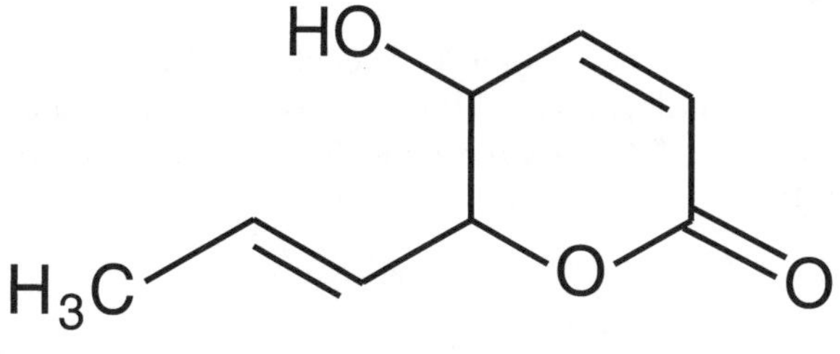

(+)-Phomalactone

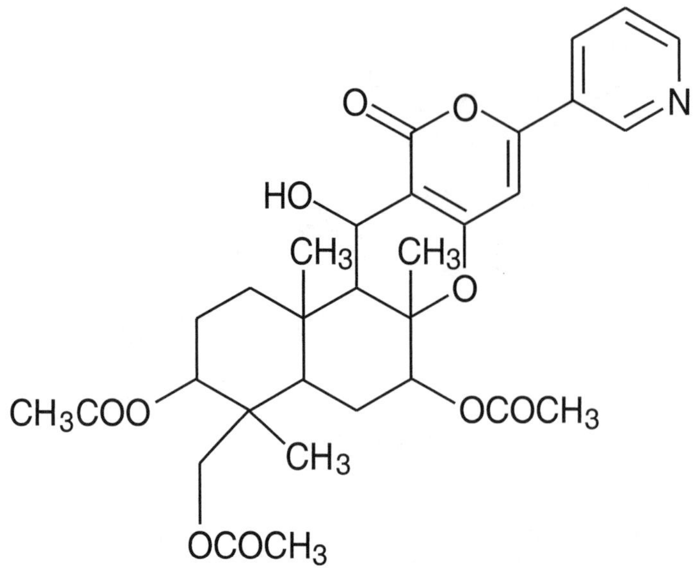

Pyripyropene

reported to be a highly effective inhibitor of acyl- coenzyme A cholesterol acyltransferase from rat liver (Kim *et al.*, 1994). Because cholesterol metabolism is also important in *H. zea* (Nes *et al.*, 1997), the mode of action in *H. zea* is likely to be similar to that for rats.

Radarins

Radarins were isolated from sclerotia of *Aspergillus sulphureus* (Laakso *et al.*, 1992a). Radarins A-D produced a 52.7, 0, 17.1 and 0 percent RGR in *H. zea* larvae at 100 ppm oral treatment, respectively (Laakso *et al.*, 1992a).

Ramulosin derivatives

Isocoumarin ramulosin derivatives were isolated from the spruce fungal endophyte *Canoplea elegantula* (Findlay *et al.*, 1995a). Compounds 1 and 3 produced 36 and 32 percent mortality of *C. fumiferana* at 5.6 and 13.7 mg in diet, respectively (Findlay *et al.*, 1995a).

Rubratoxins

Penicillium rubrum and *P. purpurogenum* (Turner and Aldridge, 1983) produce rubratoxins. Rubratoxin A and B produced LD_{50}s of 18 and 200 ppm, respectively,

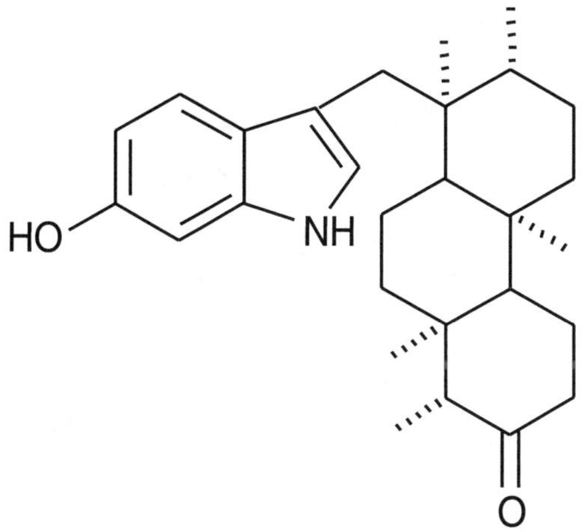

Radarin A

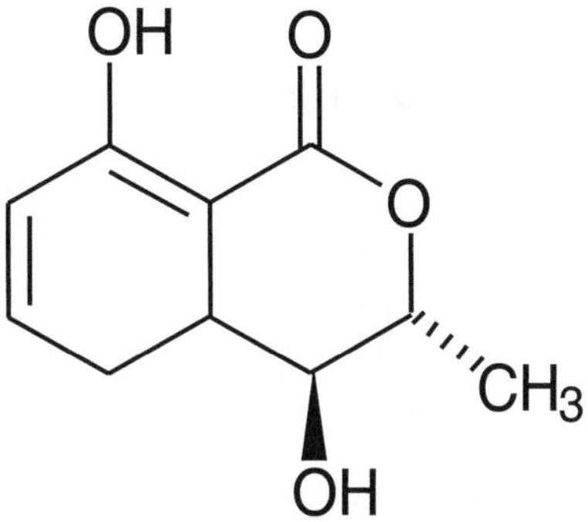

Ramulosin Derivative

against *L. sericata* larvae (Cole and Rolinson, 1972). Rubratoxin B at 1000 ppm caused a 94, 23 and 20 percent reduction in growth rates of *T. confusum, L. serricorne*, and *A. megatoma*, respectively (Wright *et al.*, 1980). Rubratoxin B produced up to 68 percent mortality of *E. kuehniella* at 100 ppm orally (Wright and Harein, 1982). Rubratoxins had little effect on *D. melanogaster* (Reiss, 1975).

Rugulosin

Rugulosin has been isolated from *Penicillium* spp., *Myrothecium verrucaria* and the balsam fir endophyte *Hormonema dematioides* (Calhoun *et al.*, 1992). The ED_{50} for *D. melanogaster* was 27.6 μg/ml (Dobias *et al.*, 1980). Rugulosin incorporated into diets at 9.2 μmol killed 14 per cent of *C. fumiferana* (Calhoun *et al.*, 1992).

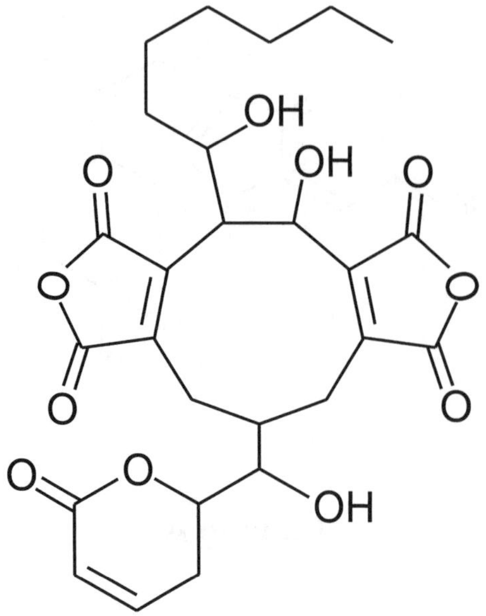

Rubratoxin B

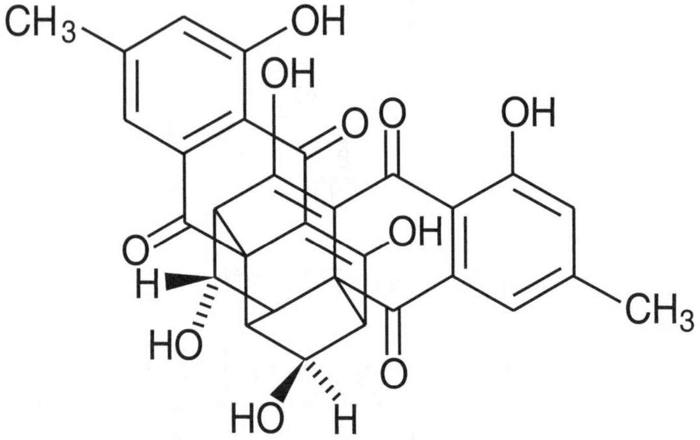

Rugulosin

Shearanines

Shearanines A-C were isolated from the ascostromata of *Eupenicillium shearii* (Belofsky *et al.*, 1995). Shearanines A and B were present in the ascostromata at 1500–1700 ppm. Shearanine A, B, and C produced an 89, 94 and 33 per cent RGR, respectively for *H. zea*, and a 69, 50 and 19 per cent RFR, respectively, for

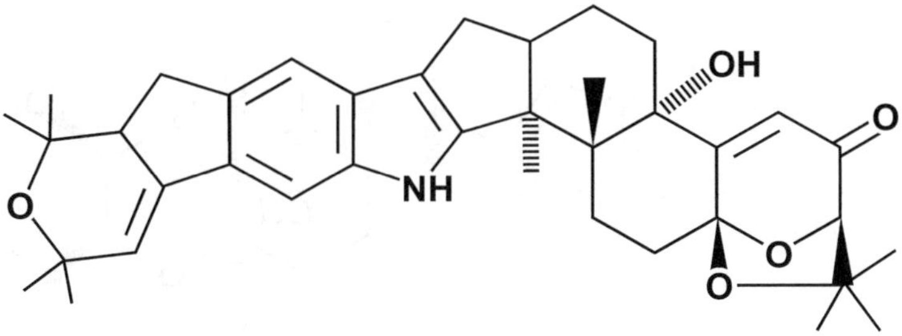

Shearinine A

C. hemipterus larvae at 100 ppm orally (Belofsky *et al.*, 1995). Shearinine A also caused 80 per cent RGR for *H. zea* when applied topically at 2 μg/2 mg insect. Shearinine B produced an 85 per cent reduction in leaf damage by *S. frugiperda* when 50 μg was applied to a 75 mg cotton-leaf disk (Belofsky *et al.*, 1995).

Sterols

A sterol sulfate was produced at 1500 ppm on corn grits by *Fusarium graminearum* (Vesonder *et al.*, 1990). It was tested as a sterol source, for ecdysteroidal effects, and general toxic effects against caterpillars and beetles (Dowd *et al.*, 1992a). The compound could apparently not be used as sterol source by *H. zea* or *S. frugiperda*, and had no detectable ecdysteroidal effects on either insect species. The sterol sulfate had no effect on feeding by adult *C. hemipterus* at 4000 ppm, but in sterol-free diet it caused a 50 per cent RFR and a 82.7 per cent RGR of *C. hemipterus* larvae (assays used second instar larvae) relative to sterol-free diet (Dowd *et al.*, 1992a). It caused

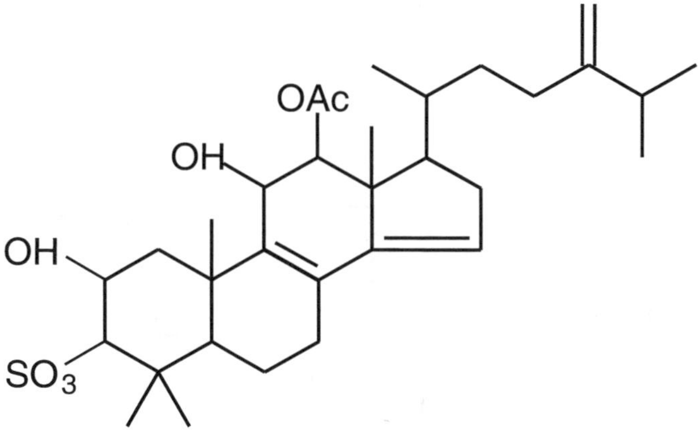

***Fusarium* Sterol Sulfate**

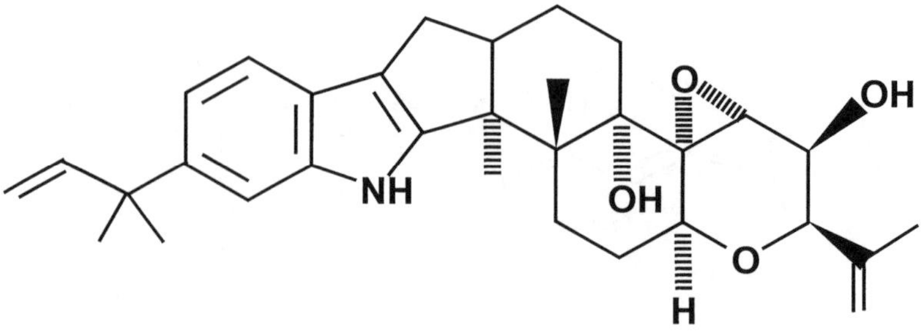

Sulpinine A

17.7 and 85.4 per cent RGR at 250 and 2500 ppm sterol diets, respectively against *S. frugiperda* larvae (Dowd *et al.*, 1992).

Sulpinines

Sulpinines were isolated from sclerotia of *Aspergillus sulphureus* (Laakso *et al.*, 1992b). Sulpinine A, B, and C produced 96.0, 87.2 and 26.5 per cent RGR in *H. zea* at 100 ppm oral treatment, but only sulpinine C caused 32 per cent RFR effect on *C. hemipterus* adults. Topically, sulpinine A produced a 54.5 per cent RGR at 2 μg/mg for *H. zea* larvae (Laakso *et al.*, 1992a,b).

Tenuazonic acid

Tenuazonic acid, α-acetyl, γ-*sec*-butyltetramic acid, is produced by *Alternaria alternata, A. tenuissima, A. longipes* and *Piricularia oryzae* (Turner, 1971; Turner and Aldridge, 1983). Tenuazonic acid produced an LC_{50} of 120 μg/ml orally against first instars of *L. sericata* in 48 h (Cole and Rolinson, 1972). Tenuazonic acid is reported as a protein synthesis inhibitor (Gottlieb and Shaw, 1967).

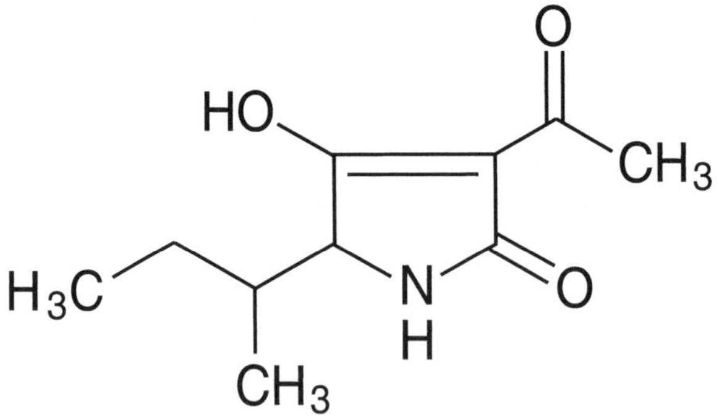

Tenuazonic Acid

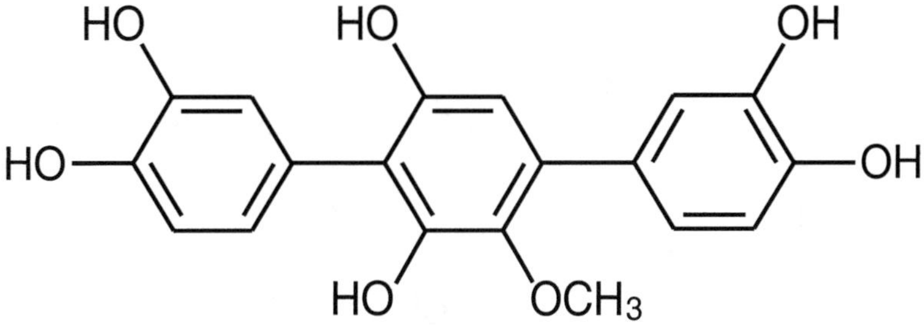

Terphenyllin Derivative

Terphenyls

The compounds 3,3″-dihydroxy-6′-desmethylterphenyllin, 3′-demethoxy-6′-desmethyl-5′-methoxycandidusin B and 6′-desmethylcandidusin B were isolated from the sclerotia of *Penicillium raistrickii* (Belofsky *et al.*, 1998a). The terphenyllin derivative produced a 28 per cent RGR for *H. zea* larvae at 500 ppm, but the other compounds were relatively inactive at this concentration (Belofsky *et al.*, 1998a).

Territrems

Territrems have been isolated from *Aspergillus tereus* (Ling *et al.*, 1979, 1984; Peng *et al.*, 1985). Orally at 250 ppm, territrems A, B, B′ and C produced no significant mortality of *H. zea*, but RGRs were 89.0, 54.1, 13.7 and 0 per cent, respectively (Dowd *et al.*, 1992b). Topically, territrem A was also most active, producing a 20 per cent RGR at 10 mg/g insect. Compared to paraoxon, the territrems were less active

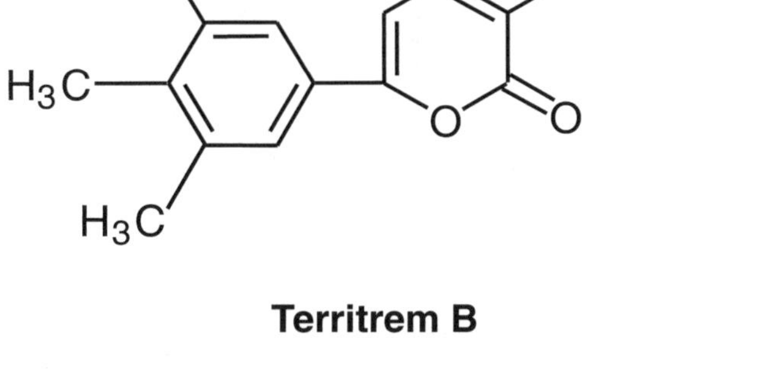

Territrem B

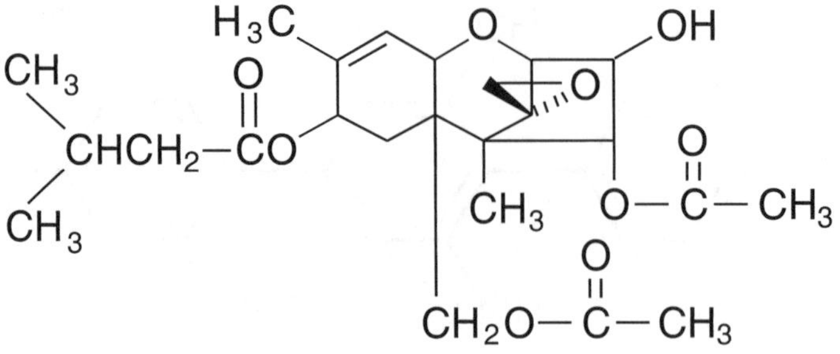

T-2 Toxin

orally, but they were as, or more active than paraoxon in inhibiting *H. zea* head acetylcholinesterase (with I_{50}s ranging from 4.7×10^{-8} to 1.1×10^{-7} M for the territrems) (Dowd *et al.*, 1992b).

Trichothecenes

Trichothecenes are a large group of fungal metabolites produced primarily by *Fusarium* species and their relatives (Turner and Aldridge, 1983). They include several mycotoxins, such as T2 toxin, deoxynivalenol and diacetoxyscirpenol. Due to the extensive amount of testing on insects, information can be divided in different insect groups.

In coleopterans, for instance, a compound isolated from *Myrothecium roridum* (later identified as verrucarin - see Wright *et al.*, 1981) reduced feeding by 81 per cent and caused 60 per cent mortality when introduced to the preoral cavity of *E. varivestis* at 0.75 μl/insect (Kishaba *et al.*, 1962). T-2 toxin at 100 ppm nearly doubled the larval development period for *T. confusum* (Wright *et al.*, 1976). T-2 toxin caused 30 per cent enhanced egg production of *T. confusum* for several weeks at 100 ppm (Wright *et al.*, 1974). There was an interaction between dietary protein content and T-2 toxin in the larvae of *Tenebrio molitor* Linnaeus (Davis and Schiefer, 1982). Efficiency of food conversion was relatively constant, but efficiency of protein conversion was influenced by con-

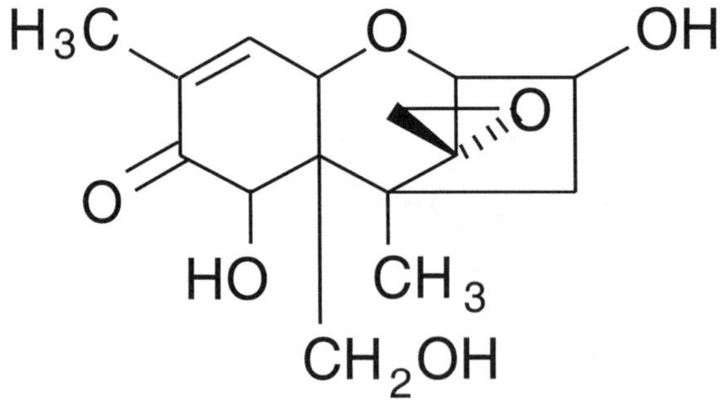

Deoxynivalenol

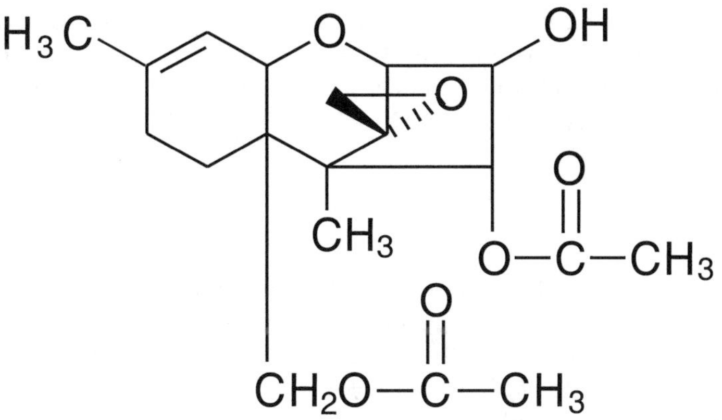

H_3C … OH … O … O … CH_3 … $O-C-CH_3$ … O … $CH_2O-C-CH_3$

Diacetoxyscirpenol

centrations of T-2 (Davis and Schiefer, 1982). Some stimulation (65%) of ATPase activity in larvae, and 25 per cent depression of ATPase activity in adults of *T. confusum* due to T-2 has been reported (Veam and Wright, 1973). The sap beetle *C. hemipterus* is able to hydrolytically degrade monoacetoxyscirpenol about 10 times faster than *S. frugiperda* or *H. zea* (Dowd and Van Middlesworth, 1989).

Some studies among lepidopterans reveal that several trichothecenes produced by *Fusarium graminearum* have been tested against *H. zea* and *S. frugiperda* at 25 ppm orally (Dowd *et al.*, 1989). For *H. zea*, 8- hydroxycalonectrin, deoxynivalenol, 7,8-dihydroxycalonectrin, 9,10-dihydrodeoxynivalenol (not naturally occurring), and sambucinol produced 52.2, 53.2, 47.0, 39.9 and 27.1 per cent RGRs, respectively, at 25 ppm oral administration of these compounds (Dowd *et al.*, 1989). For *S. frugiperda*, 8-hydroxycalonectrin, deoxynivalenol, 7,8- dihydroxycalonectrin, 9,10-dihydrodeoxynivalenol and sambucinol produced 71.2, 85.2, 50.5, 38.1 and 16.7 per cent RGRs, respectively, when given at 25 ppm orally (Dowd *et al.*, 1989). These compounds have also been tested in combination (see synergism section). An extract of a strain of *F. sporotrichoides* originally isolated from a spruce budworm, *C. fumiferana*, containing mostly T-2, HT-2 and neosolaniol (ca. 1:4:2) killed 70 per cent of spruce budworm larvae at 40 ppm (Strongman *et al.*, 1990).

Diacetoxyscirpenol and neosolaniol (*F. sambucinum*) produced 48 and 46 per cent "antifeedant" activity, respectively, against *G. mellonella* at 50 ppm (Mule' *et al.*, 1992). Increasing T-2 concentrations obtained from *Fusarium sporotrichoides* caused a sigmoidal response in the inhibition of glucose-6-phosphate dehydrogenase in *B. mori* (Shkaruba, 1976). T-2 toxin at 0.5 per cent caused 50.1 per cent mortality to *G. mellonella* larvae, and increased the diameter of Malpighian tubules (Lysenko and Vititinev, 1976). The trichothecenes are generally recognized as protein synthesis inhibitors (Joffe, 1986). Treatment with trichothecenes at 25–250 ppm orally can induce oxidative, conjugating, and hydrolytic enzymes in *H. zea* and/or *S. frugiperda* (Dowd, 1990).

Similarly, among dipterans the LC_{50}s for trichothecin and trichothecolone were ca. 100 μg/ml for *L. sericata* larvae (Cole and Rolinson, 1972). The LC_{50} for diacetoxyscirpenol for the same larvae was 7.5 μg/ml (Cole and Rolinson, 1972). A series of seventeen naturally produced 12,13-epoxytrichothecenes were tested for activity against *A. aegypti* larvae (Grove and Hosken, 1975). High activity was dependent on the presence of the 12,13 epoxide ring and ester group. Trichodermin was the most selective towards *A. aegypti* (vs. HeEp2 cells toxicity and rat oral LD_{50}s) (Grove and Hosken, 1975).

Tolypin

Tolypin (no structure available) is a thermostable, water-soluble metabolite apparently different from cyclosporines produced by *Tolypcladium cylindrosporum* and *T. inflatum* (Weiser and Matha, 1988b). Crude extracts are active against *G. mellonella, D. melanogaster*, and *Culex pipiens* Linnaeus causing 50 per cent mortality of larvae at 26 ppm after 24 h (Weiser and Matha, 1988b).

Tubingensins

Tubingensin A was isolated from the sclerotia of *Aspergillus tubingensis* (TePaske *et al.*, 1989b). It had a minimal effect on *H. zea*, producing 11 per cent mortality at 125 ppm orally (TePaske *et al.*, 1989b). Tubingensin B was isolated from the sclerotia of *A. tubingensis* (TePaske *et al.*, 1989c). It produced a minimal toxicity of 10 per cent in *H. zea* at 125 ppm after oral administration (TePaske *et al.*, 1989c).

Verruculogen and relatives

Verruculogen has been isolated from *Penicillium* and *Aspergillus* spp. (Turner and Aldridge, 1983). At 25 ppm orally, it produced a 79.8 and 79.9 per cent RGRs for *H. zea* and *S. frugiperda*, respectively (Dowd *et al.*, 1988). At 0.1 ppm in diets, verruculogen produced tremors in *B. mori* (Hayashi *et al.*, 1991). Verruculogen is known to inhibit GABA receptors in rat brain and *Torpedo* electric organ (Gant *et al.*, 1987), so interaction with GABA systems in insects is likely.

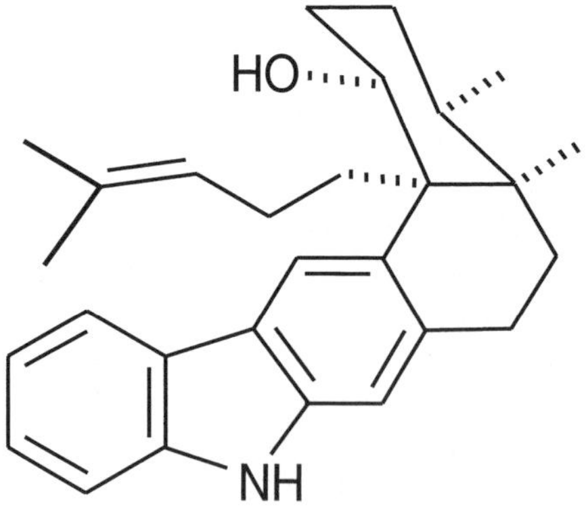

HO
NH

Tubingensin A

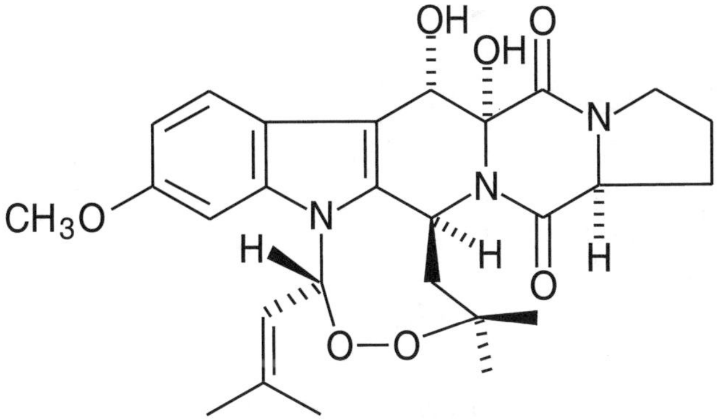

Verruculogen

The related compound, fumitremorgen A has been isolated from *Aspergillus fumigatus* (Turner and Aldridge, 1983) and the ascostromata of *Eupenicillium crustaceum* (Wang *et al.*, 1995). In contrast to verrucologen, it had no effect on *H. zea* at 50 ppm (Wang *et al.*, 1995).

Versimide

Versimide, methyl α-(methyl succinimido) acrylate, is produced by *Aspergillus versicolor* (Cole and Rolinson, 1972). It produced 100 per cent knockdown of adult *D. melanogaster* in contact toxicity assays at 5 $\mu g/cm^2$ in 4 h (Cole and Rolinson, 1972).

Viomellein

Aspergillus melleus, A. sulphureus, Penicillium citreoviride, Penicillium viridicatum and *Microsporum cookei* (Turner and Aldridge, 1983) produce viomellein. Viomellein

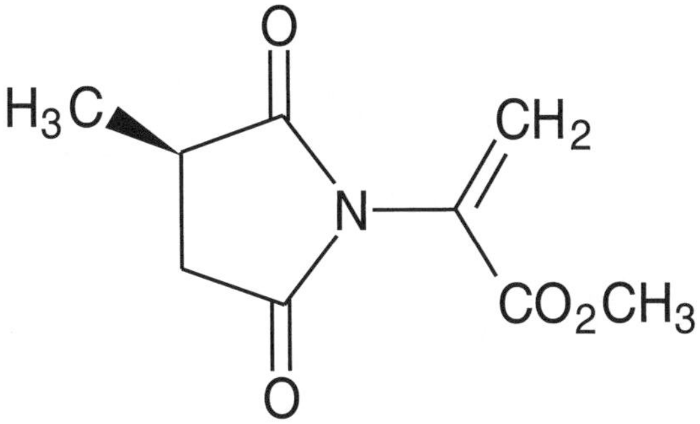

Versimide

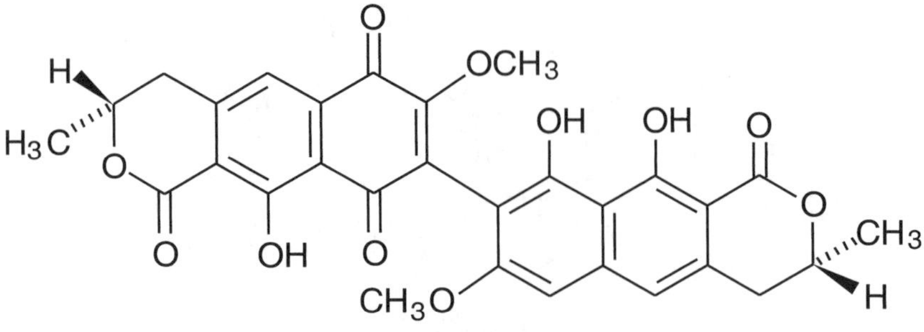

Viomellein

caused a 67 per cent reduction in feeding by *D. melanogaster* and caused 30 per cent mortality of *S. littoralis* at 10 ppm (Patterson *et al.*, 1987).

Xanthones

The compound 1,3,5,6-tetrahydroxy-8-methylxanthone was isolated from the sclerotia of *Penicillium raistrickii* and had unspecified activity against *H. zea* (Belofsky *et al.*, 1998a).

Zearalenone

Zearalenone is produced by several species of *Fusarium* (Turner and Aldridge, 1983). Zearalenone at 1000 or 10,000 ppm produced nonsignificant increases in egg production (ca. 15%) of *T. confusum* (Wright *et al.*, 1974, 1976), but higher increases were noted in other studies, depending on the composition of the food source (Harein *et al.*, 1971). Stimulation of ATPase activity was also noted when *T. confusum* was fed zearalenone (Veam and Wright, 1973). At 250 ppm in diets, zearalenone produced a 34.9 per cent RGR for *S. frugiperda* (Dowd, 1992b). Zearalenone killed 65 per cent of *T. putrescentiae* at 1 ppm after 7 days (Rodriguez *et al.*, 1980). Zearalenone fed at levels of 1000 ppm or 10,000 ppm accumulated at 8 and 14 ppm in *T. confusum* and *Alphitobius diaperinus*

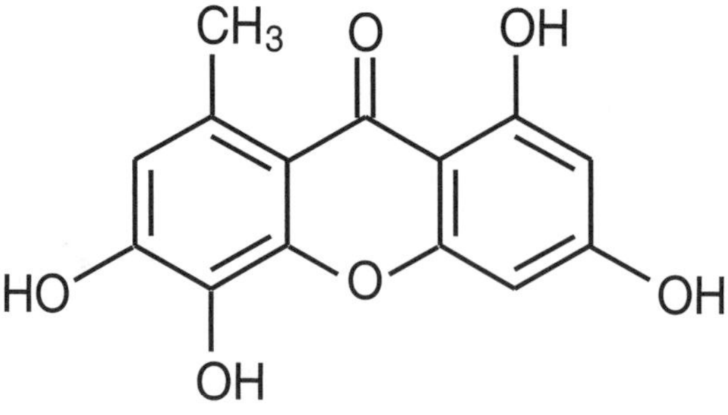

Xanthone Derivative

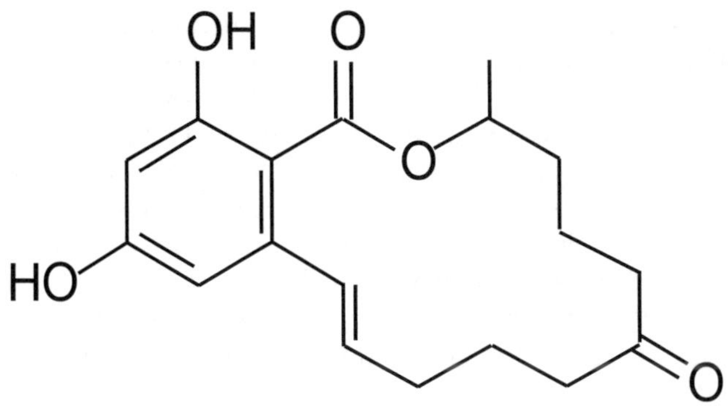

Zearalenone

(Panzer), respectively (Eugenio *et al.*, 1970). Although sequestration appeared to occur, no metabolic changes were apparent (Eugenio *et al.*, 1970).

Antiinsectan cyclic peptides

α-amanitin

α-amanitin has been isolated from mushrooms of *Amanita phalloides* and specifically inhibits nuclear RNA synthesis in eucaryotes (Fuchs and Fong, 1976), including *C. erythrocephala* when injected at 0.5 μg/larva (Shaaya and Clever, 1972). It inhibited

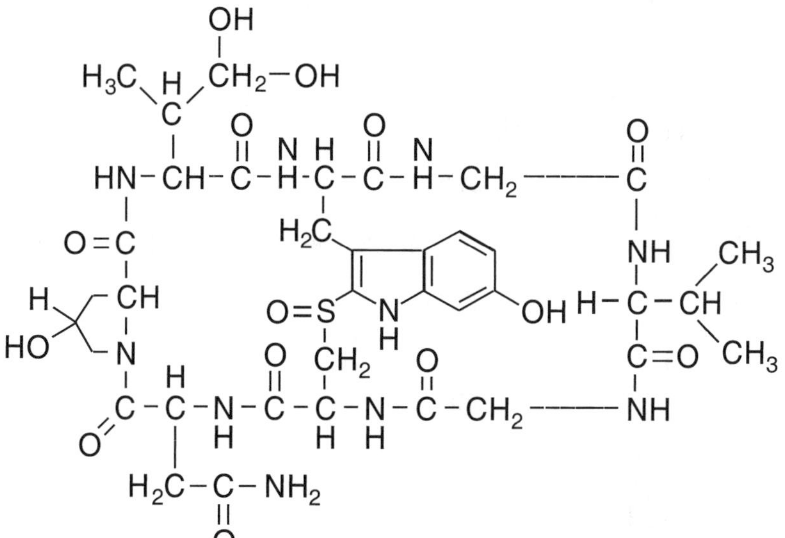

α-Amanitin

dopa decarboxylase activity in *A. aegypti*, but primarily inhibited trypsin synthesis (Fuchs and Fong, 1976). Actinomycin D, another inhibitor of RNA synthesis, had the same effect (Fuchs and Fong, 1976). Injection of 1 μg of α-amanitin into *Calliphora vicina* Robineau-Desvoidy also inhibited induction of DOPA decarboxylase (Shaaya and Sekeris, 1971). The relative toxicity of different amatoxins has been reviewed (Wieland, 1986). Inhibitory efficacy towards RNA-polymerase II of *D. melanogaster* with α-amanatin analogs indicated great reductions in activity with hexyl or longer groups (Wieland, 1986). Fruit feeding *Drosophila* are severely affected by 50 ppm of α-amanitin compared to mushroom feeding species (Jaenike *et al.*, 1983). Resistant strains of *Drosophila* have RNA polymerase II that is less sensitive to α-amanitin (Jaenike *et al.*, 1983).

Aspochracein

Aspochracein is a cyclic tripeptide (N-methyl-L-valyl-N-methyl-L-alanyl-α-capryl-L-ornithine) isolated from *Aspergillus ochraceus* (Myokei *et al.*, 1969). When injected into *B. mori* at 15 μg/g it induced paralysis followed by mortality (Myokei *et al.*, 1969).

Bassianolide

Bassianolide has been isolated from *Beauveria bassiana* and *Verticilliium lecanii* (Suzuki *et al.*, 1977; Kanoka *et al.*, 1978). When administered orally at 13 ppm to *B. mori*, 100 per cent mortality resulted (Suzuki *et al.*, 1977), while at 8 ppm in diet, 80 percent mortality after 6 days was observed (Kanoka *et al.*,1978). However, when larvae were injected with up to 10 μg, only 10 per cent mortality resulted after 1 week. Flacid paralysis (atony) was noted at 4 ppm (Kanaoka *et al.*, 1978).

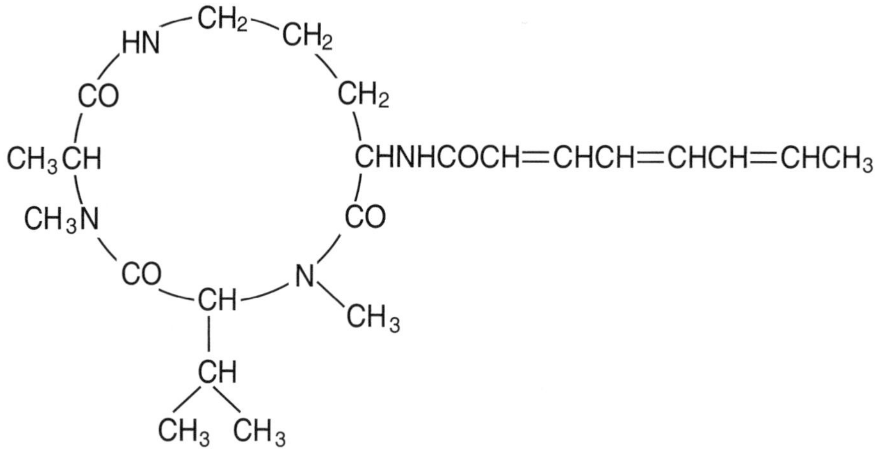

HN—CH_2—CH_2
CO　　CH_2
CH_3CH　　CHNHCOCH=CHCH=CHCH=CHCH$_3$
CH_3N　　CO
CO—CH—N—CH_3
CH
CH_3 CH_3

Aspochracein

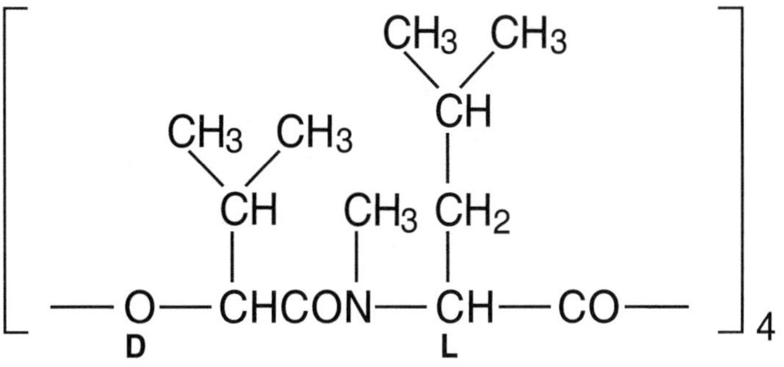

Bassianolide

Beauvericin

Beauvericin is produced by *Beauveria bassiana, Paecilomyces fumoso-roseus, Polyporus sulphureus* (Grove and Pople, 1980) and *Fusarium* spp. (Krska *et al.*, 1996). Beuvericin was initially found to be toxic to mosquito larvae (Hamil *et al.*, 1969). Beauvericin produced 39 and 86 per cent mortality of *A. aegypti* at 10 and 20 ppm, respectively (Grove and Pople, 1980). Beauvericin also caused 60 per cent knockdown, but little long term mortality at 5 μg/fly for *C. vicina* (Grove and Pople, 1980).

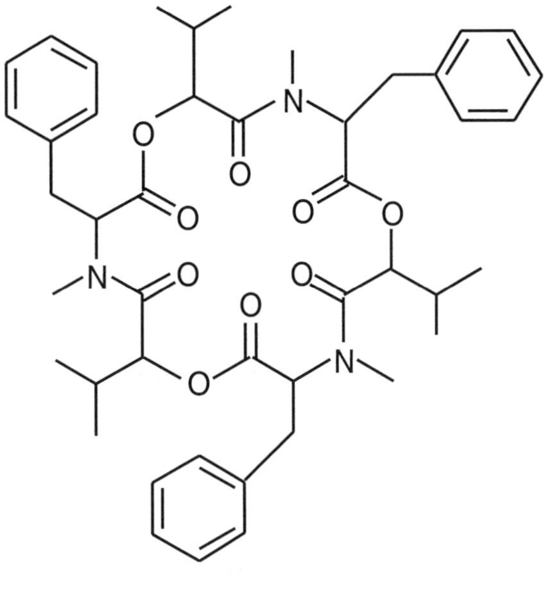

Beauvericin

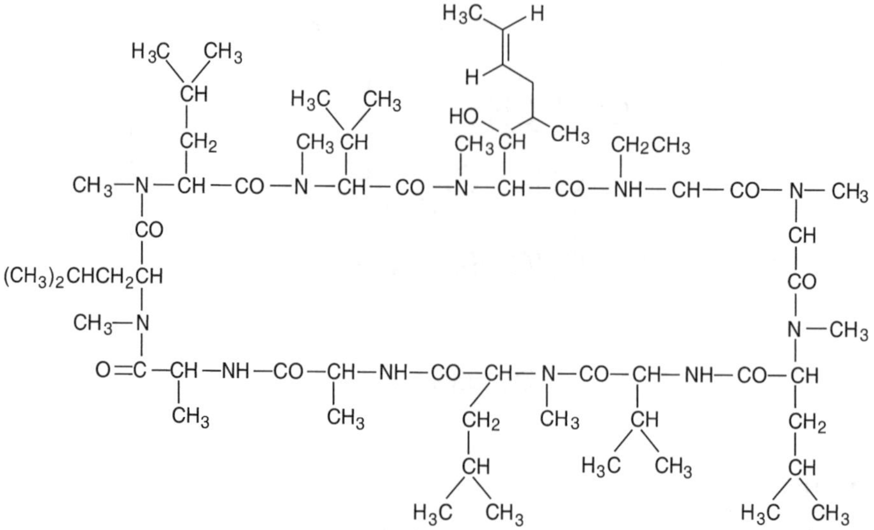

Cyclosporin A

Cyclosporins

Cyclosporins are produced by *Tolypocladium inflatum* (= *Trichoderma polysporum*), *Cylindrocarpon lucidum* and *Fusarium* (Barath *et al.*, 1974; Turner and Aldridge, 1983). Cyclosporins A, B and C produced LD_{50}s of ca. 0.6, 3.0, and 1.2 ppm, respectively, for *Culex pipiens autogenicus* larvae (Weiser and Matha, 1988a). However, the cyclosporines tested were inactive when injected into *G. mellonella* (Weiser and Matha, 1988b). The ED_{50} of cyclosporin (ramihyphin A) for *D. melanogaster* was 5.5 μg/ml (Dobias *et al.*, 1979).

Destruxins

Destruxins A and B were originally described from *Oospora destructor (Metarhizium anisopliae)* and *Aspergillus ochraceus* (Kodaira, 1961, 1962; Roberts, 1969; Tamura *et al.*, 1964). An analog of destruxin A, roseotoxin B, has been isolated from *Trichothecium roseum* (Richard *et al.*, 1969). Although it is similar in activity against insects to destruxin A (Dowd unpublished), the LC_{100} mouse ip for roseotoxin B (166 mg/kg) (Cole and Cox, 1981) is about 100 times less than that for destruxin A (1.35 mg/kg) (Kodaira, 1961). At 25 ppm orally, roseotoxin B caused 38.7 per cent mortality and a 99.7 per cent RGR for *H. zea*, and 100 per cent mortality of *S. frugiperda* (Dowd *et al.*, 1988). The LD_{50}s for destruxin A and B injected into fifth instar *B. mori* were 0.28 and 0.34 μg/g, respectively (Kodaira, 1961). Destruxins A4 and A5 were isolated from an unidentified species of *Aschersonia*, an insect pathogen (Krasnoff *et al.*, 1996). LC_{50} values were 41 and 52 ppm in diet against *D. melanogaster* for destruxins A4 and A5, respectively (Krasnoff *et al.*, 1996). Destruxin E at low doses (1×10^{-8} - 2.5×10^{-9} M) inhibited *G. mellonella* baculovirus colonization of *B. mori* cells (Quiot *et al.*, 1980).

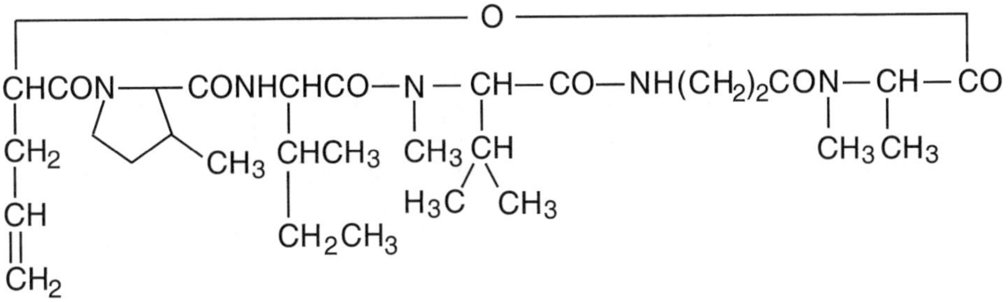

Roseotoxin B

Early experiments with destruxin A and B injected into *B. mori* fifth instars caused larval paralysis, but larvae also recovered within 24 h (Kodaira, 1962). Histological examinations of *G. mellonella* larvae indicated alterations of midgut epithelial cells, Malpighian tubules, hypodermis, brain and ganglion cells, haemocytes and muscles (Vey and Quiot, 1985). Destruxins A and B together also blocked the encapsulation response of hemocytes (Vey and Quiot, 1985). Additional effects of destruxins that have been reviewed (Clarkson and Charnley, 1996) include depolarization of muscle membranes, inhibiting ATPase, and inhibiting acidification of cellular compartments. Inhibition of ATPase may account for most of the effects of destruxins (Clarkson and Charnley, 1996). However, enhanced silk production in *H. zea* by sublethal levels of roseotoxin B (Dowd, unpublished) suggest interactions with neuroreceptors similar to those reported for neuroactive compounds that stimulate saliva production in ticks (Stich *et al.*, 1993). Destruxin E appears to be detoxified by conversion to a diol in *Locusta migratoria* (Linnaeus)(Cherton *et al.*, 1991).

Efrapeptins

Efrapeptins have been isolated from the insect pathogen *Tolypocladium niveum* (Krasnoff and Gupta, 1991). The LC_{50} against *L. decemlineata* for efrapeptins D and F was 18.9 and 8.4 ppm, respectively, for solutions applied to leaves (Krasnoff and Gupta, 1991). The efrapeptins are also known inhibitors of various ATPases, and I_{50}s

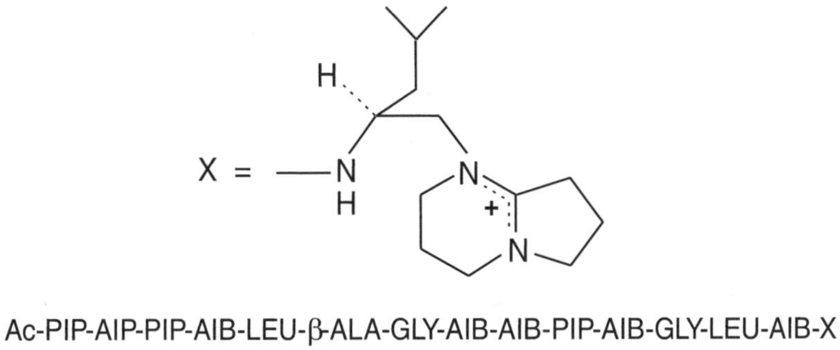

Ac-PIP-AIP-PIP-AIB-LEU-β-ALA-GLY-AIB-AIB-PIP-AIB-GLY-LEU-AIB-X

Efrapeptin D

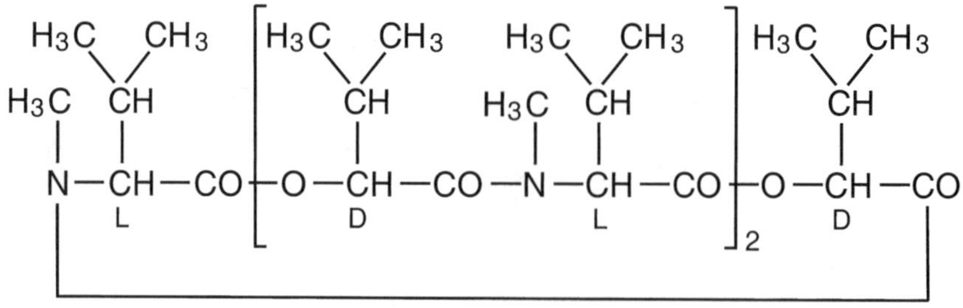

H3C
CH3
H3C
CH
N—CH—CO
L
O—CH—CO—N—CH—CO
D
L
2
O—CH—CO
D

Enniatin B

to *M. domestica* mitochondrial ATPase ranged 14.7–100 ng/ 14.7 μg protein for the five (C-F) tested (Krasnoff and Gupta, 1991).

Enniatins

Enniatins have been reported from *Fusarium* spp. (Turner, 1971) such as *F. lateritiium* (Grove and Pople, 1980), *F. avenaceum* (Strongman *et al.*, 1988), and *F. sambucinum* (Mule' *et al.*, 1992). Enniatin A caused 32 per cent mortality of adult *C. erythrocephala* after 48 h when injected at 5 μg/fly and caused 37 per cent mortality at 20 ppm after 48 h to *A. aegypti* larvae (Grove and Pople, 1980). A mixture of enniatins B, B_1, and A_1 at 5:3.5:1 produced 39 per cent mortality of adult *C. vicina* at 5 μg/fly after 2 days, and produced 31 per cent mortality of *A. aegypti* larvae at 44 ppm after 3 days (Grove and Pople, 1980). Enniatin A/A_1 produced 58 per cent mortality of *C. fumiferana* at 400 ppm (Strongman *et al.*, 1988). Enniatin B was inactive against *G. mellonella* at 100 ppm orally (Mule' *et al.*, 1992).

Isariins

Isariins were isolated from *Isaria felina* (Baute *et al.*, 1981). When 20 μg was injected into *G. mellonella*, 3-day mortalities for isariin and isariin B–D were 0, 0, 50 and 60 per cent, respectively (Baute *et al.*, 1981).

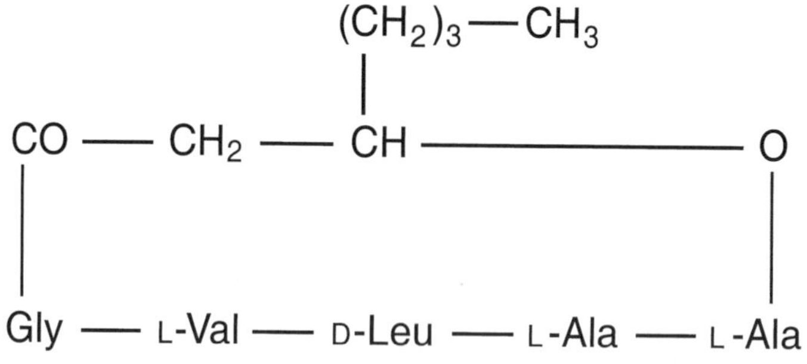

(CH2)3—CH3
CO — CH2 — CH — O
Gly — L-Val — D-Leu — L-Ala — L-Ala

Isariin D

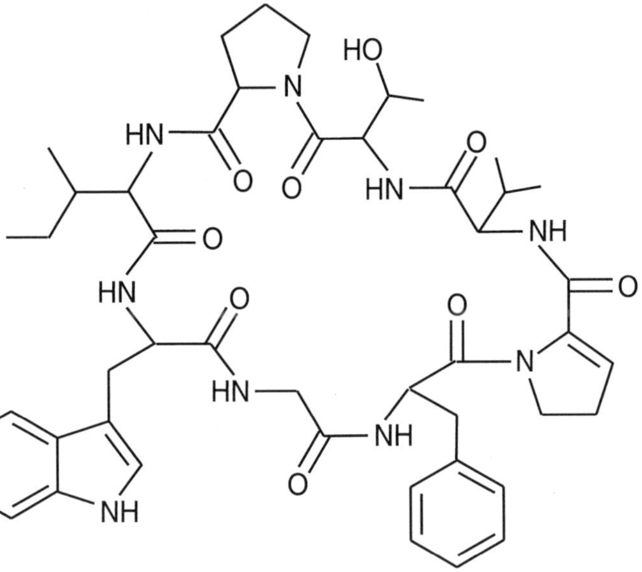

Shearamide A

Shearamide A

Shearamide A is cyclic octapeptide occurring primarily in the ascostromata of *Eupenicillium shearii* (Belofsky *et al.*, 1998b). It produced a 35 per cent RGR for *H. zea* and a 38 per cent RFR for *C. hemipterus* adults at 100 ppm orally (Belofsky *et al.*, 1998b).

Antiinsectan proteins

α-amylase inhibitors

The α-amylase inhibitors HOE-467 (tendamistat), HAIM and PAIM have been isolated from *Streptomyces tendae* (Markwick *et al.*, 1996). HOE-467 is a 74 amino acid, ca. 8 kDa protein with two disulfide bridges (CAS 1999). HOE-467 did not show any significant activity against α-amylase from several Lepidoptera, although at 5–260 μm reductions in activity of ca 15–20 per cent were noted for *Phthorimaea operculella* (Zeller) (Markwick *et al.*, 1996).

Bacillus sphaericus toxins

The *B. sphaericus* toxins have been reviewed (Porter *et al.*, 1993). There are two types of toxins, a binary one that comes from highly toxic strains, and a single toxin (SSII-1) that comes from less toxic strains and is produced during the vegetative growth phase (Porter *et al.*, 1993). The nature of these toxins is complicated, as multiple peptides are present

CARDIFF UNIVERSITY ★ PRIFYSGOL CAERDYDD ★

and some activation appears to occur due to proteolytic processing in the insect gut. The binary toxin consists of two proteins with apparent molecular weights of 43 and 56 kDa, but respective genes code for 41.9 and 51.4 kDa proteins (Baumann *et al.*, 1988; Bourgouin *et al.*, 1990). The toxins are mainly active against *Culex* and *Aedes* mosquitoes (Bourgouin *et al.*, 1990). The two components can form channels in planar lipid bilayers (Potvin *et al.*, 1998). Trypsin and chymotrypsin activate the toxin, and a 1:1 mixture has an LC_{50} of 0.49 ppm for *C. pipiens* (Baumann *et al.*, 1988).

From the lower toxicity strains, an initial toxin gene coding for a 100 kDa protein (SSII-1) was identified (Thanabalu *et al.*, 1991). Upon removal of the signal sequence, a 97 kDa toxin (Mtx 21) is formed that has an LC_{50} of 15 ppb to *Culex quinquefasciatus* Say larvae (Thanabalu *et al.*, 1992). Mosquito gut extract digested the 97kDA toxin to form a 27k Da peptide that had homology with ADP ribosyltransferase toxins (Thanabalu *et al.*, 1992). Other kDa derivatives could ADP-ribosylate two proteins from *C. quinquefasciatus* cell extracts (Thanabalu *et al.*, 1993). Further study indicated both the 27 kDa and 70 kDA proteolysis products were necessary for full activity (Thanabalu *et al.*, 1993). An additional toxin-encoding gene (31.8 kDa) producing protein Mtx-2 was isolated from a strain of *B. sphaericus* (Thanabalu and Porter, 1996). The LC_{50} for the purified protein was 0.32 ppm for *C. quinquefasciatus* (Thanabalu and Porter, 1996). Another gene, coding a 35.8 kDa protein called Mtx3 was also toxic to *C. quinquefasciatus* (Liu *et al.*, 1996b).

Bacillus thuringiensis toxins – see Chapter 3

Beauveria sulfurescens toxic glycoprotein

A protein (TF2) was isolated from *Beauveria sulfurescens* that had a molecular weight of 29 kDa and appeared to be a glycoprotein, without proteolytic activity (Mollier *et al.*, 1994). The LD_{50} by injection was 0.9 μg/ml (20 μl injected) for one toxin (TF1) and 0.1 μg/ml for the other (TF2) (Mollier *et al.*, 1994).

Chitinases

Microbial-derived chitinase activity that is active or synergistic with other toxic proteins has been described (Kramer *et al.*, 1997), but cited reports do not often list the chitinase source. The chitinase from *Serratia marcescens* has a molecular weight of ca. 36,000 and a pH optimum of 8.5 to 9.0 (Lysenko, 1976). When administered via injection to *G. mellonella*, the *S. marcescens* chitinase had an LD_{50} of 1.3–3.0 units/larva (Lysenko, 1976). When incubated in solutions containing 5000 ppm of chitinase from *Streptomyces griseus*, the peritrophic membrane of *Orgyia pseudotsugata* (McDunnough) developed small holes or eroded pits (Brandt *et al.*, 1978).

Cholesterol oxidase

A cholesterol oxidase (52.5 kDa) isolated from *Streptomyces* was toxic to *Anthonomus grandis* Boheman (Purcell *et al.*, 1993). The LC_{50} was 6 ppm after 6 days and 1.5 ppm after 16 days (Greenplate *et al.*, 1995). Growth retardation of 86, 46 and 30 per cent for *Heliothis virescens* (Fabricius), *Ostrinia nubilalis* (Hubner) and *Pectinophora gossypiella* (Saunders) occurred, respectively, at 100 ppm (Purcell, 1997). Midguts from treated *A. grandis* had disrupted epithelia (Purcell *et al.*, 1993). A cholesterol oxidase from *Pseudomonas fluorescens* caused 100 per cent mortality of *A. grandis* at 44 ppm (Purcell *et al.*, 1993). A cholesterol oxidase gene cloned from *Streptomyces* sp. produced a protein toxic to *A. grandis* when expressed in *E. coli* (Corbin *et al.*, 1994). Expression of a cholesterol oxidase gene cloned from *Streptomyces* sp. also has been reported in tobacco cells (Corbin *et al.*, 1994; Cho *et al.*, 1995). Cholesterol oxidation was often associated with reduced alkaline phosphatase activity of brush border membranes of different insect species, which was a good predictor for explaining the relative susceptibility to cholesterol oxidase (Shen *et al.*, 1997).

Clostridium toxin

Parasporal inclusion bodies from *Clostridium bifermentans* serovar *malaysia* have larvicidal activity against *Anopheles stephensi* Liston (Charles *et al.*, 1990). Activity may be associated with proteolytic activation (Charles *et al.*, 1990). Three proteins of 16, 18 and 66 kDA were isolated from the bacteria cultures and their presence was associated with toxicity of the broth to *A. stephensi* (Nicolas *et al.*, 1993). A gene for the 66 kDa toxin was cloned from this bacterium, and preparations were toxic to *A. aegypti*, *C. pipiens*, and *A. stephensi* (Barloy *et al.*, 1996).

Hirsutellins

Hirsutellin A has been isolated from cultures of *Hirsutella thompsonii* (Mazet and Vey, 1995). The apparent molecular weight of purified protein was 15,000, and the isoelectric point was at pH 10.5 (Mazet and Vey, 1995). Hirsutellin A acted very slowly, causing 100 percent mortality at 800 μg in 15 days, and 100 per cent mortality at 200 μg in 30 days, when injected into *G. mellonella* larvae (Mazet and Vey, 1995). The compound was more active against *A. aegypti*, 20 ppm solution produced 80.9 per cent mortality in 24 h (Mazet and Vey, 1995). In contact toxicity assays, a concentration of 32 μg/ml killed ca 55 percent of the mite *Phyllocoptruta oleivora* (Ashmead) in 48 h (Omoto and McCoy, 1998). A preparation from *Hirsutella thompsonii* var. *thompsonii* had a molecular weight of ca. 16,000 and an isoelectric point of 10.4 (Liu *et al.*, 1996a). A purified preparation was more selective towards cells of *S. frugiperda* than other noninvertebrate cell sources, and had ribosome inactivating activity (Liu *et al.*, 1996a). Cloning and sequencing of the gene for this protein indicated a molecular weight of 14,159, and a sequence unique from other fungal ribosomal inactivating proteins (Boucias *et al.*, 1998).

Lectins

Although many fungi produce lectins, those tested so far have limited activity against insects (Dowd, unpublished).

Phospholipases

Phospholipase C semipure preparations from different microorganisms have been tested against insects (Lysenko, 1972 a,b, 1973). The LD_{50}s of phospholipase C preparations from *Bacillus cereus, Clostridium perfringens, Corynebacterium ovis*, and *Pseudomonas chloraphis* when injected into *G. mellonella* were 0.04, 0.08, 0.015 and 1.65 units per larva, respectively (Lysenko, 1972 a,b, 1973). The most active preparation from *C. ovis* has the lowest molecular weight (15–17 kDa), the highest isoelectric point (pH 9.5 and a relatively basic pH optimum 8.4) compared to the other sources (Lysenko, 1973). The pH optima for the other sources were 7.0–7.5, and the molecular weights ranged from 24 kDa (*B. cereus*) and 32–34 kDa (*C. perfringens*) to 54 kDa (*P. chloroaphis*) (Lysenko, 1972 a,b, 1973). The isoelectric points of phospholipases from the other organisms were from 6.3 to 6.9 (Lysenko 1972 a,b, 1973). Some of these phospholipases caused hemocyte rupture, which was presumably due to hydrolysis of membrane phospholipids (Lysenko, 1974).

Photorhabdus luminescens toxins

The bacteria, *Photorhabdus luminescens*, associated with insect pathogen i.e. Heterorhabditid nematodes, produce several toxins (Bowen *et al.*, 1998). A purified complex of products (25 to 207 kDa) from cloned genes was toxic to *Manduca sexta* (Johannsen) (Bowen *et al.*, 1998).

Proteases

Toxic proteases were isolated from *Pseudomonas aeruginosa* (Lysenko and Kucera, 1968). Toxicity to *G. mellonella* larvae due to injected proteases ($X10^{-3}$ total protease units per larva) was 110 for protease B (mw 17 kDa, 139 tu/mg), 36 for protease C (mw 26 kDa, 216 tu/mg), ca. 500 for protease D (mw 13 kDa, 24.2 tu/mg) and 70–80 for protease E (mw 40 kDa, 97 tu/mg) (Lysenko and Kucera, 1968). Based on pH optima, proteases B and C were neutral (pH 7.5), protease D semialkaline (pH 8.0) and protease E alkaline (pH 9.0) (Kucera and Lysenko, 1968). Other toxic proteases were isolated from *Serratia marcescens* (Kaska *et al.*, 1976). The metaloprotease had a pH optimum of 7.5 and the serine protease had a pH optimum of 10.9. Both proteases had molecular weights of 37 kDa. When tested in a mixture by injection, the LD_{50} for *G. mellonella* was 0.078 units per larva (Kaska *et al.*, 1976). *Mamestra brassicae* (Linnaeus) was about 100-fold more sensitive to the proteases tested (Lysenko and Kucera, 1971).

Toxic proteases have also been isolated from *Metarhizium anisopliae* (Kucera, 1980). The LD_{50}s for *G. mellonella* were 80 and 1000 ppm for proteases P1 and P2, respectively (Kucera, 1980). Protease P1 had a pH optima of ca. 6.5, the pI was 7.25, the molecular weight was 35 kDa, and it was inhibited by phenylmethylsulfonyl fluoride: the author suggested it was a serine protease (Kucera, 1980). Protease P2 had a pH optima of ca. 9.0, a pI above 9.0, a molecular weight of 71 kDa, and was inhibited by 4-chloromecuribenzoate: the author suggested it was a thiol protease (Kucera, 1980). Toxic proteases have also been isolated from *Beauveria bassiana*

(Kucera and Samsinakovas, 1968). A lower molecular weight protease had a pH optimum near 6.5, and a higher molecular weight protease had a pH optima near 9.0 (Kucera and Samsinakovas, 1968).

It is uncertain what relationship these proteases have to the proteases that are induced when entomopathogens grow on cuticle. Hydrolytic enzymes associated with insect pathogenesis, include proteases, lipases, chitinases, and N-acetyl β-glucosamidases (Khachatourians, 1991; St. Leger, 1995). It appears prior reports of toxic proteases may involve proteases involved in cuticle penetrations, but the reason for their acute toxicity when injected does not seem totally consistent with a role only related to simple cuticle penetration.

Restrictocin

Restrictocin is produced by *Aspergillus restrictris* (Brandhorst *et al.*, 1996). It is a ribosomal inactivating protein (phosphodiesterase) that selectively cleaves this bond in ribosomal RNA (Jimenez and Vasquez, 1985). Other RIPs such as α-sarcin and mitogillin (also produced by *Aspergillus* spp.) are closely related structurally (Jimenez and Vasquez, 1985). Restrictocin occurs in the several thousand ppm range on spore forming structures of *A. restrictus* and will cause toxicity and inhibit feeding of beetles and caterpillars (Brandhorst *et al.*, 1996). At 1000 ppm, a relevant natural level, restrictocin killed 62.5 per cent of *S. frugiperda* larvae and 38.5 per cent of *Carpophilus freemani* larvae but produced no significant mortality to *H. zea* (Brandhorst *et al.*, 1996). In choice assays, at 1000 ppm feeding by adult *Sitophilus zeamais* Motschulsky was reduced by 3-fold, and at 2500 ppm, feeding by adult *C. freemani* was reduced by 2-fold (Brandhorst *et al.*, 1996). Differences in toxicity to *S. frugiperda* and *H. zea* appear to be due to differences in proteolytic degradation (Brandhorst *et al.*, 1996).

Vegetative insecticidal proteins (VIPs)

VIPs are produced during the vegetative growth phase of bacteria, and have been isolated from *Bacillus cereus* (Warren, 1997). The initial proteins (80 kDa and 45 kDa molecular weight) isolated from strain AB78 were active against Diabrotictes beetles, more so against *Diabrotica virgifera virgifera* LeConte and *D. longicornis* (Say) than *D. undecimpunctata howardi* Barber but not other Chyrsomelid or other beetles, caterpillars or mosquitoes (Warren, 1997). Genes for a 52,000 (vip2A(a)) and 100,000 (vip1A(a)) protein occur, both of which are required in combination for maximum activity. The LC_{50} for the combination was 0.04 ppm for newly hatched *D. virgifera virgifera* (Warren, 1997). Homologs of these two proteins have also been found from *B. thuringiensis* var. *tenebrionis* (Warren, 1997). An additional protein, vip 3A(a) has been isolated from a strain of *B. thuringiensis*, and has a molecular weight of ca. 80,000 (Estruch *et al.*, 1996; Warren, 1997). This protein is active against several caterpillar species, and a cloned product produced over 95 per cent mortality of *Agrotis ipsilon* (Hufnagel) larvae at 0.07 ppm (Warren, 1997). The vip3A protein produces gut paralysis at 4 ng/cm^2, and gut epithelium cell lysis at 40 ng/cm^2 (Yu *et al.*, 1997). Another VIP is produced by *B. sphaericus* (see above).

Uncharacterized proteins

An unknown, probably proteinaceous material (based on heat lability and other properties) produced by *Entomophthora coronata* and *E. apiculata*, caused death of *G. mellonella* and resulted in blackening of the hemolymph (Prasertphon, 1967).

Synergism

Chelators/ionophores such as kojic acid and fusaric acid are widespread in fungi (Turner and Aldridge, 1983). Both kojic acid and fusaric acid have been reported as synergists of cooccurring compounds (Dowd, 1988a,b, 1989b). They are known to inhibit relevant enzymes such as unspecific monooxygenases (Dowd, 1988b) and peroxidases/phenoloxidases (Bossi, 1960; Dowd, 1988b, 1994, 1999). Other combinations of *Penicillium* (Dowd, 1989a) and *Fusarium* (Dowd *et al.*, 1989) metabolites can result in synergistic effects on insects. Peramine can also inhibit oxidative detoxifying enzymes and act as a synergist (Dubis *et al.*, 1992). A synergistic interaction has been suggested for cyclosporins and tolypin (Weiser and Matha, 1988a). It is likely that synergistic interactions between co-occurring metabolites and proteins is widespread, and based on defensive or offensive (i.e. insect pathogen) strategies.

Synergistic interactions with proteins that are directly toxic, or that act to facilitate penetration of other proteins or bioactive compounds may be found to have greater importance as time goes on. There are some examples where combinations of chitinases and Bt toxins are synergistic (Kramer *et al.*, 1997). Presumably, the chitinase breaks down a barrier such as the peritrophic membrane to allow penetration of a more active protein. Fungal pathogens that penetrate the insect cuticle appear assisted by cuticle-degrading enzymes such a chitinases, lipases, and proteases (e.g Khachatourians, 1991). These combinations of proteins would potentially be of value in baits, insect pathogens, or transgenic plants for degrading the internal cuticular linings and peritrophic membrane of the insect to allow penetration of plant secondary metabolites or proteins.

Compounds targeting multiple target sites are also produced by the same organism. For example, *Aspergillus flavus* produces aflatoxins, which interfere with DNA replication, cyclopiazonic acid, which inhibits calcium transport ATPase, tremorgenic compounds that appear to act at GABAnergic sites, kojic acid that inhibits detoxifying enzymes and insect and plant defensive enzymes, and aflavanines that inhibit NADH oxidase (see prior discussion). This appears to be another natural example of an evolutionary polygenic strategy for preventing resistance of target insects. This multigenic strategy again provides a guideline for using microorganism-derived secondary metabolites and/or proteins for protecting plants or for engineering more effective insect pathogens.

However, insects have apparently developed the ability to deal with various defensive compounds, as discussed previously in terms of relative resistance and metabolism. A fairly unique twist on this resistance has also been reported. Inhibitors of toxic proteases from *Metarhizium anisopliae* (Kucera, 1980) have been extracted from *G. mellonella* (Kucera, 1982). Highest levels occurred in the hemolymph compared to the gut or fat body (Kucera, 1982). The inhibitor was more active against the *Metarhizium* serine protease, and also had some activity against the sulfhydryl protease (Kucera, 1982).

Future directions/commercial prospects

Some of the compounds discussed have been commercialized, such as the avermectins, spinosins, the Bt crystal proteins, cholesterol oxidases and VIP toxins for use in transgenic plants. The reasons for commercialization have been described previously. Key among these is very high biological activity against insects, with significant mortality at the 10 ppm range orally or 0.001 μg/insect range topically. Selectivity is also important. With the continued high cost (multimillion-dollar process) of development and registration of insecticides and transgenic plants, bioactive compounds must pass several tiers of testing for successful commercialization.

New sources for compounds are a key issue. Ecological approaches have resulted in high rates of discovery (Claydon and Grove 1982; Wicklow *et al.*, 1994) in comparison to commercial screening (Heisey *et al.*, 1988; Purcell, 1997; Sparks *et al.*, 1999). However, none of the ecologically discovered compounds have yet been commercialized, presumably due to low rates of activity relative to the prescribed criteria. Looking at particularly novel or untried sources has been suggested as a strategy (Sparks *et al.*, 1999). Similar effort is required to identify active proteins, with figures of 400 (Warren, 1997) to 10,000 (Purcell, 1997) culture tests needed to find something with commercial viability. All commercial products presently are produced by or derived from *Bacillus* or *Streptomyces* and were identified through random screening. Discovery from ecologically remote sources may result in a lesser chance that the target insect will have preadapted metabolic systems. The authors work has indicated insects that feed preferentially on fungi, such as *Carpophilus* sap beetles, have nearly always been more resistant than plant feeding caterpillars to compounds isolated from fungi. A similar line of logic potentially exists for plant proteins. For example, the toxicity of endosperm derived maize-derived ribosomal inactivating proteins (an N-glucosidase) was least toxic to an insect that feeds on the dry kernels (Indian meal moth, *P. interpunctella*) and greatest to cabbage looper, *T. ni*, that does not feed on maize at all (Dowd *et al.*, 1998). Again, the more novel the protein from an ecological standpoint, the less likely an insect has or can readily adapt to it. This logic may help explain the successful insect control agents that have predominantly come from soil isolates, although the proteins produced by insect pathogens are an obvious exception.

Because the detoxifying systems, especially of major pests, have been increasingly well studied, it is now possible to predict what types of moieties are subject to detoxification. Possibly, the more "alien" a source, and hence compound, can be found, the less likely an insect is able to cope with it or develop resistance to it, and microbial-derived compounds have great potential for this property. However, insect metabolism of structural moieties similar to those presumably preadapted to in plants have been reported for α-amanitin, avermectins, aflatoxins, destruxins, trichothecenes, griseofulvin, and proteins (see prior discussion on individual compounds). Microorganisms not encountered by a particular insect may be a better source of longer lifespan, effective materials, but the rate of discovery would likely to be lower than ecologically driven approaches.

Resistance theory generally assumes where multiple target sites are hit, such as by combinations of compounds, resistance is slower to develop (e.g. Gould, 1998). As discussed earlier, multiple toxins produced by same species can have different modes of action, as well as act synergistically. Recent indications of non-direct activity by reactive oxygen generating enzymes such as peroxidases or polyphenoloxidases, that apparently have multiple modes of action and target sites, suggest resistance may be more difficult for insects to develop due to multiple genes that must be altered (e.g. Dowd and Lagrimini, 1997). Fungi produce similar oxidative enzymes such as tyrosinases and laccases. Production of the colored quinones of mushrooms when damaged through a similar enzymatic mechanism suggests potentially similar mechanisms that can be exploited in fungi as well.

Single pathway controllers or regulatory gene alterations may result in production of multiple new bioactive compounds as described for A,C,P and R genes in maize (e.g. Neuffer *et al.*, 1997). Cloning of polyketide complexes (e.g. Clarkson and Charnley, 1996; Reynolds, 1998) allows for this new potential in microbial or perhaps plant application as well. Different combinations of polyketide synthase "modules", module domains, and post polyketide "tailoring" enzymes can potentially be recombined to produce new metabolites (Reynolds, 1998).

Combinatorial chemistry is a method by which lead structures can more rapidly be derivatized to more potent, selective compounds, previously impractical by conventional chemical means (e.g. Borman, 1999). There are diverse opinions on how to manage the volume of data potentially generated by this approach. Some suggestions include identifying an appropriate target site to screen against initially, computer-aided design to "prescreen electronically", and enrichment techniques associated with specific binding (Borman, 1999). Similar logic is being used to optimize biologcially active proteins, including a "fuzzy" PCR to randomly evolve proteins from a starting source with potentially enhanced or novel host range. The ability to understand secondary and tertiary protein structure allows for optimization and stabilization for activity against insects. A related example of this concept is increasing the G+C content from 37 percent in the native Bt crystal protein gene to 65 percent for expression in maize so that higher levels of expression occur (e.g. Carozzi and Koziel, 1997).

New techniques in molecular biology allow for isolation of DNA, and potentially genes, from the estimated 99.9 per cent of microorganisms that cannot be cultured (Rouhi, 1999). Quotes from industry indicated that screening rates of close to 10,000 clones per second have already been reached (Rouhi, 1999). In addition to individual genes that code proteins with biological activity, gene pathways composed of up to 100 genes are now theoretically possible (Rouhi, 1999).

Acknowledgements

I thank past and present collaborators for the opportunity to investigate the activity of microbial compounds against insects, J.B. Gloer, T.C. Sparks, and G.W. Warren for comments on prior drafts of this manuscript, and R. Sylvester for preparation of the numerous figures. Omission of any relevant prior studies was not intentional.

References

Addor, R.W., Babcock, T.J., Black, B.C., Brown, D.G., Diehl, R.E., Furch, J.A. *et al.* (1992) Insecticidal pyrroles: Discovery and overview. In D.R. Baker, J.G. Fenyes, J.J. Steffens (eds.), *Synthesis and Chemistry of Agrochemicals*, American Chemical Society, Washington, D.C., pp.283–297.

Al-adil, K.M., Kilgore, W.W., and Painter, R.R. (1973) Effects of orally ingested aflatoxin B_1 on nucleic acids and ribosomes of house fly ovaries. *Toxicol. Appl. Pharmacol.*, **26**, 130–136.

Anderson, J.F. (1966) Anomalous development of the cuticle of mosquitoes induced by griseofulvin. *J. Econ. Entomol.*, **59**, 1476–1482.

Ando, K. (1983) How to discover new antibiotics for insecticidal use. In N. Takahashi, H. Yoshioka, T. Misato and S. Matsunaka (eds.), *Pesticide Chemistry: Human Welfare and the Environment. Volume 2, Natural Products*, Pergamon Press, New York, pp. 253–259.

Aoki, A., Fukuda, R., Nakayabu, T., Ishibashi, K., Takeichi, T., and Ishida, M. (1976) Antibiotic substances B-41, their production and their use as insecticides and acaricides. U.S. Patent #3,984,564

Arnason, T., Towers, G.H.N., Philogene, B.J.R., and Lambert, J.D.H. (1983) The role of natural photosensitizers in plant resistance to insects. In P.A. Hedin (ed.), *Plant Resistance to Insects*, American Chemical Society, Washington, D.C., pp. 139–151.

Ayer, W.A., Altena, I., and Browne, L.M. (1990) Three piperazinediones and a drimane diterpenoid from *Penicillium brevi-compactum. Phytochemistry*, **29**, 1661–1665.

Barath, Z., Barathova, H., Betina, V., and Nemec, P. (1974) Ramihyfins – antifungal and morphogenic antibiotics from *Fusarium* sp. *Folia Microbiol.*, **19**, 507–511.

Barloy, F., Dele'cluse, A., Nicolas, L., and Lecadet, M.-M. (1996) Cloning and expression of the first anaerobic toxin gene from *Clostridium bivermentans* subsp. *malaysia*, encoding a new mosquitocidal protein with homologies to *Bacillus thuringiensis* delta-endotoxins. *J. Bacteriol.*, **178**, 3099–3105.

Bauer, K., Bischoff, E., Hugo, H.V., Berg, D., and Kraus, P. (1981) In H.W. Wegler (ed.), *Chemie der Pflanzenschutz und Schadlingsbekampfungsmittel*. Springer Verlag, Berlin, pp. 215–328.

Baumann, L., Broadwell, A.H., and Baumann, P. (1988) Sequence analysis of the mosquitocidal toxin genes encoding 51.4 and 41.9 kilodalton proteins from *Bacillus sphaericus* 2362 and 2297. *J. Bacteriol.*, **170**, 2045–2050.

Baute, R., Deffieux, G., Merlet, D., Baute, M.-A., and Neveu, A. (1981) New insecticidal cyclodepsipeptides from the fungus *Isaria felina*. I. Production, isolation and insecticidal properties of isariins B, C, and D. *J. Antibiotics*, **34**, 1261–1265.

Beard, R.L., and Walton, G.S. (1969) Kojic acid as an insecticidal mycotoxin. *J. Invert. Pathol.*, **14**, 53–59.

Belofsky, G.N., Gloer, J.B., Wicklow, D.T., and Dowd, P.F. (1995) Antiinsectan alkaloids: Shearinines A-C and a new paxilline derivative from the ascostromata of *Eupenicillium shearii. Tetrahedron*, **51**, 3959–3968.

Belofsky, G.N., Gloer, K.B., Gloer, J.B., Wicklow, D.T., and Dowd, P.F. (1998a) New *p*-terphenyl and polyketide metabolites from the sclerotia of *Penicillium raistrickii. J. Nat. Prod.*, **61**, 1115–1119.

Belofsky, G.N., Gloer, J.B., Wicklow, D.T., and Dowd, P.F. (1998b) Shearamide A: A new cyclic peptide from the ascostromata of *Eupenicillium shearii. Tetrahedron Lett.*, **39**, 5497–5500.

Bently, H.R., Cunningham, K.G., and Spring, F.S. (1951) Cordycepin, a metabolic product from cultures of *Cordyceps militaris* (Linn.) Link. Part II. The structure of cordycepin. *J. Chem. Soc.*, 2301–2305.

Bertram, G., Schleithoff, L., Zimmermann, P., and Wessing, A. (1991) Bafilomycin A1 is a potent inhibitor of urine formation by Malpighian tubules of *Drosophila hydei*: is a vacuolar-type ATPase involved in ion and fluid secretion? *J. Insect Physiol.*, **37**, 201–209.

Besl, H., and Blumreisinger, M. (1983) Die Eignung von *Drosophila melanogaster* zur Untersuchung der Anfälligkeit Höherer Pilze gegenüber Madenfra-beta. *Zeit. Mykol.*, **49**, 165–170.

Beutler, J.A., and der Marderosian, A.H. (1981) Chemical variation in *Amanita. J. Nat. Prod.*, **44**, 422–431.

Birch, A.J., and Wright, J.J. (1969) The brevianamides: A new class of fungal alkaloid. *J. Chem. Soc. Chem. Commun.*, 644–645.

Bolter, C.J., and Latoszek-Green, M. (1997) Effect of chronic ingestion of the cysteine proteinase inhibitor E-64, on Colorado potato beetle gut proteinases. *Entomol. exp. appl.*, **83**, 295–303.

Borman, S. 1999. Reducing time to drug discovery. *Chem. Eng. News*, **77**, 33–48.

Bossi, R. (1960) Über die Wirkung der Fusarinsaüre auf die Polyphenoloxydase. *Phytopath. Z.*, **37**, 273–316.

Bottger, G.T., Yerington, A.P., and Gertler, S.I. (1951) Synthetic organic compounds as insecticides. *U.S. Dept. Agr. Bur. Entomol. Plant. Quarantine.*, E-826 pp. 23.

Boucias, D.G., Farmerie, W.G., and Pendland, J.C. (1998) Cloning and sequencing of cDNA of the insecticidal toxin hisrsutellin A. *J. Invert. Pathol.*, **72**, 258–261.

Bourgouin, C., Dele'cluse, A., de la Torre, F., and Szulmajster, J. (1990). Transfer of the toxin protein genes of *Bacillus sphaericus* into *Bacillus thuringiensis* subsp. *israelensis* and their expression. *Appl. Environ. Microbiol.*, **56**, 340–344.

Bowen, D., Rocheleau, T.A., Blackburn, M., Andreev, O., Golubeva, E., Bhartia, R. *et al.* (1998) Insecticidal toxins of the bacterium *Photorhabdus luminescens. Science*, **280**, 2129–2132.

Brandhorst, T., Dowd, P.F., and Kenealy, W.R. (1996) The ribosome inactivating protein restrictocin deters insect feeding on *Aspergillus restrictus. Microbiology*, **142**, 1551–1556.

Brandt, C.R., Adang, M.J., and Spence, K.D. (1978) The peritrophic membrane: Ultrastructural analysis and function as a mechanical barrier to microbial infection in *Orgyia pseudotsugata. J. Invert. Pathol.*, **32**, 12–24.

Brillinger, G.U. (1979) Metabolic products of microorganisms 181. Chitin synthase from fungi, *Coprinus cinereus*, a test model fro substances with insecticidal properties. *Arch. Microbiol.*, **121**, 71–74.

Buchi, G., Klaubert, D.H., Shank, S.M., Weinreb, S.M., and Wogan, G.N. (1971) Structure and synthesis of kotanin and desmethyl kotanin, metabolites of *Aspergillus glaucus. J. Org. Chem.*, **36**, 1143–1147.

Bull, D.L. (1986) Toxicity and pharmacodynamics of avermectin in the tobacco budworm, corn earworm and fall armyworm (Noctuidae: Lepidoptera). *J. Agric. Food Chem.*, **34**, 74–78.

Calhoun, L.A., Findlay, J.A., Miller, J.D., and Whitney, N.J. (1992) Metabolites toxic to spruce budworm from balsam fir needle endophytes. *Mycol. Res.*, **96**, 281–286.

Camazine, S., and Lupo, A.T., Jr. (1984). Labile toxic compounds of the Lactarii: the role of the lacticiferous hyphae as a storage depot for precursors of pungent aldehydes. *Mycologia*, **76**, 355–358.

Camazine, S.M., Resch, J.E., Eisner, T., and Meinwald, J. (1983) Mushroom chemical defense: pungent sesquiterpenoid dialdehyde antifeedant to opossum. *J. Chem. Ecol.*, **9**, 1439–1447.

Carozzi, N.B., and Koziel, M.G. (1997) Transgenic maize expressing a *Bacillus thuringiensis* insecticidal protein for control of European corn borer. In N. Carozzi and M. Koziel (eds.), *Advances in Insect Control: The Role of Transgenic Plants*, Taylor and Francis, Bristol, PA, pp. 63–74.

CAS (1999) Tendamistat. #86595-25-0 [on line] Sci. Finder Database, American Chemical Society, Washington, D.C.

Charles, J.-F., Nicholas, L., Sebald, M., and deBarjac, H. (1990). *Clostridium bifermentans* serovar *malaysia*: sporulation, biogenesis of inclusion bodies and larvicidal effect on mosquito. *Res. Microbiol.*, **141**, 721–733.

Chen, M.-S., Johnson, B., Wen, L., Muthukrishnan, S., Kramer, K.J., Morgan, T.D. *et al.* (1992) Rice cystatin-bacterial expression, purification, cysteine proteinase inhibitory activity and insect growth suppressing activity of a truncated form of the protein. *Prot. Expr. Purif.*, **3**, 41–49.

Cherton, J.-C., Lange, C., and Mulheim, C. (1991) Direct *in vitro* and *in vivo* monitoring of destruxins metabolism in insects using internal surface reversed-phase high-performance liquid chromatography. I. Behavior of E destruxin in locusts. *J. Chromatog.*, **566**, 511–524.

Cho, H.-J., Choi, K.-P., Yamashita, M., Mirikawa, H., and Murooka, Y. (1995) Introduction and expression of the *Streptomyces* cholesterol oxidase gene (ChoA), a potent insecticidal protein active against boll weevil larvae, in tobacco cells. *Appl. Microbiol. Biotechnol.*, **44**, 133–138.

Ciegler, A. (1976) Mycotoxins as insecticides. In J.D. Briggs, (ed.), *Biological Regulation of Vectors. The Saprophytic and Aerobic Bacteria and Fungi. A Conference Report*, U.S. Dept. H.E.W., Washington, D.C., pp. 135–150.

Clark, J.M., Argentine, J.A., Lin, H., and Gao, X.Y. (1992) Mechanisms of abamectin resistance in the Colorado potato beetle. In C.A. Mullin and J.G. Scott (eds.), *Molecular Mechanisms of Insecticide Resistance*, American Chemical Society, Washington, D.C., pp. 247–263.

Clarkson, J.M., and Charnley, A.K. (1996) New insights into the mechanisms of fungal pathogenesis in insects. *Trends Microbiol.*, **4**, 197–203.

Claydon, N. (1978) Insecticidal secondary metabolites from entomogenous fungi: *Entomophthora virulenta. J. Invert. Pathol.*, **32**, 319–324.

Claydon, N., and Grove, J.F. (1978) Metabolic products of *Entomophthora virulenta. J. Chem. Soc. Perkin Trans. I*, 171–173.

Claydon, N., and Grove, J.F. (1982) *Fusarium* as an insect pathogen. *Symp. British Mycol. Soc.*, **7**, 117–128.

Claydon, N., Grove, J.F., and Pople, M. (1977) Insecticidal secondary metabolic products from the entomogenous fungus *Fusarium solani. J. Invert. Pathol.*, **30**, 216–223.

Claydon, N., Grove, J.F., and Pople, M. (1979) Insecticidal secondary metabolic products from the entomogenous fungus *Fusarium larvarum J. Invert. Pathol.*, **33**, 364–367.

Cohen, E., and Casida, J.E. (1982) Properties and inhibition of insect integumental chitin synthetase. *Pestic. Biochem. Physiol.*, **17**, 301–306.

Cole, M., and Rolinson, G.N. (1972) Microbial metabolites with insecticidal properties. *Appl. Microbiol.*, **24**, 660–662.

Cole, R.J., and Cox, R.H. (1981) *Handbook of Toxic Fungal Metabolites*, Academic Press, New York.

Conca, E., DeBernardi, M., Fronza, G., Girometta, M.A., Mellerio, G., Vidari, G. *et al.* (1981) Fungal metabolites 10. New chromenes from *Lactarius fuliginosus* and *Lactarius picinus* Fries. *Tetrahedron Lett.*, **22**, 4327–4330.

Corbin, D.R., Greenplate, J.T., Wong, E.Y., and Purcell, J.P. (1994) Cloning of an insecticidal cholesterol oxidase gene and its expression in bacteria and in plant protoplasts. *Appl. Environ. Microbiol.*, **60**, 4239–4244.

Cunningham, K.G., Hutchinson, S.A., Manson, W., and Spring, F.S. (1951) Cordycepin, a metabolic product from cultures of *Cordyceps militaris* (Linn.) Link. Part I. Isolation and characterization. *J. Chem. Soc.* 2299–2300.

Cutler, H.G. (1987) Japanese contributions to the development of allelochemicals. In P.A. Hedin (ed.), *Allelochemicals: Role in Agriculture and Forestry*, American Chemical Society, Washington, D.C., pp. 23–38.

Dähn, U., Hagenmaier, H., Höhne, H., König, W.A., Wolf, G., and Zähner, H. (1976) Nikkomycin, ein neuer Hemmstoff der chitinsynthese bei Pilzen. *Arch Microbiol.*, **107**, 143–160.

Daniewski, W.M., Gumulka, M., Przesmycka, D., Ptaszyn'ska, K., Bloszyk, E., and Droz'dz', B. (1995) Sesquiterpenes of *Lactarius* origin, antifeedant structure-activity relationships. *Phytochemistry*, **38**, 1161–1168.

Daniewski, W.M., Gumulka, M., Pankowska, E., Ptaszyn'ska, K., Bloszyk, E., Jacobsson, U. *et al.* (1993a) 3,8-ethers of lactarane sesquiterpenes. *Phytochemistry*, **32**, 1499–1502.

Daniewski, W.M., Gumulka, M., Ptaszyn'ska, K., Skibicki, P., Bloszyk, E., Droz'dz', B. *et al.* (1993b) Antifeedant activity of some sesquiterpenoids of the genus *Lactarius* (Agaricales: Russulaceae). *Eur. J. Entomol.*, **90**, 65–70.

Davis, G.R.F., and Schiefer, H.B. (1982) Effects of dietary T-2 toxin concentrations fed to larvae of the yellow mealworm at three dietary protein levels. *Comp. Biochem. Physiol.*, **73C**, 13–16.

DeAmicis, C.V., Dripps, J.E., Hatton, C.J., and Karr, L.L. (1997) Physical and biological properties of the spinosyns: Novel macrolide pest-control agents from fermentation. In P.A. Hedin, R.M. Hollingsworth, E.P. Masler, J. Miyamoto and D.G. Thompson (eds.), *Phytochemicals for Pest Control*, American Chemical Society, Washington, D.C., pp. 144–154.

DeGuzman, F.S., Gloer, J.B., Wicklow, D.T., and Dowd, P.F. (1992) New diketopiperazine metabolites from the sclerotia of *Aspergillus ochraceus. J. Nat. Prod.*, **55**, 931–939.

DeGuzman, F.S., Brus, D.R., Rippentrop, J.M., Gloer, K.B., Gloer, J.B., Wicklow, D.T. *et al.* (1994) Ochrindoles A-D: New Bis-indolyl benzenoids from the sclerotia of *Aspergillus ochraceus* NRRL 3519. *J. Nat. Prod.*, **57**, 634–639.

Dellwig, H., Kurz, J., Pflüger, W., Schedel, M., Vobis, G., and Wünsche, C. (1988) Rodaplutin, a new peptidylnucleoside from *Nocardioides albus. J. Antibiotics*, **41**, 1145–1147.

Dev, S., and Koul, O. (1997) *Insecticides of Natural Origin*, Harwood Academic Publishers gmbh, Amsterdam, 365 pp.

Dickinson, C., and Lucas, J. (1979) *The Encyclopedia of Mushrooms*, Crescent Books, New York.

Dobias, J., Nemec, P., and Brtko, J. (1977) The inhibitory effect of kojic acid and its two derivatives on the development of *Drosophila melanogaster. Biolo'gia (Bratislava)*, **32**, 417–421.

Dobias, J., Betina, V., and Nemec, P. (1980) Insekticidna aktivita ramihyfin A, citrininu a rugulozinu. *Biologia (Bratislava)*, **35**, 431–434.

Dowd, P.F. (1988a) Synergism of aflatoxin B_1 toxicity with the co-occurring fungal metabolite kojic acid to two caterpillars. *Entomol. exp. appl.*, **47**, 69–71.

Dowd, P.F. (1988b) Toxicological and biochemical interactions of the fungal metabolites fusaric acid and kojic acid with xenobiotics in *Heliothis zea* (F.) and *Spodoptera frugiperda* (J.E. Smith). *Pestic. Biochem. Physiol.*, **32**, 123–134.

Dowd, P.F. (1989a) Toxicity of naturally occurring levels of the *Penicillium* mycotoxins citrinin, ochratoxin A and penicillic acid to the corn earworm, *Heliothis zea*, and the fall armyworm, *Spodoptera frugiperda* (Lepidoptera: Noctuidae). *Environ. Entomol.*, **18**, 24–29.

Dowd, P.F. (1989b) Fusaric acid: A secondary fungal metabolite that synergizes toxicity of cooccurring host allelochemicals to the corn earworm, *Heliothis zea* (Lepidoptera). *J. Chem. Ecol.*, **15**, 249–254.

Dowd, P.F. (1990) Responses of representative midgut detoxifying enzymes from *Heliothis zea* and *Spodoptera frugiperda* to trichothecenes. *Insect Biochem.*, **20**, 349–356.

Dowd, P.F. (1992a) Detoxification of mycotoxins by insects. In C.A. Mullin and J.G. Scott (eds.), *Molecular Mechanisms of Insecticide Resistance*, American Chemical Society, Washington, D.C., pp. 264–275.

Dowd, P.F. (1992b) Insect interactions with mycotoxin-producing fungi and their hosts. In D. Bhatnagar, E.B. Lillehoj and D.K. Arora (eds.), *Handbook of Applied Mycology. Volume 5: Mycotoxins in Ecological Systems*, Marcel Dekker, New York, pp. 137–155.

Dowd, P.F. (1993) Toxicity of the fungal metabolite griseofulvin to *Helicoverpa zea* and *Spodoptera frugiperda. Entomol. exp. appl.*, **69**, 5–11.

Dowd, P.F. (1994) Enhanced maize (*Zea mays* L.) pericarp browning: associations with insect resistance and involvement of oxidizing enzymes. *J. Chem. Ecol.*, **20**, 2777–2803.

Dowd, P.F. (2000) Relative inhibition of insect phenoloxidase by cyclic fungal metabolites.(In Press)

Dowd, P.F., and VanMiddlesworth, F.L. (1989) *In vitro* metabolism of the trichothecene 4-monoacetoxyscirpenol by fungus and non-fungus feeding insects. *Experientia*, 45, 393–395.

Dowd, P.F., and Miller, O.K. (1990) Insecticidal properties of *Lactarius fuliginosus* and *Lactarius fumosus. Entomol. exp. appl.* **57**, 23–28.

Dowd, P.F., and Vega, F.E. (1996) Enzymatic oxidation products of allelochemicals as a potential direct resistance mechanism against insects: Effects on the corn leafhopper *Dalbulus maidis. Nat. Toxins*, **4**, 85–91.

Dowd, P.F., and Lagrimini, L.M. (1997) The role of peroxidase in host insect defenses. In N. Carozzi and M. Koziel (eds.), *Advances in Insect Control: The Role of Transgenic Plants*, Taylor and Francis, Bristol, PA, pp. 195–223.

Dowd, P.F., Cole, R.J., and Vesonder, R.F. (1988) Toxicity of selected tremorgenic mycotoxins and related compounds to *Spodoptera frugiperda* and *Heliothis zea. J. Antibiotics*, **41**, 1868–1872.

Dowd, P.F., Miller, J.D., and Greenhalgh, R. (1989) Toxicity and interactions of some *Fusarium graminearum* metabolites to caterpillars. *Mycologia*, **81**, 646–650.

Dowd, P.F., Burmeister, H.R., and Vesonder, R.F. (1992a) Antiinsectan properties of a novel *Fusarium*-derived sterol sulfate. *Entomol. exp. appl.*, **63**, 95–100.

Dowd, P.F., Mehta, A.D., and Boston, R.S. (1998) Antiinsectan activity of the maize ribosomal inactivating protein. *J. Agric. Food Chem.*, **46**, 3775–3779.

Dowd, P.F., Peng, F.C., Chen, J.W., and Ling, K.H. (1992b) Toxicity and anticholinesterase activity of the fungal metabolites territrems to the corn earworm, *Helicoverpa zea. Entomol. exp. appl.*, **65**, 57–64.

Dubis, E.N., Brattsten, L.B., and Dingan, L.B. (1992) Effects of the endophyte associated alkaloid peramine on southern armyworm microsomal cytochrome P450. In C.A. Mullin and J.G. Scott (eds.), *Molecular Mechanisms of Insecticide Resistance*, American Chemical Society, Washington, D.C., pp. 125–136.

Edmonds, H.S., Gatehouse, L.N., Hilder, V.A., and Gatehouse, J.A. (1996) The inhibitory effects of the cysteine protease inhibitor, oryzacystatin, on digestive proteases and on larval survival and development of the southern corn rootworm (*Diabrotica undecimpunctata howardi*). *Entomol. exp. appl.*, **78**, 83–94.

Endo, A., and Kuroda, M. (1976) Citrinin, an inhibitor of cholesterol biosynthesis. *J. Antibiotics*, **29**, 841–843.

Estruch, J.J., Warren, G.W., Mullins, M.A., Nye, G.J., and Koziel, M.G. (1996) Vip3A, a novel *Bacillus thuringiensis* vegetative insecticidal protein with a wide spectrum of activities against lepidopteran insects. *Proc. Nat. Acad. Sci.*, **93**, 5389–5394.

Eugenio, C., De las Casas, E., Harein, P.K., and Mirocha, C.J. (1970) Detection of the mycotoxin F-2 in the confused flour beetle and lesser mealworm. *J. Econ. Entomol.*, **63**, 412–415.

Eugster, C.H. (1969) Chemie der Wirkstoffe aus dem Fliegenpilz (*Amanita muscaria*). In L. Zechmeister (ed.), *Progress in the Chemistry of Organic Natural Products, Volume 27*, Springer Verlag, New York, pp. 261–321.

Findlay, J.A., Buthelezi, S., Lavoie, R., and Peña-Rodriguez, L. (1995a) Bioactive isocoumarins and related metabolites from conifer endophytes. *J. Nat. Prod.*, **58**, 1759–1766.

Findlay, J.A., Buthelezi, S., Li, G., Seveck, M., and Miller, J.D. (1997a) Insect toxins from an endophytic fungus from wintergreen. *J. Nat. Prod.*, **60**, 1214–1215.

Findlay, J.A., Li, G., and Johnson, J.A. (1997b) Bioactive compounds from an endophytic fungus from eastern larch (*Larix laricina*) needles. *Can. J. Chem.*, **75**, 716–719.

Findlay, J.A., Li, G., and Penner, P.E. (1995b) Novel diterpenoid insect toxins from a conifer endophyte. *J. Nat. Prod.*, **58**, 197–200.

Fisher, M.H. (1997) Structure-activity relationships of the avermectins and milbemycins. In P.A. Hedin, R.M. Hollingsworth, E.P. Masler, J. Miyamoto and D.G. Thompson (eds.), *Phytochemicals for Pest Control*, American Chemical Society, Washington, D.C., pp. 220–238.

Fuchs, M.S., and Fong, W.-F. (1976) Inhibition of blood digestion by α-amanitin and actinomycin D and its effect on ovarian development in *Aedes aegypti. J. Insect Physiol.*, **22**, 465–471.

Gallagher, R.T., Clardy, J., and Wilson, B.J. (1980a) Aflatrem, a tremorgenic toxin from *Aspergillus flavus. Tetrahedron Lett.*, **21**, 239–242.

Gallagher, R.T., Finer, J., Clardy, J., Leutwiler, A., Weibel, F., Acklin, W., *et al.* (1980b) Paspalinine, a tremorgenic metabolite from *Claviceps paspali* Stevens et Hall. *Tetrahedron Lett.*, **21**, 235–238.

Gallagher, R.T., McCabe, T.T., Hirotsu, K., and Clardy, J. (1980c) Aflavinine, a novel indole-mevalonate metabolite from tremorgen-producing *Aspergillus flavus. Tetrahedron Lett.*, **21**, 243–246.

Gant, D.B., Cole, R.J., Valdes, J.J., Eldefrawi, M.E., and Eldefrawi, A.T. (1987) Action of tremorgenic mycotoxins on $GABA_A$ receptor. *Life Sci.*, **41**, 2207–2214.

Gershezon, J., and Croteau, R. (1991) Terpenoids. In G.A. Rosenthal and M.R. Berenbaum (eds.), *Herbivores, Their Interactions with Secondary Plant Metabolites, 2nd Edition, Volume I The Chemical Participants*, Academic Press, New York, pp. 165–220.

Glasby, J.S. (1976) *Encyclopedia of Antibiotics*. John Wiley & Sons, New York.

Glennon, R.A. (1987) Central serotonin receptors as targets for drug research. *J. Med. Chem.*, **30**, 1–12.

Gloer, J.B. (1995a) Antiinsectan natural products from fungal sclerotia. *Acc. Chem. Res.*, **28**, 343–350.

Gloer, J.B. (1995b) The chemistry of fungal antagonism and defense. *Can. J. Bot.*, **73** (Suppl. 1), S1265–S1274.

Gloer, J.B., and Truckenbrod, S.M. (1988) Interference competition among coprophilous fungi: production of (+)-isoepoxydon by *Poronia punctata. Appl. Environ. Microbiol.*, **54**, 861–864.

Gloer, J.B., Rinderknecht, B.L., Wicklow, D.T., and Dowd, P.F. (1989) Nominine: A new insecticidal indole diterpene from the sclerotia of *Aspergillus nomius. J. Org. Chem.*, **54**, 2530–2532.

Gloer, J.B., TePaske, M.R., Sima, J.S., Wicklow, D.T., and Dowd, P.F. (1988) Antiinsectan aflavinine derivatives from the sclerotia of *Aspergillus flavus. J. Org. Chem.*, **53**, 5457–5460.

Goegelman, R.T., and Ondeyka, J.G. (1988) Derivatives of paraherquamide isolated from a fermentation broth active as antiparasitic agents. European Patent Application #88202603.2.

Gottlieb, D., and Shaw, P.D. (1967) *Antibiotics Volume I: Mechanism of Action*. Springer Verlag, New York.

Gould, F. (1998) Sustainability of transgenic insecticidal cultivars: Integrating pest genetics and ecology. *Annu. Rev. Entomol.*, **43**, 701–26.

Greenhalgh, R., Meirer, R.M., Blackwell, B.A., Miller, J.D., Taylor, A., and ApSimon, J.W. (1984a) Minor metabolites of *Fusarium roseum* (ATCC 28114). *J. Agric. Food Chem.*, **32**, 1261–1264.

Greenhalgh, R., Meirer, R.M., Blackwell, B.A., Miller, J.D., Taylor, A., and ApSimon, J.W. (1984b) Minor metabolites of *Fusarium roseum* (ATCC 28114) part 2. *J. Agric. Food Chem.*, **34**, 115–118.

Greenplate, J.T., Duck, N.B., Pershing, J.C., and Purcell, J.P. (1995) Cholesterol oxidase: an oostatic and larvicidal agent active against the cotton boll weevil, *Anthonomus grandis. Entomol. exp. appl.*, **74**, 253–258.

Grove, J.F., and Hosken, M. (1975) The larvacidal activity of some 12,13-epoxytrichothece-9-enes. *Biochem. Pharmacol.*, **24**, 959–962.

Grove, J.F., and Pople, M. (1979) Metabolic products of *Fusarium larvarum* Fuckel. The fusarentins and the absolute configuration of monocerin. *J. Chem. Soc. Perkin Trans. I.*, 2048–2051.

Grove, J.F., and Pople, M. (1980) The insecticidal activity of beauvericin and the enniatin complex. *Mycopathologia*, **70**, 103–105.

Grove, J.F., and Pople, M. (1981) The insecticidal activity of some fungal dihydroisocoumarins. *Mycopathologia*, **76**, 65–67.

Hall, C., Wu, M., Crane, F.L., Takahashi, H., and Tamura, S. (1966) Piericidin A: a new inhibitor of mitochondrial electron transport. *Biochem. Biophys. Res. Comm.*, **25**, 373–377.

Hamil, R.L., Higgins, C.E., Boaz, H.E., and Gorman, M. (1969) The structure of beauvericin, a new depsipeptide antibiotic toxic to *Artemia salina. Tetrahedron Lett.*, **49**, 4255–4258.

Hanada, K., Tamai, M., Yamagishi, M., Ohmura, S., Sawada, J., and Tanaka, I. (1978) Isolation and characterization of E-64, a new thiol protease inhibitor. *Agric. Biol. Chem.*, **42**, 523–528.

Hanessian, S., DeJongh, D.C., and McCloskey, J.A. (1966) Further evidence on the structure of cordycepin. *Biochem. Biophys. Acta.*, **117**, 480–482.

Harein, P.K., De las Casas, E., Eugenio, C., and Mirocha, C.J. (1971) Reproduction and survival of confused flour beetles exposed to metabolites produced by *Fusarium roseum* var. *graminearum. J. Econ. Entomol.*, **64**, 975–976.

Harold, F.M., Altendorf, H., and Hirata, H. (1974) Probing membrane transport mechanisms with ionophores. *Ann. N.Y. Acad. Sci.*, **235**, 149–160.

Harries, F.H. (1967) Fecundity of female codling moth treated with novobiocin and other antibiotics. *J. Econ. Entomol.*, **60**, 7–10.

Hayashi, H., and Sakaguchi, A. (1998) Okaramine G, a new okaramine congener from *Penicillium simplicissimum. Biosci. Biotechnol. Biochem.*, **62**, 804–806.

Hayashi, H., Murao, S., and Arai, M. (1991) Verruculogen as a convulsive factor against silkworm, from *Penicillium simplissimum. Chem. Express.*, **6**, 989–992.

Hayashi, H., Takiuchi, K., Murao, S., and Arai, M. (1989) Structure and insecticidal activity of new indole alkaloids, okaramines A and B, from *Penicillium simplicissimum. Agric. Biol. Chem.*, **53**, 461–469.

Hayashi, H., Asabu, Y., Murao, S., and Arai, M. (1995) New okaramine congeners, okaramines D, E, and F, from *Penicillium simplicissimum. Biosci. Biotech. Biochem.*, **59**, 246–250.

Hayashi, H., Asabu, Y., Murao, S., Nakayama, M., and Arai, M. (1993a) Penitrem A, as a convulsive factor against silkworm, from *Penicillium simplicissimum* AK-40. *Chem. Express*, **8**, 177–180.

Hayashi, H., Asabu, Y., Murao, S., Nakayama, M., and Arai, M. (1993b) A new congener of penitrems, 6-bromopenitrem E, from *Penicillium simplicissimum* AK-40. *Chem. Express*, **8**, 233–236.

Heisey, R.M., Huang, J., Mishra, S.K., Keller, J.E., Miller, J.R., Putnam, A.R. *et al.* (1988a) Production of valinomycin, an insecticidal antibiotic, by *Streptomyces griseus* var. *flexipertum* var. *nov. J. Agric. Food Chem.*, **36**, 1283–1286.

Heisey, R.M., Mishra, S.K., Putnam, A.R., Miller, J.R., Whitenack, C.J., Keller, J.E. *et al.* (1988b) Production of herbicidal and insecticidal metabolites by soil microorganisms. In H. Cutler (ed.), *Biologically Active Natural Products*, American Chemical Society, Washington, D.C., pp. 65–78.

Huang, H.T., and Shapiro, M. (1971) Insecticidal activity of microbial metabolites. In D.J.D. Hockenhull (ed.), *Progress in Industrial Microbiology.*, CRC Press, Cleveland, OH, pp. 79–112.

Huang, J.-K., Dowd, P.F., Keudell, K.C., Klofenstein, W.E., Wen, L., Bagby, M.O. *et al.* (1996) Biotransformation of saturated monohydroxyl fatty acids to 2-tetrahydrofuranyl acetic acid derivatives: Mechanism of formations and the biological activity of 5-*n*-hexyl-tetrahydrofuran-2-acetic acid. *J. Am. Oil Chem. Soc.*, **73**, 1465–1469.

Hughes, P.B., Dauterman, W.C., and Montoyama, N. (1989) Inhibition of growth and development of tobacco hornworm (Lepidoptera: Sphingidae) larvae by cyromazine. *J. Econ. Entomol.*, **82**, 45–51.

Hunt, D.A., and Treacy, M.F. (1998) Pyrrole insecticides: A new class of agriculturally important insecticides functioning as uncouplers of oxidative phosphorylation. In I. Ishaaya and D. Degheele (eds.), *Insecticides with Novel Modes of Action*, Springer, New York, pp. 138–151.

Isogai, A., Horii, T., Suzuki, A. Murakoshi, S. , Ikeda, K., Sato, S. *et al.* (1975) Isolation and identification of nigragillin as a insecticidal metabolite produced by *Aspergillus niger. Agr. Biol. Chem.*, **39**, 739–740.

Isogai, A., Sakuda, S., Matsumoto, S., Ogura, M., Furihata, K., Seto, H. *et al.* (1984) The structure of leucanicidin, a novel insecticidal macrolide produced by *Streptomyces halstedii. Agric. Biol. Chem.*, **48**, 1379–1381.

Jaenike, J., Grimaldi, D.A., Sluder, A.E., and Greenleaf, A.L. (1983) α-Amanitin tolerance in mycophagous *Drosophila. Science*, **221**, 165–167.

Jansson, R.K., and Dybas, R.A. (1998) Avermectins: Biochemical modes of action, biological activity and agricultural importance. In I. Ishaaya and D. Degheele (eds.), *Insecticides with Novel Modes of Action*, Springer, New York, pp. 152–167.

Janzen, D.H. (1977) Why fruits rot, seeds mold and meat spoils. *Am. Nat.*, **111**, 691–713.

Jarvis, J.L., Guthrie, W.D., and Lillehoj, E.B. (1984) Aflatoxin and selected biosynthetic precursors: Effects on the European corn borer in the laboratory . *J. Agric. Entomol.*, **1**, 354–359.

Jawetz, E., Melnick, J.L., and Adelberg, E.A. (1982) *Review of Medical Microbiology*, LANGE Medical Publications, Los Altos, CA

Jiminez, A., and Vasquez, D. (1985) Plant and fungal protein and glycoprotein toxins inhibiting eukaryote protein synthesis. *Annu. Rev. Microbiol.*, **39**, 649–672.

Joffe, A.Z. (1986) *Fusarium Species: Their Biology and Toxicology*, Wiley, New York.

Jones, N.D., Chaney, M.O., Chamberlain, J.W., Hamill, R.L., and Chen, S. (1973) Structure of A204A, a new polyether antibiotic. *J. Am. Chem. Soc.*, **95**, 3399–3400.

Kanaoka, M., Isogai, A., Murakoshi, S., Ichinoe, M., Suzuki, A., and Tamara, S. (1978) Bassianolide, a new insecticidal cyclodepsipeptide from *Beauveria bassiana* and *Verticillium lecanii. Agric. Biol. Chem.*, **42**, 629–635.

Kaška, M., Lysenko, O., and Chaloupka, J. (1976) Exocellular proteases of *Serratia marcescens* and their toxicity to larvae of *Galleria mellonella. Folia Microbiol.*, **21**, 465–473.

Khachatourians, G.C. (1991) Physiology and genetics of entomopathogenic fungi. In D.K. Arora, L. Ajello and K.G. Mukerji (eds.), *Handbook of Applied Mycology, Volume 2: Humans, Animals and Insects*, Marcel Dekker, New York, pp. 613–663.

Kim, Y.K., Tomoda, H., Nishida, H., Sunazuka, T., Obata, R., and Omura, S. (1995) Pyripyropenes, novel inhibitors of acyl-Co-A: cholesterol acyltransferase produced by *Aspergillus fumigatus*. II. Structure elucidation of pyripyropenes A, B, C, and D. *J. Antibiot.*, **47**, 154–162.

Kishaba, A.N., Shankland, D.L., Curtis, R.W., and Wilson, M.C. (1962) Substances inhibitory to insect feeding with insecticidal properties from fungi. *J. Econ. Entomol.*, **55**, 211–214.

Kodaira, Y. (1961) Toxic substances to insects, produced by *Aspergillus ochraceus* and *Oospora destructor. Agric. Biol. Chem.*, **25**, 261–262.

Kodaira, Y. (1962) Studies on the new toxic substance to insects, destruxin A and B, produced by *Oospora destructor*. Part I. Isolation and purification of destruxin A and B. *Agric. Biol. Chem.*, **26**, 36–42.

Koga, D., Isogai, A., Sakuda, S., Matsumoto, S., Suzuki, A., and Koseki, K. (1986) The structure of allosamidin, a novel insect chitinase inhibitor, produced by *Streptomyces* sp. *Agric. Biol. Chem.*, **51**, 471–476.

Konno, K. (1995) Biologically active components of poisonous mushrooms. *Food Rev. Int.*, **11**, 83–107.

Kramer, K.J., Muthukrishnan, S., Johnson, L., and White, F. (1997) Chitinases for insect control. In N. Carozzi and M. Koziel (eds.), *Advances in Insect Control: The Role of Transgenic Plants*, Taylor and Francis, Bristol, PA, pp. 185–193.

Krasnoff, S.B., and Gupta, S. (1991) Identification and directed biosynthesis of efrapeptins in the fungus *Tolypocladium geodes* Gams (Deuteromycotina: Hyphomycetes). *J. Chem. Ecol.*, **17**, 1953–1962.

Krasnoff, S.B., and Gupta, S. (1994) Identification of the antibiotic phomalactone from the entomopathogenic fungus *Hirsutella thompsonii* var. *synnematosa. J. Chem. Ecol.*, **20**, 293–302.

Krasnoff, S.B., Gibson, D.M., Belofsky, G.N., Gloer, K.B., and Gloer, J.B. (1996) New destruxins from the entomopathogenic fungus *Aeschrsonia* sp. *J. Nat. Prod.*, **59**, 485–489.

Kretschmer, A., Dorgerloh, M., Deeg, M., and Hagenmaier, H. (1985) The structures of novel insecticidal macrolides: Bafilomycins D and E and oxohygrolidin. *Agric. Biol. Chem.*, **49**, 2509–2511.

Krska, R., Schumacher, R., Grasserbauer, M., and Scott, P.M. (1996) Determination of the *Fusarium* mycotoxin beauvericin at μg/kg levels in corn by high performance liquid chromatography with diode-array detection. *J. Chromatog.*, **746A**, 233–238.

Kubo, I., Kim, M., Wood, W.F., and Kaoki, H. (1986) Clitocine, a new insecticidal nucleoside from the mushroom *Clitocybe inversa. Tetrahedron Lett.*, **27**, 4277–4280.

Kubo, M., Kato, Y., Morisaka, K., Inamori, Y., Nomoto, K., Takemoto, T. *et al.* (1981) Insecticidal activity of streptothricin antibiotics. *Chem. Pharm. Bull.*, **29**, 3727–3730.

Kucera, M. (1980) Proteases from the fungus *Metarhizium anisopliae* toxic for *Galleria mellonella* larvae. *J. Invert. Pathol.*, **35**, 304–310.

Kucera, M. (1982) Inhibition of the toxic proteinases from *Metarhizium anisopliae* by extracts of *Galleria mellonella* larvae. *J. Invert. Pathol.*, **40**, 299–300.

Kucera, M., and Lysenko, O. (1968) The mechanism of pathogenicity of *Pseudomonas aeruginosa*. V. Isolation and properties of the proteinases toxic for larvae of the greater wax moth *Galleria mellonella* L. *Folia Microbiol. (Prague)*, **13**, 288–294.

Kucera, M., and Samsinakova, A. (1968) Toxins of the entomophagous fungus *Beauveria bassiana. J. Invert. Pathol.*, **12**, 316–320.

Kuhn, D.G. (1997) Structure-activity relationships for insecticidal pyrroles. In: P.A. Hedin, R.M. Hollingsworth, E.P. Masler, J. Miyamoto and D.G. Thompson (eds.), *Phytochemicals for Pest Control*, American Chemical Society, Washington, D.C., pp. 195–205.

Kuo, S.-C., and Lampen, J.O. (1974) The action of antibiotics on enzyme secretion in yeasts: Studies with cytochalasin A. *Ann. N.Y. Acad. Sci.*, **235**, 137–148.

Laakso, J.A., Gloer, J.B., Wicklow, D.T., and Dowd, P.F. (1992a) Radarins A-D: New antiinsectan and cytotoxic indole diterpenoids from the sclerotia of *Aspergillus sulphureus. J. Org. Chem.*, **57**, 138–141.

Laakso, J.A., Gloer, J.B., Wicklow, D.T., and Dowd, P.F. (1992b) Sulpinines A-C and secopenitrem B: New antiinsectan metabolites from the sclerotia of *Aspergillus sulphureus. J. Org. Chem.*, **57**, 2066–2071.

Laakso, J.A., Gloer, J.B., Wicklow, D.T., and Dowd, P.F. (1993) A new penitrem analog with antiinsectan activity from the sclerotia of *A. sulphureus. J. Agric. Food Chem.*, **41**, 973–975.

Laakso, J.A., Narske, E.D., Gloer, J.B., Wicklow, D.T., and Dowd, P.F. (1994) Isokotanins A-C: New bicoumarins from the sclerotia of *Aspergillus alliaceus. J. Nat. Prod.*, **57**, 128–133.

Lillehoj, E.B. (1992) Aflatoxin: Genetic mobilization agent. In D. Bhatnagar, E.B. Lillehoj and D.K. Arora (eds.), *Handbook of Applied Mycology. Volume 5: Mycotoxins in Ecological Systems*, Marcel Dekker, New York, pp. 1–22.

Ling, K.H., Yang, C.K., and Peng, F.T. (1979) Territrems, termorgenic mycotoxins of *Aspergillus terreus. Appl. Environ. Microbiol.*, **37**, 355–357.

Ling, K.H., Liou, H.H., Yang, C.M., and Yang, C.K. (1984) Isolation, chemical structure, acute toxicity and some physicochemcial properties of territrem C from *Aspergillus terreus. Appl. Environ. Microbiol.*, **47**, 98–100.

Lipsky, R.H., and Silverman, S.J. (1987) Effects of mycophenolic acid on detection of glial filaments in human and rat astrocytoma cultures. *Cancer Res.*, **47**, 4900–4904.

Liu, J.-C., Boucias, D.G., Pendland, J.C., Liu, W.-Z., and Maruniak, J. (1996a) The mode of action of hirsutellin A on eukaryotic cells. *J. Invert. Pathol.*, **67**, 224–228.

Liu, J.-W., Porter, A.G., Wee, B.Y., and Thanabalu, T. (1996b) New gene from nine *Bacillus sphaericus* strains encoding highly conserved 35.8-kilodalton mosquitocidal toxins. *Appl. Environ. Microbiol.*, **62**, 2174–2176.

Lysenko, M.A., and Vititnev, I.V. (1976) Effect of mycotoxins on the histological structure of the malpighian tubules of the caterpillar *Galleria mellonella* L. *Nauk. Pr. Ukr. Sil's'kogospod Akad.*, **161**, 100–103.

Lysenko, O. (1972a) Pathogenicity of *Bacillus cereus* for insects. I. Production of phospholipase C. *Folia Mcirobiol.*, **17**, 221–227.

Lysenko, O. (1972b) Pathogenicity of *Bacillus cereus* for insects. II. Toxicity of phospholipase C for *Galleria mellonella. Folia Microbiol.*, **17**, 228–231.

Lysenko, O. (1973) Phospholipase C of *Pseudomonas chlorapis* and its toxicity for insects. *Folia Microbiol.*, **18**, 125–132.

Lysenko, O. (1974) Bacterial exoenzymes toxic for insects; proteinase and lecithinase. *J. Hyb. Epidem. Microbiol. Immunol.*, **18**, 347–352.

Lysenko, O. (1976) Chitinase of *Serratia marcescens* and its toxicity to insects. *J. Invert. Pathol.*, **27**: 385–386.

Lysenko, O., and Kucera, M. (1968) The mechanism of pathogenicity of *Pseudomonas aeruginosa*. VI. The toxicity of proteinases for larvae of the greater wax moth, *Galleria mellonella* L. *Folia Microbiol. (Prague)*, **13**, 295–299.

Lysenko, O., and Kucera, M. (1971) Micro-organisms as sources of new insecticidal chemicals: Toxins. In H.D. Burgess and N.W. Hussey (eds.), *Microbial Control of Insects and Mites*. Academic Press, New York, pp. 205–227.

Magnusson, G., Thore'n, S., and Wickberg, B. (1972) Fungal extractives I. Structure of a sesquiterpene dialdehyde from *Lactarius* by computer simulation of the NMR spectrum. *Tetrahedron Lett.* **13**, 1105–1108.

Magnusson, G., Thore'n, S., and Drakenberg, T. (1973) Structure of a novel sesquiterpene dialdehyde from *Lactarius* by spectroscopic methods. *Tetrahedron*, **29**, 1621–1624.

Mantle, P.G., and Weedon, C.M. 1994. Biosynthesis and transformation of tremorgenic indole diterpenoids by *Penicillium paxilli* and *Acremonium lolii. Phytochemistry*, **36**, 1209–1217.

Mantle, P.G., Burt, S.J., MacGeorge, K.M., Bilton, J.N., and Sheppard, R.N. 1990. Oxidative transformation of paxilline in sheep bile. *Xenobiotica*, **20**, 809–821.

Markwick, N.P., Laing, W.A., Christeller, J.T., Reid, S.J., and Newton, M.R. (1996) α-Amylase activities in larval midgut extracts from four species of Lepidoptera (Tortricidae and Gelechiidae): response to pH and to inhibitors from wheat, barley, kidney bean and *Streptomyces. J. Econ. Entomol.*, **89**, 39–45.

Matsumoto, S., Sakuda, S., Isogai, A., and Suzuki, A. (1984) Search for microbial insect growth regulators: L-alanosine as an ecdysis inhibitor. *Agric. Biol. Chem.*, **48**, 827–828.

Matsuoka, K., Higuchi, T., Maeshima, M., and Nakamura, K. (1997) A vacuolar-type H^+-ATPase in a non-vacuolar organelle is required for the sorting of soluble vacuolar protein precursors in tobacco cells. *Plant Cell.*, **9**, 533–546.

Mazet, I., and Vey, A. (1995) Hirsutellin A, a toxic protein produced *in vitro* by *Hirsutella thompsonii. Microbiology.*, **141**, 1343–1348.

McInerney, B.V., Gregson, R.P., Lacey, M.J., Akhurst, R.J., and Gregson, R.P. (1991) Biologically active metabolites from *Xenorhabdus* spp. Part 1. Dithiolopyrrolone derivatives with antibiotic activity. *J. Nat. Prod.*, **54**, 774–784.

Miller, J.D. (1986) Toxic metabolites of epiphytic and endophytic fungi of conifer needles. In N.J. Fokkema and J. van den Heuvel (eds.), *Microbiology of the Phyllosphere*. Cambridge University Press, New York, pp. 223–231.

Mishima, H. (1983) Milbemycin: A family of macrolide antibiotics with insecticidal activity. In N. Takahashi, H. Yoshioka, T. Misato and S. Matsunaka (eds.), *Pesticide Chemistry, Human Welfare and the Environment. Volume 2. Natural Products*, Pergamon Press, New York, pp. 129–134.

Mishra, S.K., Keller, J.E., Miller, J.R., Heisey, R.M., Nair, M.G., and Putnam, A.R. (1987) Insecticidal and nematicidal properties of microbial metabolites. *J. Indust. Microbiol.*, **2**, 267–276.

Mizuno, T. (1995) Bioactive biomolecules of mushrooms: Food function and medicinal effect of mushroom fungi. *Food Rev. Int.*, **11**, 7–21.

Mollier, P., Lagnel, J., Fournet, B., Ai:oun, A., and Riba, G. (1994) A glycoprotein highly toxic for *Galleria mellonella* larvae secreted by the entomopathogenic fungus *Beauveria sulfurescens. J. Invert. Pathol.*, **64**, 200–207.

Morishita, A., Ishikawa, S., Ito, Y., Ogawa, K., Takada, M., Yaginuma, S. *et al.* (1994) AM4299 A and B, novel thiol protease inhibitors. *J. Antibiotics*, **47**, 1065–1068.

Mrozik, H. (1994) Advances in research and development of avermectins. In P.A. Hedin, J.J. Menn and R.M. Hollingworth (eds.), *Natural and Engineered Pest Management Agents.*, American Chemical Society, Washington D.C., pp. 54–73.

Mule', G., D'Ambrosio, A., Logrieco, A., and Bottalico, A. (1992) Toxicity of mycotoxins of *Fusarium sambucinum* for feeding in *Galleria mellonella. Entomol. exp. appl.*, **62**, 17–22.

Murao, S., Hayashi, H., Takiuchi, K., and Arai, M. (1988) Okaramine A, a novel indole alkaloid with insecticidal activity from *Penicillium simplicissimum. Agric. Biol. Chem.*, **52**, 885–886.

Murdock, L.L., Shade, R.E., and Pomeroy, M.A. (1988) Effect of E-64, a cysteine proteinase inhibitor, on cowpea weevil growth, development and fecundity. *Environ. Entomol.*, **17**, 467–469.

Murdock, L.L., Brookhart, G., Dunn, P.E., Foard, D.E., Kelley, S., Kitch, L. *et al.* (1987) Cysteine digestive proteinases in Coleoptera. *Comp. Biochem. Physiol.*, **87B**, 783–787.

Myokei, R., Sakurai, A., Chang, C.-F., Kodaira, Y. Takahashi, N., and Tamura, S. (1969) Structure of aspochracin, an insecticidal metabolite of *Aspergillus ochraceus. Tetrahedron Lett.*, **9**, 695–698.

Nagasawa, H., Isogai, A., Suzuki, A., and Tamura, S. (1979) ^{13}C-NMR spectra and stereochemistry of isoechinulins A, B and C. *Agric. Biol. Chem.*, **43**, 1759–1763.

Nagasawa, H., Isogai, A., Ikeda, K., Sato, S., Murakoshi, S., Suzuki, A. *et al.* (1975) Isolation and structure elucidation of a new indole metabolite from *Aspergillus ruber. Agric. Biol. Chem.*, **39**, 1901–1902.

Nair, M.G., Chandra, A., and Thorogood, D.L. (1993) Griseulin, a new nitro-containing bioactive metabolite produced by *Streptomyces* spp. *J. Antibiotics*, **46**, 1762–1763.

Nair, M.G., Chandra, A., and Thorogood, D.L. (1995) Nematicidal and mosquitocidal aromatic nitro compounds produced by *Streptomyces* spp. *Pestic. Sci.*, **43**, 361–365.

Nair, M.G., Putnam, A.R., Mishra, S.K., Mulks, M.H., Taft, W.H., Keller, J.E. *et al.* (1989) Faeriefungin: A new broad-spectrum antibiotic from *Streptomyces griseus* var. *autotrophicus. J. Nat. Prod.*, **52**, 797–809.

Nawrot, J., Bloszyk, E., Harmatha, J., Novotny, L., and Drozdz, B. (1986) Action of antifeedants of plant origin on beetles infesting stored products. *Acta Entomol. Bohemoslov.*, **83**, 327–335.

Nes, W.D., Lopez, M., Zhou, W., Guo, D., Dowd, P.F., and Norton, R.A. (1997) Sterol utilization and metabolism by *Heliothis zea. Lipids*, **32**, 1317–1323.

Neuffer, M.G., Coe, E.H., and Wessler, S.R. (1997) *Mutants of Maize*, Cold Spring Harbor Laboratory Press.

Nicolas, L., Charles, J.-F., and deBarjac, J. (1993) *Clostridium bifermetans* serovar *malasia*: Characterization of putative mosquito larvacidal proteins. *FEMS Microbiol Lett.*, **113**, 23–28.

Nozawa, K., Nakajima, S., Kawai, K., and Udagawa, S. (1988) Isolation and structure of indole terpenes, possible biosynthetic intermediates to the tremorgenic mycotoxin paxillines. *J. Chem. Soc. Perkin Trans I.*, 2607–2610.

Oh, H., Gloer, J.B., Wicklow, D.T., and Dowd, P.F. (1998a) Arenarins A-C: New cytotoxic fungal metabolites from the sclerotia of *Aspergillus arenarius. J. Nat. Prod.*, **61**, 702–705.

Oh, H., Swenson, D.C., Gloer, J.B., Wicklow, D.T., and Dowd, P.F. (1998b) Chaetochalasin A: A new bioactive metabolite from *Chaetomium brasiliense. Tetrahedron Lett.*, **39**, 7633–7636.

Oishi, H., Hosokawa, T., Okutomi, T., Suzuki, K., and Ando, K. (1969) Pesticidal activity of aureothin. *Agric. Biol. Chem.*, **33**, 1790–1791.

Oishi, H., Sugawa, T., Okutomi, T., Suzuki, K., Hayashi, T., Sawada, M. *et al.* (1970) Insecticidal activity of macrotetrolide antibiotics. *J. Antibiotics*, **23**, 105–106.

Omoto, C., and McCoy, C.W. (1998) Toxicity of purified fungal toxin hirsutellin A to the citrus rust mite *Phyllocoptruta oleivora* (Ash.). *J. Invert. Pathol.*, **72**, 319–322.

Pansa, M.C., Natalizi, G.M., and Bettini, S. (1973) Toxicity of valinomycin on insects. *J. Invert. Pathol.*, **22**, 148–152.

Patterson, R.R.M., Simmonds, M.S.J., and Blaney, W.M. (1987) Mycopesticidal effects of characterized extracts of *Penicillium* isolates and purified secondary metabolites (including mycotoxins) on *Drosophila melanogaster* and *Spodoptera littoralis. J. Invert. Pathol.*, **50**, 124–133.

Peng, F.C., Ling, K.H., Wang, Y., and Lee, G.H. (1985) Isolation, chemical structure, acute toxicity and some physicochemical properties of territrem B' from *Aspergillus terreus. Appl. Environ. Entomol.*, **49**, 721–723.

Pittindrigh, B.R., Huesing, J.E., Shade, R.E., and Murdock, L.L. (1997) Effects of lectins, CRY1A/CRY1B Bt delta-endotoxin, PAPA, protease and alpha-amylase inhibitors, on the development of the rice weevil, *Sitophilus oryzae*, using an artificial seed bioassay. *Entomol. exp. appl.*, **82**, 201–211.

Porter, A.G., Davidson, E.W., and Liu, J.-W. (1993) Mosquitocidal toxins of bacilli and their genetic manipulation for effective biological control of mosquitoes. *Microbiol. Rev.*, **57**, 838–861.

Potvin, L., Charles, J.-F., Berry, C., Humphreys, M.J., Coux, F., Menestrina, G. *et al.* (1998) *Bacillus sphaericus* binary toxin and its individual components form channels in planar lipid bilayers. Abstract, *VIIth International Colloquium on Invertebrate Pathology and Microbial Control,, IVth International Conference on Bacillus thuringiensis*. pp.47

Prasertphon, S. (1967) Mycotoxin production by species of *Entomophthora. J. Invert. Pathol.*, **9**, 281–282.

Purcell, J.P. (1997) Cholesterol oxidase for the control of boll weevil. In N. Carozzi and M. Koziel (eds.), *Advances in Insect Control: The Role of Transgenic Plants*, Taylor and Francis, Bristol, PA, pp. 95–108.

Purcell, J.P., Greenplate, J.T., Jennings, M.G., Ryerse, J.S., Pershing, J.C., Sims, S.R. *et al.* (1993) Cholesterol oxidase: a potent insecticidal protein active against boll weevil larvae. *Biochem. Biophys. Res. Commun.*, **196**, 1406–1413.

Questel, D.D., and Gertler, S.I. (1952) Laboratory tests of toxicity of some organic compounds to the European corn borer, 1947–1952. *U.S. Dept. Agric. Bur. Entomol. Plant Quarantine*, # E-840, 12 pp.

Quiot, J.-M., Vey, A., Vago, C., and Païs, M. (1980) Action antivirale d'une mycotoxine. E'tude d'une toxine de l'hyphomyce'te *Metarhizium anisopliae* (Metsch) sorok en culture cellulaire. *C.R. Acad. Sc. Paris.*, **291**, 763–765.

Rao, H.G.R., and De las Casas, E. 1974. Effect of metabolites of *Chaetomium* on *Tribolium confusum. Proc. NCB Entomol. Soc. Am.*, **29**, 152–153.
Rebeiz, C.A., Juvick, J.A., and Rebeiz, C.C. (1988) Porphyric insecticides. I. Concept and phenomenology. *Pestic. Biochem. Physiol.*, **30**, 11–27.
Rebeiz, C.A., Juvick, J.A., Rebeiz, C.C., Bouton, C.E., and Gut, L.J. (1990) Porphyric insecticides. 2. 1,10 Phenanthroline, a potent porphyric insecticide modulator. *Pestic. Biochem. Physiol.* **36**, 201–207.
Reese, J.C. (1977) The effects of plant biochemicals on insect growth and nutritional physiology. In P.A. Hedin (ed.), *Host Plant Resistance to Pests*, American Chemical Society, Washington, D.C., pp. 129–152.
Reis, J. (1975) Insecticidal and larvicidal activities of mycotoxins aflatoxin B1, rubratoxin B, patulin, and diacetoxyscirpenol towards *Drosophila melanogaster. Chem. Biol. Interact.*, **10**, 339–342.
Reynolds, K.A. (1998) Combinatorial biosynthesis: Lessons learned from nature. *Proc. Natl. Acad. Sci.*, **95**, 12744–12746.
Richard, J.L., Engstrom, G.W., Pier, A.C., and Tiffany, L.H. (1969) Toxigenicity of *Trichothecium roseum* Link: Isolation and partial characterization of a toxic metabolite. *Mycopath. Mycol. Appl.*, **39**, 231–240.
Riley, R.T., and Showker, J.L. (1991) The mechanism of patulin's cytotoxicity and the antioxidant activity of indole tetramic acids. *Toxicol. Appl. Pharmacol.*, **109**, 108–126.
Roberts, D.W. (1969). Toxins from the entomogenous fungus *Metarhizium anisopliae*. Isolation of destruxins from submerged cultures. *J. Invert. Pathol.*, **14**, 82–88.
Roberts, D.W. (1981). Toxins of entomopathogenic fungi. In H.D. Burges (ed.), *Microbial Control of Pests and Plant Diseases 1970–1980*, Academic Press, New York, pp. 441–464.
Rodriguez, J.G., Potts, M.F., and Rodriguez, L.D. (1980) Mycotoxin toxicity to *Tyrophagus putrescentiae*, an arthropod associated with fungi in stored food products or stored grain. *J. Econ. Entomol.*, **73**, 282–284.
Rouhi, A.M. (1999) Chemistry from unknown microbes. *Chem. Engin. News*, **77**, 21–23.
Rowan, D.D., and Gaynor, D.L. (1986) Isolation of feeding deterrents against Argentine stem weevil from ryegrass infected with the endophyte *Acremonium loliae. J. Chem. Ecol.*, 112, 647–658.
Rowan, D.D., Hunt, M.B., and Gaynor, D.L. (1986) Peramine, a novel insect feeding deterrent from ryegrass infected with the endophyte *Acremonium loliae. J. Chem. Soc. Chem. Commun.*, 935–936.
Rowan, D.D., Dymock, J.J., and Brimble, M.A. (1990) Effect of fungal metabolite peramine and analogs on feeding and development of Argentine stem weevil (*Listronotus bonariensis*). *J. Chem. Ecol.*, **16**, 1683–1695.
Roylance, J.T., Hill, N.S., and Agee, C.S. (1994) Ergovaline and peramine production in endophyte-inected tall fescue: independent regulation and effects of plant and endophyte genotype. *J. Chem. Ecol.*, **20**, 2171–2183.
Sakuda, S., Isogai, A., Matsumoto, S., and Suzuki, A. (1987) Search for microbial insect growth regulators II. Allosamidin, a novel insect chitinase inhibitor. *J. Antibiotics*, **40**, 296–300.
Sakuda, S., Isogai, A., Matsumoto, S., Suzuki, A., and Koseki, K. (1986) The structure of allosamidin, a novel insect chitinase inhibitor, produced by *Streptomyces* sp. *Tetrahedron Lett.*, **27**, 2475–2478.
Sargent, J.E. (1974) The effect of ochratoxin A and other metabolites of *Aspergillus ochraceus* Wilhelm on stored product insects. *Proc. North Central Branch. Entomol. Soc. Am.*, **29**, 155.
Sasikala, C., Ramana, C.V., and Rao, P.R. (1994) 5-Aminolevulinic acid: A potential herbicide/insecticide from microorganisms. *Biotechnol. Prog.*, **10**, 451–459.
Sassa, T., and Yoshikoshi, H. (1983) New terpene-linked cyclohexenone epoxides, macrophorin A,B, and C, produced by the fungus causing *Macrophoma* fruit rot of apple. *Agric. Biol. Chem.*, **47**, 187–189.
Sattelle, D.B. (1990) GABA receptors of insects. *Adv. Insect Physiol.*, **22**, 1–113.
Sato, N., Horiuchi, T., Hamano, M., Sekine, H., Chiba, S., Yamamoto, H. *et al.* (1996) Kojistatin A, a new cysteine protease inhibitor produced by *Aspergillus oryzae. Biosci. Biotech. Biochem.*, **60**, 1747–1748.
Scherertz, P.C., Mills, R.R., and Llewellyn, G.C. (1978) Comparative dietary patterns and body weight changes in adult male American cockroaches fed pure aflatoxin B1. *Bull. Environ. Contam. Toxicol.*, **20**, 687–695.
Sebesta, K., Farkas, J., and Horska, K. (1981) Thuringiensin, the *beta*-exotoxin of *Bacillus thuringiensis*. In H.D. Burges (ed.), *Microbial Control of Pests and Plant Diseases 1970–1980*, Academic, New York, pp. 249–281.
Shaaya, E., and Clever, U. (1972) *In vivo* effects of α-amanitin on RNA synthesis in *Calliphora erythrocephala. Biochim. Biophys. Acta.*, **272**, 373–381.
Shankland, D.L. (1976) The nervous system: Comparative physiology and pharmacology. In C.F. Wilkinson (ed.), *Insecticide Biochemistry and Physiology*. Plenum Press, New York, pp. 229–270.
Shen, Z., Corbin, D.R., Greenplate, J.T., Grebenok, R.J., Galbraith, D.W., and Purcell, J.P. (1997) Studies on the mode of action of cholesterol oxidase on insect midgut membranes. *Arch. Insect Biochem. Physiol.*, **34**, 429–442.
Shkaruba, N.G. (1976) Glucose 6-phosphate dehydrogenase activity of the silkworm (*Bombyx mori* L.) under the effect of sevin and fusariotoxin. *Nauk. Pr. Ukr. Sil's'kogospod. Akad.*, **161**, 95–7.

Siegel, M.R., Latch, G.C.M., Bush, L.P., Fannin, F.F., Rowan, D.D., Tapper, B.A. *et al.* (1990) Fungal endophyte-infected grasses: alkaloid accumulation and aphid response. *J. Chem. Ecol.*, **16**, 3301–3315.

Sparks, T.C., Thompson, G.D., Kirst, H.A., Hertlein, M.B., Larson, L.A., Worden, T.V. *et al.* (1998) Biological activity of the spinosins, new fermentation derived insect control agents, on tobacco budworm (Lepidoptera: Noctuidae) larvae. *J. Econ. Entomol.*, **91**, 1277–1283.

Sparks, T.C., Thompson, G.D., Kirst, H.A., Hertlein, M.B., Mynderse, J.S., Turner, J.R. *et al.* (1999) Fermentation derived insect control agents: The spinosyns. In F.R. Hall and J.J. Menn (eds.), *Biopesticides: Use and Delivery*. Humana Press, Ttowa, NJ, pp. 171–188.

Spindler, K.-D., and Spindler-Barth, M. (1994) Inhibition of chitinolytic enzymes from *Streptomyces griseus* (Bacteria), *Artemia salina* (Crustacea), and a cell line from *Chironomus tentans* (Insecta) by allosamidin and isoallosamidin. *Pestic. Sci.*, **40**, 113–120.

Springer, J.P., Clardy, J., Well, J.M., Cole, R.J., and Kirksey, J.W. (1975) The structure of paxilline, a tremorgenic metabolite of *Penicillium paxilli* Barner. *Tetrahedron Lett.*, **16**, 2531–2534.

Staub, G.M., Gloer, J.B., Wicklow, D.T., and Dowd, P.F. (1992) Aspernomine: A cytotoxic antiinsectan metabolite with a novel ring system from the sclerotia of *Aspergillus nomius. J. Am. Chem. Soc.*, **114**, 1015–1017.

Staub, G.M., Gloer, K.B., Gloer, J.B., Wicklow, D.T., and Dowd, P.F. (1993) New paspalinine derivatives with antiinsectan activity from the sclerotia of *Aspergillus nomius. Tetrahedron Lett.*, **34**, 2569–2572.

Stich, R.W., Sauer, J.R., Bantle, J.A., and Kogan, K.M. (1993) Detection of *Anaplasma mirginale* (Rickettsiales: Anaplasmataceae) in secretagogue induced oral secretions of *Dermanocenter andersoni* (Acari: Ixodidae) with the polymerase chain reaction. *J. Med. Entomol.*, **30**, 789–794.

St. Leger, R. (1995) The role of cuticle-degrading proteases in fungal pathogenesis of insects. *Can. J. Bot.* (*Suppl. 1*), S1119–S1125.

Stonard, R.J., Ayer, S.W., Kotyk, J.J., Letendre, L.J., McGary, C.I., Nickson, T.E. *et al.* (1994) Microbial secondary metabolites as a source of agrochemicals. In P.A. Hedin, J.J. Menn, and R.M. Hollingworth (eds.), *Natural and Engineered Pest Management Agents*. American Chemical Society, Washington, D.C., pp. 25–36.

Strong, L., and Brown, T.A. (1987) Avermectins in insect control and biology: a review. *Bull. Ent. Res.*, **77**, 357–389.

Strongman, D.B., Strunz, G.M., and Yu, C.-M. (1990) Trichothecene mycotoxins produced by *Fusarium sporotrichoides* DAOM 197255 and their effects on spruce budworm *Choristoneura fumiferana. J. Chem. Ecol.*, **16**, 1605–1609.

Strongman, D.B., Strunz, G.M., Gigue're, P., Yu, C.-M., and Calhoun, L. (1988) Enniatins from *Fusarium avenaceum* isolated from balsam fir foliage and their toxicity to spruce budworm larvae, *Choristoneura fumiferana* (Clem.) (Lepidoptera: Tortricidae). *J. Chem. Ecol.*, **14**, 753–764.

Suzuki, A., Kanaoka, M., Isogai, A., Murakoshi, S., Ichinoe, M., and Tamura, S. (1977) Bassianolide a new insecticidal cyclodepsipeptide from *Beauveria bassiana* and *Verticillium lecanii. Tetrahedron Lett.*, **25**, 2167–2170.

Takahashi, N., Suzuki, A., Kimura, Y., Miyamoto, S., Tamura, S., Mitsui, T. *et al.* (1968) Isolation, structure and physiological activities of piericidin B, natural insecticide produced by a *Streptomyces. Agric. Biol. Chem.*, **32**, 1115–1112.

Takase, S., Kawai, Y., Uchida, I., Tanaka, H., and Aoki, H. (1984) Structure of amauromine, a new alkaloid with vasodialating activity produced by *Amauroascus* sp. *Tetrahedron Lett.*, **25**, 4673–4676.

Takiguchi, Y., Mishima, H., Okuda, M., Terao, M., Aoki, A., and Fukuda, R. (1980) Milbemycins, a new family of macrolide antibiotics: fermentation, isolation and physico-chemical properties. *J. Antibiotics*, **33**, 1120–1127.

Tamura, S., Kuyama, S., Kodaira, Y., and Higashikawa, S. (1964) The structure of destruxin B, a toxic metabolite of *Oospora destructor. Agric. Biol. Chem.*, **28**, 137–138.

Tamura, S., Takahashi, N. , Miyamoto, S., Mori, R., Suzuki, S., and Nagatsu, J. (1963) Isolation and physiological activities of piericidin A, a natural insecticide produced by *Streptomyces. Agric. Biol. Chem.*, **27**, 576–582.

TePaske, M.R., Gloer, J.B., Wicklow, D.T., and Dowd, P.F. (1989a) Three new aflavinines from the sclerotia of *Aspergillus tubingensis. Tetrahedron*, **16**, 4961–4968.

TePaske, M.R., Gloer, J.B., Wicklow, D.T., and Dowd, P.F. (1989b) Tubingensin A: An antiviral carbazole alkaloid from the sclerotia of *Aspergillus tubingensis. J. Org. Chem.*, **54**, 4743–4746.

TePaske, M.R., Gloer, J.B., Wicklow, D.T., and Dowd, P.F. (1989c) The structure of tubingensin B: A cytotoxic carbazole alkaloid from the sclerotia of *Aspergillus tubingenis. Tetrahedron Lett.*, **30**, 5965–5968.

TePaske, M.R., Gloer, J.B., Dowd, P.F., and Wicklow, D.T. (1990) Aflavazole: an antiinsectan carbazole metabolite from the sclerotia of *Aspergillus flavus. J. Org. Chem.*, **55**, 5299–5301.

TePaske, M.R., Gloer, J.B., Wicklow, D.T., and Dowd, P.F. (1991) Leporin A: An antiinsectan *N*-alkoxypyridone from the sclerotia of *Aspergillus leporis. Tetrahedron Lett.*, **32**, 5687–5690.

TePaske, M.R., Gloer, J.B., Wicklow, D.T., and Dowd, P.F. (1992) Aflavarin and B-aflatrem: New anti-insectan metabolites from the sclerotia of *Aspergillus flavus. J. Nat. Prod.*, **55**, 1080–1086.

Thanabalu, T., and Porter, A.G. (1996) A *Bacillus sphaericus* gene encoding a novel type of mosquitocidal toxin of 31.8 kDa. *Gene*, **170**, 85–89.

Thanabalu, T., Hindley, J., and Berry, C. (1992) Proteolytic processing of the mosquitocidal toxin from *Bacillus sphaericus* SSII-1. *J. Bacteriol.*, **174**, 5051–5056.

Thanabalu, T., Berry, C., and Hindley, J. (1993) Cytotoxicity and ADP-ribosylating activity of the mosquitocidal toxin from *Bacillus sphaericus* SSII-1: Possible roles of the 27- and 70-kilodalton peptides. *J. Bacteriol.*, **175**, 2314–2320.

Thanabalu, T., Hindley, J., Jackson-Yap, J., and Berry, C. (1991) Cloning, sequencing, and expression of a gene encoding a 100-kilodalton mosquitocidal toxin from *Bacillus sphaericus* SSII-1. *J. Bacteriol.*, **173**, 2776–2785.

Trial, H. Jr., and Dimond, J.B. (1979) Emodin in buckthorn: a feeding deterrent to phytophagous insects. *Can. Entomol.*, **111**, 207–212.

Turner, W.B. (1971) *Fungal Metabolites*. Academic Press, New York.

Turner, W.B., and Aldridge, D.C. (1983) *Fungal Metabolites II*, Academic Press, New York.

Veam, E.V., and Wright, V.F. (1973) Effect of F-2 and T-2 mycotoxins on the ATPase enzyme system of *Tribolium confusum. Tribolium Inform. Bull.*, **16**, 110.

Vesonder, R.F., Burmeister, H.R., and Tjarks, L.W. (1990) 4,4,24-trimethylcholesta-8,14,24(28)-trien-2-a,3B,11a,12B-tetrol 12-acetate, 3-sulfate from *Fusarium graminearum. Phytochemistry*, **29**, 2263–2265.

Vey, A., and Quiot, J.-M. (1985) Personal communication cited in Ferron, P., Fungal control. In G.A. Kerkut and L.I Gilbert (eds.), *Comprehensive Insect Physiology, Biochemistry and Pharmacology. Volume 12. Insect Control*. Pergamon Press, New York, pp. 313–346.

Wang, H., Gloer, J.B., Wicklow, D.T., and Dowd, P.F. (1995) Aflavanines and other antiinsectan metabolites from the ascostromata of *Eupenicillium crustaceum* and related species. *Appl. Environ. Microbiol.*, **61**, 4429–4435.

Warren, G.W. (1997) Vegetative insecticidal proteins: Novel proteins for control of corn pests. In N. Carozzi and M. Koziel (eds.), *Advances in Insect Control: The Role of Transgenic Plants*. Taylor and Francis, Bristol, PA, pp. 109–121.

Weiland, T. (1986) *Peptides of Poisonous Amanita Mushrooms*. Springer-Verlag, New York.

Weiser, J., and Matha, V. (1988a) The insecticidal activity of cyclosporines on mosquito larvae. *J. Invert. Pathol.*, **51**, 92–93.

Weiser, J., and Matha, V. (1988b) Tolypin, a new insecticidal metabolite of the fungi of the genus *Tolypocladium. J. Invert. Pathol.*, **51**, 94–96.

Wells, J.M., Cole, R.J., and Kirksey, J.W. (1975) Emodin, a toxic metabolite of *Aspergillus wentii* isolated from weevil-damaged chestnuts. *Appl. Microbiol.*, **30**, 26–28.

Wessing, A., Bertram, G., and Zierold, K. (1993) Effects of bafilomycin A1 and amiloride on the apical potassium and proton gradients in *Drosophila* Malpighian tubules studied by X-ray microanalysis and microelectrode measurements. *J. Comp. Physiol. Biochem. Syst. Environ. Physiol.*, **163**, 452–462.

Westley, J.W., Liu, C.-M., Sello, L.H., Evans, R.H., Troupe, N., Blount, J.F. *et al.* (1984) The structure and absolute configuration of the 18-membered macrollide lactone antibiotic X-4357B (concanamycin A). *J. Antibiotics.*, **37**, 1738–1749.

Whyte, A.C., Gloer, J.B., Wicklow, D.T., and Dowd, P.F. (1996) Sclerotiamide: A new member of the paraherquamide class with potent antiinsectan activity from the sclerotia of *A. sclerotiorum. J. Nat. Prod.*, **59**, 1093–1095.

Wicklow, D.T. (1984) Ecological approaches to the study of mycotoxigenic fungi. In H. Kurata and Y. Ueno (eds.), *Toxigenic Fungi, Their Toxins and Health Hazards*, Elsevier, Amsterdam, pp. 33–43.

Wicklow, D.T. (1988) Metabolites in the coevolution of fungal chemical defence systems. In K.A. Pyrozynski and D.L. Hawksworth (eds.), *Coevolution of Fungi with Plants and Animals*, Academic Press, New York, pp. 173–201.

Wicklow, D.T., and Cole, R.J. (1982) Tremorgenic indole metabolites and aflatoxins in sclerotia of *Aspergillus flavus*: an evolutionary perspective. *Can. J. Bot.*, **60**, 525–528.

Wicklow, D.T., Dowd, P.F., and Gloer, J.B. (1994) Antiinsectan effects of *Aspergillus* metabolites. In K.A. Powell, H. Renwick and J.F. Peabody (eds.), *The Genus Aspergillus*. Scientc. Tech. Publishers, Madison, WI, pp. 93–114.

Wicklow, D.T., Dowd, P.F., TePaske, M.R., and Gloer, J.B. (1988) Sclerotial metabolites of *Aspergillus flavus* toxic to a detritivorous maize insect *Carpophilus hemipterus*, Nitidulidae. *Trans. Br. Mycol. Soc.*, **91**, 433–438.

Wicklow, D.T., Dowd, P.F., Alfatafta, A.A., and Gloer, J.B. (1996) Ochratoxin A: an antiinsectan metabolite from the sclerotia of *Aspergillus carbonarius* NRRL 369. *Can. J. Microbiol.*, **42**, 1100–1103.

Wicklow, D.T., Joshi, B.K., Gamble, W.R., Gloer, J.B., and Dowd, P.F. (1998) Antifungal metabolites (momorden, monocillin IV and cerebrosides) from *Humicola fuscoatra* Traaen NRRL 22980, a mycoparasite of *Aspergillus flavus* sclerotia. *Appl. Environ. Microbiol.*, **64**, 4482–4484.

Wolfson, J.L., and Murdock, L.L. (1987) Suppression of larval Colorado potato beetle growth and development by digestive proteinase inhibitors. *Entomol. exp. appl.*, **44**, 235–240.

Woo, J.-T., Ono, H., and Tsuji, T. (1995) Cathestatins, new cysteine protease inhibitors produced by *Penicillium citrinum. Biosci. Biotech. Biochem.*, **59**, 350–352.

Wright, V.F., and Harein, P.K. (1982) Effects of some mycotoxins on the Mediterranean flour moth. *Environ. Entomol.*, **11**, 1043–1045.

Wright, V.F., Harein, P.K., and de las Casas, E. (1974) The effects of the mycotoxins zearalenone (F-2) and T-2 toxin on *Tribolium confusum. Proc. North Central Branch, Entomol. Soc. Am.*, **29**, 152.

Wright, V.F., de las Casas, E., and Harein, P.K. (1976) The response of *Tribolium confusum* to the mycotoxins zearalenone (F-2) and T-2 toxin. *Environ. Entomol.*, **5**, 371–374.

Wright, V.F., de las Casas, E., and Harein, P.K. (1980) Evaluation of *Penicillium* mycotoxins for activity in stored-product coleoptera. *Environ. Entomol.*, **9**, 217–221.

Wright, V.F., Vesonder, R.F., and Ciegler, A. (1981) Mycotoxins and other fungal metabolites as insecticides. In D. Kurstak (ed.), *Microbial and Viral Pesticides*. Marcel Dekker, New York, pp. 559–583.

Yaginuma, S., Asahi, A., Mirishita, A., Hayashi, M., Tsujino, M., and Takada, M. (1989) Isolation and characterization of new thiol protease inhibitors estatins A and B. *J. Antibiotics.*, **62**, 1362–1369.

Yokota, T., Sakurai, A., Iriuchijima, S., and Takahashi, N. (1981) Isolation and ^{13}C NMR study of cyclopiazonic acid, a toxic alkaloid produced by muscardine fungi *Aspergillus flavus* and *Aspergillus oryzae. Agric. Biol. Chem.*, **45**, 53–56.

Yoshida, S., and Takahashi, N. (1978) Piericidins: Naturally occurring inhibitors against mitochondrial respiration. *Heterocycles.*, **10**, 425–467.

Yu, C-G., Mullins, M.A., Warren, G.W., Koziel, M.G., and Estruch, J.J. (1997) The *Bacillus thuringiensis* vegetative insecticidal protein vip3A lyses midgut epithelium cells of susceptible insects. *Appl. Environ. Microbiol.*, **63**, 532–536.

MICROBIAL CONTROL OF INSECT PESTS: ROLE OF GENETIC ENGINEERING AND TISSUE CULTURE

3

K. Narayanan
Project Directorate of Biological Control, PB No. 2491, Hebbal, Bangalore 560 024, India

Introduction

The use of insect pathogens, as one of the major components in biological control of crop pests for integrated pest management is gaining general acceptance as a realistic goal and has stimulated research on their understanding and effective utilization. Although various entomopathogens and their products are currently used to control insect pests which offer certain advantages over conventional chemical insecticides, they also have certain disadvantages especially the lack of speed of kill in the case of insect viruses that has limited their application. With the advent of recombinant-DNA (r-DNA) technology, an opportunity has emerged in alleviating certain commercial short-comings of pathogens and fostering the creation of new generation biopesticides. Modification of insect viruses especially through genetic engineering is anticipated to increase their effectiveness greatly. This chapter discusses the existing information on various biotechnological approaches, viz. the role of genetic manipulations of various insect pathogens (especially insect viruses under insect/tissue/cell culture environment). It also focuses on molecular approach like r-DNA that has been applied in genetic improvement of various microbial control agents, viz. insect viruses, bacteria, fungi, nematodes and protozoa for increasing their efficiency. For the conventional use of various entomopathogens for control of crop pests several reviews are available (Granados and Federici, 1986; Jayaraj *et al.*, 1989; Kurstak, 1982; Tanada and Kaya, 1993; Cunningham, 1995; Seema Mishra, 1998).

Research on insect viruses, especially baculovirus, have helped in understanding of the molecular biology and their utilization as efficient vectors. This has become possible ever since the expression of human interferon (Smith *et al.*, 1983; Maeda *et al.*, 1984) and *Escherichia coli* β galactosidase (Pennock *et al.*, 1984) systems have become available. Driven by a pressing need for safer and more environmentally compatible insecticides accelerated by the ever-increasing problem of insecticide resistance, progress in developing improved insect microbes/microbial based products has been rapid. Many of the predictions and possibilities offered by the advent of recombinant DNA technology (Miller *et al.*, 1983; Kirschbaum, 1985) and insect tissue culture (Mitsuhashi and Maramorosch, 1964; Vaughn *et al.*, 1977) are extant. Further improvement in existing microbial pesticides and genetically modified organisms, therefore, is inevitable to make way for their commercial application.

Viruses

Baculoviridae, Polydnaviridae and Ascoviridae are three families of viruses specific to insects and related invertebrates. Viruses belonging to the sub-family Eubaculovirinae (Francki *et al.*, 1991; Mayo and Pringle, 1997) differ from that of plant and animal viruses wherein the infective virions are occluded in many sided occlusion bodies called "polyhedra", which afford environmental stability to virions. Out of 1100 hosts record covering 13 insect orders belonging to 12 viral families world-wide (Martignoni and Iwai, 1986), India has 33 host records covering 2 insect orders belonging to 4 viral families. Insect viruses especially baculoviruses [comprising nuclear polyhedrosis virus (NPV) and granulosis virus (GV)] are mostly specific viruses and they can only replicate in members of invertebrate hosts, principally Lepidoptera which not only constitute the major agricultural crop pests but also forms the youngest of the insect orders. No baculovirus has been described that infects and replicates in mammalian cells or plant cells or in other invertebrates. Their safety is well documented and accepted by FAO (1973) and hence frequently used as microbial pesticides. Insect baculoviruses, which are currently used to control agricultural crop pests, offer certain advantages over chemical insecticides like species specificity, stability, environmental safety, etc. In some respects the advantage of species specificity of baculoviruses renders them inappropriate for the control of multitudes of pests. Paradoxically, the specificity of insect viruses is both to their advantage and disadvantage. Species specificity of certain baculoviruses is considered an asset, by way of conserving beneficial insects and killing the targeted pests.

To improve the efficacy of viral insecticides, several strategies of both conventional and genetic engineering approaches of insect viruses that take into consideration of viral pathogenesis, are under development. The conventional approach includes mainly strain selection. For example, it has been known for some time that different isolates of same viral species from different geographical locations can vary significantly in efficacy against the same target pest. NPVs isolated from different populations of *Spodoptera frugiperda* (J.E. Smith)(fall armyworm) in Louisiana have been shown to vary by 16-fold in their LC_{50} against this pest (Fuxa *et al.*, 1988). More recently, it has been determined that isolates of an NPV that occurs commonly in Thailand in populations of *S. frugiperda* are 10-fold more effective against populations of *S. exigua* larvae in California. Though the basis for these differences is not known, these findings suggest that natural variations in pathogenic populations and virulence of viruses isolated from different geographical areas should receive more attention towards the selection of isolates that are used as viral insecticides. Through conventional approach of screening of field isolates from different agroclimatic regions, increasing the viral activity through serial passage *in vivo* and direct application of conventional genetics through chemical mutagens so as to generate variants of baculoviruses may yield good results (Wood *et al.*, 1981; Vyas, 1991; Kolondy-Hirsch and Van Beek, 1997). The r-DNA technology offers the possibility of developing entirely new biological insecticides having broader host range with increased speed of kill based on the construction of genetically modified microorganisms that retain the advantages of classical biological control agents and suffer fewer of their drawbacks.

Strategies for engineering baculovirus insecticides

The three major aims in the engineering of baculovirus insecticides are (i) to increase the speed of action or at least to induce early cessation of feeding, (ii) to enhance the virulence to control the older larvae at lower doses, and (iii) to extend host specificity in order to have commercial potential. A number of strategies that have been tried to attain these goals as well as for efficient utilization of various insect pathogens in microbial control of insect pests have been discussed below. Some of these strategies are an outcome of extensive knowledge of molecular biology of baculoviruses and their replication in insect cells. Their efficient use as expression vectors by way of expressing various insect specific neurotoxin genes and behaviour-modifying genes in order to hasten the death of the host or feeding inhibition are also an outcome of such knowledge. In many instances increased speed of action and enhanced virulence albeit not an extended host range can be achieved with a single strategy.

At the outset, some aspects of the biology of baculoviruses have been described in brief in order to appreciate the methods employed to genetically engineer baculoviruses and to enable the subsequent discussion of their genetic manipulations and consequences towards its effective utilization. By understanding what role each virus gene plays in virus growth and survival; we have the possibility of increasing the efficacy of the virus by modifying or even deleting viral genes which counteract viral efficacy as pesticides. Further, to appreciate the methods employed to genetically engineered baculoviruses, it is necessary to first describe some aspects of their basic biology. For more detailed information on the molecular biology and genetic improvement of baculoviruses the readers may refer to the earlier reviews (Doerfler and Bohm, 1986; Granados and Federici, 1986; Bonning and Hammock, 1996; Possee *et al.*, 1997).

Molecular biology of insect baculoviruses

Baculoviruses have a genome of double stranded, co-valently closed circular DNA of approximately 100–130 kilobase pairs (kbp). This is packed into rod shaped nucleocapsid, which are enveloped by a lipid-protein membrane to form the virion. Based on the structural criteria, the family of Baculoviridae is divided into three sub-groups. Subgroup A consists of nuclear polyhedrosis viruses (NPV), in which virions are occluded either singly (SNPV) or in multiples (MNPV) with an intra-nuclear paracrystalline matrix formed by the single viral encoded polyhedrin protein. The polyhedral protein has a molecular weight of 29 kilo daltons and is highly conserved among different baculoviruses. These occlusions called polyhedra, are bigger (0.5–5.0 μ) in size with silicaceous framework, which protect the virus from environmental hazards such as ultraviolet rays and desiccation, and they are responsible for the horizontal transmission from insect to insect under field conditions.

They have a unique bi-phasic life cycle (differing from most of other DNA animal virus groups) which involves the temporarily regulated expression of two morphologically and functionally different viruses, i.e. they differ in their protein composition, morphology, and tissue tropism and the roles in the viral life cycle. They have genetically similar viral phenotypes, viz. budded virus (BV) or extracellular virus (ECV) in the first phase and occluded virus (OV) in the later phase of the infection. The primary and natural infection starts when the insect ingests polyhedra contaminated plants. The occlusions or polyhedra dissolve in the alkaline environment of the insect gut juices and by some enzymatic degradation, releasing the virions, i.e. polyhedral derived virus (PDV) that invade and replicate in the midgut epithelial cells-especially in the columnar epithelial cells of insect midgut by fusion with microvilli.

The nucleocapsids are then transported to the nucleus where uncoating of the viral DNA occurs followed by gene expression and viral DNA replication. Progeny nucleocapsids are observed as assemblages within and around dense virogenic stroma. Some progeny nucleocapsids bud through the nuclear membrane and into the plasma membrane but apparently lose the nucleus derived envelope in the cytoplasm. These nucleocapsids then bud through the cytoplasmic membrane into the haemocoel acquiring the budded virus envelope that contains the virus encoded spike glycoproteins (gp64). These virions of the budded virus phenotype appear to be specialized for the secondary infection beyond the midgut. They also seem to be responsible for vertical transmission from cell to cell and other insect tissues since the open circulatory system of the insect provides easy access to other tissues of the insect for the budded virus. Though the infection of different larval tissues occurs in a sequential manner, recently Engelhard *et al.* (1994) has hypothesised that the viruses use the tracheal system of the insect as a conduit. A second group of progeny nucleocapsids becomes enveloped within the nucleus by *de nova* assembled envelope. These virions are subsequently occluded within polyhedrin proteins that crystallise around them. Synthesis of budded virus and occluded virus is temporarily regulated and baculovirus genes are transcribed in a regulated cascade involving four phases of transcription, viz. early, delayed early, late and very late phase (Blissard and Rohrmann, 1990; Doerfler and Bohm, 1986; Granados and Federici, 1986).

Upon infection of cells, viral genes of the immediate early class are transcribed by host factors (including RNA polymerases, products of these genes, and possibly some host factors), which turn on (trans-activate) an array of delayed early genes, including virus-encoded RNA and DNA polymerases, and turn off some host functions. Transcriptions of these delayed early genes obviously start before the onset of DNA replications. Both immediate early and delayed early gene expression may continue throughout the later stages of infection. Late genes are switched on concurrently with the onset of DNA replication, and delayed early genes promote their expression. The late genes code for structural proteins of the virus particles. Genes of each of these four classes are not clustered but randomly distributed throughout the genome. The hierarchical nature of baculovirus gene regulation is similar to that of other large DNA viruses like adeno-herpesviruses. Baculoviruses, however, have an additional, unique class of very late genes, which code for proteins that are involved in the late stages of virus infection and polyhedral inclusion bodies morphogenesis. Two of these very late genes, coding for polyhedrin and a protein of 10 kDa called *p10* are hyper expressed. Late after infection mutations or deletion in these very late genes do not affect ECV production and this forms the basis for the use of baculoviruses as vectors for the expression of foreign genes.

Very late in the infection of baculovirus, two proteins, viz. polyhedrin and protein (*p10*) are abundantly synthesized. The only role for polyhedrin appears to be polyhedral inclusion body (PIBs) formation, which is irrelevant or not essential for the production or multiplication of *in vitro* infective non-occluded viruses. Foreign gene can, therefore, be inserted in place of the polyhedrin gene eliminating its function. The resulting viral genome will initially direct the production of non-occluded virus and with the time course of polyhedrin protein gene expressed, go on to produce product(s) as directed by the inserted foreign DNA.

Baculovirus vector construction

The large size of the baculovirus genome makes it difficult to ligate foreign DNA directly into the genome, co-transfection of insect cells with wild type DNA and a transfer vector containing the foreign gene construct recombinant virus. The majority of these vectors contain a high copy number plasmid, containing sequence from AcMNPV that include the promoter of the polyhedrin gene and flanking sequences. The foreign gene is inserted downstream of the polyhedrin promoter. Co-transfection of such vectors with wild type DNA results in homologous recombination in the flanking virus sequences transferring the foreign gene into the virus genome, in place of the polyhedrin gene. It means that the foreign DNA replaces the polyhedrin gene coding region, but remains under the control of the polyhedrin gene promoter to produce polyhedrin negative virus.

Screening of recombinants

In the case of allelic replacement of the polyhedrin gene, recombinant viruses are usually recognised by the absence of PIBs using light microscope (Smith *et al.*, 1983).

Sometimes they are recognised by the presence of foreign gene using Southern Blot hybridization analysis, its mRNA species (Northern analysis) or recombinant protein (using immuno-fluorescens or immuno blot analysis). Alternative strategies involve the use of marker genes in the transfer vector, such as an additional polyhedral gene or the bacterial "*lacZ*" gene placed immediately downstream to the polyhedrin locus (Zuidema *et al.*, 1990). These markers facilitate positive selection schemes for recombinants. The insertion of foreign genes into the *p10* locus markers is more complicated because phenotypic markers for the absence of *p10* are lacking. A new generation of transfer vector is available which facilitates the screening for recombinant using β-galactosidase producing a "*lacZ*" gene as a marker (Vlak *et al.*, 1990). There are several reasons and advantages for using baculoviruses as expression system for expressing foreign genes of our interest for improving the baculovirus biopesticides (Bishop and Possee, 1990; Luckow and Summers, 1988; Maeda, 1989a, b; Miller, 1988; Summers, 1989; Wood and Granados, 1991).

Advantages of using baculovirus expression vector

Though excellent prokaryotic host vector systems are currently available the success of baculovirus as an eukaryotic vector system owes much to several factors. As stated earlier baculoviruses are non-pathogenic to vertebrates or plants and do not employ transformed cells or transforming elements as some of the mammalian expression systems do and hence safer to handle and easy to work with. Both promoters of polyhedrin and *p10* protein are remarkably strong with clear temporal control of expression providing maximum potential for protein production as evidenced by the fact that the gene products constitute the bulk of the total infected-cell protein by the very late phase of infection. There is a large body of evidence to suggest that nuclear and cytoplasmic proteins are usually expressed to very high levels by recombinant baculoviruses (up to 200 mg/l of infected cells) (McCorrel and King, 1997). They are highly conserved and neither polyhedrin nor *p10* is essential for replication in cell culture. The genes are expressed late in the infection cycle, allowing the production of even cytotoxic products with little effect on viral replication. So baculoviruses are said to be helper independent viruses. The polyhedrin also serves as an easy marker for insertion of foreign gene. Hence, visualisation, detection or selection of polyhedrin negative recombinant is made easy by light microscope using plaque assay technique. Further one of the important advantages of baculovirus recombinant (vector) is the replication competence of the resulting recombinants. With regard to the safety of the recombinant DNA experiments, NPVs lacking the polyhedrin gene have the advantage of being easily inactivated in the field.

The large sized (100–130 kD), covalently closed, circular, double stranded DNA genome of baculoviruses makes them amenable to all nucleic acid technologies applied to mammalian viral systems. The extendible rod shaped nucleocapsid allows it to accommodate an extra DNA segment including host transposons (Jehle and Vlak, 1996) (theoretically up to 100 kbp), unlike that of other eukaryotic Simian virus (SV40) vector, which has a limitation up to 5 kbp size for foreign DNA accommodation. It is the most powerful and efficient vector and the best alternative as of today to any other eukaryotic (SV40, Vaccinia virus) and lower eukaryotic (yeast) vectors because of its efficient gene product expression without any serious instability problems. Unlike other prokaryotic systems (bacteria) where the requirement of passenger DNA gene free of intervening sequence (introns) is a must [because of the possible instability of the messenger ribonucleic acid (mRNA) transcript or translated protein product], baculovirus vectors are the best preferred expression vectors under insect cell environment for the expression of any type of foreign gene products due to efficient gene splitting mechanism. Hence a wide variety of genes from viruses, fungi, bacteria, plants, and animals have been abundantly expressed in insect cells infected with baculovirus expression vectors (Luckow, 1991).

Recently, Hoffman *et al.* (1995), Boyce and Bucher (1996) and Shoji *et al.* (1997) have shown the use of baculovirus as efficient vector for the delivery of foreign genes into various mammalian cells like human and rat hepatocytes, Hela, CPK and COS 7. Further, the availability of insect *in vitro* cell culture system and working with safe and non-oncogenic or transformed insect cells with various associated advantages (dealt under insect cell culture) has enabled in-depth study of certain baculoviruses with considerable simplification of their genetic manipulation. Thus it is evident that a foreign gene, which could increase the speed of kill either by disabling or inactivation, can be placed in the baculovirus genome. This upon infection into the natural population of insects/cell system via the virus, can augment their control over and above that of the virus alone. However, the identification of a foreign gene effective for insect control is crucial for the construction of recombinant baculovirus insecticide, which should be selectively toxic to insects, safe and effective. Most genes chosen for introduction will code for a protein that falls into any of several differing classes mentioned above (Keeley and Hayes, 1987; Kirschbaum, 1985).

The existing information on strategies/approaches to increase the efficacy of the NPV for the biological control of insect pests has been reviewed (Bishop, 1994; Maeda, 1995; Miller, 1995; Wood, 1995; Shuler *et al.*, 1995; Choudhary, 1996; Narayanan, 1996b; Cory and Hails, 1997; Thiem, 1997). These studies are based on the concept of genetic engineering of viruses to clone the new genes deleterious to the insects using insect cell environment. Such approaches vis-à-vis the specific genes are discussed in following sections.

Expression of foreign toxic genes

Insect specific neurotoxin genes

Because the insect nervous system has been the most common and vulnerable target site for the conventional chemical pesticide arsenal, viz. organochlorines, organophosphates, carbamates, and pyrethroids, the search for alternative pest control agents has also been focused on natural toxins that specifically affect the insect nervous system. Arthropod neurotoxins represent a powerful chemical adaptation of venomous animals, which almost instantly paralyse their insect preys by specifically interfering with normal functioning of the host nervous system. The neurotoxins act by blocking synaptic transmission or by inhibiting the activity of an ion channel. Because of the degree of specificity, natural toxins of the insect nervous systems have become extremely valuable in designing biopesticides, including recombinant baculoviruses. Since these toxins are not active topically, use of baculovirus as a delivering vehicle has been envisaged. As such recombinant baculoviruses perform the dual functions of efficiently expressing biologically active toxins and serving as vehicles for efficient delivery of the toxins to insect targets. At the same time, the insecticidal activities of these toxins substantially enhance the speed of kill of baculoviruses.

Scorpion venom toxin gene

Carbonell *et al.* (1988) have explored the possibility of using recombinant DNA technology to incorporate a synthetic gene encoding an insect specific paralytic neurotoxin found in the venom of Middle Asian sub-species *Buthus eupeus* into (AcMNPV) genome with the aim of halting insect feeding more rapidly. Though they could express polyhedrin/toxin fusion gene that could yield substantial levels of fusion protein without a limitation by transcription or RNA stability, they failed to detect paralytic activity either due to protein instability or inefficient translation. The venom of North African (Algerian) scorpion, *Androctonus australis* (Linnaeus) contains specific neurotoxin. The neurotoxic protein AaIT has been shown to bind to the sodium channel proteins and acts by causing specific modifications in the sodium (Na+) conductance of neuron, producing a pre-synaptic excitary effect leading to paralysis and death which is the characteristic symptom consistent with sodium channel blocking. A dose of 50 mg/kg of AaIT neurotoxin when administered to mice produced no effect. It exclusively affects the insect nervous system and it is toxic only when injected into the body. AaIT effect is thus similar to pyrethrum insecticides and could be an excellent candidate for improvement of the efficacy of baculovirus insecticides.

Stewart *et al.* (1991) have constructed a recombinant baculovirus derived from AcMNPV containing the above insect specific neurotoxin aiming to improve the virulence of AcMNPV by reducing the survival time of the infected insects with concomitant reduction in host plant damage. The improved insecticidal activity of the recombinant virus demonstrated by the reduction in both median survival time (ST_{50}) and median lethal dose (LD_{50}) value in *Trichoplusia ni* (Hubner) larvae when compared to wild AcMNPV (ST_{50}, 113 h) was obvious as a decrease of 25 per cent was recorded. Further, the recombinant virus treated plants consistently showed less damage to the leaves than the individuals infected with AcMNPV. In host range studies, with recombinant virus containing scorpion gene, no significant change in host range was identified between the parent and recombinant virus eventhough the expressed toxin can affect a variety of insect species (Possee *et al.*, 1993). This is because the host range is specified by the virus and not by the introduced toxin.

Maeda *et al.* (1991) have cloned the synthetic AaIT gene in silkworm *Bombyx mori* (Linnaeus) NPV expression system to express the toxic protein. The protein was secreted into the haemolymph that caused cessation of feeding and early mortality (60 h post infection) than with wild virus infected larvae which died after 96 h post infection. In all these studies recombinant virus was polyhedron negative and the larva had to be provided with the recombinant baculoviruses by injection rather than via feeding. However, in order to have field application value, baculovirus should be orally infectious. McCutchen *et al.* (1991) were able to develop a polyhedron positive recombinant AcMNPV which carried the same AaIT gene under the control of *p10* promoter. This could increase the speed of kill by about 30 per cent.

It is thus evident from the above studies that the differences could occur which can be attributed to the source of toxin gene and/or the sequence used and the vector/or the host insect system that has been tried. For example, Carbonell *et al.* (1988) have used the toxin from *B. eupeus* in AcMNPV expression system under *S. frugiperda* cell environment whereas Maeda *et al.* (1991) have used BmNPV expression system using *B. mori* cell line.

Predatory mite toxin gene

The straw itch mite, *Pyemotes tritici* (Lagreze-Fussat and Montagne) possesses an extremely toxic venom used primarily to paralyze insect prey. Before feeding, the female mite immobilises host by injecting the venom from paired glands, apparently located in the basal part of the pedipalps (Weiser and Slama, 1964). The venom is extremely potent since a single female has been observed to paralyse an insect 17,000 times its own weight. The venom does not exhibit host specificity since it is toxic to a wide variety of hosts as well as non-host species. Paralysis is irreversible through disruption of neuro-muscular functions, and does not affect the respiratory mechanism.

Tomalsky and Miller (1992) could increase the paralytic effect of the novel neurotoxin encoded to ×34 whose molecular weight is 27 kDa. It is commonly known as *txp-1* isolated from the predatory mite, *P. tritici.* They also increased its effectiveness to broaden the host range of insects and make it particularly attractive to use in viral pesticide development. It is significant that *txp-1* does not immediately kill but rather paralyses the host. Thus the virus can replicate under these conditions and the yield of virus is sufficient to allow its production in insect larvae, which is the current conventional method for large-scale virus pesticide production. In contrast to the NPV expressing a scorpion toxin, the one expressing a mite toxin may possibly receive public acceptance as a protective viral pesticide readily, because of the absence of the stigma and apprehension associated with proteins derived from scorpion venom. Recently, Popham *et al.* (1997) have improved the *Helicoverpa zea* (Boddie) NPV by way of inserting mite toxin gene and expressed the toxin during infection resulting in 50 per cent mortality within 40h after virus treatment. This is said to be the fastest acting recombinant baculovirus reported to date.

Predatory spider toxin gene

Spider venoms are heterogeneous mixtures of natural insecticides, which paralyse insects. The continued search for toxins suitable for use in bioengineered pesticides has resulted in the recent discovery of a new toxin, Nps-901 from the spider, *Diguetia canites.* When injected, this toxin induced paralysis that was specific to insects and cloning into a baculovirus enhanced the virus insecticidal activity (Krapecho *et al.*,

1995). Recently, Hughes *et al.* (1997) expressed another spider toxin gene from *Tegenaria agrestis* (Walckenaer) in addition to earlier *D. canities* toxin gene. The recombinant viruses expressing insect specific toxin genes from *T. agrestis* and *D. canities*, designated *vAcTalTX-1* and *vAcDTX 9.2*, respectively, significantly reduced both FT_{50} and ST_{50} in three lepidopteran pests. Reductions in feeding times compared to the wild-type virus ranged from 16 to 39 per cent with vAcTalTX-1 and 30 to 40 per cent with vAcDTX 9.2. Reductions in survival time were lower ranging from 18 to 33 per cent with vAcTalTX-1 and 9 to 24 per cent with vAcDTX 9.2. While vAcTalTX-1 tends to kill faster than vAcDTX 9.2; vAcDTX 9.2 induced feeding inhibition faster than vAcTalTX-1 suggesting that it would be more effective in reducing crop damage.

Parasitic venom gene

The recent study by Quistad *et al.* (1994) on the purification and characterization of insecticidal toxin from the venom of the parasitic wasp, *Bracon hebetor* Say has shown that the toxin paralyses insects and has a generalised effect on most of the lepidopteran species. In future there is a possibility of using this toxin (Brh) as an insecticide through baculovirus vector system. It has shown 400-fold higher biocidal activity (0.03 mg/g) against *Spodoptera* when compared to scorpion toxin LD_{50} (13 mg/g) and 100-fold higher activity against *Galleria* when compared to mite toxin (LD_{50} 0.3 mg/g).

Bacillus thuringiensis (Bt δ-endotoxin gene)

Usually, the recombinant virus expressing the toxin or as a matter of fact any toxic gene should be particularly useful in agricultural pest control programme, i.e. it should induce rapid cessation of pest feeding to limit the crop damage. The endotoxin produced by *Bacillus thuringiensis* and its other species is an outstanding example of this type.

During sporulation of *B. thuringiensis*, a Gram-positive, rod shaped aerobic bacterium produces crystalline inclusions containing the toxins. Insecticidal properties of delta endotoxin gene of *B. thuringiensis* and its other strains are very well known (Hofte and Whitely, 1989). The presence of active toxin within the insect gut causing anorexia is known for decades now (Heimpel and Angus, 1959). The toxin is shown to generate pores in the cell membrane leading to disruption of the osmotic balance, cell lysis-a process called colloid osmotic lysis, and leakage of the gut content into haemocoel. Then the insect quickly ceases feeding and dies. The gene for the protein may be either encoded on plasmids or on the bacterial chromosomal DNA. In the case of *B. thuringiensis* subsp. *kurstaki* HD-173, the gene is encoded by a 75 kbp plasmid. The toxin encodes a 130 kDa protein, which is cleaved proteolytically by the gut protease to an active toxin of 68 kDa in the insect gut.

The idea of cloning Bt toxin gene in the NPV genome was contemplated simultaneously by different scientists in different laboratories using different viral vectors. Merryweather *et al.* (1990) have inserted delta endotoxin gene from *B.t.* subsp. *kurstaki* HD 173 into AcMNPV using both polyhedron and *p10* promoters. Analysis of infected cell extracts showed that the delta endotoxin was expressed in insect cells as 130 k, 62 k, 42 k proteins, and synthesis was at peak 18h-post infection. When extracts from the cells infected with polyhedrin negative virus were fed to *T. ni* larvae, feeding by the insect was inhibited and death occurred that was inconsistent with virus infection.

In bioassays with second instar *T. ni* larvae the dosage required to kill 50 per cent of the larvae (LD_{50}) and the time required to kill 50 percent of the larvae (LT_{50}) with Bt recombinant and wild type viruses were not significantly different. The polyhedrin-positive virus has an LD_{50} about 2-fold higher than that of unmodified AcMNPV. This suggests that the endotoxin produced in the insect does not have any effect. This is not surprising because the protein was expressed as protein and remained intracellular. In this form it would be inaccessible to the gut lumen and thus unable to have any effect. The reason may be the insertion and expression of the truncated (active) toxin sequence rather than protoxin (Possee *et al.*, 1991; Wood and Granados, 1991). Martens *et al.* (1990) cloned the complete insecticidal crystal protein gene cryIA (b) of *B. thuringiensis* subsp. *aizawi* 7-21 into AcMNPV in place of polyhedrin gene. *S. frugiperda* (Sf) cells infected with recombinant virus produced biologically active endotoxin crystal protein. Infected cell extracts inhibited feeding of the large white cabbage butterfly, *Pieris brassicae* (Linnaeus). The difference in active expression of Bt crystal protein toxin by Martens *et al.* (1990) when compared to attempts of Merryweather *et al.* (1990) may be due to highly virulent Bt strains used. For instance, Bt subsp. *aizawi* isolated by Kalfon and de Barjac (1985) when compared to earlier Bt subsp. *kurstaki* used by Merryweather *et al.* (1990) was comparatively less virulent to most of the noctuid lepidopteran.

Even though Merryweather *et al.* (1990) were not able to express the Bt gene product in an active way, the important feature of the study was the use of *p10* promoter which has been positioned upstream of the polyhedra gene and produced a polyhedron plus recombinant virus with Bt cloned. The insect viruses generally protect themselves from the vagaries of the environment through encapsulation in a polyhedral coat during the field application for pest control, and the presence of polyhedral inclusion is a must for the survival of the virus in the field. If polyhedral gene is replaced then the progeny virus will quickly perish when liberated after the death of the host. Hence recombinant virus with polyhedron plus may be better than virus minus polyhedron. These studies clearly provide basis for strategy in which Bt crystal protein gene can be used for future baculovirus pathogenicity. However, the studies both by Ribeiro and Crook (1993) and by Martens *et al.* (1995) by way of expressing full-length, truncated and mature forms of crystal protein genes from *B.t* subsp. *kurstaki* into a baculovirus have shown that the recombinant viruses do not increase the insecticidal activity. The reasons being that (i) the protoxins produced are inactive and not likely to be activated *in vivo*. (ii) secretion of Bt protoxins is poor, (iii) production of the mature toxins results in cytotoxicity, and (iv) crystals of the protein are mainly produced in the fat body and not in the midgut cells or haemolymph, where the toxic effects of Bt are mainly produced. However, it is important to note that in all the earlier studies (Merryweather, 1990; Martens *et al.*, 1990; Pang *et al.*, 1992), the recombinant viruses were engineered to express proteins rather than activated toxins. Before Bt toxins can be ruled out as candidates for improving baculovirus efficacy, attempt should be made to express activated toxins, as they occur in native toxins and fuse with signal sequences that permit them to be secreted into the haemolymph or into the midgut cells.

Baculovirus expression of hormone genes

Neurohormones are the master regulatory hormones of insects and affect critical physiological processes that include moulting, metamorphosis, reproduction and general homeostasis and kill or debilitate the treated insects. Sensitive neuroendocrine events include hormone synthesis, hormone secretion and degradation. Further, neurohormones are proteins and as such are intrinsically unstable and unsuited for application in the environment due to light, heat, and micro-organisms. Also, it is unlikely that natural neurohormones could gain entry into exposed insects and they would be digested to their constituent amino acids by gut proteases, if consumed orally. So the use of insect viruses as highly efficient cloning-expression vectors for neurohormone genes comes to the rescue for their efficient utilization in biological control of insect pests. Following few sections will explain the advantage of using baculovirus as an expression vector for some neuropeptides and will discuss existing important information on the use of genetic engineering to increase the efficiency of NPV through cloning certain behaviour modifying genes which are deleterious to the insect.

It is evident that a foreign gene, which could greatly increase the speed of kill either by disabling or inactivating the larvae by causing feeding deterrence early, can be placed in the baculovirus genome. This upon infecting the natural population of insects/cell system via the virus can augment their control over and above that of the virus alone. However, the identification of a foreign gene effective for insect control is crucial for the construction of recombinant baculovirus insecticide, which should be safe, effective and selectively toxic to insects. Most genes chosen for introduction will code for a protein that falls into any of several different classes mentioned above. (Keeley and Hayes, 1987; Kirschbaum, 1985).

Eclosion hormone gene

The insect development is characterised by a series of moults, which allow growth and metamorphosis. Moulting in insects is a complex process that involves apolysis, secretion of the new cuticle and shedding or ecdysis of the old cuticle. The eclosion hormone triggers shedding of the old cuticle. Recently, Eldridge *et al.* (1991) expressed eclosion hormone in *Manduca sexta* (Johannsen) through the AcMNPV baculovirus expression system with high level of biological activity. This can pave way for further development of this unique insect neurohormone.

It is thus evident that viruses can produce quantitatively the higher levels of hormones and enzymes and use of insect gene presents less of a problem in assessing the safety of such recombinant viruses, since the genes are not "foreign" to the target insects. The investigators speculate that this type of hormone biosynthesis, may be due to the interference of UDP-glucosyl transferase gene present in the control of virus itself. They also suggest the future line of work in overcoming *in vitro* instability and low and late expression of JHE, so as to effectively use this technique for the development of genetically engineered insecticides

Diuretic hormone gene

Water balance in insect is controlled by its intake and excretion in response to changes in the insect environment. Diuretic and anti-diuretic hormones are considered to play important roles in maintaining the water balance in insects. This balance might be disrupted if a recombinant virus produced elevated levels of either of the hormone in infected larvae. Maeda (1989b) expressed a synthetic diuretic hormone gene of the tobacco hornworm, in a recombinant baculovirus, viz. *B. mori* NPV. When silkworm larvae were injected with the resulting BmNPV, the diuretic hormone was expressed causing a strong alteration in the larval fluid metabolism. The haemolymph of *B. mori* larvae infected with recombinant BmNPV was decreased by 30 per cent with an increased mortality in comparison with wild type NPV infected larvae which required five to six days to achieve the effect of a similar level. These observations demonstrate that genetically engineered NPV has the ability to change the water metabolism and increase the mortality of the larvae. Here the infection of larvae was by BV rather than by feeding occlusion bodies, and analysis of the true biological properties of the virus *in vivo* remains to be done.

Baculovirus expression of enzyme genes

Juvenile hormone esterase gene

The regulatory process associated with the metamorphosis of insect larvae into pupae is also suitable target for the action of recombinant baculoviruses. The metamorphic changes of insect caterpillars into pupae, and ultimately into adults are regulated by the juvenile hormone (JH). A reduction in the titre of JH early in the late instar has been shown to initiate metamorphosis and lead to cessation of feeding behaviour. The reduction in JH is associated with a drastic increase in the levels of juvenile hormone esterase (JHE) (Bonning and Hammock, 1994).

Considering the above concept, Hammock *et al.* (1990) cloned *JHE* gene into the transfer vector of AcMNPV and produced the recombinant NPV carrying the gene and observed about 5-fold reduction in growth in the cabbage looper *T. ni* larvae infected with recombinant NPV compared to control. The feeding efficiency also decreased in recombinant NPV-infected first instar larvae. However, infection of late instar larvae showed no significant difference in feeding or growth. Various explanations for this could be: (i) In later larval stages the levels of JHE produced may be unable to overcome hormone biosynthesis; (ii) The production of a virus gene-encoded ecdysteroid UDP-Glucosyl transferase may reduce the effects of JHE; and (iii) JHE is extremely unstable *in vivo* when produced by the recombinant virus, or in its natural form.

The lack of effect on the later larval instars reduces the prospects for the successful use of this enzyme in a recombinant baculovirus insecticide. It is not likely that insect pests could be consistently targeted at the neonate stage. However, it is encouraging that an insect gene can be used to modify the efficacy of a baculovirus insecticide. Undoubtedly, studies are required to explain how to increase levels of JHE in virus-infected insects and how to improve the stability of the enzyme *in vivo* by way of preventing the inactivation of protein.

Ecdysteroid UDP-glucosyltransferase gene (egt)

Ecdysteroid UDP-glucosyltransferase (*EGT*) is produced by AcMNPV and maintains the insects in an actively feeding state throughout the infections by blocking moulting. The deletion of the gene encoding EGT was envisaged to accelerate the virus-induced mortality by allowing the infected larvae to begin moulting, and resulting in feeding cessation. For instance, *S. frugiperda* larvae infected with egt-negative virus fed less and died more rapidly than those infected with wild type virus. Thus an egt-negative baculovirus by itself or an "*egt*" negative baculovirus over-expressing genes such as JHE can be used for the development of enhanced baculovirus insecticide. However, wild type viruses with "*egt*" gene cannot be used for over-expressing such genes since they cannot be expected to have any significant effect due to the inhibition of ecdysis by "*egt*".

It has been calculated that the baculovirus genome of AcMNPV contains approximately 150 putative genes that have been identified (Ayres *et al.*, 1994). The function of more than 30 genes has been either confirmed or speculated. O'Reilly and Miller (1989) and Barret *et al.* (1995) have identified a gene that interferes with ecdysteroid UDP-glycosyltransferase of the host. This particular gene of viral origin has been identified in AcMNPV genome and shown to interfere with the normal development of the host insects which catalyses the transfer of glucose from UDP-glucose to ecdysteriods, the insect moulting hormones. In fall armyworm, *S. frugiperda* expression of "*egt*" gene allowed the virus to interfere with the normal insect development so that moulting was blocked in infected larvae and most of the fourth instar larvae infected with the wild type AcMNPV did not moult to the fifth instars. However, all the larvae infected with an "*egt*" minus mutant moulted. The characteristics of impending larval-pupal moults, viz. feeding cessation, wandering and spinning were noted with "*egt*" minus infected larvae but no data were presented regarding any alteration in the pesticidal properties of the "*egt*" mutant virus. Thus the "*egt*" minus baculoviruses by themselves or with foreign gene inserts such as JHE gene may be useful in the development of enhanced viral pesticides.

The interaction between a virus and its host is the result of long coevolutionary history. The deletion of viral genes affects virulence, yield or persistence. However, the deletion of particular genes can lead to increased virulence and improved insecticidal action, for example, the removal of the OB envelope gene results in increased virulence due to more efficient dissolution of OB in the alkaline system.

O'Reilly and Miller (1991) made a novel approach to the engineering of improved baculovirus pesticides. Instead of involving the expression of some foreign gene, here they deleted the "*egt*" gene, which significantly improves the pesticide characteristic of AcMNPV. Larvae infected with "*egt*" deletion mutant display considerable reduced feeding and early mortality as compared to uninfected larvae. This technique will pave the way for easy registration of genetically improved viral pesticides by pesticide regulatory agencies, since one of the safety concerns involved with generation of recombinant viral pesticides is the introduction of a foreign gene which may alter their properties in some undesirable way. The recent study by Sarvari *et al.* (1990) has shown the expression of foreign gene and toxic gene products through insect baculovirus expression system for their effective utilization with high virulence properties.

Baculovirus expression of activator genes

Enhancin genes

The nuclear polyhedrosis viral genome contains a number of genes with varied functions. One of the viral genes that affect the pesticidal activity of the baculovirus is viral enhancing factor (VEF) or "enhancin" gene of *T. ni* granulosis virus (GV) (Derksen and Granados, 1988). The presence of "enhancin" gene has been reported in GV of *Helicoverpa armigera* (Hubner) and *Mythimna unipuncta* (Haworth) (Roelvink *et al.*, 1995) and in NPV of *Lymantria dispar* (Linnaeus) (Bischoff and Slavicek, 1997). The VEF/enhancin has biological properties similar to *M. unipuncta* and is GV synergistic factor (Sf), a component of the polyhedra (granules of GV) (Tanada, 1985). Following dissolution of polyhedra in the midgut of host larvae the VEF disrupts the integrity of the peritrophic membrane lining the midgut and inflicts severe damage to this membrane. Thus the removal of this barrier significantly increases the efficiency of viral infections.

Recently, Wang and Granados (1997b) have identified an invertebrate intestinal mucin (IIM) from a lepidopteran insect *T. ni*, similar to the mucus layer found in mammals, protecting the digestive tract from microbial infections. It has been shown that *T. ni* granulosis virus (TnGV) has evolved a novel strategy due to the presence of "enhancin" to overcome intestinal mucinous defense barrier against micro-organisms due to mucin degrading activity, which increases the susceptibility of larvae to NPV infection (Wang and Granados, 1998; Peng *et al.*, 1999). Federici and Stern (1990) reported that GV of great leaf skeletoniser, *Harrisina brillians* Barnes and McDunnough replicates extensively in the midgut epithelium and sheds the virus into faeces of their host. This implies that there is a future possibility of understanding the molecular mechanism underlying the production of viral occlusion bodies in midgut epithelium. Once this mechanism is known it would be easier to engineer NPVs, GVs, EPV (entomopox virus) and other viruses to produce effective viral insecticides and potential insect control agents. The over-expression of and/or increased occlusion of enhancin VEF in polyhedra or granules would significantly reduce the amount of field inoculum required to achieve high rate of infection. Since baculovirus "enhancin" protein genes are virus coded they can be better utilized to improve the efficacy of viral pesticides in future.

Using spindles associated with entomopox virus

Recent evidence indicates that the spindles associated with the spheroids of *M. separata* (Walker) entomopox virus can enhance the effectiveness of a nuclear polyhedrosis virus

(Xu and Hukuhara, 1992), by reducing the LD_{50} of the NPV of *P. separata*, infecting the same host by as much as 9000 times. Similarly, Mitsuhashi *et al.* (1998) have reported, 10,000 times more sensitivity of *B. mori* larvae when inoculated with spindles associated with entomopoxvirus of *Anomala cuprea* (Hope) than the control. Recently, Narayanan (1998d) in his preliminary study characterised the spindles associated with baculovirus of *Galleria mellonella* (Linnaeus) and has shown its enhanced activity than the control by way of increasing the infectivity of NPV of *S. litura* when inoculated against *S. litura*. Since baculovirus "spindle" protein gene is virus coded, it can be better utilized to improve the efficacy of viral pesticides in future, and should clarify the mechanism of enhancement.

Improvement of insecticidal effects

Although baculoviruses infect over 600 insect species (Martignoni and Iwai, 1986), individual isolates normally show a limited host range and infect only closely related species. Thus, baculoviruses in general have narrow host specificity. This property makes baculoviruses potentially very useful biocontrol agents for the control of insect pests. To increase their usefulness as effective biological insecticides, it is desirable to alter these viruses genetically to be more virulent and to infect a broader host range. Konda and Maeda (1991) have isolated recombinant virus through heterologus recombinant with a wider host range than the parental AcMNPV and BmNPV both *in vitro* and *in vivo*. Mori *et al.* (1992) developed a hybrid baculovirus derived from AcMNPV and BmNPV which had a wider host range and replicated and produced polyhedra both in *S. frugiperda* cells (Sf-21), and cells and the larvae of the silkworm, *B. mori*.

As discussed earlier baculovirus early, late and very late genes are expressed in a temporally regulated cascade. Early genes are transcribed by the host RNA polymerase II and late and very late genes are transcribed by a virus-specific RNA polymerase. Baculovirus late gene expression is tightly linked to DNA replication. Some of the baculovirus genes identified to date that influence the host range are *P143* (putative DNA helicase) gene, apoptotic suppresser genes (*p35* and *iap*), and some of the cell line late transcriptional factors like *lef-7, hrf-1, hcf-1*, etc.

Helicase gene (p 143)

Lu and Carstens (1991) in case of AcMNPV have showed the putative DNA helicase gene p143 as essential for DNA replication. Similarly, Ahrens and Rohrmann (1996) reported the sequence of helicase gene in *Orgyia pseudotsugata* (McDunnough). Being essential for DNA replication, the baculovirus helicase gene has also been shown to influence viral host range. *B. mori* NPV (BmNPV) and AcMNPV infect *B. mori* and *S. frugiperda* cells, respectively, but cannot replicate in the heterologous cell line. However, in co-transfection experiments AcMNPV recombinants, containing a fragment of the BmNPV helicase gene instead of the corresponding AcMNPV sequence, were produced that were capable of replicating both in *B. mori* and

S. frugiperda cells. Recently, Argaud *et al.* (1998) have shown that amino acid changes at position 564 and 577 are required to kill *B. mori* larvae.

Apoptotic and anti-apoptotic gene (p 35)

A specific gene *p35*, which is required for AcMNPV late gene expression and virus DNA replication in Sf-21 cells, was identified as being responsible for blocking the apoptotic response (Clem *et al.*, 1991). Apoptosis-a distinct type of programmed cell death, is a phenomenon evolved as a primitive viral defense in certain vertebrate animals and invertebrates lacking humoral immunity to function as antiviral defense mechanism, is gaining importance in cellular defense against viral infection (Narayanan, 1996a). Apoptosis may be defined as "a process where the cells die in a controlled manner in response to specific stimuli, apparently following an intrinsic programme". It is an active process of cellular self-destruction with distinctive morphological and biochemical features. For more details on apoptosis and its role in the microbial control of insect pests the readers may refer to the recent review by Narayanan (1998a).

Cellular apoptosis may be induced by a variety of different extracellular and intracellular stimuli and apoptosis during viral infection is a general response of insect cells to virus infection, which can differ depending upon the cell types. Apoptosis is a genetically controlled process. Clem *et al.* (1991) have reported that genes, which are responsible for inducing apoptosis have not been observed in insect baculoviruses. However, inducement of apoptosis by "*p35*" deficient mutant (vAcAnh) of AcMNPV may cause cell death in some insect cells (Sf-21 cell line derived from *S. frugiperda* and BmN4 cell line derived from *B. mori*) or may not in others (Tn-368 cells derived from *T. ni*). Recently, Pang *et al.* (1998) used *H. armigera* single embedded NPV (HaSNPV) against the *T. ni* cell line, BTI-Tn-5BI-4 (Hi5) and recorded apoptotic cell death. The same group have recently observed that the HaSNPV-induced apoptosis in Hi5 cells was completely suppressed by co-infection with *T. ni* multicapsid nucleopolyhedrovirus (TnMNPV) and partially blocked by TnMNPV p35 expression alone (Dai *et al.*, 1999).

Though apoptosis was first observed in Sf-21 cells, later on it was demonstrated that it was not only Sf-21 cells which are unique in their apoptotic response to infection by vAcAnh, the Bmn4 cell line also undergoes apoptosis when infected with vAcAnh. On the other hand, *p35* mutant of AcMNPV cannot induce apoptosis in Tn-368 cells (Clem *et al.*, 1991). Prikhod'Ko and Miller (1996) have demonstrated that transient expression of the AcMNPV "ie-i" gene induced apoptosis in Sf-21 cells but not in Tn-368 cells just like *p35* deficient mutant (vAcAnh) of AcMNPV virus. Palli *et al.* (1996) too have found similar cell type-dependent effects. They found that AcMNPV replicates and produces occlusion bodies in FPMI-Cf-70, an ovarian cell line of spruce budworm, *Choristoneura fumiferana* (Clemens), wherein it induced apoptosis and failed to replicate in Cf-203, a mid-gut cell line of the same insect. Chejanovsky and Gershberg (1995) have found inducement of apoptosis in *Spodoptera littoralis* (Boisduval) by wild type AcMNPV. These observations suggest that apoptosis may also play a role in determining the host range of baculovirus, however, the phenomenon remains specific to cell types.

Autographa californica (Speyer) multiple embedded nuclear polyhedrosis virus (AcMNPV), carried apoptosis gene *p35* in addition to "*iap*" gene, that works in conjunction with "*iap*" to inhibit apoptosis in a wider range of insects. This could explain the relatively broad host range exhibited by AcMNPV compared to other baculoviruses with a narrower host range. This is evident from the study of *O. pseudotsugata* NPV (though it is closely related to AcMNPV) which lacks *p35* gene, therefore, possessing narrower host range as compared to AcMNPV (Clem and Miller, 1994). However, *iap* homology (*Op-iap* gene) that functions to inhibit apoptosis has been identified in this case (Birnbaum *et al.*, 1994). Nucleotide sequence analysis of the *B. mori* nuclear polyhedrosis virus (BmNPV) genome revealed the existence of a gene homologous to the *p35* gene of *A. californica* MNPV (AcMNPV), which has been shown to prevent virus-induced apoptosis (Kamitha *et al.*, 1993). Codling moth of apple, *Cydia pomonella* (Linnaeus) granulosis virus, "*iap*" gene (inhibitor of apoptosis) was isolated (Cp-iap) (Crook *et al.*, 1993). Occurrence of a homologue of Cp-iap and Ac-iap in the genome of an insect iridovirus by sequence homology (Handermann *et al.*, 1992) suggests that these DNA containing viruses do have the means of circumventing apoptosis.

Baculoviruses are generally host specific such as *Helicoverpa* NPV and *O. pseudotsugata* NPV. Infection by virus is limited to a single species or few closely related species of insects. AcMNPV, on the other hand, has a broader host range than many baculoviruses both *in vitro* and *in vivo*, infecting at least 33 species of lepidopteran larvae in 10 families as well as more than 25 different cell lines (Groner, 1986).

From the above discussion we can conclude that in baculoviruses two different genes, *p35* and "*iap*", are capable to overcome apoptosis, "*iap*" being the primary apoptosis inhibiting gene carried by baculoviruses. However, some baculoviruses, such as AcMNPV, have acquired additional gene (*p35*) that works in conjunction with "*iap*" to inhibit apoptosis in a wider range of insects. This may be the reason for the relatively broad host range exhibited by AcMNPV compared to other baculoviruses. On the other hand, *O. pseudotsugata*, though closely related to AcMNPV baculovirus, lacks *p35* gene, thereby showing the narrower host range (Clem and Miller, 1994). It appears that the co-evolution of cellular defensive strategies and viral offensive strategies are finely tuned. If it is understood as to which genes play what roles in which tissues in a species, the disarming of insect defenses at both the cellular and organismal levels will become easier. Accordingly, information necessary to control or to modify host range properties of the virus will be achieved (Miller, 1995).

Late expression factor gene (*lef-7*)

Morris *et al.* (1994) have identified a sequence of single activity gene *lef-7*, commonly called as late expression factor gene. In replication assays, *lef-7* stimulated DNA replication in Sf-21 cells, but was dispensable in TN-368 cells. In virus infected cells, *lef-7* was required for DNA replication in Sf-21 cells and for stimulating DNA synthesis in Tn-368 cells, which has similarity with wild type AcMNPV system. However, *lef-7* deletion viruses are significantly less infectious than AcMNPV, as seen in *S. frugiperda* and *T. ni* larvae.

Host range factor gene (*hrf-1*)

The broad spectrum AcMNPV does not infect either gypsy moth larvae or its cell line IPLB-ld 6527. Recently, Thiem *et al.* (1996) have identified a host range factor (*hrf-1*) gene that promotes AcMNPV replication in IPLB-ld 6527 cell system. Co-transfection AcMNPV and plasmid carrying the LdMNPV gene into IPLB-ld 6527 cells results in AcMNPV replication.

Host cell specific factor gene (*hcf-1*)

Lu and Miller (1995) identified a gene, which had differential effect on late gene expression in two different cell lines, the Tn-368 and Sf-21, and named it as host cell factor (*hcf-1*) gene. They (Lu and Miller, 1996) also characterized the effects of deleting the AcMNPV *hcf-1* genes (*hcf-1* mutant) on the virus infection of different *T. ni* cell lines, and *S. frugiperda* and *T. ni* larvae. These *in vitro* and *in vivo* studies have demonstrated that *hcf-1* has species as well as tissue specificity and is important for AcMNPV virulence in *T. ni* larvae. The existence of such species-specific factors in baculovirus genomes implies that the ability of the viruses to infect alternate species may be an important aspect of their natural life history. Manipulation of such genes may allow the control of their distribution in nature (Lu and Miller, 1995).

Characterization of baculovirus genes

Cysteine protease (*v. cath*) gene

Ohkawa *et al.* (1994) and Slack *et al.* (1995) have reported a functionally active cysteine protease gene encoded by *B. mori* NPV and AcMNPV with a sequence homology to those of Papain family of cysteine proteases reported for alphaviruses, coronaviruses and tobacco etch viruses. They have the characteristics of the mammalian lysomal protease, cathepsin L. The role of cysteine protease (*v-cath*) gene is to breakdown the infected host tissues and facilitate the horizontal transmission of virus. Infection of insects with recombinant viruses containing an inactivated *v-cath* gene results in insects with an altered appearance (after death) that fails to disintegrate normally. Thus *v-cath* appears to facilitate the release of occlusion from insect after death. Cathepsins are general cysteine proteinases and *v-cath* is produced in wild type virus infections, the activity of *v-cath* may be problematic in the expression or purification of some proteins from baculovirus expression systems by way of either degrading the foreign products after or during production of foreign proteins. However, *v-cath* is not an essential gene (Ohkawa *et al.*, 1994; Slack *et al.*, 1995), the use of a recombinant of baculovirus which the protease gene has eliminated, i.e. use of *v-cath* (-) mutant may improve conditions for expression by reducing the degradation of over expressed foreign products.

Chitinase gene

Unlike vertebrates, which have extensive exposure of epithelial cells to the external environment, insects have extensive protection of their epithelial tissue. The chitinous cuticle of the insect covers virtually all-external surface, even extending through the foregut, hindgut and tracheal tubes, constituting the first line of passive defense in insects. Hawtin *et al.* (1995) and Thomas *et al.* (1998) have identified a functional chitinase gene (*chiA*) in the genome of AcMNPV and expressed its endo and exochitinase activity in insect cell system. The chitinase extracted from AcMNPV infected cells hydrolyzed more than those from bacterial, fungal or plant substrates used in the enzyme assays. This demonstrates the broad substrate range in viral system. The median time for mortality of fourth instar larvae of *S. frugiperda* infected with a recombinant virus containing chitinase gene was approximately 20h shorter than that for insects infected with a wild-type virus (Gopalakrishnan *et al.*, 1995). The above studies showed that insect chitinase has potential to enhance insecticidal activity of entomopathogens. Such engineered baculoviruses possessing the chitinolysis activity, in addition to their standard infectivity, should be more effective in the field against pests, rather than directly applying chitinase based insecticidal spray formulations (Mazzone, 1987).

25K FP gene

Just like virus encoded chitinase and protease involvement in the liquefaction of virus infected insects (Hawtin *et al.*, 1995), recently Katsuma *et al.* (1999) have identified the *25K* gene from *B. mori* NPV as an additional gene involved in the post-mortem host degradation. It alters cell metabolism and /or host cell gene expression, in addition to it's role in lowering of polyhedron production, i.e. few polyhedra phenotypes (FP) (Beames and Summers, 1989). Thus *25K* gene was involved in the complex phenotype induction by virus infection and was important for virus transmission. Without *25K* gene, host degradation followed by efficient dispersal of polyhedra would not occur.

P74 gene

A few gene products have been implicated in the insecticidal activity of baculoviruses or in their infectivity or pathogenicity. The *p74* gene, located downstream from the *p10* gene, is involved in establishing lethal fusion infection in columnar gut cells, despite the absence of prominent viral spike glycoprotein (gp64) which is usually present in budded virus. Hong *et al.* (1994) and Thielman *et al.* (1996) have demonstrated that AcMNPV encoded *p74* is essential for establishing lethal infection by oral inoculation with ODV but not by intrahaemocoelic inoculation with budded virus. Deletion of this gene abolishes infection of insects whereas the overproduction of *p74* enhances infectivity five-fold (Carstens and Lu, 1990).

Other strategies for engineering baculovirus

Anti-sense RNA strategy

Although attempts to develop baculovirus based insecticides by insertion of genes encoding neurotoxins, enzymes, etc. as detailed above have met with some success, several reports have pointed out difficulties in correct reaction or processing of peptides introduced into baculovirus genome. Another concern is the possibility of development of insect resistance to foreign protein. Recently, Lee *et al.* (1997) described an alternative strategy by way of using strong virus promoter to produce an excess of antisense transcripts which is complementary to the endogenous mRNA of a host gene where protein product is presumed essential for normal larval growth and development. The consequent block in the translation of an essential protein would be expected almost immediately to halt the normal insect physiology.

Accordingly, they have shown the existence of "*c. myc*" like protein in *S. frugiperda* similar to human "*c. myc*" protein which has an important regulating function including prolification, arrest, differentiation and death. They inserted human "*c. myc*" gene downstream from the polyhedrin promoter AcMNPV and tested against *S. frugiperda* larvae. The results of the bioassay showed that the antisense construction stopped feeding as soon as the polyhedrin driven transcripts accumulated and subsequently death of the larvae occurred. Antisense inhibition of host gene was previously observed in larvae infected with recombinant baculoviruses containing juvenile hormone esterase gene under the polyhedrin promoter or *p10* promoters (Roelvink *et al.*, 1992). Thus the anti-sense strategy will be effective in virus insecticide production in future.

It has been found that expression of bacterial and eukaryotic genes can be blocked by the production of antisense RNA. A copy of the coding sequence of the target gene is engineered in the opposite orientation downstream from a strong promoter. The antisense RNA thus produced is complimentary to the "sense" transcript and the two strands will anneal to form double stranded RNA that cannot be translated into protein. This strategy has been successfully used to alter flower colouring (Van der Krol *et al.*, 1988) and to suppress virus replication in plants (Powell *et al.*, 1989). Important insect genes coding for metabolic or regulating enzymes or hormones could be engineered in antisense orientation behind a baculovirus promoter. The expression of this antisense gene in infected insects could then block essential host functions. A baculovirus model system has been designed to explore the potential of the strategy.

In insects similar strategy has been derived recently to halt larval growth and feeding. An anti-sense gene fragment complimentary to the m-RNA of a host gene whose protein product, presumed to be essential for larval growth and development, is used. As the two transcripts (sense and anti-sense) would remain bound to each other, there would be a block in the translation of an essential protein and normal insect physiology would be altered. This anti-sense approach was found to be efficient and larvae stopped feeding almost immediately after the appearance of the polyhedrin promoter driven transcript strand.

Baculovirus expression of maize mitochondrial protein

A maize mitochondrial protein involved in cytoplasmic male sterility and disease susceptibility has been expressed in AcMNPV (Korth and Lewis, 1993). This protein, URF13, is hydrophilic and binds tightly to the membrane of cells. Injection of larvae of the cabbage looper, *T. ni*, with polyhedrin-negative recombinant virus resulted in a

40 per cent decrease in lethal time compared with that of the wild-virus. However, URF13 decreased the ability of AcMNPV to produce polyhedrin when polyhedrin-positive constructs were made.

Baculovirus expression of insect virus surface protein

As stated earlier, one of the commercial disadvantages of the baculovirus pathogen is the host specificity of the virus. This is mainly because of the presence of specific midgut binding domain in the virus. Since the primary objective of the biotechnology research is to extend the host range nature of the virus and to increase the speed of kill, Sivasubramanian *et al.* (1991) cloned an insect virus surface protein gene of wide host spectrum with strong midgut binding domain of Bt delta endotoxin. Using the chimeric clones they have produced a viral vector with wider host range and increased toxicity.

Baculovirus expression of insect's antibacterial proteins

As a part of a survival strategy, insects have evolved numerous and effective defence mechanisms to resist infection caused by various microorganisms like fungi, bacteria, protozoa and even viruses. The defense mechanism in insects, in general, is broadly classified into two groups. The first one is non-specific immune system, which consists of structural and passive barrier like cuticle, gut, physio-chemical properties and peritrophic membrane. The second one is specific immune system involving cellular and humoral immunity. In cellular immunology, mechanisms such as phagocytosis and encapsulation are operative. In humoral immunity, responses in insects involve synthesis and release of several well characterised antibacterial immunoproteins like cecropins, attacins, diptericin and defensins against several gram positive and gram negative bacterial pathogens (Boman and Hultmark, 1987; Hori and Watanabe 1980; Morishima *et al.*, 1990). Sun Shao-Cong *et al.* (1990) have identified hemolin which belongs to the immunoglobulin superfamily anti-bacterial proteins. The antibacterial nature of gut contents of *S. litura* (Govindarajan *et al.*, 1975) and partial characterisation of haemolymph bacterial proteins and lysomes from certain insects like *B.mori* and *G. mellonella* has been reported (Powning and Davidson 1973; Sethuraman *et al.*, 1993; Abraham *et al.*, 1995).

These insect antibacterial proteins are the best-characterized invertebrate antibacterial factors. They have counter parts in mammals and are used against parasites that cause diseases like malaria, chagas and leishmanias (Kaaya *et al.*, 1987). Cloning and expression of some of these antibacterial protein genes for the future designing and management of certain insect vectors causing human and animal diseases, hold much promise. Such an approach needs to be followed not only in mammalian arthropod disease vectors (Miller *et al.*, 1987; Narayanan, 1998b) but also in insect pests including silkworm against various diseases. In this connection it is imperative to mention that Yamada *et al.* (1990) expressed sarcotoxin 1A, an antibacterial protein against most of the Gram-negative bacteria from the fleshfly, *Boettoperisca peregrina* (Robineau-Desvoidy) using BmNPV expression system. The recombinant sarcotoxin

1A was secreted into the culture medium and being unstable in this system because of the presence of viral encoded cysteine protease, addition of cysteine proteinase inhibitor P-chloromercuri-benzene-sulphonic acid (PCMBS) considerably prevented the sarcotoxin 1A-degradation. This is a very useful system to study recombinant sarcotoxin and its utilization.

Baculovirus expression of polydna virus (PDV)

Certain female parasitic wasps in the families of Ichneumonidae and Braconidae carry the particles containing double stranded circular, multiple DNA virus called Ichnovirus and Bracovirus. These viruses are introduced through the cuticle during the oviposition of parasitoids. These are essential for the survival of their progeny in their habitual host. In the absence of these virus particles, the parasitoid egg is recognised as foreign and encapsulated by host blood cells, i.e. insect haemocytes, especially by granulocytes, whereas in the presence of virus the parasitoid is not encapsulated (Edson *et al.*, 1981). Though most of the baculoviruses promote their own survival by suppressing apoptosis of host cells, as is the case of *Microplitis demolitor* polydna virus (MdPDV), they are likely to be transferred vertically. MdPDVs promote their own survival by inducing apoptosis of host immune cells, which would otherwise kill the developing *M. demolitor* eggs.

In general, larvae of lepidopteran insects become increasingly resistant to baculovirus infection as they age (Narayanan, 1979; Engelhard and Volkman, 1995). Such developmental resistance has been reported in many species of lepidopteran larvae challenged with baculoviruses including *H. armigera* with its own baculovirus. The susceptibility of late 5th instar *H. armigera* (which is immune) to HaNPV after parasitisation by a tachinid *Eucelatoria* sp. breaking the maturation immunity (Narayanan, 1980a) suggests the possibility of presence of some unknown factors/ particles or viruses similar to polydna of hymenopteran insects. They probably might alter the immune system by inducing apoptosis in the key immune granulocytes, which mediate encapsulation in the beginning, and leave many other putative replacement cells (which are not infected owing to the absence of free polydna virus 24–36 h post parasitisation) as is observed for HaNPV infection by way of inhibiting apoptosis (Strand, 1994). The immune suppression of *Drosophila melanogaster* Meigen, for the parasitisation of *Leptospilina heterotoma* (Thomson), a cynipid wasp reported by Rizki and Rizki (1990) suggests that insect groups other than braconids and ichneumonids can also alter the host developmental immune response.

Larvae of many dipteran parasitoids especially tachinids maintain contact with outside air by attaching their posterior spiracles to the host's tracheal system or a hole in the integument (Askew, 1971). In many instances, the larvae of these parasitoids turn the immune response of their host to their own advantage by building a respiratory funnel. The presence of such a respiratory funnel allows the developing parasitoid continuous access to fresh air through the host's tracheal system or hole in the host's integument. Engelhard *et al.* (1994) have discovered that the tracheal system is the major conduit for baculovirus movement through infected hosts. The finding that the larvae can be infected directly via the tracheal system has profound implications for the use of baculoviruses as pest control agents. There is availability of other species of *Eucelatoria*, viz. *Eucelatoria rubentis*, which has 4 times broader host range than *E. bryani* (Reitz and Adler, 1996) and can be utilized for the purpose. The incorporation of polydna virus genes with immuno suppressant activity into the genome of HaSNPV might lower resistance in grown up larvae of *H. armigera*. This may enable this pest to be controlled even in the grown up stage with a recombinant HaSNPV (Washburn *et al.*, 1996) by allowing *Campoletis sonorensis* (Cameron), an ichneumonid parasite to oviposit into *H. zea* larva highly resistant to AcMNPV.

Mechanism of baculovirus expression

Mode of action of *Bacillus thuringiensis* (Bt) δ-endotoxin

The toxic action of Bt *in vivo* is known to be dependent on the binding of activated toxin to the external surface of midgut microvilli. Here the toxin appears to bind to specific proteins (Hofmann *et al.*, 1988) and then intercalate, forming trans-membrane cation pores (Knowles and Ellar, 1987) that lead to cell death. However, it has been shown recently that the Cry1A and Cry 3A toxins could be inserted into planar lipid bi-layers that have no protein receptors (Stalin *et al.*, 1990). Moreover, recent evidence from patch clamp studies of the action of the Cry IC toxin on Sf cells indicates that this Bt toxin may act inside the cell and is capable of inserting into the cell membrane from the cytoplasmic side (Schwartz *et al.*, 1991). Thus by using a baculovirus expression vector, it should be possible to circumvent the midgut microvillus barrier and express different forms of Bt toxins within cells, and to determine whether they are toxic and if so where they act. Such studies, provided they use activated toxin, might enable the intoxication mechanism to be separated from the binding action of the molecule and clarify whether specificity is determined solely at the level of the midgut epithelium.

Mode of action of optical brighteners

Among the several drawbacks that have hampered the effective use of baculoviruses for pest management, loss of infectivity in the environment due to ultraviolet (UV) radiation and protracted time for death are the major ones. Recently, Dougherty *et al.* (1996) have described the use of optical brightener as a radiation protectant for the gypsy moth, *Lymantria dispar* (Linnaeus) MNPV, enhancing the residual activity upto 214-fold. While radiation protection and enhancing effect of optical brightener such as M2R are well documented for several baculoviruses and their hosts, very little is known about the enhancement mechanism. Recently, Washburn *et al.* (1998) have used a reporter gene recombinant of AcMNPV (AcMNPV-hsp 70/*lacZ*). They used AcMNPV to express *lacZ* as a reporter gene under the Heat shock protein (Hsp) promoter of *Drosophila* in order to investigate the enhancing effect of M2R on pathogenesis, mortality and death time in larvae of *T. ni* and *H. virescens*. They found that stilbene derived optical brightener M2R enhanced AcMNPV pathogenesis by blocking the sloughing of infected primary target cells in the midgut, thereby countering developmental resistance and increasing mortality.

Developmental resistance

Larvae of lepidopteran insects commonly become increasingly resistant to baculovirus infections as they age. The mechanism responsible for this developmental resistance is not known, but the phenomenon does not occur if the rival inoculum is administered intra-haemocoelically instead of orally, which is the natural route of infection. This observation indicates that the factors mediating developmental resistance are operative during infection of the primary target tissue, the larval midgut, and not during subsequent systemic infection. To learn more about the mechanism of developmental resistance, Engelhard and Volkman (1995) orally inoculated four cohorts of fourth instar larvae with a recombinant of *A. californica* M nuclear polyhedrosis virus expressing a reporter gene. While these cohorts differed only by a few hours in age, they found increasing resistance to infection in successively older cohorts. By assessing the presence and location of infected cells at intervals during the first 48 hours after inoculation, they identified two key factors relevant to the resistance pattern among the developmental cohorts. These factors were (i) an age-dependent rate of establishing and/or sloughing infected midgut cells, and (ii) the ability of fourth instar *T. ni* to completely clear infection of the midgut epithelium by ecdysis to the fifth instar. Kirkpatrick *et al.* (1998) have studied the physiological basis of developmental resistance to AcMNPV in fifth instar larvae of *H. virescens* using a recombinant of AcMNPV carrying a "*LacZ* " reporter gene and they suggested that the developmental resistance was partly hormone mediated.

Host range of multiple embedded nuclear polyhedrosis virus

As stated earlier, the infection cycle of baculovirus includes two distinct phenotypes, an occluded form, called occluded derived virus (ODV) and bud form, called budded virus (BV). ODV is embedded in a crystalline protein structure (the occlusion), which helps to protect the infective virions from environmental degradation. In the nucleoplolyhedrosis virus, ODV is composed of virions containing either a single nucleocapsid (SENPV) or one to many nucleocapsids (MENPV).

To gain insight into the biological relevance of these two different packaging strategies, Washburn *et al.* (1999), in their time course experiments, using ODV fractions of AcMNPV carrying a β-galactisidase reporter gene (AcMNPV. Hsp70/lacZ), demonstrated clearly that infection initiated by ODV-M moved more rapidly into secondary target cells of tracheal epidermis of *T. ni* than did infections initiated by ODV-S, thereby accelerating the onset of irreversible systemic infections. This indicates wider host range of MENPVs than SENPVs.

Molecular techniques for better utilization of baculoviruses

Virus characterisation through "REN" analysis

Using restriction endonuclease (REN) analysis, precise description of various virus isolates and identification of high level of genotypic variations in baculovirus population is possible. It is useful to have base-line information and to find out the fate of field released baculoviruses and to see if there is occurrence of any mutation.

Identification of infection processes

Baculoviruses are genetically designed to follow the infection process. Recent studies by Engelhard *et al.* (1994) and Kirkpatrick *et al.* (1994) using AcMNPV expressing "*lacZ*" as a reporter gene have demonstrated that the larval tracheal system appears to be the major conduit for virus. Recently, Locke (1997) has shown that the caterpillars have specialised parts of tracheal system, which function as lungs, providing blood cells (haemocytes) with oxygen via haemocytes to the trachea instead of conventional path of trachea going to the tissues. These findings have profound implications for the use of baculovirus as pest control agents because infection of trachea can be initiated through the spiracles via topical applications. This will help in escaping the midgut barrier as well as for the subsequent true systemic infection of almost all organs that are bathing in the haemocoel with open circulatory system. To understand the host immune response using a combination of a highly refractory host (*H. zea* to AcMNPV) and AcMNPV expressing "*lacZ*" they have shown that infected cells became encapsulated with insect haemocytes and were subsequently cleared.

Identification of latency (or) persistency of virus

Normally horizontal transmission of insect baculoviruses is through oral feeding of foliage contaminated with occlusion bodies of insect virus. And vertical transmission from adult to progeny is both trans-ovum and trans-ovarial process. Baculovirus epizootics often decimate host insect population not only in the field but also in the laboratory due to latent infection. So the development of simple and highly sensitive method for the detection of the nuclear polyhedrosis virus in large scale culture populations not only of insect pests but also other useful insects like honey bee, silk worm, etc. is a must. Noguchi *et al.* (1994) have proposed a tool for practical diagnosis of the NPV of *B. mori* in co-operative rearing using polymerase chain reaction (PCR) at the DNA level. Using this technique they demonstrated the presence of a possible latent virus in a culture of the cabbage moth, *Mamestra brassicae* (Linnaeus) which was stressed by infection with a second baculovirus. Similar studies are needed to show the extent and role of such persistent infections in natural population, which is going to cause viral epizootics. Lupiani *et al.* (1999) have developed a PCR assay technique to detect the presence of *H. zea* reproductive virus (previously called as gonad specific non-occluded virus) in various geographical populations. This suggests its usefulness as highly specific, sensitive, and rapid way of detecting the presence of *H. zea* reproductive virus. It has also been envisaged that PCR technique could help in the study of the host range, tissue specificity and incidence of this virus in wild populations of corn earworm.

Another technique useful for this purpose is "fluorescence" technique as has been studied by using "green fluorescent protein (GEP)" gene from jelly fish or "luciferin" gene from firefly. Chao *et al.* (1996) have transferred the green fluorescent protein (GFP) gene from the jelly fish *Aequorea victoria* into a typical baculovirus, the AcMNPV, thus providing an easy visible marker for detecting infected insects. The above technique can be used to predict infection process in insects (Barret *et al.*, 1998) and the dispersal and/or long term persistence of the recombinant viruses before their widespread use, which may adversely affect the environment.

Insect luciferase gene from firefly, *Photinus pyralis* was expressed in Sf-21 cells using a baculovirus vector (Hasnain and Nakhai, 1990) and in *B. mori* cells and larvae (Vikas *et al.*, 1995). Jha *et al.* (1990) have expressed luciferase gene both in *T. ni* and *S. littoralis* larvae through a recombinant AcMNPV. Since luciferase was not secreted into the haemolymph but remained in the body tissue, it was suggested that expression of luciferase could be used as an excellent reporter enzyme to study virus infection, dissemination, and host range determination. It could also monitor the release of recombinant in the environment when used as a biocide. The above studies indicate that luciferase may be used as reporter gene expression in insects, identification and selection of recombinant baculoviruses. Similarly, bacterial luciferase derived from a fusion of the *lux A* and *lux B* gene of *Vibrio harveyi* has also been expressed at very high levels in caterpillars like cabbage looper, *T. ni* and saltmarsh caterpillar, *E. acrea* and Sf-21 cells (Richardson *et al.*, 1992).

Field trials with genetically improved/modified baculoviruses

Field testing of genetically modified AcMNPV was performed between 1986–1989. The first virus used contained a small, unique genetic marker to facilitate monitoring. The second virus contained another genetic marker, but also had a polyhedrin gene deletion preventing the production of virus occlusion bodies (polyhedron-negative), which protect the virus in the environment. The third virus is also polyhedron-negative but contained bacterial beta-galactosidase gene, under the control of a virus gene promoter, to serve as an innocuous foreign reporter gene. These experiments demonstrated that the small-scale field trials with recombinant baculoviruses presented minimal risk to the environment (Bishop *et al.*, 1988). First field trial of a genetically improved multiple embedded nuclear polyhedrosis virus of looper, *A. californica* (AcMNPV) that expressed the insecticidal toxin (A) derived from the venom of the scorpion *A. australis* was carried out in United Kingdom by Cory *et al.* (1994). They also demonstrated that recombinant virus expressing insect selective toxin kills *T. ni* larvae more rapidly (reducing the time to kill the larvae by 30%) than the wild type virus, resulting in improved crop protection by way of reduction in crop damage (67–80%). There is evidence that secondary transmission (horizontal transmission) is lower in insects treated with recombinant virus. This may be due to the lower yields of polyhedra since genetically modified AcMNPV lacks *p10* gene as suggested by Mulock and Faulkner (1997). Field testing of an *egt*-deleted recombinant AcMNPV was approved by the U.S. Environmental Protection Agency and the first field test was performed (Miller, 1995).

Safety of genetically improved/modified baculoviruses

Effect on non-target beneficial arthropods

One concern associated with genetically improved recombinant viruses is their effects on non-target insects associated with pests (Coghlan, 1994). In this connection Heinz *et al.* (1995) evaluated the direct effects of wild-type (AcMNPV) and a recombinant AcMNPV (AcAaIT) on three beneficial insects. The recombinant NPV expresses an insect selective neurotoxin, AaIT, which was isolated from the scorpion, *A. australis*. Two generalist predators, *Chrysoperla carnea* (Stephens) and *Orius insidiosus* (Say) were not adversely affected by feeding on larvae of *H. virescens* infected with AcAaIT. Similarly, no adverse effects were detected in the honey bee, *Apis mellifera* Linnaeus, when injected with wild-type or recombinant NPVs. McNitt *et al.* (1995) have also observed no difference between the social wasp, *Polistes metricus* when they were fed uninfected *S. frugiperda* larvae and larvae infected with toxin expressing viruses. Results from these studies may provide a foundation upon which potential risks associated with genetically engineered NPVs may be evaluated on a limited scale in greenhouse or field experiments.

Host range, persistence and ecology of viruses

Ecology and environmental impact evaluation of recombinant baculovirus insecticides have recently been reviewed by Richards *et al.* (1998). Genetically improved viruses act as biological insecticides and not as biological control agents that will become permanent establishment. To address the issue of persistence and assessment of the random virus dissemination through genetically modified and improved virus, field trials using crippled recombinant virus via deletion of the polyhedrin gene with genetically improved virus have been conducted (Bishop *et al.*, 1988; Cory *et al.*, 1994). The above studies revealed that there was no difference in host range for the genetically crippled virus in comparison with parent virus. Recently, D'Amico *et al.* (1999) have provided a fast, simple and inexpensive identification method for studying the ecology of genetically engineered gypsy moth *L. dispar* NPV by way of inserting β-galactosidase marker gene and have shown the movement of virus both in time and space.

Insect cell culture

One area of great interest for the development of insect pest control methods, through the use of obligate pathogens like insect virus and protozoan in microbial control is by the application of insect tissue culture for their mass production. At present several baculoviruses are produced commercially by the insectory method, using artificial diets with little or no sterility precautions (Kompier *et al.*, 1988; Maramarosch, 1987). The recent advances in laboratory culturing of insect cells might offer a satisfactory alternative method for large-scale production of not only insect virus but also obligate pathogens like insect protozoans. The advantages of *in vitro* multiplication rather than *in vivo* are several, viz. free from other viral microbial contamination, comparatively easy since insect cells can be cloned and stored, and provide stable and uniform production of obligate pathogens. In fact, the development of baculovirus as expression vector for foreign gene expression and production of an array of proteins of agricultural, veterinary, medical and pharmaceutical importance has renewed the interest in the study of insect cell cultures (Luckow, 1991; Volkman, 1995; Davis, 1995; Narayanan, 1995b).

In countries like Australia, UK and USA, *in vitro* studies have already produced important new information on the biochemistry, viral replicative process and the virus genetics (Bilimora *et al.*, 1986; Granados and Hashimoto, 1989). *In vitro* studies have also shown the complexity of the nuclear polyhedrosis replicative cycle. Though pioneering work on the development of dipteran mosquito cell lines for arbovirus work has been reported by Singh (1967) virtually no work has been done on insect cell culture especially on lepidopteran insects for the purpose of insect viruses in India. However, preliminary attempts were made to grow cells from embryonic tissue of potato tuber moth, *Phthorimaea operculella* (Zeller) (Pant *et al.*, 1977) and primary cell cultures from ovarian tissue of *Locusta migratoria* (Linnaeus) and *Schistocerca gregaria* (Forskal) (Raina and Khurad, 1988).

Narayanan (1993a) has successfully established the primary haemocyte culture of *S. litura* and infected the primary haemocyte culture with haemolymph obtained from larva already infected with nuclear polyhedrosis virus by *per os*. This revealed typical cytopathic effect (CPE) by way of enlargement of cell nucleus, granulations around nuclear region and polyhedral formation. The future possibility of *in vitro* replication of *S. litura* NPV in alien cell line of *S. frugiperda* grown under serum and serum free medium containing 16 m of $Alcl_3$ by inoculating with filtered supernatant medium from infected primary haemocyte culture of *S. litura* has also been envisaged (Narayanan, 1993b). *S. frugiperda* cell line was successfully cultured using neo-natal calf serum/bovine serum instead of using usual 10 per cent foetal bovine serum (Narayanan, 1994b). Temporary storage of cells (5–15°C) over a period of one month as well as long term cryo-preservation of cells under liquid nitrogen was successfully attempted without affecting the viability of cells. Recently, Sundeep and Pant (1998) have established a continuous cell line from the larval haemocytes of *H. armigera*. They have also reported the susceptibility of this cell line to AcMNPV and SlNPV. A recent report (Anonymous, 1999) has shown establishment of continuous cell lines both from the larval and pupal ovaries of *S.litura* and their susceptibility both to homologous SlNPV and heterologous AcMNPV. However, the susceptibility of cell lines developed from larval and pupal ovaries to alien broad-spectrum AcMNPV warrants detailed investigations on the susceptibility of larvae to AcMNPV.

Need for insect cell culture

As on today, it seems that insect cell culture system could play an important role in production of baculoviruses, control of dormant pests and quarantine pests, production of polydna viruses and hybridoma work. From an Indian perspective, baculoviruses of two important national pests, *H. armigera* and *S. litura*, are being produced through conventional diet surface contamination technique. This is labour intensive, time consuming, needs spacious insect rearing facilities and is not easily accessible for automation. It also does not satisfy the commercial requirement of dependability, consistency, yield, quality and cost. Thus it would be worth while to use insect culture techniques for large scale production of insect viruses. Such cultures will be important in case of seasonal pests as well , which are effectively controlled by the NPV (Jayaraj *et al.*, 1989; Narayanan, 1985c,d)

In the life cycle of certain parasitic hymenopteran insects, they require as an integral component, the participation of polydna viruses (Braco and Ichno viruses) which replicate in the ovary of certain females. Some physiological and immunological effects on the parasitic larvae have been attributed to the presence of polydna viruses (Edson *et al.*, 1981). For example, inhibition of phenol oxidase activity has been shown in parasitized larvae of *T. ni.* Similarly, it has been reported that the virus component of the calyx fluid of *Campoletis* sp. is capable of inhibiting encapsulation of parasitic eggs by the host, *Heliothis* sp. (Stoltz, 1986), thereby facilitating endoparasite protection against host-immune defenses (Summers and Dill-Hajj, 1995). The ichno virus infection in an established gypsy moth cell line (Kim *et al.*, 1996) and expression of polydna virus gene product using a poly histidine baculovirus vector (Soldevila *et al.*, 1997) suggests the importance of insect tissue culture for mass multiplication and use of polydna virus to overcome the abrogation of their defense mechanism, that increases the efficiency of biological control of crop pests both with parasites and pathogens in an integrated manner.

The development and use of 'Hybridoma' technique in insect culture by way of producing monoclonal antibodies has been suggested, because it has been proven to be easy, sensitive and rapid (Maramorosch, 1987). Once particular monoclonal antibody preparations are characterized, they can be used as standardized reagents. Monoclonal antibodies will be important in (i) baculovirus identification, (ii) baculovirus taxonomy, (iii) study of the genome organisation between baculovirus, (iv) study of baculovirus replication *in vitro*, and (v) isolation of specific m-RNA from infected cells by immuno precipitation of the ribosome complex. Among the analytical techniques that have been perfected during last decade are the Western-blot ELISA, combined with the use of monoclonal anti-sera, which appear to be the most valuable for baculovirus polypeptide characterization and comparative studies between isolates (McCarthy and Gettig, 1986).

Advantages of using insect cell culture system

The development of baculovirus as expression vector for foreign gene expression has renewed the interest in the study of insect, especially lepidopteran cell lines. Insect cells can be maintained at room temperature (26–28°C) without pH indicators and trypsin treatments. There is no requirement for CO_2. Further, generation time is generally short (16–18 h) and plaque assaying is easy. Insect cells can be cultured/maintained in serum free synthetic media (Hink, 1991; Vaughn and Weiss, 1991). This has the associated advantage of low cost, less batch variation in quality and free from contaminating microorganisms, which not only helps in the *in vitro* mass production of insect viral biopesticides but also in the downstream processing of protein products. Ten-fold efficiency of transfection of baculovirus DNA into insect cells by using cationic lipids, such as lipofection (GIBCO/BRL) is possible because of the formation of the liposomes.

The availability of permissive cell lines that can be grown as monolayer and suspension culture (Vaughn and Weiss, 1991) has made scaling up process easy with existing systems engaged in the production of virus for veterinary/medical vaccines with/without major modifications. All post-translation modifications like glycosylation (N and "O"-glycosylation), fatty acid acylation (palmitylation and myristylation) nuclear-transport, C-terminal amidation, disulphide bond formation, signal peptide together with other proteolytic cleavage and formation of tertiary structure with clear unfolding of expressed proteins are performed in an authentic manner under insect cell environment. Most of the recombinant proteins that are expressed are soluble, and antigenically, immunologically, and functionally similar to their authentic counterparts (Luckow, 1991; Holzman, 1995; Overton and Kost, 1995).

Need for indigenous insect cell culture

Most of the baculovirus expression vector work that has been carried out for the production of an array of foreign gene products of agriculture/veterinary/medical and pharmaceutical importance has been done with Sf-9 cells from the ovarian tissue of fall armyworm, *S. frugiperda*. In fact, the absence of indigenous insect cell culture system for many native nuclear polyhedrosis virus associated insects, hampers the genetic manipulation and development for having own indigenous baculovirus expression system in many countries.

It is well documented that insect viruses do not appear to replicate well in reproductive organs like ovaries and testes unlike that of other favourable tissue like fat body which is one of the primary sites of virus multiplication in the insect. Some of the post translational modifications of foreign proteins expressed through baculovirus insect cell culture environment like "C" terminal amidation enzyme are found in the fat bodies of lepidopteran insects (Choudhary *et al.*, 1992). Electron microscopic studies have shown that polyhedra formed within *S. frugiperda* cells are often many but empty polyhedra. Recent studies have shown considerable increase in level of expression of foreign gene products and biological activity in various cell lines other than Sf-9 cell lines (Atkinson *et al.*, 1990). These studies indicate the importance of insect cell lines, if developed indigenously.

Hink *et al.* (1991) and Wickham *et al.* (1992) have found that individual cell lines differed in their ability to synthesise secreted versus non-secreted proteins, presumably due to the differing machinery that each cell line possessed. Flipsen *et al.* (1993) have shown relatively high expression of polyhedrin and *p10* in *S. exigua* midgut cells than through the conventional *S. frugiperda* ovarian cell system suggesting the possible role of cellular factors in the expression. McKenna *et al.* (1997) have established three new *T. ni* cell lines, Tn-Aa14, Tn-4b and Tn-4b31, from embryonic tissue in a commercially viable serum free medium, Ex-cell 400. The cell lines are highly susceptible to wild type and recombinant viruses and have the capacity to express the foreign genes, at levels exceeding those of the earlier cell lines Tn-5B1-4 (Hi5 cells) developed by Granados *et al.* (1994) and Wickham *et al.* (1992). Hence, much can be learnt in future by studying the levels of the expression and the protein processing, targeting and transport of many recombinant proteins in other insect cell lines and cell lines derived from other tissues and other susceptible host specimens available indigenously for better understanding of the phenomenon.

Efforts are needed in future for *in vitro* culture of insect viruses and other obligate pathogens using serum free media for large-scale production for insect pest management programme, using air-lift fermentors (King *et al.*, 1992; Maiorella *et al.*, 1988). Further, development of stably transformed insect cells (Jarvis *et al.*, 1990) coupled with immediate early viral promoters will go a long way for continuous expression of insect specific toxins and other behaviour modifying gene products. Use of hollow fibre bioreactors and micro-encapsulation technique will alleviate the problem of maintaining high cell densities for obtaining the high product concentrations. With our recent understanding of the genetics and cell biology of apoptosis, there is a tremendous scope for the manipulation of cells by transfecting anti-apoptotic genes like *bcl-2.* Supplementing the culture medium with appropriate survival factors it is possible to enhance the robustness and survival of cells in culture in future.

Bacterial pathogens

Among many bacterial pathogens, which are used for the control of crop pests, Gram-positive, rod-shaped, aerobic, spore forming and crystalliferous bacteria *Bacillus thuringiensis* commonly called Bt is the extensively studied bacterium. The insecticidal toxin synthesised is called parasporal body (known as crystal), which is proteinaceous in nature tightly packed by hydrophobic bonds and disulphide bridge (Anandakumar *et al.*, 1996). Though biology and genetics (Bulla *et al.*, 1978; Aronson, 1986) and molecular biology of Bt (Hofte and Whiteley, 1989; Yoshisue *et al.*, 1995) have been well studied elsewhere, virtually no work has been done on the molecular biology and microbial genetics of Bt in India. Meenakshi and Jayaraman (1979) have studied the formation of crystal proteins during sporulation. Sivamani and Rajendran (1992) have successfully demonstrated a protoplast fusion between *B.t.* var. *kurstaki* and a native root colonizer Rhizobacterium, *Pseudomonas fluorescens.* This shows the potential for increasing the habitat to which Bt toxin might be applied even though the expression of the insecticidal crystal protein (ICP) gene in *P. fluorescens* hybrid is lower than in the parent.

Mode of action of Bt

Upon ingestion of insoluble prominent crystal inclusions of Bt which are released upon lysis during its stationery phase; are solubilised in the gut, which has high pH of (more than 9) releasing protein called delta endotoxin. These protoxins of 130 kDa are activated by the gut trypsin like proteases which typically cleave about 500 amino acids from

C-terminus and 28 amino acids from the N-terminus leaving a 65–55 kDa protease resistant toxic active core comprising the N-terminal half of the protoxin. The resultant proteinase resistant toxic fragments bind to specific receptors in the target tissue. These receptors are tentatively identified as 120–180 kDa glycoproteins. After binding to a specific border located on the brush border membrane (BBM) of columnar epithelial cells, the toxin inserts irreversibly into the plasma membrane of the cells. The next step involves the formation of spores in the plasma membrane, which disturbs osmotic equilibrium (maintained by the cells) by pumping ions in to the extra cellular medium. Pore formation in the columnar epithelial membrane renders the cell's volume regulation mechanism ineffective. Accordingly, the cell swells and ultimately bursts by a process known as "colloid osmotic lysis". This leads to the distruption of integrity and finally death of insect from starvation and septicaemia.

Insecticidal crystal proteins (ICPs) of Bt are grouped into four major classes: Lepidoptera specific (CryI); Lepidoptera-Diptera specific (CryII); Coleoptera specific (CryIII) and Diptera specific (CryIV). Among the lepidopterous specific ICPs, six different types, viz. Cry IA (a), Cry IA (b), Cry IA (c), Cry IB, Cry IC and Cry ID have been recognized (Hofte and Whiteley, 1989). Two new classes of toxin genes, viz. Cry V and Cry VI toxic against Lepidoptera, Coleoptera and nematodes respectively have been subsequently added to the above classification (Feitelson *et al.*, 1992). All the above classes of Bt proteins have extremely short half-life when applied topically due to inactivation by ultraviolet rays. An improved method to deliver the Bt toxic protein was developed by genetically improving the genes, which encode their proteins directly into plants. The major benefits of this system are economic, environmental, and qualitative. In addition to the reduced input costs to the farmer, the transgenic plants provide season-long protection independent of weather condition, effective control of burrowing insects difficult to reach with spores and control at all the stages of insect development. Crickmore *et al.* (1996) have suggested a revised nomenclature for the Bt *cry* genes, which is solely based on amino acid identity where in the Roman numerals have been exchanged for Arabic numerals, and the parentheses are removed, thus, *cry*IA(a) becomes *cry*1Aa.

Need for genetic manipulation and genetic engineering

The screenings of field isolates of entomogenous bacteria for their insecticidal properties and further strain improvement through conventional genetic approach have yielded good results. It is evident from the recent isolation of *B. thuringiensis* subsp. *tenebrionis* (Kreig *et al.*, 1983) and *Bacillus thuringiensis* subsp. *sandiego* (Herrnstadt *et al.*, 1986) for the control of coleopteran pests. *B. thuringiensis* subsp. *aizawi* (Mallapadidam, 1992; Sanchis *et al.*, 1989) for the control of noctuid insects like *Spodoptera* and *Heliothis* spp. complex (which were otherwise refractive to earlier Bt strains) and other Bt strains which are more potent to dipterans including certain plant and animal parasitic nematodes are also on record (Feitelson *et al.*, 1992). Further, recent advances of using bacterial pathogens as such for effective microbial control of insect pests include the use of evapo-retardants, photo-protectants, use of bait and encapsulation technique, etc. for increasing its persistence/stability under field conditions. However, the above conventional approaches for increasing the field persistence and residual activity of Bt have not given many encouraging results. This implies the need for newer approaches. Accordingly, the advent of recombinant DNA technology offers the possibility of developing entirely new Bt or microbial agents based on the construction of genetically modified microorganisms. Such organisms may retain the advent of classical control agents and suffer fewer of their drawbacks like narrow specificity, short shelf life, low potency, presence of viable spores, etc. thereby providing protection to any industry which ventures into new research and development expenditure in this area. Hence, utilization of recombinant DNA technique to insert *B. thuringiensis* endotoxin gene into various prokaryotes such as leaf and root colonizing bacteria *Clavibacter xyli*, and *Rhizobium* sp. to obtain transgenic bacteria as well as to have a "Biopacked" *Pseudomonas* containing Bt endotoxin crystals. Similarly, such insertions in eukaryotes like baculovirus including plants to have transgenic plants and transgenic baculovirus are some of the biotechnological approaches that have been tried for the effective utilization of bacterial pathogens by way of increasing their persistence and efficacy (Narayanan, 1994a, 1995a, 1996b, 1997; Anandakumar *et al.*, 1996).

It has been observed that the activity of a Bt strain depends on the activities and relative amounts of individual endotoxins produced and plasmid borne nature of Bt endotoxin genes. Thus it is possible to construct better Bt strains using standard genetic techniques of plasmid curing and conjugate plasmid transfer. Recombinant DNA technique promises even more precise control since it should be possible to construct strains that produce only the desired endotoxin in the desired proportion. Hence, the following biotechnological approaches that have been applied towards increasing the efficiency of Bt have been presented in brief.

Conventional strain improvement by conjugation

The conventional strain improvement can be achieved by conjugation or by transduction. Since most of the cry genes are plasmid borne, the plasmid-coded genes can be exchanged via conjugation process among the Bt strains producing new Bt insecticides with different spectra of toxicity. Recently, an Ecogen scientist combined beetle active with caterpillar active Bt protein in the same strain by conjugating *B.t. kurstaki* with *B.t. tenebrionis* to produce hybrid strain EG2424.

Transduction, on the other hand, is a transfer of bacterial DNA between cells (intra-inter-serotype) via transducing phage particles. Kalman *et al.* (1995) have placed *cry1Ca* gene from *B.t. aizawi* into the chromosome of two *B.t. kurstaki* strains using transducing bacteriophage thus producing strains with a broader insecticidal spectrum.

Although the above conventional genetic manipulation techniques have the capability to increase potency and control specificity, they are limited by several factors due to which it is not possible to construct strains containing only the desired delta endotoxin genes. For example, in the case of conjugational transfer to create a novel Bt, not all Bt toxins are located on transferable plasmids. The toxin protein with useful insecticidal activity may be synthesized in low amounts. The plasmid incompatibility could also be a problem. Conjugation is not easily controllable in the laboratory, therefore, it limits the number of toxins (generally two to three) that can be present in the final stage. However, tools of genetic engineering like recombinant DNA technique can be used to overcome all the limitations of the classical genetics.

Enhancing the Bt toxin activity in plants through recombinant DNA technique

Recombinant DNA (r-DNA) technology provides the tools for developing safe, efficient and cost-effective microbial agents. It is now possible to combine the best traits of several different organisms into a single strain, expressing δ-endotoxins that exhibit enhanced insecticidal activity, longer residual activity and broader host range. Thus the feasibility of using Bt for insect control has been increased by advances in recombinant DNA technology, which facilitated cloning of toxin genes and their expression in plants and other organisms. As a result recombinant DNA technology provides enormous potential for advances in insect control.

Reason for Bt toxic gene as attractive candidate for transgenics

Most natural plant defenses against insects-such as alkaloids, tannins and terpenes are products of complex metabolic process and multi-enzyme pathways, and are therefore, difficult to engineer into new plant species when compared with Bt δ-endotoxin, which is comparatively straight-forward process. Similarly, there are non-Bt proteins, which interfere with nutritional need vis-à-vis development of the insect, such as proteinase and amylase inhibitors and chitinase proteins. In spite of small size, abundance and stability of the proteinase inhibitor, insects have proven to be flexible enough to alter the proteinase composition in their midgut to overcome the inhibitor produced by transgenic plants. So is also the case with chitinase though it targets peritrophic membrane, which is the main internal defense barrier. LC_{50} ranges of Bt are from only 50 to 500 ng/ml of diet. Hence, Bt δ-endotoxins are particularly attractive candidate for genetic engineering. Bt toxins are extraordinary lethal to certain pests. For instance based on amounts used in agricultural applications molecules of Bt toxin are 80,000 times more potent than organophosphates and 300 times more potent than pyrethroids (Feitelson *et al.*, 1992). Moreover, Bt causes rapid cessation of pest feeding (called anorexial effect) thereby limiting crop damage.

Insertion of Bt endotoxin into prokaryotes

Since expressing Bt "cry" genes in organisms that are stable in the environment may be a useful approach to overcome the problem of limited field stability of Bt preparation, several groups have been successful in transferring Bt endotoxin "cry" genes into an alternate host organism to substantially change the means of application of the insecticides. Watrud *et al.* (1985) have introduced a cloned Bt endotoxin gene into the corn root colonizing bacterium *Pseudomonas fluorescens*. This, in principle is to deliver the insect crystals beneath the surface of the soil in areas where root feeding insects do the most damage. Similarly, Stone *et al.* (1989) reported the construction of a recombinant *P. fluorescens* that is toxic to tobacco budworm. An ICP gene from *B.t.* subsp. *kurstaki* (HD-1) was cloned into a plant colonizing bacterium *P. cepacia* to protect tobacco plants from tobacco budworms (Stock *et al.*, 1990). Cloning of the lepidopterous specific gene into a *Pseudomonas* to protect N_2 fixing root nodules of pigeon pea from soil dwelling pests was attempted by Nambiar *et al.* (1990). Gene encoding a 65 kD toxin from *B. thuringiensis* subsp. *tenebrionis* was cloned into *Rhizobium leguminosarum* (Skot *et al.*, 1990). Bezdicek *et al.* (1991) introduced the *cry3* gene into *R. leguminosarum* and *R. meliloti* so as to reduce the feeding damage by the nodule-feeding insects, *Sitona lineatus* on pea plant and *S. hispidulus* on alfalfa.

Though expressing Bt "cry" genes in organisms that are stable in the environment may be useful approach to overcome the problem of limited field stability of Bt preparations, the registration of a live genetically engineered product posed insurmountable challenge to the regulating agencies during product development and registration. Hence, expression of the Bt endotoxin using dead cells of *P. fluorescens* is thought of and is a recent concept. Apart from *Pseudomonas* being a non-sporulating organism unlike Bt and providing the added advantage of directing the cells' energy towards the production of more toxin than would be produced by a natural Bt. The recombinant *Pseudomonas* can be killed during production thus providing a source of recombinant but non-viable bacterial insecticide. Thus, these products are environmentally benign because the microorganism cannot spread from the site of application or transfer its genetic information. Thus, this technology has distinct advantages in terms of environmental compatibility, efficacy and registration of genetically modified organisms. Thus, the micro encapsulation of Bt crystal, i.e. "biopacking" by *Pseudomonas* cell wall, the process called "cell cap" will protect the endotoxin from the environmental factors. The two commercial Mycogens cell cap products, viz. MVP and M-Trak are those wherein the protein biotoxin genes of Bt are expressed in *P. fluorescens* (Pf) cell. In the process of biopacking (i) The gene coding for the δ-endotoxin protein is removed from the Bt cell. (ii) The gene is incorporated into a plasmid that contains genetic information to allow for expression of the protein. (iii) The plasmid is inserted into a Pf cell. The recombinant cells are grown in aerobic, submerged culture fermentation, and at an optimum stage of the growth process they are induced to express the protein toxin. Unlike traditional Bt, cell cap products do not produce spores, more of the Pf cells energy can, therefore, be directed to the expression of protein. A chemical fixative is added to the complete fermentation broth to rapidly kill the bio-toxin-containing Pf and simultaneously to stabilize the cells. The stabilization process strengthens the cell wall by cross-linking while inactivating proteolytic enzymes that can degrade the bio-toxin. The process results in an active, stable, bio-toxin encapsulated within a non-viable cell. These products are environmentally benign because the microorganism cannot spread from the site of application or transfer its genetic information (Panetta, 1993).

Another concept in using *P. fluorescens* is to express the Bt endotoxin at high levels when it carries the Bt on a recombinant plasmid since unlike Bt, *Pseudomonas* do not sporulate. The recombinant *Pseudomonas* can be killed by a proprietary chemical treatment that cross link the bacterial cell wall to yield a non-viable encapsulated bacterium surrounding the crystal protein during production thus providing a source of recombinant but non-viable bacterial insecticide. Thus the micro-encapsulation of Bt crystal, i.e. "Biopacking" protects the endotoxin from the environmental factors (Gelernter, 1990). Yet another approach attempted by Crop Genetic International is introducing gene into an endophytic bacterium, which colonizes in the xylem of plants and provides a type of systemic immunity against susceptible insect pests. A biopesticide developed in this category has recently been named as "Incide".

Another category is of leaf colonizing bacteria (phylloplane bacteria). The introduction of ICP genes into the plant-associated microorganisms appears to be the most viable alternative. The process being much quicker and more practicable than producing transgenic plants, which have certain limitations such as the range of the toxin genes that can be usefully expressed and avoid development of resistance to ICP by the insect larvae. Dimock *et al.* (1988) earlier attempted to clone the Bt toxin gene into the phylloplane bacteria *Clavibacter xyli.* Turner *et al.* (1991) introduced the cry *1Ac* gene into another plant endophyte, *Cyanodontis.* Using the conjugational approach, Bora *et al.* (1994) transferred the *cry1Aa* gene of Bt into *B. megatherium*, which resides in the cotton phyllosphere. Increased production of insecticidal proteins with concomitant protection against *H. armigera* was reported by Kalman *et al.* (1995). Transfer of an insecticidal protein gene of Bt on to plant colonising *Azosprillum* that may be used to control root feeding insects has also been attempted (Ananthakumar *et al.*, 1996.). Recently, Sekar (unpublished) from Madurai Kamaraj University, Tamil Nadu, has isolated an organism-strain RSI-identified as a *Bacillus* sp., which is capable of colonizing in cotton leaves and found it to be an excellent colonizer of cotton phyllosphere. Into this he has introduced *cry1Aa* gene of Bt by conjugal transfer. He has also shown that transconjugation colonizes cotton plants for a prolonged period and it also protects the plant from the attack of *Helicoverpa* for more than 30 days.

Insertion of Bt δ-endotoxin into eukaryote-plants

The insertion can be achieved via truncated toxin. Plants have been transformed and regenerated that express the Bt crystal proteins as is in the case of tobacco (Barton *et al.*, 1987; Vaeck *et al.*, 1987) and tomato (Fischhoff *et al.*, 1987) for the control of *Manduca sexta* (Johannsen), *H. zea* and *H. virescens.* This has been achieved by utilizing a native coding sequence for the *cry1A* genes fused with constitutive promoter (35SCaMV promoter) operable in plants using *Agrobacterium*-mediated T-DNA transfer vector with no detectable expression. It is also known that the proteins are activated by the insect gut proteases, which typically cleave some 500 amino acids from the C-terminus of 130 kDa protoxins and amino acids from the N-terminus, leaving a 65–55 kDa protease resistant active core comprising the N-terminal half of the protoxins. Following this scientific knowledge truncated version of the native genes was used in plant transformation studies (Warren *et al.*, 1992).

Another procedure is using synthetic gene or by optimising codon usage. To circumvent the problem of low expression from the native cry genes in plants, synthetic genes have been constructed using different strategies. Eliminating potential poly adenylation sites, eliminating potential transcriptional stalling sites, eliminating mRNA instability signals, eliminating potential intron splice sites and by optimising translation by altering codon usage are well known. These strategies alter the overall guanine-cytosine (G and C) content of the genes, thereby presenting themselves a preferable way to the new host. Native cry gene has a low G+C content (37%) whereas plants tend to have a high G+C (60–70%). In addition, synthetic Bt genes are prerequisite to obtain expression of crop genes in monocot crops such as corn and rice (Koziel *et al.*, 1993). For example, the native *cry1Ab* gene transferred into rice produced no detectable protein, but the rice optimized *cry1Ab* codon transgenic *Japonica* lines showed 10–50 per cent mortality against the striped stem borer and 45–55 percent mortality against leaf folder (Fujimoto *et al.*, 1993). Hence, several laboratories have constructed partially or completely modified *cry* genes that resulted in significant improvements in gene expression in cotton (Perlak *et al.*, 1990), tomato (Perlak *et al.*, 1991), and potato (Perlak *et al.*, 1993). Anandakumar *et al.* (1998) have constitutively expressed the Cry1Ab protein in transgenic brinjal by introducing a synthetic gene otpimised for plant codon usage.

The insertions can also be achieved by using different promoters. Continuous gene expression in all plant parts raises the risk of pests developing resistance in addition to yield reduction as the plant directs more resources than necessary to its defense. Hence, efforts were directed to concentrate the toxin expression in these parts where insect feeds. In this connection Koziel *et al.* (1993) have developed the first transgenic maize containing *cry1Ab* maize under the control of tissues specific promoters such as maize pith preferred promoters, maize pollen promoter and maize phosphoenol pyruvate carboxylase (PEPC) promoter which conferred resistance against European corn borer, *Ostrinia nubilalis* (Hubner). One of the approaches to circumvent the problem of consitutive expression is to temporarily (i.e. stage specific) express the gene using the chemically responsible promoters. A fusion gene construct of *cry1Ab* with PR-la promoter was introduced into tobacco via. *A. tumifaciens* and when the resultant transgenic lines were tested against tobacco hornworm larvae, they were found to confer significant resistance.

Plastid transformation is another approach of insertion. The transcriptional and translational machinery of the plastid is prokaryotic in origin and its genome is relatively AT-rich. Thus circumventing the circuitous route of *cry* gene codon modification could demonstrate expression nearly 3–5 per cent of total soluble leaf protein of wild type *cry1Ac* in transplatomic tobacco lines using this technique (McBride *et al.*, 1995).

Perlak *et al.* (1990) have expressed truncated forms of the insect control protein genes of *B. thuringiensis* var. *kurstaki* HD-1 (*cry1Ab*) and Hd-73 (*cry1Ac*) in cotton plants at levels that provided effective control of cabbage looper, *T. ni*; beet armyworm, *S. exigua* and cotton bollworm, *H. zea*. Fujimoto *et al.* (1993) have introduced a truncated delta endotoxin gene, *cry1Ab* of *B. thuringiensis* into a *Japonica* rice. Transgenic plants efficiently expressed the toxic gene and bioassay studies had indicated that transgenic plants were more resistant to two major pests of rice; viz. rice stemborer, *Chilo suppressalis* (Walker) and rice leaf folder, *Cnaphalocrocis medinalis* (Guenee). Koziel *et al.* (1993) introduced a synthetic gene encoding a truncated version of the Cry IAb protein derived from *B. thuringiensis* into immature embryos of an elite line of maize using microprojectile bombardment. Plants expressing high levels of the insecticidal protein exhibited excellent resistance to repeated heavy infestations of European corn borer (*O. nubilalis*). Parrot *et al.* (1994) have evaluated the soybean plants transgenic for a native *cry1Ab* gene from *B. thuringiensis* var. *kurstaki* HD-1 against velvet bean caterpillar, *Anticarsia gemmatalis* (Hubner) and found reduced feeding correlated with the presence of the transgene.

Insertion of Bt endotoxin gene into eukaryotes' insect baculovirus

The cloning the Bt toxin into the baculovirus genome done in different laboratories using insect baculoviruses as viral vectors has already been dealt under section on viruses.

Expansion of insecticidal host range of Bt

The expansion of insecticidal host range of Bt could be accomplished by using protease (trypsin) inhibitor gene. MacIntosh *et al.* (1990) have shown the enhanced activity of insect control proteins from *Bacillus thuringiensis* var. *kurstaki* against their target insect pests of tobacco, boll worm and other lepidopteran and coleopteran insects by several serine protease inhibitors from soybeans. They also genetically improved tobacco plants expressing a protease inhibitor fused to a truncated *B. thuringiensis* insect control protein and have shown the levels of activity enhancement similar to those seen with purified inhibitors.

Secondly, co-expressing coleopteran and lepidopteran toxic genes could be utilised for expansion. Lereclus *et al.* (1992) have shown through their novel approach the expansion of insecticidal host range of Bt via *in vitro* genetic recombination. This has been achieved by inserting the sequence IS232 to deliver *cry3A* gene into an isolate producing Cry1A toxin resulting in a strain, which displayed insecticidal activity against both Lepidoptera and Coleoptera.

In Vitro bioassay of Bt toxins

Research efforts to improve formulations of these organisms, to increase the toxicity spectrum of various isolates or to understand the mechanism of action of toxins produced by these *Bacilli* rely heavily on bioassays against target pests. Several forms of bioassays of insects have been developed to measure the effect of various toxic entities of Bt. Parameters like mortality or decreased feeding when larvae were forced-fed or via now known to be surface contamination technique of food plant, diet surface contamination or diet incorporation technique, etc. are the common methods. The choice of test insect, life stage assayed, method of administration and environmental conditions may all affect the result of bioassay. Even *in vitro* bioassays of Bt crystal proteins against certain insect cell lines have proved useful, which demonstrated the high degree of response specificity (Johnson, 1989). In these assays, the activity of toxin preparations may be measured by histopathological observations, by vital staining or by quantitating cell lysis, by assaying ATP or other intracellular enzymes like LADH.

Hence, development of insect culture (Johnson, 1989; Maramorosch, 1987) system for bioassays seems to be a necessity.

The tobacco cut worm, *S. litura* is an economically important polyphagous insect and known to be resistant to some Bt strains based on histological, X-ray and physiological studies. There was no anorexient effect and no change in anatomical structure of gut and pH of blood and gut contents. In certain cases the gut contents showed antibacterial activity (Govindarajan *et al.*, 1975). Thus, further basic studies concerning molecular biology and genetics of Bt are necessary if one has to understand the affinity of the various Bt toxins, their binding to the brush border membrane of the receptors, and the mode of action/resistance mechanism. To create a novel Bt, which is safe to silkworm but highly toxic to other insect pests by way of manipulating various toxic domains is another aspect of consideration and is possible through *in vitro* bioassay systems.

Management of insect resistance to transgenic Bt

Resistance to Bt in insects has been demonstrated in the laboratory and in the field in case of *P. xylostella*. Underlying mechanism of resistance indicates 50-fold reduction in the affinity of the *Cry1Ab* toxin to the brush border membrane of the midgut in the case of *M. unipuncta* and 200-fold in the case of *P.xylostella* (Bruce *et al.*, 1996). In the case of Bt transgenics, resistance development in target pests is due to the constitutive expression of a single cry gene. The following are the various strategies proposed theoretically, lacking any supporting experimental data.

Using refugia

Facilitating the survival of susceptible insects by way of growing non-transgenic plants along with transgenic plants in a definite ratio, is one of the best approaches to slow down resistance development (McGaughey and Whalon, 1992). It delays the development of insect resistance to Bt introduced crops by providing susceptible insects for mating with resistant insects and thus creating a gene mix.

Using gene pyramiding

This is based on the presumption that almost unlimited number of different Bt toxins are available in nature and that resistance can be managed by using these in various mixture, mosaic, rotational or sequential systems. Recently, Chakrabarti *et al.* (1998) have reported the synergism of *Cry1Ac* with *Cry1F* toxin, where EC $_{50s}$ of *Cry1Ac* toxin were 13 times lower due to the presence of *cry1F*. It suggests that the toxins *Cry1Ac* and *Cry1F* can be expressed together in transgenic crop plants for future effective control of *H. armigera* and also for resistance management strategy. However, caution has to be exercised in future research since already extensive cross resistance among different Bt toxins has been reported in the case of *P. xylostella* (Tabashinik *et al.*, 1990) and in the laboratory populations of *M. unipuncta* (McGaughey and Johnson, 1992).

Toxin dose acquisition

A high dose of Bt, which consistently kills heterozygotes along with untreated refuges as a potential means of managing resistance development in transgenic plants was advocated by Denholm and Rowland (1992). This approach maintains constitutive and continuous exposes of Bt toxins in transgenic plants, which is sufficient to kill the heterozygotes in a population.

Targeted delivery

Continuous and constitutive expression of Bt toxic genes may result in selection pressure, tissue specific (use of stem, root, boll, pod or seed), stage specific (vegetative or reproductive), and inducible expression by way of using wound specific promoters along with chemical spray like salicylic acid to induce gene expression will aid in delaying the resistance development in insect pests (Williams *et al.*, 1992).

Second generation toxic genes for transgenic plants

Recently, Estruch *et al.* (1996) characterized the novel insecticidal proteins produced by certain Bt isolates in logarithmic stage of bacterial growth called as vegetative insecticidal proteins (VIP). Bowen *et al.* (1998) have characterized four insecticidal toxins from the bacterium *Photorhabdus luminescence* encoded by toxin complex loci *tca, tcb, tcc* and *tcd*, representing the second generation of insecticidal *trans*-genes that will complement the novel Bt δ endotoxin in future.

Safety of Bt

The safety record of Bt preparation containing spores and crystals is impeccable. There have been no reports of harm associated with its use for pest control. The Bt toxin crystals are toxic to lepidopteran larvae but are non-toxic to all animals, plants and all insects other than lepidopterans. Melin and Cozzi (1990) concluded that although there are reports of feeding deterrence and prey avoidance by some predators and parasitoids of Bt affected organism when applied at recommended dose, the Bt spore-crystal complex has been shown to have minimal effects. Prior to early 1970, some *B. thuringiensis* spore preparations contained toxic substance known as beta-exotoxin. This toxin is considered as structural analogue of ATP with an inhibitory effect on RNA synthesis (Lecadet and de Barjac, 1981). Sharma and Sahu (1977) have shown the mutagenic potentialities of exotoxin by inhibiting spindle formation and cytokinesis, and induces micronuclei, chromocentric nuclei and minor deviation in spindle activity in *Allium cepa*. While USA and Canada now require that commercial Bt preparation be tested to assure that it is free from β-exotoxin, while in India Bt strains producing β-exotoxin are being freely marketed. There are some adverse reports where Bt preparation splashed in eyes resulted in corneal ulcer (Samples and Buettner, 1982). Coghlan (1998) reported that a Bt serotype of H-34-*konkurian* caused a nasty infection when strain was put into wounds in mice with weakened immunosystems. Therefore, caution is being exercised when Bt is being advocated as a spray for the control of onchocerciasis (river blindness) in West Africa where most of the people are prone to immuno suppressive infections, such as measles in childhood, malaria and of course AIDS (Dixon, 1994). Most of Bt toxins are plasmid borne. Since mutations can lead to loss in plasmids, therefore, can trigger *B. thuringiensis* into acrystalliferous *B. cerus*, which is a very close relative of deadly anthrax *Bacillus*.

Safety of the activated toxin

Since activated Bt toxin damages susceptible insect cells after binding to a receptor on the epithelial cell membrane, binding studies should be performed to determine if mammalian tissues accessible to the toxin have receptors as there are some similarities in the tissues of two widely different organisms. Insect epithelial tissue is composed of two major cell types; a columnar cell layer with brush border membrane (BBM) resembling that of human intestine; and a unique goblet cells, containing larger vacuolar cavity (Anandakumar and Sharma, 1994). Bonfanti *et al.* (1992) have reported the resemblance of structural organization and molecular architecture of the midgut brush border of *M. sexta* with that of intestinal epithelium of vertebrates. It has been shown that the Cry1A and Cry3A toxins can insert into planar lipid bilayers that have no protein receptors (Stalin *et al.*, 1990). Moreover, recent evidence from patch clamp studies of the action of the Cry1C toxin on Sf cells indicates that this toxin may act inside the cell and is capable of inserting into the cell membrane from the cytoplasmic side (Schwartz *et al.*, 1991). It shows that the possibility of circumventing the midgut microvillus barrier and expression of different forms of Bt toxins within the cells exist.

The results of most of the toxin binding studies may not be relevant or pertinent since Bt endotoxin fusion mechanism comes from infection of insect cells (mostly derived from embryonic tissue) in insect cell culture, where the pH is near neutral and the temperature is constant at 27°C. Then the pertinent question is how δ-endotoxin fuses with BBM of columnar epithelial "gut" cells in the highly alkaline midgut environment. It is a well known fact that individual cell lines differ in their ability to synthesise secreted versus non-secreted proteins due to the different machanisms that each cell line possessed (Hink *et al.*, 1991, Wickham *et al.*, 1992). Further, it is commonly argued that proteins including activated δ-endotoxins are broken down to their constitutional amino acids in the acidic mono-gastric intestinal tract of human beings, horse, and poly-gastric animals. But *Clostridium botulinum* toxin is an exception to the tenet where large proteins (over 100 kD) resist many proteases. Hence studies should be performed to determine whether activated Bt toxins resist proteases in the human gut. Generally the toxicity of cry toxins appears to be correlated with receptor number, and the receptor affinity, i.e. the higher the receptor concentration the greater the toxicity. However, a recent study by Wolfersbarger (1990) shows that there is no correlation between receptor affinity and toxicity in gypsy moth *L. dispar*, therefore, a straight correlation between toxicity and toxin binding according to either receptor affinity or receptor concentration may not always be a generally applicable concept.

In view of the above facts it looks that comprehensive examination of mammalian tissues is a must to show empirically that activated Bt toxins will not damage mammalian cells.

Safety of transgenic Bt

There are number of important considerations which must be taken into account before deciding whether to release a Bt transformed crop into the environment. These include (i) assessment of the risk of the gene flow into native habitats and wild relatives of the crop (ii) whether non-target species are at risk from the toxin. Even though there is no evidence that native Bt δ-endotoxin is hazardous, activated delta endotoxin as expressed by Bt plants should be carefully tested as a new agent because of numerous reasons.

The transgenic Bt proteins are produced in high levels throughout the green tissue of transgenic plants and depending on the promoter, also in pollen, kernels, roots, and other non-green parts of the plants, unlike that of typical Bt commercial formulations as pesticidal sprays covering foliage in a variable way. Moreover, transgenic Bt proteins expressed during most of the growing period of the plants to target pests may also effect non-target insects which will in turn influence the natural enemies, unlike Bt commercial preparations where residues will decay rapidly under field condition. Further, in insect tolerant Bt transgenic plants, solubility and proteolytic processing are bypassed because only the soluble toxin core of the cry proteins is expressed in modified and truncated forms that differ from the crystalline Bt proteins present in commercial preparations.

On the whole no adverse effects on selected beneficial organisms are reported (Dogan *et al.*, 1996; Pilcher *et al.*, 1997; Shuler *et al.*, 1999). However, recently Hilbeck *et al.* (1998) have observed 62 percent immature mortality of the common predator *C. carnea* when it consumed its prey, the transgenic corn-fed European corn borer. Very recently, Losey *et al.* (1999) have shown only 56 percent survival of monarch butterfly, *Danaus plexippus* (Linaeus) when it was fed with pollen collected from Bt corn, thereby showing the Bt corn plant expressing the Bt toxin in pollen.

This is no surprise for monarch larvae, being susceptible to transgenic Bt since it is also a lepidopteran and Bt is being considered as lepidoptericide. However, the expression of Bt toxin in pollen, and the susceptibility of non-targeted lepidopterans which already face a loss of habitat, is having profound implications for the conservation of monarch butterflies.

The primary strategy for delaying insect resistance to transgenic Bt plants is to provide refuges of host plants that do not produce Bt toxins. But there is a recent finding of Liu *et al.* (1999) which states that a resistant strain of cotton pink bollworm larvae, *P. gossypiella* on Bt cotton takes longer time (5.7 days) to develop than susceptible larvae on non-Bt cotton. This means they are less likely to interbreed, which may aid the spread of resistance through non-random mating of the population. Hence this developmental asynchrony must be considered in efforts to sustain the refuge strategy in future.

Fungal pathogens

More than 700 species of fungi, mostly Deuteromycetes from about 90 genera, are pathogenic to insects (Charnley, 1989). Some of the genera that have been most intensively investigated for mycoinsecticides are *Beauveria, Metarhizium, Verticillium, Hirsutella, Poecoelomyces, Nomuraea, Aschersonia, Erynia*, etc. The first two genera have been identified from 700 and 300 species of insects, respectively and also they have been used on a large scale over a number of years in other countries. In India, there is a large body of literature available on the occurrence, cultivation and insect pathogenesis of *Metarhizium anisopliae* (Gopalakrishnan and Narayanan, 1988a; Sundara Babu, 1980), *Beauveria bassiana* (Gopalakrishnan and Narayanan, 1988a; Jayaramaiah and Veeresh, 1983) and *Verticillium lecanii* (Easwaramoorthy and Jayaraj, 1985) and *Nomuraea rileyi* (Gopalakrishnan and Narayanan, 1988a; Gopalakrishnan and Narayanan, 1988b).

Among the various entomopathogenic fungi, the white muscardine fungus, *B. bassiana* and green muscardine fungus, *M. anisopliae* have been the subject of intensive study in recent years. However, very limited information is available regarding the molecular biology and genetic manipulations. Though entomopathogenic fungi have many potential advantages such as safety to both user and non-target organisms, easy integration within biological control programme with naturally occurring materials and prolonged pest control from a single application are certain technical constraints like the slow speed of kill. The potential for rational strain improvement is limited due to ignorance of the molecular and biochemical basis of pathogenicity and the absence of sexual cycle in fungi. However, recent demonstrations of transformations in a number of filamentous fungi (Fincham, 1989) have indicated that molecular cloning techniques could be used to investigate the pathogenicity determinants of entomopathogenic fungi, isolate genes coding for such specific pathogenicity determinants and produce organisms with enhanced virulence.

At present *M. anisopliae* seems to be the most appropriate fungus for use in pest control. There are two putative pathogenicity/virulence determinants, viz. endoprotease "prI" (a chymoelastase protease enzyme predominantly produced during infection, which assists in penetration of this fungus by softening the cuticle) and dextruxins "dtx" (an insecticidal compound) so as to produce organism with enhanced virulence. Recently, a group at Bath University have demonstrated the DNA mediated transformation of *M. anisopliae* to benomyl resistance using cosmid p50 which harbours beta-tubulin gene from *Neurospora crassa* and "benA3" gene from *Aspergillus nudilans*, respectively (Bernier *et al.*, 1989; Staples *et al.*, 1988). The transformants were mitotically stable when sub-cultured on non-selective agar and retained the ability to infect and kill the larvae of *M. sexta*, which will pave the way for concurrent use of this fungus with fungicide on the crop when needed. The recent development of species specific DNA probes (Hegedus and Khachatourians, 1993), the use of RFLP (restriction fragment length polymorphism) (Pipe *et al.*, 1989) and the use of monoclonal antibodies (Ke *et al.*, 1990) provide a practical approach to the detection of *B. bassiana* and *M. anisopliae* isolates. So the future genetic manipulation and protoplast fusion technique will offer much scope to have an isolate or strain with increased virulence. However, the detailed account of entomopathogenic fungi in IPM system can be seen in Chapter 7 of this volume.

Protozoan pathogens

Entomopathogenic protozoans have been recognised as important organisms in the natural regulation of the population of certain insects. Although entomopathogenic protozoan species occur in all the major sub-groups of protozoa the majority occurs in the order Microsporidia. Protozoans have been little considered as applied microbial control agents because entomophilic species in general cause chronic or debilitative infections in a narrow range of hosts except *Vairimorpha* sp. Most importantly they can not compete with more virulent pathogens such as bacteria and fungi.

The microsporidia are a large and diverse group of parasitic protozoan whose separation has been difficult taxonomically. Usually life cycle and morphological characteristics of the stages have been used in grouping of families, but identification of species remains a constraint even at the ultra-structure level. In India, no work has been carried out on the molecular/genetic aspects of entomophilic protozoans except some preliminary relationship studies between host and pathogens are known (Narayanan, 1985a, 1987, 1988a; Narayanan and Jayaraj, 1979; Rabindra, 1981).

The effective utilization of protozoan pathogens lies in the basic research at molecular level characterization. This can be achieved by using pulse field electrophoresis to resolve the chromosomal DNA molecules (Munderloh *et al.*, 1990) in *Nosema*. SDS-PAGE electrophoresis technique can be used for determining intra-generic relativeness among *Vairimorpha* isolates (Moore and Brooks, 1993) for proper identification of various microsporidians. Thus, instead of relying on the spore morphology and utilization of insect tissue culture for their mass production, genetic manipulation could help in effective utilization of these organisms as long-term microbial control agents.

Nematodes

Although nematodes are more evident as injurious parasites of many animals and plants, and those species which limit the population of agriculturally important insect pests, they do so by causing reduced fecundity, delayed development, behavioural inhibition or host death. Among these, Steinernematids and Heterorhabditids nematodes are effectively used for the control of several insect pests (Narayanan and Gopalakrishnan, 1987; Poinar, 1981) and are formidable biological control agents. However, environmental factors, especially high temperature may reduce their efficacy. To overcome this hindrance for successful biological control Shapiro *et al.* (1997) demonstrated the potential of using conventional approach of hybridization to genetically improve entomopathogenic nematodes. A trait for heat tolerance was transferred from 185 strains, which were heat tolerant to the HP88, a commercial strain of *H. bacteriophora*. The transfer was accomplished by allowing them to make the hybrid. Nature of the progeny was confirmed using a marker mutant of the HP88 strain and by back crossing the entomopathogenic nematodes. The nematode acts as a syringe to inoculate susceptible insect larvae with the bacterium, which then multiplies and kills the insect host.

The nematode feeds on the bacteria and multiplies in the insect carcass and nematode progeny carrying bacterial inoculate infects additional insects (Poinar, 1981). Conventionally, these nematodes are being mass produced using *in vivo* technique (Dutky *et al.*, 1964; Narayanan, 1988b) which poses some problem because of the existence of phase variation executed by the *Xenorhabdus* bacterium which is symbiotically associated with these nematodes.

Xenorhabdus nematophilus and *Photorhabdus luminescence* are the respective bacteria key to the culture of Steinernematids and Heterorhabditids. Phase variation is common in *Xenorhabdus* sp. The primary phase, which is the phase naturally occurring in the infective stage nematode provides better growth condition for infective nematodes than the secondary phase although either phase may be carried within the infective stage nematode. Infectives formed in the presence of both phases contain only the primary phase (Aukhurst, 1986). The primary phase symbionts are of major importance in the nematode bacterial association. Either phase can convert a variety of media into suitable nutrients for the nematodes so that they can be mass-produced *in vitro*. However, the primary phase is regularised for the maximisation of mass production that is necessary for commercial scale. Only the primary form is isolated from the infective juvenile stage, but this form tends to be unstable and the production of the secondary form is greatly enhanced (Woodring and Kaya, 1988). In case of entomophilic nematodes the need to identify the gene(s) responsible for retention of primary form of mutualistic bacterial pathogens like *X. nematophilus* and *X. luminescence* associated with certain Steinernematids and Heterorhabditids will in future help in developing the cheap biopesticides on a large scale based on these organisms.

The poor storage stability of these nematodes, particularly at higher temperatures, is one factor that has inhibited wider use of entomopathogenic nematodes. Attempts have been made to improve high-temperature stability by isolating heat tolerant strains (Glaser *et al.*, 1996) and classical hybridization studies (Shapiro *et al.*, 1997). However, it so happens that apart from enhancing the beneficial traits, the selective breeding increases the other traits like increased penetration and high reproductive capacity but reduced the nematode fitness by decreasing the storage stability (Gaugler *et al.*, 1990). Recently, Hashimi *et al.* (1998) transformed *H. bactereophora* nematode with heat shock protein hsp 70a gene, (from the free living nematode, *Caenorhabditis elegans*) through recombinant DNA technique and found that transgenic nematodes were 18 times more tolerant of heat shock than wild type nematodes. According to Wilson *et al.* (1999) the genetically engineered *H. bacteriophora* with hsp 70a gene, had improved heat tolerant character without affecting other important characteristics like infectivity, reproductive capacity, etc. and the transgenic nematode was found to be safe to mice when injected intraperitoneally.

Recently, Bowen *et al.* (1998) have characterised four insecticidal toxins from the bacterium *Photorhabdus* (pht) encoded by toxin complex loci *tca, tcb, tcc* and *tcd.* They also showed that *P. luminescens* toxins like *tca* and *tcd* are as potent as the δ-endotoxins of Bt when administered orally to *M. sexta.* They further suggested that in future alterations or co-deployment of pht and Bt toxins would prolong the effective life of biological insecticides by delaying the evolution of resistance to either component alone.

Actinomycetes

A novel class of compounds called avermectins (macrocyclic lactones) isolated from fermentation of the soil actinomycetes, *Streptomycetes avermitilis* from Japan have shown nematicidal, acaricidal and insecticidal activity (Lasota and Bybas, 1991). These have been described in Chapter 2 in detail.

Destruction of symbionts

In contrast to all other earlier strategies, which involved the use of application of various microbes for the control of insect pests, the removal and destruction of endosymbionts may significantly contribute to control certain insect pests. The possibility of insect control especially the dipteran pests, which invariably harbour the microbes in their gut seems plausible by way of destroying their symbionts. It has been observed that removal of symbionts has detrimental effect on the host especially in the marked reduction on insect growth, difficulty in moulting, deranged metamorphosis, reduction or loss in reproductive capacity and death in severe cases (Chinnarajan *et al.*, 1972; Tanada and Kaya, 1993). The recent approach is to transform the Bt toxin gene constitutively by using transposon vector to the *Pseudomonas* sp. of certain fruit flies.

Conclusions

It is obvious from above discussion that various biotechnological approaches have led to the development of more effective, faster acting baculoviruses, bacteria, fungi, protozoan and nematodes that could provide a novel agent for the pesticide arsenal including a unique mode of action for combating resistant or selected pests. They apparently have also influenced animal/human health by way of better management of arthropod vectors. Further, it is evident from the above examples that genetic manipulations of insect pathogens like bacteria, virus, fungus, etc. can be modelled for specific or broad range insect control with an aim to form a "Super organism" with virulence, pathogenicity, host range, persistence, and mass production.

Already success has been achieved in gene technology application in certain advanced countries like USA and UK using genetically improved recombinant baculovirus for the control of crop pests under field conditions (Bishop *et al.*, 1988; Cory *et.al.*, 1994; D'Amico *et al.*, 1999). Some work has been carried out using foreign baculoviruses as expression vector for the production of vaccines, pharmaceuticals, genetic markers, etc. for diagnosis purposes using BmNPV as expression vector for the production and *B. mori* larvae as biofactory (Nakhai *et al.*, 1991; Naik and Shaila, 1991; Reddy *et al.*, 1997). Some success of using indigenous baculoviruses like *Heliothis* (Narayanan, 1980a,b), *Spodoptera* (Ramakrishnan, 1992), *Chilo* spp. (Easwaramoorthy and Santhalakshmi, 1988) and *Oryctes rhinocerous* (Linnaeus) (Mohan, 1985) has been achieved in India in order to compensate for the high costs of chemical insecticides. Obviously, development of selective recombinant insecticides under insect cell environment should augment any integrated pest management programme by reducing the impact on non-target species including beneficial insects. Consequently, resurgence of primary pests and outbreak of secondary pests will be minimal.

Majority of work that has been carried out so far relates to the baculovirus of *A. californica* and only on one cultured cell line derived from *S. frugiperda* which is of heterologous origin. Further, AcMNPV, which is more often used in recombinant work is originally isolated from alfalfa looper *A. californica*, which is cross infective against more than 46 other insects belonging to different species, genera and families (Groner, 1986). The recent report on the susceptibility of *B. mori* cell lines developed from larval and pupal ovaries to alien broad spectrum AcMNPV warrants detailed studies on the susceptibility of *B. mori* larvae to AcMNPV especially in the states where sericulture is being commercially practiced (Anonymous, 1999). Although most of the development and application of insect cell-based expression systems is the result of intense research into the study of AcMNPV, recent efforts to develop expression systems based on a variety of other insect viruses have been initiated and show promising signs of success. One reason for using viruses other than AcMNPV, as genetic tools is to facilitate the fundamental studies of gene expression in a wider variety of host species. A second motivating factor is to develop novel expression systems that can be used on a commercial scale and is not restricted by existing licensing and patent issues. However, in India some preliminary work has been done towards molecular characterisation of certain indigenous baculoviruses (Anuradha and Ramakrishnan, 1993; Mathad *et al.*, 1991; Mohan and Gopinathan, 1991; Narayanan, 1991; Ramakrishnan, 1992; Vikas and Gopinathan, 1996; Satyanarayanan Sriram *et al.*, 1999). In fact, certain basic studies on the genome analysis are required to understand cross-infective systems (Jayaraj *et al.*, 1989; Narayanan, 1985b), and chance mutations, if any (Narayanan, 1994b). Hence, the need of the hour is "Back to the Basic" approach to initiate more studies on the various aspects of basic molecular studies on different indigenous entomopathogens and to initiate work on the genetic improvement of indigenous insect baculoviruses and bacterial pathogens. However, the ecology of the baculoviruses has been regarded as a black box (Coghlan, 1994). There is explosive increase in the availability of basic information on the genome of AcMNPV (Ayres *et al.*, 1994) and BmNPV (Gomi *et al.*, 1999) as well as the information that both AcMNPV and BmNPV systems provide high level expression of foreign genes. Because of their 90 per cent nucleotide sequence identified with well conserved genes (Kamita *et al.*, 1993 cited by Du *et al.*, 1999); it is worthwhile to probe into remaining 10 per cent which distinguish the species specificity of BmNPV with that of broad host spectrum activity of AcMNPV. This knowledge can be utilized in future manipulation of indigenous baculoviruses like HaNPV and SlNPV for their better commercial exploitation. Research is being pursued to find out the 1.5 per cent genome difference between humans and chimpanzee. This is being done to distinguish between two legged man with less body hair, well developed sound box and highly susceptible to AIDS disease and four legged chimpanzee full of body hair, with no larynx and mostly free from or less susceptible to AIDS disease (Gibbons, 1998).

A promising technique towards the genetic improvement of indigenous baculoviruses is the use of very early promoter of baculoviruses as vectors (Jarvis and Summers, 1992; Jarvis *et al.*, 1990). For instance, neurotoxins especially from indigenous parasite, *B. hebetor* venom toxin has shown 400-fold biocidal activity against *Spodoptera* when compared to scorpion toxin. Other insect behaviour modifying gene products under stable indigenous insect cell environment especially in midgut cells (Flipsen *et al.*, 1993) could possibly increase the efficacy of viral pesticides by accelerating the onset of morbidity or paralysis. This blocks feeding and finally kills the target pest while retaining polyhedral production which is essential for the organismal transmission in the field populations.

Gene technology with its range, variety and sophistication comparable to nuclear and space research, is the password to contemporary research in "microbial control of insect pests". It is a "boon" or "panacea" if applied with integrity, openness and in a spirit of healthy competition (Rangarajan and Padmanaban, 1996; Narayanan, 1996c). The recent creation of a mammalian species in the laboratory, a sheep named "Dolly" (Wilmut *et al.*, 1997) and cows "Molly" and "Polly" has evoked widespread interest and concern in all sections of people. Unscrupulous and uncontrolled research aimed at rapid gains will undoubtedly make it a "ban" or "peril" which we should not allow to happen.

Acknowledgement

The author is grateful to Dr S.P. Singh, Project Director, Project Directorate of Biological Control, Bangalore for providing the facilities.

References

Abraham, E.G., Nagaraju, J., Salunke, D., Gupta, H.M., and Datta, R.K. (1995) Purification and partial characterization of an induced antibacterial protein in the silkworm, *Bombyx mori*. *J. Invertebr. Pathol.*, **65**, 17–24.

Ahrens, C.H., and Rohrmann, G.F. (1996) The DNA polymerase and helicase genes of a baculovirus of *Orgyia pseudotsugata*. *J. Gen. Virol.* **77**, 825–837.

Anandakumar, P., and Sharma, R.P. (1994) Genetic engineering of insect-resistant crop plants with *Bacillus thuringiensis* crystal protein genes *J. Plant Biochem. Biotechnol.*, **3**, 3–8.

Anandakumar, P., Sharma, R.P., and Malik, V.S. (1996) The insecticidal proteins of *Bacillus thuringiensis*. *Adv. Appl. Microbiol.* **42**, 1–44.
Anandakumar, P., Mandaokar, A., Sreenivasu, K., Chakrabarti, S.K., Sharma, S.R., Bisaria, S. *et al.* (1998) Insect resistant transgenic brinjal plants. *Mol. Breed.*, **4**, 33–37.
Anonymous (1999). Development of lepidopteran cell lines for baculovirus studies. National Institute of Virology, Pune. 8 pp.
Anuradha, S., and Ramakrishnan, N. (1993). Genome analysis of NPV of *Amsacta albistriga* (Walker). *Proc. Int. Conf. Biotechnol. in Agriculture and Forestry*, Indian Agricultural Research Institute, New Delhi.
Argaud, O., Croizier, L., Lopez-Ferber, M., and Croizier, G. (1998). Two key mutations in the host-range specificity domain of the *p143* gene of *Autographa californica* nucelopolyhedro virus are required to kill *Bombyx mori* larvae. *J. Gen. Virol.*, **79**, 931–935.
Aronson, A.I. (1986) *Bacillus thuringiensis* and related insect pathogens. *Microbiol. Rev.*, **50**, 1–24.
Askew, R.R. (1971) *Parasitic Insects*. Elsevier, New York, 316 pp.
Atkinson, A.E., Early, F.G.P., Beadle, D.J., and King, L.A. (1990) Expression and characterization of the chick nicotinic acetylcholine receptor and sub-unit in insect cells using a baculovirus vector. *Eur. J. Biochem.* **192**, 451–458.
Aukhurst, R.J. (1986). Recent advances in *Xenorhabdus* research. In R.A. Samson, J.M. Vlak and D. Peter (eds.), *Fundamentals and Applied Aspects of Invertebrate Pathology*, Foundation 4th Int. Colloq. Invertebr. Pathol., Washington DC, pp. 304–307.
Ayres, M.D., Howard, S.C., Kuzio, J., Lopez-Ferber, M., and Possee, R.D. (1994). The complete DNA sequence of *Autographa californica* nuclear polyhedrosis virus. *Virology*, **202**, 586–605.
Barrett, J.W., Brownwright, A.J., Primavera, M.J., and Palli, S.R. (1998). Studies on the nucleopolyhedrosis infection process in insects by using the green fluorescence proteinase reporter. *J. Virol.*, **72**, 3377–3382.
Barret, J.W., Krell, P.J., and Arif, B.M. (1995). Characterization, sequencing and phylogeny of the ecdysteroid UDP-glycosyl transferase gene from two distinct nuclear polyhedrosis viruses isolated from *Choristoneura fumiferana*. *J. Gen. Virol.*, **76**, 2447–2456.
Barton, K.A., Whiteley, H.R., and Ning-Sunh, Yang (1987). *Bacillus thuringiensis* δ-endotoxin expressed in transgenic *Nicotiana tabacum* provides resistance to lepidoptera insects. *Plant Physiol.*, **85**, 1103–1109.
Beames, B., and Summers, M.D. (1989) Location and nucleotide sequence of the 25K protein missing from baculovirus few polyhedra (FP) mutants. *Virology*, **168**, 344–353.
Bernier, L., Cooper, R.M., Charnley, A.K., and Clarkson, J.M. (1989) Transformation of entomopathogenic fungus *Metarhizium anisopliae* to benomyl resistance. *FEMS Microbiol. Lett.*, **60**, 161–166.
Bezdicek, D.F., Quinn, M.A., and Kahn, M. (1991) Genetically engineering *Bacillus thuringiensis* genes into Rhizobia to control nodule feeding insects. USDA CRIS Report No. 9131200.
Bilimora, S.L., Lin, H.S., and Reinihp, A.J. (1986) Gene expression of baculovirus in semi permissive insect cell lines. In R.A. Samson, J.M. Vlak and D. Peter (eds.), *Fundamentals and Applied Aspects of Invertebrate Pathology*, Foundation 4th Int. Colloq. Invertebr. Pathol., Washington DC, pp. 51–54.
Birnbaum, M.J., Clem, R.J., and Miller, L.K. (1994) An apoptosis-inhibiting gene from a Nuclear Polyhedrosis Virus encoding a polypeptide with Cys/His sequence motifs. *J. Virol.*, **68**, 2521–2528.
Bischoff, D.S., and Slavicek, J.M. (1997) Molecular analysis of an enhancing gene in the *Lymantria dispar* nulcear polyhedrosis virus. *J. Virol.*, **71**, 8133–8140.
Bishop, D.H.L. (1994) Biopesticides. *Curr. Opin. Biotech.*, **5**, 307–311.
Bishop, D.H.L., and Possee, R.D. (1990) Baculovirus expression vectors. *Adv. Gene Technol.*, **10**, 55–72.
Bishop, D.H.L., Entwistle, P.F., Cameron, I.R., Allen C.J., and Possee, R.D. (1988). Field trials of genetically-engineered baculovirus insecticides. In "M. Sussman, C.H.Colins, F.A. Skinner and D.E. Stewart-Tull (eds.), *The Release of Genetically Engineered Micro-Organisms*, Academic Press, New York, pp. 143–179.
Blissard, G.W., and Rohrmann, G.F. (1990) Baculovirus diversity and molecular biology. *Ann. Rev. Entomol.*, **35**, 127–155.
Boman, H.G., and Hultmark, D. (1987) Cell-free immunity in insects. *Ann. Rev. Microbiol.*, **41**, 103–126.
Bonfanti, P., Colombo, A., Heintzelman, M.B., Mooseker, M., and Camatini, M. (1992) The molecular architecture of an insect midgut brush border cytoskeleton. *Eur. J. Cell Biol.*, **57**, 298–307.
Bonning, B.C., and Hammock. B.D. (1994) Insect control by use of recombinant baculoviruses expressing juvenile hormone esterase. *Am. Chem. Soc. Symp. Series.*, **551**, 368–383.
Bonning, B.C., and Hammock. B.D. (1996) Development of recombinant baculoviruses for insect control. *Ann. Rev. Entomol.*, **41**, 191–210.
Bora, R.S., Murty, M.G., Shenbagarathai, R., and Sekar, V. (1994) Introduction of a Lepidoptera-specific insecticidal crystal protein gene of *Bacillus thuringiensis* subsp. *kurstaki* by conjugal transfer to a

Bacillus megaterium strain that persists in the cotton phyllosphere. *Appl. Environ. Microbiol.*, **60**, 214–222.

Bowen, D., Rocheleau, T.A., Blackburn, M., Andreev, O., Golubeva, E., Bhartia, R. *et al.* (1998). Insecticidal toxins from the bacterium *Photorhabdus luminescens. Science*, **280**, 2129–2132.

Boyce, F., and Bucher, N.L.R. (1996) Baculovirus-mediated gene transfer into mammalian cells. *Proc. Natl. Acad. Sci.* U.S.A., **93**, 2348– 2352.

Bruce, E.T., Francies, R.G., and Naomi, F. (1996) Resistance to *Bacillus thuringiensis* in *Plutella xylostella.* In T.M. Brown (ed.), *Molecular Genetics and Evolution of Pesticide Resistance*, American Chemical Society, Washington DC, pp. 130–140.

Bulla, L.A. Jr., Costilow, R.N., and Sharpe, E.S. (1978) Biology of *Bacillus popilliae. Adv. Appl. Microbiol.*, **23**, 1–18.

Carbonell, L.F., Hodge, M.K., Tomalski, M.D., and Miller, L K. (1988) Synthesis of a gene coding for an insect specific scorpion neurotoxin and attempts to express it using baculovirus vectors. *Gene*, **73**, 409–418.

Carstens, E.B., and Lu, A. (1990) Nucleotide sequence and transcriptional analyis of the Hind III P region of a temperature sensitive mutant of *Autographa californica* NPV. *J. Gen. Virol.*, **71**, 3035–3040.

Chakrabarti, S.K., Mandaokar, A.D., Anandakumar, P., and Sharma, R.P. (1998) Synergistic effect of Cry1Ac and Cry1F δ-endotoxins of *Bacillus thuringiensis* on cotton bollworm, *Helicoverpa armigera. Curr. Sci.*, **75**, 663–664.

Chao, Y.C., Chen, S.L., and Li, C.F. (1996) Pest control by fluorescence. *Nature*, **380**, 396–397.

Charnley, A.K. (1989). Mycoinsecticides: Present use and future prospects. *Progress and Prospects in Insect Control, BCPC Monograph*, **43**, 165–181.

Chejenovsky, N., and Gershburg, E. (1995) The wild type *Autographa californica* nuclear polyhedrosis virus induces apoptosis of *Spodoptera littoralis* cells. *Virology*, **209**, 519–525.

Chinnarajan, A.M., Jayaraj, S., and Narayanan, K. (1972) Destruction of endosymbionts with oxytetracycline and sulphanilamide in the gourd fruit fly *Dacus cucurbitae* Coq. (Trypetidae : Diptera). *Hindustan Antibiot. Bull.*, **15**, 16–22.

Choudhary, P.V. (1996) Virus-mediated approaches to insect pest control: Potential of genetically engineered baculovirus. In R.K. Upadhyay, K.G. Mukerji and R.L. Rajak (eds.), *IPM System in Agriculture, Vol 2, Biocontrol in Emerging Biotechnology*, Aditya Books Private Ltd., New Delhi, pp. 124–154.

Choudhary, P.V., Kamita, S., and Maeda, S. (1992) Expression of foreign genes in insect larvae using baculovirus vectors. In C. Richardson and N.J. Clifton (eds.), *Methods of Molecular Biology* : *Protocols for Baculovirus Expression*, Humana Press, Clifton, New Jersey, pp. 1–34.

Clem, R.J., and Miller, L.K. (1994) Induction and inhibition of apoptosis by insect viruses. *Apoptosis II* : *The Molecular Basis of Apoptosis in Disease*, Cold Spring Harbour Laboratory Press, pp. 89–110.

Clem, R.J., Fechheimer, M., and Miller, L.K. (1991) Prevention of apoptosis by a baculovirus gene during infection of insect cells. *Science*, **254**, 1388–1390.

Coghlan, A. (1994) Will the scorpion gene run wild? *New Scient.*, **142**, 14–15.

Coghlan, A. (1998) Friend turns to foe: A useful bacterium can develop a taste for blood. *New Scient.*, **158**, 7.

Cory, J.S., and Hails, R.S. (1997) The ecology of biosafety of baculoviruses. *Curr. Opin. Biol.*, **8**, 323–327.

Cory, J.S., Hirst, M.L., Williams, T., Halls, R.S., Goulson, D., Green, B.M. *et al.* (1994) Field trials of genetically improved baculovirus insecticide. *Nature*, **370**, 138–140.

Crickmore, N., Zeigler, D.R., Feitelson, J., Schnepf, E., Vanrie, J. Lambert, B. *et al.* (1996) Revision of the nomenclature for the *Bacillus thuringiensis* pesticidal crystal proteins. *Microbiol. Mol. Biol. Rev.*, **62**, 807–813.

Crook, N.E., Clem, R.J., and Miller, L.K. (1993) An apoptosis-inhibiting baculovirus gene with zinc finger-like motif. *J. Virol.*, **67**, 2168–2174.

Cunningham, J.C. (1995). Baculoviruses as microbial insecticides. In R. Reuveni (ed.), *Novel Approaches to Integrated Pest Management*, Lewis Publishers, Boca Raton, Florida, pp. 261–292.

Dai, X., Shi, X., Pang, Y., and Su, D. (1999) Prevention of baculovirus-induced apoptosis of BTI-Tn-5B1-4 (Hi5) cells by the p35 gene of *Trichoplusia ni* multicapsid nucleopolyhedrovirus. *J. Gen. Virol.*, **80**, 1841–1845.

D'Amico, V., Elkinton, J.S., Podgwaite, J.D., Slavicek, J.M., McManus, M.L., and Burand, J.P. (1999) A field release of genetically engineered gypsy moth (*Lymantria dispar* L) nuclear polyhedrosis virus (LdNPV). *J. Invertebr. Pathol.*, **73**, 260–268.

Davis, A.H. (1995). "Baculophage", a new tool for protein display. *Biotechnology*, **13**, 1046.

Denholm, I., and Rowland, M.W. (1992) Tactics for managing pesticide resistance in arthropods: Theory and practice *Ann. Rev. Entomol.*, **37**, 91–112.

Derksen and Granados, R.R. (1988) Alteration of a Lepidoptera peritrophic membrane by a baculovirus and enhancement of viral infectivity. *Virology*, **167**, 242–250.

Dimock, M.B., Beach, R.M., and Carlson, P.S. (1988) Endophytic bacteria for the delivery of crop protection agents. In D.W. Roberts and R.R. Granados (eds.), *Proc. Conf. Biotechnology, Biological Pesticides and Novel Plant-Pest Resistance for Insect Pest Management*, Boyce Thompson Institute, Ithaca, USA, pp. 88–92.

Dixon, B. (1994) Keeping an eye on *Bacillus thuringiensis. Biotechnology* ,**12**, 435.

Doerfler, W., and Bohm, P. (1986) Molecular biology of baculoviruses. *Curr. Topic Microbiol. Immunol.*, **131**, 51–68.

Dogan, E.B., Berry, R.E., Reed, G.L., and Rossignol, P.A. (1996) Biological parameters of convergent lady beetle (Coleoptera: Coccinellidae) feeding on aphids (Homoptera: Aphididae) on transgenic potato. *J. Econ. Entomol.*, **89**, 1105–1108.

Dougherty, E.M., Guthrie, K.P., and Shapiro, M. (1996). Optical brightners provide baculovirus activity enhancement and UV radiation protection. *Biol. Contr.*, **7**, 71–74.

Du, Q., Lehavi, D., Faktor, O., Qi, Y., and Chejanovsky, N. (1999) Isolation of an apoptosis suppressor gene of the *Spodoptera littoralis* nucleoplyhedrovirus. *J. Virol.*, **73**, 1278–1285.

Dutky, S.R., Thompson, J.V., and Cantwell, G.E. (1964) A technique for the mass propagation of the DD-136 nematode. *J. Insect Pathol.*, **6**, 417–422.

Easwaramoorthy, S., and Jayaraj, S. (1985) Use of *Verticillium lecanii* Zimm. in the microbial control of coffee green bug *Coccus viridis* (Green). In S. Jayaraj (ed.), *Microbial Control and Pest Management*, Tamil Nadu Agricultural University, Coimbatore, pp. 136–140.

Easwarmoorthy, S., and Santhalakshmi, G. (1988) Efficacy of granulosis virus in the control of shoot borer, *Chilo infuscatellus* Snellen. *J. Biol. Contr.*, **2**, 26–28.

Edson, K.M., Vinson, S.B., Stoltz, D.B., and Summers, M.D. (1981) Virus in a parasitoid wasp suppression of the cellular immune response in the parasitoids host. *Science*, **211**, 582–583.

Eldridge, R., Horodyski, F.M., Morton, D.B., O'Reilly, D.R., Truman, J.W., Riddiford, L.M. *et al.* (1991) Expression of eclosion hormone gene in insect cells using baculovirus vectors. *Insect Biochem.*, **21**, 341–351.

Engelhard, E.K., and Volkman, L.E. (1995) Developmental resistance in fourth instar *Trichoplusia ni* orally inoculated with *Autographa californica* M nuclear polyhedrosis virus. *Virology*, **209**, 384–389.

Engelhard, E.K., Kam-Morgan, L.N.W., Washburn, J.O., and Volkman, L.E. (1994) The insect tracheal system: A conduit for the systemic spread of *Autographa californica* M nuclear polyhedrosis virus. *Proc. Natl. Acad. Sci.*, USA, **91**, 3224–3227.

Estruch, J.J., Warren, G.W., Mullins, M.A., Nye, G.J., Craig, J.A., and Koziel, M.G. (1996) VIP 3A, a novel insecticidal protein with a wide spectrum of activities against lepidopteran insects. *Proc. Natl. Acad. Sci.*, **93**, 5389–5394.

FAO (1973) Use of insect viruses in integrated pest control. *Plant Prot. Bull.*, **5**, 142–143.

Federici B.A., and Stern, V.M. (1990) Replication and occlusion of a granuosis virus in larval and adult midgut epithelium of Western great leaf skeletonisor, *Harrisina brillians. J. Invertebr. Pathol.*, **56**, 401–414.

Feitelson, J.S., Payne, J., and Kim. L. (1992) *Bacillus thuringiensis*; Insects and beyond. *Biotechnology*, **10**, 171–175.

Fincham, J.R. S. (1989) Transformation in fungi. *Microbiol. Rev.*, **53**, 148–170.

Fischhoff, D.A., Bowdish, K.S., Perlak, F.J., Pamela, G.M., McCornvick, S.M., Niedermeyer, J.G. *et al.* (1987) Insect tolerant transgenic tomato plants. *Biotechnology*, **5**, 807–813.

Flipsen, J.T.M., Valent, J.W.M., Goldbach, R.W., and Vlak, J.M. (1993) Expression of polyhedrin and p^{10} in the midgut of AcMNPV-infected *Spodoptera exigua* larvae, an immunoelectron microscopic investigation. *J. Invertebr. Pathol.*, **61**, 17–23.

Francki, R.J.B., Fauquet, C.M., Knudson, D.L., and Brown, F. (1991) Classification and nomenclature of viruses. *Fifth Report of the Interrnational Committee on Taxonomy of Viruses, Arch. and Virol., Suppl.*2, Springer Verlag, New York, pp. 45.

Fujimoto, H., Itoh, K., Yamamoto, M., Kyozuka, J., and Shimamoto, K. (1993) Insect resistant rice generated by introduction of a modified delta endotoxin gene of *Bacillus thuringiensis. Biotechnology*, **11**, 1151–1155.

Funakoshi, M., and Aizawa, K. (1989) Antiviral substance in the silkworm gut juice against a nuclear polyhedrosis virus of the silkworm, *Bombyx mori. J. Invertebr. Pathol.*, **53**, 135–136.

Fuxa, J.R., Mitchell, F.L., and Richter, A.R. (1988) Resistance of *Spodoptera frugiperda* (Lep: Noctuidae) to a nuclear polyhedrosis virus in the field and laboratory. *Entomophaga*, **33**, 55–63.

Gaugler, R., Campbell, J.F., and McGuire, T. (1990) Fitness of a genetically improved entomopathogenic nematode. *J. Invertebr. Pathol.*, **56**, 106–116.

Gelernter, W.D. (1990) MVP Tm bioengineered, bioencapsulated product for the control of lepidopteran larvae. In *Abstracts of the Vth International Colloquium on Invertebrate Pathology and Microbial Control*, Adelaide, Australia, pp. 14.

Gibbons, A. (1998) Which of our genes make us human? *Science*, **281**, 1432–1434.

Glaser, I., Kozodoi, E., Hashmi, G., and Gaugler, R. (1996) Biological characteristics of the entomopathogenic nematode *Heterorhabditis bacteriophora* IS-5: A heat tolerant nematode from Israel. *Nematologica*, **42**, 481–492.

Gomi, S., Majima, K., and Maeda, S. (1999) Sequence analysis of the genome of *Bombyx mori* nucleopolyhedrovirus. *J. Gen. Virol.*, **80**, 1323–1337.

Gopalakrishnan, C., and Narayanan, K. (1988a). Occurrence of two entomopathogens *Metarhizium anisopliae* (Metchinikoff) Sorokin var. *minor* Tulloch and *Nomuraea rileyi* (Farlow) Samson on *Heliothis armigera* Hubner (Noctuiidae: Lepidoptera). *Curr. Sci.*, **57**, 867–868.

Gopalakrishnan, C., and Narayanan, K. (1988b) Occurrence of entomopathogenic fungi *Nomuraea rileyi* (Farlow) Samson on *Acontia graellsi* F. (Noctuiidae: Lepidoptera) and *Beauveria bassiana* (Balsamo) Vuill. on *Myllocerous subfaciatus* G. (Curculionidae: Coleoptera). *Indian J. Biocontrol*, **2**, 58–59.

Gopalakrishnan, B., Muthukrishnan, S., and Kramer, K.J. (1995) Baculovirus-mediated expression of a *Manduca sexta* chitinase gene; properties of the recombinant protein. *Insect Biochem. Molec. Biol.*, **25**, 255–265.

Govindarajan, R., Jayaraj, S., and Narayanan, K. (1975) Observation on the nature of resistance in *Spodoptera litura* F. (Noctuidae: Lepidoptera) to infection by *Bacillus thuringiensis* Berliner. *Indian J. expt. Biol.*, **13**, 548–550.

Granados, R.R., and Federici, B.A. (eds.) (1986) *Biology of Baculoviruses Vol. I. Biological Properties and Molecular Biology and Vol. II. Practical Application for Insect Control.* CRC Press, Boca Raton, Florida, USA.

Granados, R.R., and Hashimoto, Y. (1989) Infectivity of baculovirus in cultured cells. In J. Mitsuhashi (ed.), *Invertebrate Cell System and Applications, Vol.2*, CRC Press Inc., Boca Raton, Florida, USA, pp. 3–13.

Granados, R.R., Li, G., Derkson, A.C.G., and MaKenna, K.A. (1994) A new insect cell line from *Trichoplusia ni* (BTI-Tn-5B1-4) sucesptible to *Trichoplusia ni* single enveloped nuclear polyhedrosis virus. *J. Invertebr. Pathol.*, **64**, 260–266.

Groner, A. (1986) Specificity and safety of baculoviruses. In R.R. Granados and B.A. Federici (eds.), *Biology of Baculoviruses, Vol. I, Biological Properties and Molecular Biology*, CRC Press Inc., Boca Raton, Florida, USA, pp. 177–202.

Hammock, B.D., Bonning, B.C., Possee, R.D., Hanzlik, T.N., and Maeda, S. (1990) Expression and effects of the juvenile hormone esterase in a baculovirus vector. *Nature*, **344**, 458–461.

Handermann, M., Schnitzler, P., Rosen-Wolff. A., Raab, K., Sonntag, K.C., and Darai, G. (1992) Identification and mapping of origins of DNA replication within the DNA sequences of the genome of insect iridescent virus type 6. *Virus Genes*, **6**, 19.

Hashmi, S., Hashmi, G., Glaser, I., and Gaugler, G. (1998) Thermal response of *Heterorhabditis bacteriophora* transformed with *Caenorhabditis elegans* hsp70 encoding gene. *J. Exp. Zool.*, **281**, 164–170.

Hasnain, S.E.and Nakhai, B. (1990) Expression of the gene encoding firefly luciferase in insect cells using a baculovirus vector. *Gene*, **91**, 135–138.

Hawtin, R.E., Arnold, K., Ayres, M.D., Zanotto, P.M.de A., Howard, S.C., Gooday, G.W. *et al.* (1995) Identification and preliminary characterization of a chitinase gene in *Autographa californica* nuclear polyhedrosis virus genome. *Virology*, **212**, 673–685.

Hegedus, D.D., and Khachatourians, G.G. (1993) Construction of cloned DNA probes for the specific detection of the entomopathogenic fungus *Beauveria bassiana* in grasshoppers. *J. Invertebr. Pathol.*, **62**, 233–240.

Heimpel, A.M., and Angus, T.A. (1959). The site of action of crystalliferous bacteria in lepidoptera larvae. *J. Insect Pathol.*, **1**, 151–170.

Heinz, K.M., McCulchen, B.F., Herrmann, R., Parrella, M.P., and Hammock, B.D. (1995) Direct effect of recombinant nuclear polyhedrosis viruses on selected non-target organisms. *J. Econ. Entomol.*, **88**, 259–264.

Herrnstadt, C., Soares, G.G., Wilcox, E.R., and Edwards, D.L. (1986) A new strain of *Bacillus thuringiensis* with activity against coleopteran insects. *Biotechnology*, **4**, 305–308.

Hillbeck, A., Baumgartner, M., Fried, P.M., and Bigler, F. (1998) Effects of transgenic *Bacillus thuringiensis* corn-fed prey on mortality and development time of immature *Chrysoperla carnea* (Neuroptera: Chrysopidae). *Environ. Entomol.*, **27**, 480–487.

Hink, W.F. (1991) A serum free medium for the culture of insect cells and production of recombinant proteins. *In Vitro Dev. Biol.*, **27**, 397–401.

Hink, W.F., Thompson, D.R., Davidson, D.J., Mayer, A.I., and Castellino, F.J. (1991). Expression of three recombinant proteins using baculovirus vectors in 23 insect cell lines. *Biotech. Prog.*, **7**, 9–14.

Hofmann, C., Sandig, V., Jennings, G., Rudolph, M., Schlag, P. and Strauss, M. (1995) Efficient gene transfer into human hepatocytes by baculovirus vectors. *Proc. Natl. Acad .Sci.*, USA, **92**, 10099–10103.

Hofmann, C., Vanderbruggen, H., Hofte, H., Van Rie, J., Jansens, S., and Van Mellaert, H. (1988) Specificity of *Bacillus thuringiensis* δ-endotoxins is correlated with the presence of high-affinity binding sites in the brush border membrane of target insect midguts. *Proc. Natl. Acad. Sci.* USA, **85**, 7844–7848.

Hofte, H., and Whiteley, H.R. (1989) Insecticidal crystal proteins of *Bacillus thuringiensis. Microbiol. Rev.*, **53**, 242–255.

Holzman, D. (1995) Cultured insect cells are protein production resource. *Nature*, **61**, 567.

Hong, Tao, Braunagel, S.C., and Summers, M.D. (1994). Transcription, translation and cellular localization of PDV-E66: A structural protein of the PDV envelope of *Autographa californica* nuclear polyhedrosis virus. *Virology*, **204**, 210–222.

Hori, Y., and Watanabe, H. (1980). Recent advances in sericulture. *Ann. Rev. Entomol.*, **25**, 49–71.

Hughes, P.R., Wood, H.A., Brun, J.P., Simpson, S.F., Duggan, A.J., and Dybas, J.A. (1997) Enhanced bioactivity of recombinant baculoviruses expressing insect specific spider toxins in lepidopteran crop pests. *J. Invertebr. Pathol.*, **69**, 112–118.

Jarvis, D.L., and Summers, M.D. (1992) Baculovirus expression vectors. In R.E. Isaacson (ed.), *Recombinant DNA Vaccines: Rationale and Strategy*. Marcel and Dekker Inc., New York. pp. 265–291.

Jarvis, D.L., Fleming, J.A.G.N., Kovacs, G.R., Summers, M.D., and Gaurino, L.A. (1990) Use of early baculovirus promoters for continuous expression and efficient processing of foreign gene products in stably transformed lepidopteran cells. *Biotechnology*, **8**, 950–955.

Jayaraj, S., Rabindra, R.J., and Narayanan, K. (1989) Development and use of microbial agents for the control of *Heliothis* spp. (Lepidoptera; Noctuiidae) in India. *Proc. Workshop on Biological Control of Heliothis: Increasing the Effectiveness of Natural Enemies*, New Delhi, India, pp. 483–504.

Jayaramiah, M., and Veeresh, G.K. (1983) Studies on the symptom of infection caused by the new silkworm white muscardine fungus, *Beauveria bassiana* (SNcc.) Petch to different stages of white grub, *Holotrichia serrata* Fab. (Coleoptera: Carabidae). *J. Soil Biol. Ecol*, **3**, 7–12.

Jehle, J.A., and Vlak, J.M. (1996) Transposable elements in baculoviruses: Their possible role in horizontal gene transfer. In E.R. Schmidt and T. Hankeln (eds.), *Transgenic Organisms and Ecosafety*, Springer-Verlag, Berlin, pp.31–41.

Jha, P.K., Nakhai, B., Sridhar, P., Talwar, G.P., and Hasnain, S.E. (1990) Firefly luciferase, synthesized to very high levels in caterpillars infected with a recombinant baculovirus can also be used as an efficient reporter enzyme *in vivo. FEBS Letters*, **274**, 23–26.

Johnson, D.E. (1989) Speicificity of cultured insect tissue cells for the bioassay of entomocidal protein of *Bacillus thuringiensis.* In J. Mitsuhashi (ed.), *Invertebrate Cell System Applications*, CRC Press Inc., Boca Raton, Florida, pp. 85–90.

Kaya, G.P., Fly, G.C., and Boman, H.G. (1987) Insect immunity, induction of cecropin and attacin like antibacterial factors in the haemolymph of *Glossina morsitans. Insect Biochem.*, **17**, 309–315.

Kalfon, A.R., and de Barjac, H. (1985) Screening of the insecticidal activity of *B. thuringiensis* strains against the Egyptian leaf worm, *Spodoptera littoralis. Entomophaga*, **30**, 176–185.

Kalman, S., Kiehne, K.L., Cooper, N., Reynoso, M.S., and Yamamoto, T. (1995) Enhanced production of insecticidal proteins in *Bacillus thuringiensis* strains carrying an additional crystal protein gene in their chromosomes. *Appl. Environ. Microbiol.*, **61**, 3063–3068.

Kamita, S.G., Majima, K., and Maeda, S. (1993) Identification and characterization of the p 35 gene of *Bombyx mori* nuclear polyhedrosis virus that prevents virus induced apoptosis. *J. Virol.*, **67**, 455–463.

Katsuma, S., Noguchi, Y., Zhou, C.L.E., Koyayashi, M., and Maeda, S. (1999) Characterisation of the 25K FP gene of the baculovirus *Bombyx mori* nucleopolyhedrovirus : Implications for post-mortem host degradation. *J. Gen. Virol.*, **80**, 783–791.

Ke, Z., Xie, W., Wang, X., Long, Q., and Pu, Z (1990) A monoclonal antibody to *Nosema bombysis* and its use for the identification of microsporidian spores. *J. Invertebr. Pathol.*, **56**, 395–400.

Keeley, L.L., and Hayes, T.K. (1987) Speculations on biotechnology applications for insect neuro endocrine research. *Insect Biochem.*, **17**, 639–651.

Kim, M.K., Sisson, G., and Stoltz, D. (1996) Ichnovirus infection of an established gypsy moth cell line. *J. Gen. Virol.*, **77**, 2321–2328.

King, G.A., Daugulis, A.J., Faulkner, P., Bayly, D., and Goosen, M.F.A. (1992) Recombinant β -galactosidase production in serum free medium by insect cells in a 14L air-lift fermentor. *Biotechnol. Prog.*, **8**, 567–571.

Kirkpatrick, B.A., Washburn, J.O., and Volkman, L.E. (1998) AcMNPV pathogenesis and developmental resistance in fifth instar *Heliothis virescens. J. Invertebr. Pathol.*, **72**, 63–72.

Kirkpatrick, B.A., Washburn, J.O., Engelhard, E.K., and Volkman, L.E. (1994) Primary infection of insect tracheae by *Autographa californica* M nuclear polyhedrosis virus. *Virology*, **203**, 184–186.

Kirschbaum, J.B. (1985) Potential implications of genetic engineering and other biotechnologies to insect control. *Ann. Rev. Entomol.*, **30**, 51–70.
Knowles, B.H., and Ellar, D.J. (1987). Colloid-osmotic lysis is a general feature of the mechanism of action of *Bacillus thuringiensis* delta endotoxin with different specificity. *Biochim. Biophysic. Acta.*, **924**, 509–518.
Kolondy-Hirsch, D., and Van Beek, N. (1997) Selection of a morphological variant of *A. californica* NPV with increased virulence following serial passage in *Plutella xylostella. J. Invertebr. Pathol.*, **69**, 205–211.
Kompier, R., Tramper, J., and Vlak, J.M. (1988) A continuous process after the production of baculovirus using insect cell lines. *Biotechnol. Lett.*, **10**, 849–854.
Konda, A., and Maeda, S. (1991) Host range expansion by recombination of the baculoviruses *Bombyx mori* nuclear polyhedrosis virus and *Autographa californica* nuclear polyhedrosis virus. *J. Virol.*, **65**, 3625–3632.
Korth, K.L., and Lewis, C.S. (1993) Baculovirus expression of the maize mitochondrial protein URF13 confers insecticidal activity in cell cultures and larvae. *Proc. Natl. Acad. Sci.* USA, **90**, 3388–3392.
Koziel, M.G., Beland, G.I., Bowman, C., Carozzi, N.B., Crenshaw, R., Crossland, L. *et al.* (1993) Field performance of elite transgenic maize plants expressing an insecticidal protein derived from *Bacillus thuringiensis. Biotechnology*, **11**, 194–200.
Krapecho, K.J., Kral. R.M.Jr., Van Wagener, B.C., Eppler, K.G., and Morgan, T.K. (1995) Characterization and cloning of insecticidal peptides from the primitive weaving spider *Diguetia canities. Insect Biochem. Molec. Biol.*, **25**, 991–1000.
Kreig, A., Hugher, A.M., Langenbruch, G.A., and Schnetter, W. (1983) *Bacillus thuringiensis* var. *tenebrionis* a new pathotype effctive against larvae of Coleoptera. *Z. Ang. Ent.*, **96**, 500–508.
Kurstak, E. (1982) *Microbial and Viral Pesticides*. Marcell and Dekker Inc., New york, 720 pp.
Lasota, J.A., and Dybas, R.A. (1991) Avermectins, a novel class of compounds: implications for use in arthropod pest control. *Ann. Rev. Entomol.*, **36**, 91–118.
Lecadet, M.M., and de Barjac, H. (1981) *Bacillus thuringiensis* beta-exotoxin. In E.W. Davidson (ed.), *Pathogenesis of Invertebrate Microbial Diseases*, Allanheld Osmun & Co. Publishers, Totowa, New Jersey, pp. 231–295.
Lee, S.Y., Qu X., Chen, W., Poloumienko, A., Macafee, N., Morin, B. *et al.* (1997) Insecticidal activity of a recombinant baculovirus containing an antisense c-myc fragment. *J. Gen. Virol.*, **78**, 273–281.
Lereclus, D., Agassie, M., Gominet, M., and Chaufaux, J. (1995) Overproduction of encapsulated insecticide crystal proteins in a *Bacillus thuringiensis* spoOA mutant. *Biotechnology*, **13**, 67–71.
Lereclus, D., Vallade, M., Chaufaux, J., Arantes, O., and Rambaud, S (1992) Expansion of insecticidal host range of *Bacillus thuringiensis* by *in vivo* genetic recombination. *Biotechnology*, **10**, 418–421.
Liu, Yong-Biao, Tabashnik, B.E., Dennehy, T.J., Patin, A.L. and Bartlett, A.C. (1999) Development time and resistance to *B.t.* crops. *Nature*, **400**, 519.
Locke, M. (1997) Caterpillars have evolved lungs for haemocyte gas exchange. *J. Insect Physiol.*, **44**, 1–20.
Losey, J.E., Rayor, L.S., and Carter, M.E. (1999) Transgenic pollen harms monarch larvae. *Nature*, **399**, 214.
Lu, A., and Carstens, E.B. (1991) Nucleotide sequence of a gene essential for viral DNA replication in the baculovirus *Autographa californica* nuclear polyhedrosis virus. *Virology*, **181**, 336–347.
Lu, A., and Miller, L.K. (1995) Differential requirement for baculovirus late expression factor genes in two cell lines. *J. Virol.*, **69**, 6265–6272.
Lu, A., and Miller, L.K. (1996) Species-specific effects of the *hcf-1* gene on baculovirus virulence. *J. Virol.*, **70**, 5123–5130.
Luckow, V.A. (1991) Cloning and expression of heterologous genes in insect cells with baculovirus vectors. In A. Procop and R. Bajpai (eds.), *Recombinant DNA Technology and Application*, McGraw Hill, New York, pp. 97–153.
Luckow, V.A., and Summers, M.D. (1988) Trends in the development of baculovirus expression vectors. *Biotechnology*, **6**, 47–52.
Lupiani, B., Raina, A.K., and Huber, C. (1999) Development and use of a PCR assay for detection of the reproductive virus in wild populations of *Helicoverpa zea* (Lep.: Noctuidae). *J. Invertebr. Pathol.*, **73**, 107–112.
MacIntosh, S.C., Kishore, G.M., Perlak, F.J., Marrone, P.G., Stone, T.B., Sins, S.R. *et al.* (1990) Potentiation of *Bacillus thuringiensis* insecticidal activity by serine protease inhibitors. *J. Agric. Food Chem.*, **38**, 1145–1152.
Maeda, S. (1989a) Expression of foreign genes in insects using baculovirus vectors. *Ann. Rev. Entomol.*, **34**, 351–372.
Maeda, S. (1989b) Increased insecticidal effect by a recombinant baculovirus carrying a synthetic diuretic hormone gene. *Biochem. Biophys. Res. Commun.*, **165**, 1177–1183.
Maeda, S. (1995) Further development of recombinat baculovirus insecticides. *Curr. Opin. Biotechnol.*, **6**, 313–319.

Maeda, S., Kawai, T., Obinata, M., Chika, T., Horiuchi, T., Maekawa, K. *et al.* (1984) Characterization of human interferon ∞ produced by a gene transferred by a baculovirus vector in the silkworm, *Bombyx mori*. *Proc. Japan Acad.*, **60B**, 423–426.

Maeda, S., Volrath, S.L., Hanzlik, T.N., Harper, S.A., Majima, K., Maddox, D.N. *et al.* (1991) Insecticidal effects of an insect specific neurotoxin expressed by a recombinant baculovirus. *Virology*, **184**, 777–780.

Maiorella, B., Inflow, D., Stranger, A., and Harano, D. (1988) Large scale insect cell culture for recombinant protein production. *Biotechnology*, **6**, 1406–1410.

Mallapadidam (1992) Insecticidal crystal protein from *Bacillus thuringiensis* is highly toxic to *Heliothis armigera*. *J. Invertebr. Pathol.*, **59**, 109–111.

Maramorosch, K. (ed.) (1987) *Biotechnology in Invertebrate Pathology and Cell Culture*, Academic Press, San Diego.

Martens, J.W.M., Honee, G., Zuidema, D., Vankent, J.W.M., Visser, B., and Vlak, J.M. (1990) Insecticidal activity of a bacterial crystal protein expressed by a recombinant baculovirus in insect cells. *Appl. Environ. Microbiol.*, **56**, 2764–2770.

Martens, J.W.M., Knoester, M., Weijts, F., Groffen, S.J.A., Hu, Z., Bosch, D. *et al.* (1995) Characterization of baculovirus insecticides expressing tailored *Bacillus thuringiensis* Cry1A (b) crystal proteins. *J. Invertebr. Pathol.*, **66**, 249–257.

Martignoni, M.E., and Iwai, (1986) *A Catalog of Viral Diseases of Insects, Mites and Ticks*, 4th edition, USDA, Portland, PNN–195.

Mathad, S.B., Hinchigeri, S.B., and Savanurmath, C.J. (1991) Characterization of nuclear polyhedrosis virus of *Mythimna (Pseudaletia) separata*. *J. Biol. Contr.*, **5**, 23–27.

Mayo, M.A., and Pringle, C.R. (1997) Virus taxonomy-1997. *J. Gen. Virol.*, **79**, 649–657.

Mazzone, H.M. (1987) Control of invertebrate pests through the chitin pathway. In K. Maramorosch (ed.), *Biotechnology in Invertebrate Pathology and Cell Culture*, Academic Press, New York, pp. 439–450.

McBride, K.E., Svab, Z., Schaaf, D.J., Hogan, P.S., Stalker, D.M., and Maliga, P. (1995) Amplification of a chimeric *Bacillus* gene in chloroplasts leads to an extraordinary level of an insecticidal protein in tobacco. *Biotechnology*, **13**, 362–365.

McCarroll, L., and King, L.A. 1997 Stable insect cell cultures for recombinant protein production. *Curr. Opin. Biotechnol.*, **8**, 590–594.

McCarthy, W.J., and Getting, R.R. (1986) Current developments in baculovirus serology. In R.R. Granados and B.A. Federici (eds.), *The Biology of Baculoviruses Vol.1 Biological Properties and Molecular Biology*, CRC press Inc., Boca Raton, Florida, pp. 147–158.

McCutchen, B.F., Choudhary, P.V., Crenshaw, R., Maddox, D., Kamita, S.G., Alekar, N. *et al.* (1991) Development of a recombinant baculovirus expressing an insect selective neurotoxin potential for pest control. *Biotechnology*, **9**, 848–852.

McGaughey, W.H., and Whalon, M.E. (1992) Managing insect resistance to *Bacillus thuringiensis* toxins. *Science*, **258**, 1451–1455.

McGaughey, W.H., and Johnson, D.E. (1992) Indian mealmoth (Lepidoptera: Pyralidae) resistance to different strains and mixtures of *Bacillus thuringiensis*. *J. Econ. Entomol.*, **85**, 1594–1600.

McKenna, K.A., Hong, H., vanNunen, E., and Granados, R.R. (1997) Establishment of new *Trichoplusia ni* cell lines in serum-free medium for baculovirus and recombinant protein production. *J. Invertebr. Pathol.*, **71**, 82–90.

McNitt, L., Espelie, K.E., and Miller, L.K. (1995) Assessing the safety of toxin-producing baculovirus biopesticides to a non target predator, the social wasp, *Polistes metricus* Say. *Biol .Contr.*, **5**, 267–278.

Meenakshi, K., and Jayaraman, K. (1979) On the formation of crystal proteins during sporulation in *Bacillus thuringiensis*. *Arch. Microbiol.*, **120**, 9–14.

Melin, B.E., and Cozzi, E.M. (1990) Safety of non-target invertebrates of lepidopteran strains of *Bacillus thuringiensis* and their β-exotoxins. In M. Laird, L.A. Lacey and E.W. Davidson (eds.), *Safety of Microbial Insecticides*, CRC Press Inc., Boca Raton, Florida, pp. 149–167.

Merryweather, A.T., Weger, U., Harris, M.P.G., Hirst, M., Booth, T., and Possee, R.D. (1990) Construction of genetically engineered baculovirus insecticides containing the *Bacillus thuringiensis* subsp. *kurstaki* HD-73 delta endotoxin.*J. Gen. Virol.*, **71**, 1535–1544.

Miller, L.K. (1988) Baculoviruses as gene expression vectors. *Ann. Rev. Microbiol.*, **42**, 177–199.

Miller, L.K. (1995) Genetically engineered insect virus pesticides: Present and future. *J. Invertebr. Pathol.*, **65**, 211–216.

Miller, L.K., Miller, D., and Adang, M. (1983) An insect virus for genetic engineering: Developing baculovirus polyhedrin substitution vectors. In P. Lurguin and A. Kleinhofs (eds.), *Genetic Engineering in Eukaryotes*, Plenum Press, New York, pp. 89–97.

Miller, L.M., Sakai, R.K., Romans, P., Gwadz, R.W., Kantoff, P., and Coon, H.G. (1987) Stable integration and expression of a bacterial gene in the mosquito *Anopheles gambiae*. *Science*, **237**, 779–781.

Mitsuhashi, J., and Maramorosch, K. (1964). Leaf hopper tissue culture , embryonic nymphal and imaginal tissues from aseptic insects. *Contr. Boyce Thompson Inst.* **22**, 435–460.

Mitsuhashi, W., Furuta, Y., and Sato, M. (1998) The spindles of an entomopoxvirus of Coleoptera (*Anomala cuprea*) strongly enhance the infectivity of a nucleopolyhedrovirus in Lepidoptera (*Bombyx mori*). *J. Invertebr. Pathol.*, **71**, 186–188.

Mohan, K.S. (1985) Examination, isolation and field use of baculovirus of *Oryctes*. In S. Jayaraj (ed.), *Microbial Control and Pest Management*, Tamil Nadu Agricultural University, Coimbatore, pp. 107–109.

Mohan, K.S., and Gopinathan, K.P. (1991) Physical mapping of the genomic DNA of the *Oryctes rhinoceros* baculovirus . *Gene*, **107**, 343–344.

Moore, C.B., and Brooks, W.M. (1993) An evaluation of SDS-polyacrylamide gel electrophoretic analysis for the determination of intrageneric relationships of *Vairimorpha* isolates. *J. Invertebr. Pathol.*, **62**, 285–288.

Mori, H., Nakazawa, H., Shirai, N., Shibata, N., Sumida, M., and Matsubara, F. (1992) Foreign gene expression by a baculovirus vector with an expanded host range. *J. Gen. Virol.*, **73**, 1877–1880.

Morishima, I., Suginaka, S., Ueno, T., and Hirano, H. (1990) Isolation and structure of cecropins, inducible antibacterial peptides from silkworm, *Bombyx mori. Comp. Biochem. Physiol.*, 95B, 551–554.

Morris, T., Todd, J.W., Fisher, B., and Miller, L.K. (1994) Identification of *lef-7*: a baculovirus gene affecting late gene replication. *Virology*, **200**, 360–369.

Mulock, B., and Faulkner, P. (1997) Field studies of a genetical modified baculovirus, *Autographa californica* nuclear polyhedrosis virus. *J. Invertebr. Pathol.*, **70**, 33–40.

Munderloh, U.G., Kurti, T.J., and Ross, S.E. (1990) Electrophoretic characterization of chromosomal DNA from two microsporidia. *J. Invertebr. Pathol.*, **56**, 243–248.

Naik, S.S., and Shaila, M.S. (1991) Cloning and expression of rinder pest virus haemagglutin DNA in a baculovirus expression vector system. In *Abstracts of the International Conference on Virology in the Tropics*, Lucknow, India, p. 233.

Nakhai, B., Sridhar, P., Talwar, G.P., and Hasnain, S.E. (1991) Construction , purification, and characterization of a recombinant baculovirus containing the gene for alpha subunit of human chorionic gonadotropin. *Indian J. Biochem. Biophy.*, **28**, 237–242.

Nambiar, P.T.C., Ma, S.W., and Iyer, V.N. (1990) Limiting an insect infestation of N_2-fixing root nodules of the pigeon pea (*Cajanus cajan*) by engineering the expression of an entomocidal gene in its root nodules. *Appl. Environ. Microbiol.*, **56**, 2866–2869.

Narayanan, K. (1979) Studies on the nulcear polyhedrosis virus of gram pod borer, *Heliothis armigera* (Hub.) (Noctuidae: Lepidoptera). Ph.D. Thesis, Tamil Nadu Agricultural University, Coimbatore.

Narayanan, K. (1980a) Studies on the interaction between host (*Heliothis armigera* Hubn.), pathogen (nuclear polyhedrosis virus) and parasite (*Eucelatoria* sp. nr. *armigera* Coq.), Paper presented at *Third Annual Workshop of Biological Control of Crop Pests and Weeds*, Ludhiana, p. 165.

Narayanan, K. (1980b) Field evaluation of nuclear polyhedrosis virus of *Heliothis armigera* (Hbn.) on chickpea, *Cicer arietium* L. *Biological Control* of *Heliothis Workshop*, Department of Primary Industries, Tom street, Toowomba, Australia.

Narayanan, K. (1985a) Susceptibility of *Spilosoma obliqua* Walker to *Nosema* sp. *Curr. Sci.*, **54**, 487–488.

Narayanan, K. (1985b) Studies on the cross infectivity of the nuclear polyhedrosis virus of *Adisura atkinsoni* Moore (Noctuidae:Lepidoptera) . *Curr. Sci.*, **55**, 372–373.

Narayanan, K. (1985c) Use of nuclear polyhedrosis virus of *Adisura atkinsoni* Moore in the integrated control of pests on field beans. In S. Jayaraj (ed.), *Microbial Control and Pest Management*, Tamil Nadu Agricultural University, Coimbatore, pp. 87–89.

Narayanan, K. (1985d) An artificial diet for rearing *Adisura atkinsoni* Moore (Noctuidae:Lepidoptera) and *Sphaenarches anisodactylus* Walker (Pterophoridae : Lepidoptera). *Indian J. Agric. Sci.*, **55**, 300–301.

Narayanan, K. (1987) Susceptibility of *Heliothis armigera* Hubner to *Vairimorpha* sp. *Indian. J. Biocontrol*, **1**, 148–149.

Narayanan, K. (1988a) Mass production of entomogenous protozoa in pest management. Paper presented in the *Summer Institute on Techniques for Mass Production of Biocontrol Agents for Management of Pests and Diseases*, Tamil Nadu Agricultural University, Coimbatore.

Narayanan, K. (1988b) Mass propagation technique of entomophilic nematode *Steinernema feltiae* in pest management with its cost of production. Paper presented in *the Summer Institute on Techniques for Mass Production of Biocontrol Agents for Management of Pests and Diseases*, Tamil Nadu Agricultural University, Coimbatore.

Narayanan, K. (1991) Restriction map of *Spodoptera litura* nuclear polyhedrosis virus genome with endonuclease EcoRI and BstI. In T.N. Ananthakrishnan (ed.), *Emerging Trends in Biological Control of Phytophagous Insects*, Oxford & IBH Publishing Co. Pvt. Ltd., New Delhi, pp. 78–79.

Narayanan, K. (1993a). Establishment of haemocyte primary culture of *Spodoptera litura* for *in vitro* multiplication of nuclear polyhedrosis virus. *Abstract V National Symposium on Advances in Biological Control of Insect Pests*, Muzzaffarnagar, Bihar, India, p.30.

Narayanan, K. (1993b) *In vitro* attempt to infect *Spodoptera frugiperda* (Sf-9) cells with baculovirus (nuclear polyhedrosis virus) of *Spodoptera litura* (Lepidoptera : Noctuidae). *Abstract International Symposium on Virus Cell Interaction: Cellular and Molecular Responses*, Bangalore, India, p. 34.

Narayanan, K. (1994a) Role of biotechnology in effective utilization of insect pathogens in integrated pest management programme in India. In S. Goel (ed.), *Biological Control of Insect Pests*, Uttar Pradesh Zoological Society, Muzzafarnagar, India, pp. 37–47.

Narayanan, K. (1994b) Development and use of microbial control agents for the control of *Heliothis armigera* and *Spodoptera litura. Final DBT Technical Report* , Govt. of India, New Delhi, 81pp.

Narayanan, K. (1995a) Microbial control of insect pests: A cornerstone in sustainable agriculture. In. R.K. Behl, A.L. Khurana and R.C. Dogra (eds.), *Plant Microbe Interaction in Sustainable Agriculture*, Max Mueller Bhavan, New Delhi and CCS Haryana Agricultural University, Hisar, India, pp. 173–182.

Narayanan, K. (1995b) Insect tissue culture: History, development and its application towards microbial control of insect pests. *J. Appl. Zool. Res.*, **6**, 81–84.

Narayanan, K. (1996a) Insect immunity: Its role in management of pests. *IPM Sustain. Agri. Ent. Appr.*, **6**, 1–5.

Narayanan, K. (1996b) Role of biotechnology in effective utilization of insect pathogens in integrated pest management programme in India. In R.K. Upadhayay, K.G. Mukherji and R.L. Rajak (eds.), *IPM System in Agriculture, Vol.2, Biocontrol in Emerging Biotechnology*, Aditya Books Pvt. Ltd., New Delhi, pp. 153–196.

Narayanan, K. (1996c). Gene technology in microbial control of insect pests : Boon or bane? Paper presented at *GENOME 96 Workshop on Biotechnology*, Bangalore.

Narayanan, K. (1997). Biotechnological immobolization of entomopathogens in pest suppression. *Uttar Pradesh J. Zool.*, **17**, 191–199.

Narayanan, K. (1998a). Apoptosis : Its impact on microbial control of insect pests. *Curr. Sci.*, **75**, 114–122.

Narayanan, K. (1998b). Biological control of vectors of parasitic diseases. Lecture note during the *Fourth National Training Programme on Vectors and Vector Borne Parasitic Diseases*, Veterinary College, University of Agricultural Sciences, Bangalore, pp. 58–70.

Narayanan, K. (1998c). Efficient utilization of insect microbes through biotechnology: An overview. *Abstract, National Symposium on Development of Microbial Pesticides and Insect Pest Management*, Pune, p7.

Narayanan, K. (1998d). Characterization of spindle shaped inclusions associated with baculovirus of greater wax moth, *Galleria mellonella. Abstract, National Symposium on Development of Microbial Pesticides and Insect Pest Management*, Pune, p. 29.

Narayanan, K., and Gopalakrishnan, C. (1987). Effect of entomogenous nematode *Steinernema feltiae* to the prepupa and adult of *Spodoptera litura* (Noctuidae: Lepidoptera). *Indian J. Nematol.* 17, 273–276.

Narayanan, K., and Jayaraj, S. (1979) *Spodoptera litura* as a host for *Nosema* sp. *Curr. Sci.*, **48**, 219.

Noguchi, Y., Kobayashi, M., and Shimada, T. (1994) An application for the polymerase chain reaction for practical diagnosis of the nuclear polyhedrosis in large-scale culture of *Bombyx mori. J. Seric. Sci. Japan.*, **63**, 399–406.

Ohkawa, T., Majima, K., and Maeda, S. (1994) A cysteine protease encoded by the baculovirus *Bombyx mori* nuclear polyhedrosis virus. *J. Virol.*, **68**, 6619–6625.

O'Reilly, D.R., and Miller, L.K. (1989) A baculovirus blocks insect moulting by producing ecdysteroid UDP-Glycosyl transferase. *Science*, **245**, 1110–1112.

O'Reilly, D.R., and Miller, L.K. (1991) Improvement of a baculovirus pesticide by deletion of EGT gene. *Biotechnology*, **9**, 1086–1089.

Overton, L.K., and Kost, T.A. (1995) Potential application of insect cell based expression systems in the bio/pharmaceutical industry. In M.L. Schuler, H.A. Wood, R.R. Grandos and D.A. Hannor (eds.), *Baculovirus Expression Systems and Biopesticides*, Wiley-Liss Inc., New York, pp. 233–242.

Palli, S.R., Caputo, G.F., Sohi, S.S., Brownwright, A.J., Ladd, T.R., Cook, B.J. *et al.* (1996) CfMNPV blocks Ac MNPV-induced apoptosis in a continuous midgut cell line. *Virology*, **222**, 201–213.

Panetta, J.D. (1993) Engineered microbes: A cell cap system. In L. Kim (ed.), *Advanced Engineered Pesticides*, Marcel and Dekker Inc., New York.

Pang, Y., Frutos, R., and Federici, B.A. (1992) Synthesis and toxicity of full length and truncated bacterial Cry 1VD mosquitocidal proteins expressed in lepidopteran cells using a baculovirus vector. *J. Gen. Virol.*, **73**, 89–101.

Pang, Y., Dai, X., Shi, X., Long, Q., and Wang, X. (1998) Apoptosis of insect cells induced by baculovirus. *Acta Scient. Natur. Univ. Sunyat.*, **37**, 1–6.

Pant, U., Mascarenhas, A.F., and Jagannathan, V. (1977) *In vitro* cultivation of a cell line from embryonic tissue of potato tuber moth, *Gnorimoschema operculella* (Xeller). *Indian J. Expt. Biol.*, **15**, 244–245.

Parrot, W.A., All, J.N., Adang, M.J., Baily, M.J., Boerma, H.R., and Stewart, C.N. Jr. (1994) Recovery and evaluation of soybean plants transgenic for a *Bacillus thuringiensis* var. *kurstaki* insecticidal gene. *Invitro. Cell. Dev. Biol.*, **30**, 144–149.

Pennock, G.D., Shoemaker, C., and Miller, L.K. (1984) Strong and regulated expression of *Escherichia coli* galactosidase in insect cells with a baculovirus vector. *Mol. Cell Biol.*, **4**, 399–406.

Perlak, F.J., Deaton, R.W., Armstrong, T.A., Fuchs, R.L., Sims, S.R., Greenplate, J.T. *et al.* (1990) Insect resistant cotton plants. *Biotechnology*, **8**, 939–943.

Perlak, F.J., Fuchs, R.L., Dean, D.A., McPherson, S.L., and Fischhoff, D.A. (1991) Modification of the coding sequence enhances plant expression of insect control protein genes. *Proc. Natl. Acad. Sci.* USA, **88**, 3324–3328.

Perlak, F.J., Stone, T.B., Muskopf, Y.M., Peterson, L.J., Parker, G.B., McPherson, S.A. *et al.* (1993) Genetically improved potatoes. Potatoes: Protection from damage by Colorado potato beetles. *Plant Mol. Biol.*, **22**, 313–321.

Pilcher, C.D., Obrycki, J.J., Rice, M.E., and Lewis, L.C. (1997) Pre-imaginal development, survival and field abundance of insect predators on transgenic *Bacillus thuringiensis* corn. *Environ. Entomol.*, **26**, 446–454.

Pipe, N.D., Heale, J.R., Bainbridge, B.W., and Gillespie, A.T. (1989) DNA, RFLP analysis for the strains of entomopathogen *M. anisopliae*. *BCPC Monograph*, **43**, 247.

Poinar, G.O. Jr. (1981) Use of nematodes for the microbial control of insects. In H.D. Burges and N.W. Hussey (eds.), *Microbial Control of Insects and Mites*, Academic Press, London, pp. 181–204.

Popham, J.R.H., Li, Y., and Miller, L.K. (1997) Genetic improvement of *Helicoverpa zea* nuclear polyhedrosis virus as a biopesticide. *Biol. Contr.*, **10**, 83–91.

Possee, R.D. (1997) Baculovirus as expression vector. *Curr. Opin. Biotechnol.*, **8**, 569–572.

Possee, R.D., Barnett, A.L., Hawtin, R.E., and King, L.A. (1997) Engineered baculoviruses for pest control. *Pestic. Sci.*, **51**, 462–470.

Possee, R.D., Bonnig, B.C., and Merryweather, A.T. (1991) Expression of proteins with insecticidal activities using baculovirus vectors. In A. Prokop and R.K.Bajpai (eds.), *Recombinant DNA Technology and Application*, McGraw Hill, New York, pp. 234–239.

Possee, R.D., Hirst, M., Jones, L.D., and Bishop, D.H.L. (1993) Field-tests of genetically engineered baculoviruses. *BCPC Monograph No.55, Opportunities for Molecular Biology in Crop Production*, pp. 23–37.

Powell, P.A., Stark, D.M., Sanders, P.R., and Beachy, R.N. (1989) Protection against tobacco mosaic virus in transgenic plants that express tobacco mosaic virus antisense RNA. *Proc. Natl. Acad. Sci.* USA, **86**, 6949–6952.

Powning, R.F., and Davidson, W.J. (1973) Studies on insect bacteriolytic enzymes. I. Lysozyme in haemolymph of *Galleria mellonella* and *Bombyx mori*. *Comp. Biochem. Physiol.*, **45B**, 669–686.

Prikhod'Ko, E.A., and Miller, L.K. (1996) Induction of apoptosis by baculovirus transactivator IEI. *J. Virol.*, **70**, 7116–7124.

Quistad, G.B., Ngugan, K., Bernasconi, P., and Leisy, D.J. (1994) Purification and characterization of insecticidal toxins from venom glands of the parasitic wasp, *Bracon hebetor. Insect Biochem. Molec. Biol.*, **24**, 955–961.

Rabindra, R.J. (1981) The effects of *Farinocystis tribolii* on the growth and development of the flour beetle, *Tribolium castaneum. J. Invertebr. Pathol.*, **38**, 345–351.

Raina, S.K., and Khurad, A.M. (1988) Primary cultures of ovarian tissues from locusts, *Locusta migratoria* and *Schistocerca gregaria* (Orthoptera: Acrididae). In Y. Kurado, E. Kurstak and K. Maramorosch (eds.), *Invertebrate and Fish Tissue Culture*, Springer-Verlag, Berlin, pp. 239.

Ramakrishnan, N. (1992) Nuclear polyhedrosis virus of *Spodoptera litura*. An approach for its efficient use. In. T.N. Ananthakrishnan (ed.), *Emerging Trends in Biological Control of Phytophagous Insects*, Oxford & IBH Publishing Co. Pvt. Ltd., New Delhi, pp. 159–164.

Rangarajan, P.N., and Padmanaban, G. (1996) Gene therapy: Principles, practice, problems and prospects. *Curr. Sci.*, **71**, 360–368.

Reddy, G.R., Natarajan, C., and Suryanarayana, V.V.S. (1997) Expression of 2C protein gene of foot and mouth disease virus type Asia 1 in insect cells. *Abstract 12th Annual Convention and Conference of Indian Virological Society*, Gujrat Agricultural University, Anand.

Reitz, S.R., and Adler, P.H. (1996) Biology and larval taxonomy of *Eucelatoria bryani* Sabrosky and *E. rubentis* (Conquillett) (Diptera:Tachinidae). *Proc. Ent. Soc. Washington*, **98**, 625–629.

Ribeiro, B.M., and Crook, N.E. (1993). Expression of full length and truncated forms of crystal protein genes from *Bacillus thuringiensis* subsp. *kurtstaki* in a baculovirus and pathogenicity of the recombinant viruses. *J. Invertebr. Pathol.* **62**, 121–130.
Richards, A., Mathews, M., and Christian, P. (1998) Ecological considerations for the environmental impact evaluation of recombinant baculovirus insecticides. *Ann. Rev. Entomol.*, **43**, 493–517.
Richardson, C.D., Banville, M., Lalumiere, M., Vialard, J., and Meighen, E.A. (1992) Bacterial luciferase produced with rapid-screening baculovirus vectors is a sensitive reporter for infection of insect cells and larvae. *Intervirology*, **34**, 213–227.
Rizki, R.M., and Rizki, T.M. (1990) Parasitoid virus like particles destroy *Drosophila* cellular immunity. *Proc. Natl. Acad. Sci.* USA, **87**, 8388–8392.
Roelvink, P.W., Corsaro, B.G., and Granados, R.R. (1995) Characterization of the *Helicoverpa armigera* and *Pseudaletia unipuncta* granulosis enhancin genes. *J. Gen. Virol.*, **76**, 2693–2705.
Roelvink, P.W., Van Meer, M.M.M., De Kort, C.A.D., Posee, R.D., Hammock, B.D., and Vlak, J.M. (1992). Dissimilar expression of *Autographa californica* multiple nucleocapsid nuclear polyhedrosis virus polyhedrin and p10 genes. *J. Gen. Virol.*, **73**, 1481–1489.
Samples, J.R., and Buettner, H. (1982) Corneal ulcer caused by a biological insecticide (*Bacillus thuringiensis*). *Am. J. Opthalmol.*, **95**, 258–260.
Sanchis, V., Lereculus, D., Menou, G., Chaufaux, J., Lecadet, M.M., and Dedonder, R. (1989) Cloning from *Bacillus thuringiensis* subsp. *aizawi* 7.29 of a gene encoding a delta endotoxin specifically active against lepidopteran insect species of the Noctuidae family. *Israel J. Ent.*, **23**, 201–212.
Sarvari, M., Osikos, G., Sass, M., Gal, P., Schumaker, V.N., and Zavodsky, P. (1990) Ecdysteroids increase the yield of recombinant protein produced in baculovirus insect cell expression system. *Biochem. Biophys. Res. Commun.*, **107**, 1154–1161.
Satyanarayanan Sriram, Gopinathan, K.P., and Sriram, S. (1999) The potential role of a late gene expression factor *lef-2* from *Bombyx mori* NPV in very late gene transaction and DNA replication. *Virology*, **251**, 108–122.
Schwartz, J.I., Garneau, L., Masson, L., and Brousseau, R. (1991) Early response of cultured lepidopteran cells to exposure to δ-endotoxin from *Bacillus thuringiensis*; involvement of calcium and anion channels. *Biochem. Biophys. Acta*, **1065**, 250–260.
Seema Mishra (1998) Baculoviruses as biopesticides. *Curr. Sci.*, **75**, 1015–1022.
Sethuraman, B.N., Nagaraju, J., and Datta, R.K. (1993) Purification and partial characterization of antiviral protein in silkworm, *Bombyx mori. Indian J. Seric.*, **32**, 63–67.
Shapiro, D.I., Glazer, I., and Segal, D. (1997) Genetic improvement of heat tolerance in *Heterorhabditis bacteriophora. Biol. Contr.*, **8**, 153–159.
Sharma, C.B.S.R., and Sahu, R.K. (1977) Cytogenetic hazards from agricultural chemicals. A preliminary study on the responses of root meristems to exotoxin from *Bacillus thuringiensis* a constituent of a microbial insecticide, thuricide. *Mutation Res.*, **46**, 19–26.
Shoji, I., Aizaki. H., Tani, H., Ishii, K., Chiba, T., Saito, I. *et al.* (1997) Efficient gene transfer into various mammalian cells, including non-hepatic cells, by baculovirus vectors. *J. Virol.*, **78**, 2657–2664.
Schuler, T.H., Poppy, G.M., Kerry, B.R., and Denholm, I. (1999) Potential side effects of insect-resistant transgenic plants on arthropod natural enemies. *Tibtech.*, **17**, 210–216.
Shuler, M.L., Wood, H.A., Granados, R.R., and Hamer, D.A. (1995) *Baculovirus Expression Systems and Biopesticides.* Wiley-Liss Inc., New York, pp. 264.
Singh, K.R.P. (1967) Cell cultures derived from larvae of *Aedes albopictus* (Kuse) and *Aedes aegypti* L. *Curr. Sci.*, **19**, 506–508.
Sivamani, E., and Rajendran, N. (1992) Protoplast fusion between insecticidal *Bacillus thuringiensis* var. *kurstaki* HD 73 and plant growth promoting rhizobacterial strain of *Pseudomonas flourescens* PFCP. In T.N. Ananthakrishnan (ed.), *Emerging Trends in Biological Control of Phytophagous Insects*, Oxford & IBH Publishing Co. Pvt. Ltd., New Delhi, pp. 147–158.
Sivasubramanian, N., Post, C.A., and Hice, R.H. (1991) Use of baculovirus protein for developing novel protein insecticides. *Abstract International Conference on Virology in the Tropics*, Lucknow, India, p. 225.
Skot, L., Harrison, S.P., Nath, A., Mytton, L.R., and Clifford, B.C. (1990) Expression of insecticidal activity in *Rhizobium* containing the delta-endotoxin gene cloned from *Bacillus thuringiensis* subsp. *tenebrionis. Plant Soil*, **127**, 285–295.
Slack, J.M., Kuzio, J., and Faulkner, P. (1995) Characterization of V-Cath, a cathepsin L like proteinase expressed by the baculovirus *Autographa californica* multiple nuclear polyhedrosis virus. *J. Gen. Virol.*, **76**, 1091–1098.
Smith, G.E., Summer, M.D., and Fraser, M.J. (1983) Production of human -interferon in insect cells infected with a baculovirus expression vector. *Mol. Cell. Biol.*, **3**, 2156–2165.

Soldevila, A.I., Heuston, S., and Webb, B.A. (1997) Purification and analysis of a polydna virus gene product expressed using a polyhistidine baculovirus vector. *Insect Biochem. Molec. Biol.*, **27**, 201–211.

Sridhar, P., Awasthi, A.K., Azim, C.A., Burma, S., Habib, S., Jain, A. *et al.* (1994) Baculovirus vector mediated expression of heterologous genes in insect cells. *J. Biosci.*, **19**, 603–614.

Stalin, S.L., Abrams, C., and English, L. (1990) Delta-endotoxin form cation-selective channels in planar lipid bilayers. *Biochem. Biophys. Res. Comm.*, **169**, 765–772.

Staples, R.C., Legerst. R.J., Bhairi, S., and Roberts, D.W. (1988) Strategies for genetic engineering of fungal entomopathogens. In *Proceedings of Conference on Biotechnology, Biological Pesticides and Novel Plant-Pest Resistance for Insect Pest Management*, Boyce Thompson Institute, Ithaca, pp. 44–48.

Steinhaus, E.A., and Hughes, K.M. (1952) A granulosis virus of the Western grape leaf skeletoniser. *J. Econ. Entomol*, **45**, 744–745.

Stewart, L.M.D., Hirst, M., Ferber, M.L., Merryweather, A.T., Jane Cayley, P., and Possee, R.D. (1991) Construction of improved baculovirus insecticide containing an insect specific toxin gene. *Nature*, **352**, 85–88.

Stock, C.A., McLoughlin, T.J., Klein, J.A., and Adang, M.J. (1990) Expression of *Bacillus thuringiensis* crystal protein gene of *Pseudomonas cepacia* 526. *Can. J. Microbiol.*, **36**, 879–884.

Stoltz, D.B. (1986) Viruses of parasitic Hymenoptera. In R.A.Samson, J.M. Vlak and D. Peters (eds.), *Fundamentals and Applied Aspects of Invertebrate Pathology*, Foundation Fourth Int. Colloq. Invertebr. Pathol., The Netherlands, pp. 81–82.

Stone, T.B., Sims, S.R., and Marrone, P.G. (1989) Selection of tobacco budworm for resistance to a genetically engineered *Pseudomonas fluorescens* containing the delta endotoxin gene cloned from *Bacillus thuringiensis* subsp. *kurstaki. J. Invertebr. Pathol.*, **53**, 228–234.

Strand, M.R. (1994) *Microplitis demolitor* polydna virus infects and expresses in specific morphotypes of *Pseudoplusia includens* haemocytes. *J. Gen. Virol.*, **75**, 3007–3020.

Summers, M.D. (1989) Recombinant proteins expressed by baculovirus vectors. In A.L. Notkins and M.B.A. Oldstono (eds.), *Concepts in Viral Pathogenesis*, pp. 77–86.

Summers, M.D., and Dil-Hajj, S.D. (1995) Polydna virus facilitated endoparasite protection against host immune defenses. *Proc. Natl. Acad. Sci.* USA, **92**, 29–36.

Sundeep, A.B., and Pant, U. (1998) A new *Helicoverpa armigera* cell line for the multiplication of baculoviruses. *Abstracts National Symposium on Development of Microbial Pesticides and Insect Pest Management*, Pune, India.

Sun Shao-Cong, Lindstrom, I., Boman, H.G., Faye, I., and Schmidt, O. (1990) Hemolin : An insect-immune protein belonging to the immunoglobulin superfamily. *Science*, **250**, 1729–1732.

Sundara Babu, P.C. (1980) Studies on the pathogenecity of *Metarhizium anisopliae* (Metchinikoff) Sorokin var. major Tulloch in *Oryctes rhinoceros* L. Ph.D. Thesis , Tamil Nadu Agricultural University, Coimbatore.

Tanada, Y. (1985) A synopsis of studies on the synergistic property of an insect baculovirus. A tribute to Edward A. Steinhaus. *J. Invertebr. Pathol.*, **45**, 125–138.

Tanada, Y., and Kaya, H.K. (1993) *Insect Pathology*, Academic Press, New York.

Tabashnik, B.E., Cushing, N.L., Finson, A., and Johnson, M.W. (1990) Field development of resistance to *Bacillus thuringiensis* in diamondback moth (Lepidoptera: Plutellidae). *J. Econ. Entomol.*, **85**, 1671–1676.

Theilmann, D.A., Chantler, J.K., Stewart, S., Flipsen, T.M., Vlak, J.M., and Crook, N.E. (1996) Characterization of a highly conserved baculovirus structural protein that is specific for occlusion derived virions. *Virology*, **218**, 148–158.

Theim, S.M. (1997) Prospects for altering host range for baculovirus bioinsecticides. *Curr. Opin. Biotechnol.*, **8**, 317–322.

Theim, S.M., Du, X., Quentin, M.E., and Berner, M.M. (1996) Identification of a baculovirus gene that promotes *Autographa californica* nuclear polyhedrosis virus replication in a non permissive insect cell line. *J. Virol.*, **70**, 2221–2229.

Thomas, C.J., King, L.A., and Possee, R.D. (1998) Localization of a baculovirus-inducted chitinase in the insect cell endoplasmic reticulism. *J. Virol.*, **72**, 10207–10212.

Tomalski, M.D., and Miller, L.K. (1992) Expression of paralytic neurotoxin gene to improve insect baculoviruses as biopesticides. *Biotechnology*, **10**, 545–549.

Turner, J.T., Lampel, J.S., Stearman, R.S., Sundin, G.W., Gunynzki, P. and Anderson, J.J.1991) Stability of the δ-endotoxin gene from *Bacillus thuringiensis* subsp. *kurstaki* in a recombinant strain of *Clavibacter xyli* subsp. *cynodontis. Appl. Environ. Microbiol.*, **57**, 3522–3528.

Vaeck, M., Reynnerts, A., Hofte, H., Jansens, S., Beuckeleer, M.D., Dean, C. *et al.* (1987) Transgenic plants protected from insect attack. *Nature*, **328**, 33.

Van der Krol, A.R., Mol, J.N.M., and Stuitje, A.R. (1988) Modulation of eukaryotic gene expression by complementary RNA or DNA sequences. *Biotechniques*, **6**, 958–976.

Vaughn, J.L., and Weiss, S.A. (1991) Formulating media for the culture of insect cells. *Biopharm.*, **4**, 16–19.

Vaughn, J.L., Goodwin, R.H., Thompkin, G.J., and McCawley, P. (1977) The establishment of two cell lines from the insect *Spodoptera frugiperda* (Lep.: Noctuidae). *In Vitro*, **13**, 213–217.

Vikas, B.P., and Gopinathan, K.P. (1996) Characterisation of a local isolate of *Bombyx mori* nuclear polyhedrosis virus. *Curr. Sci.*, **70**, 147–153.

Vikas, B.P., Sumathy, S., and Gopinathan, K.P. (1995) Baculovirus mediated high-level expression of luciferase in silkworm cells and larvae. *Biotechniques*, **19**, 195–199.

Vlak, J.M., Schouten, A., Usmany, M., Belsham, G.J., Klinge-Roode, E.C., Maule, A.J. *et al.* (1990) Expression of cauliflower mosaic virus gene I using a baculovirus vector based upon the baculovirus *p10* gene and a novel selection method. *Virology*, **179**, 312–320.

Volkman, L.E. (1995) Baculovirus bounty. *Science*, **269**, 1834.

Vyas, H.G. (1991) Serological homology of geographic isolates of *Heliothis* baculovirus. *Abstract International Conference on Virology in the Tropics*, Lucknow, India, p. 226.

Wang, P., and Granados, R.R. (1997) An intestinal mucin is the target substrate for a baculovirus enhancin. *Proc. Natl. Acad. Sci.* USA, **94**, 6977–6982.

Wang, P., and Granados, R.R. (1998) Observations on the presence of the peritrophic membrane in larvae of *Trichoplusia ni* and its role in mimiting baculovirus infection. *J. Invertebr. Pathol.*, **72**, 57–62.

Warren, G.W., Carozzi, N.B., Desai, N., and Koziel, M.G. (1992) Field evaluation of transgenic tobacco containing a *Bacillus thuringiensis* insecticidal protein gene. *J. Econ. Entomol.*, **5**, 1651–1659.

Washburn, J.O., Kirkpatrick, B.A., and Volkman, L.E. (1996) Insect protection against viruses. *Nature*, **383**, 7667.

Washburn, J.O., Kirkpatrick, B.A., and Volkman, L.E. (1998) Evidence that the stilbene-derived optical brightener M2R enhances *Autographa californica* M nucleopolyhedrovirus infection of *Trichoplusia ni* and *Heliothis virescens* by preventing sloughing of infected midgut epithelial cells. *Biol. Contr.*, **11**, 58–69.

Washburn, J.O., Lyons, E.J., Haas-Stapleton and Volkman, L.E (1999) Multiple nucleocaosid packaging of *Autographa californica* nuclear polyhedrosis virus acccelerates the onset of sytemic infection in *Trichoplusia ni. J. Virol.*, **73**, 411–416.

Watrud L.S., Perlak, F.J., Minh-Tien, T., Kusano, K., Mayer, E.J., Miller-Wideman, M.A. *et al.* (1985) Cloning of *Bacillus thuringiensis* subsp. *kurstaki* delta endotoxin gene into the *Pseudomonas fluorescens*: Molecular biology and ecology of engineered microbial pesticide. In H.O. Halvorson, D. Pramer and M. Rogul (eds.), *Engineered Organisms in the Environment-Scientific Issues*, Am. Soc. Microbiol., Washington DC, pp. 40–46.

Weiser, J., and Slama, K. (1964) Effects of the toxin of *Pyemotes* (Acarina: Pyemotidae) in the insect prey, with special reference to respiration. *Ann. Entomol. Soc. Am.*, **57**, 479–482.

Wickham, T.J., Davis, T., Granados, R.R., Shuler, M.L., and Wood, H.A. (1992) Screening of insect cell lines for the production of recombinant proteins and infectious virus in the baculovirus expression systems. *Biotechnol. Prog.*, **8**, 391–396.

Williams, S., Friedrich, L., Dincher, S., Carozi, N., Kessman, H., Ward, E. *et al.* (1992) Chemical regulation of *Bacillus thuringiensis* delta-endotoxin expression in transgenic plants. *Biotechnology*, **10**, 540–542.

Wilmut, I., Schnieke, A.E., McWhir, J., Kind, A.J., and Campbell, K.H.S. (1997) Viable offspring derived from fetal and adult mammalian cells. *Nature*, **385**, 810–813.

Wilson, L.E., Wilkinson, N., Marlow, S.A., Possee, R.D., and King, L.A. (1997) Identification of recombinant baculovirus using green fluorescent protein as a selective marker. *Biotechniques*, **22**, 674–681.

Wilson, M., Xin, W., Hashmi, S., and Gaugler, R. (1999) Risk assessment and fitness of a transgenic entomopathogenic nematode. *Biol. Contr.*, **15**, 81–87.

Wolfersberger, M.G. (1990) The toxicity of two *Bacillus thuringiensis* delta-endotoxins to gypsy moth larvae is inversely related to the affinity of binding sites on midgut brush border membranes for the toxins. *Experientia*, **46**, 475–477.

Wood, H.A. (1995) Development and testing of genetically improved baculovirus insecticides. In M.L. Schuler, H.A. Wood, R.R. Grandos and D.A. Hamer (eds.), *Baculovirus Expression Systems and Biopesticides*, Wiley-Liss Inc., New York, pp. 91–102.

Wood, H.A., and Granados, R.R. (1991) Genetically engineered baculovirus as agents for pest control. *Ann. Rev. Microbiol.*, **45**, 69–87.

Wood, H.A., Hughes, P., Johnson, J.D., and Langridge, W.H.R. (1981) Increased virulence of *Autographa californica* nuclear polyhedrosis virus by mutagenesis. *J. Invertebr. Pathol.*, **38**, 236–241.

Woodring, J.L., and Kaya, H.K. (1988) *Steinermatid and Heterorhabtid Nematodes. A Handbook of Biology and Techniques*, Souvnier Co-operative Series Bulletin 331. Arkansas Agri. Expt. Stn., Fayetteville, Arkansas, pp. 30.

Xu, J and Hukuhara, T. (1992) Enhanced infection of a nuclear polyhedrosis virus in larvae of the army worm, *Pseudaletia separata*, by a factor in the spheroids of an entomopoxvirus. *J. Invertebr. Pathol.*, **60**, 259–264.

Yamada, K., Nakajima, Y., and Natori, S. (1990) Production of recombinant sarcotoxin 1A in *Bombyx mori* cells. *Biochem. J.*, **272**, 633–636.

Yoshisue, H., Ihara, K., Nishimoto, T., Sakai, H., and Komano, T. (1995) Cloning and characterization of a *Bacillus thuringiensis* homolog of the spo IIID gene from *Bacillus subtilis. Gene*, **154**, 23–29.

Zuidema, D., Schouten, A., Usmany, M., Maule, A.J., Belsham, G.J., Roosien, J. *et al.* (1990) Expression of cauliflower mosaic virus gene I in insect cells using a novel polyhedrin-based baculovirus expression vector. *J. Gen. Virol.*, **71**, 2201–2210.

Microbial Biopesticides Developed as Inducible Plant Defensive Systems Transgenically

4

Salvatore Arpaia[1], Giuseppe Mennella[2], Giuseppe L. Rotino[3] and Francesco Sunseri[4]

[1]Metapontum Agrobios, S.S. 106 Jonica Km 448.2, I-75010, Metaponto (MT), Italy, [2]Istituto Sperimentale per l'Orticoltura, PO Box 48, I-84094 Pontecagnano (SA), Italy, [3]Istituto Sperimentale per l'Oriticoltura, Via Paullese 28, I-26836 Montanaso Lombardo (LO), Italy, [4]Dipartimento di Biologia, Difesa e Biotecnologie Agroforestali, Università degli Studi della Basilicata, Contrada Macchia Romana, I-85100 Potenza, Italy

Introduction

After several years of intense experimental work, the use of transgenic plants as a means of plant protection has recently been adopted commercially on large scale. In 1998, 27.8 million hectares were cultivated with transgenic corn, cotton and potato plants (Moffat, 1998). The area under transgenic crops further increased to 39.9 million hectares in 1999 (C. James, personal communication). This area is more than three times when compared to 1997 (11 million ha); more than one third of the total genetically modified plants grown in 1997 were transformed to induce insect resistance (USDA/APHIS Biotechnology Permit Database) using synthetic genes derived from the bacterium, *Bacillus thuringiensis* Berl. (Bt). Other insect resistance strategies are currently being studied (see Carozzi and Koziel, 1997). Bt-expressing plants are at the moment the best known transgenic systems available for pest control.

Soon after the first transgenic plants were obtained, insect ecologists suggested some possible drawbacks of an extensive use of these resistant plants (e.g. Gould, 1988b, 1991). The main concern is the possible onset of resistance by target insects to Bt-expressing plants (see Chapter 3), and among the possible solutions proposed for avoiding or delaying such an adaptation has been the use of inducible promoters (e.g. Gould, 1988a).

This chapter will review the mechanisms of natural inducible defenses in plants, the most recent advances in *B. thuringiensis* toxin expression in transgenic plants and try to evaluate the possible effect of inducible expression of microbial toxins (with particular emphasis on Bt) in transgenic plants.

Inducible defenses in plants

In nature all plants are subject to wounding, either by herbivory, mechanical damage or a variety of other offenders. The long co-evolution between herbivores, pathogens and their food plants has led to the development of a large array of defense mechanisms in plants (Price, 1984). In contrast, other authors support the idea that the presence of defensive mechanisms and injuries by the herbivores are independent phenomena (e.g. Jermy, 1993). Among the most common plant defenses, we can single out the widespread occurrence of inducible defenses. Wounding triggers the expression of a number of defensive genes both near to the wound and far from the wounded tissues; the induced response is often very fast and can be activated by removal of small parts of plant tissues indicating an active physiological response from the plant (Baldwin, 1993). The reaction evoked by the primary invader, protects the plant against a subsequent attack brought by the same or a different pathogen or herbivore. Phytophagous insects, nematodes, pathogenic fungi, and bacteria may cause the induced reactions. A virulent form of pathogens, cultivar non-pathogenic races, and non-pathogenic bacteria may also cause it.

Several different molecules are involved in these reaction mechanisms; whilst some of these are rather general and specific, some molecules are especially important for disease resistance such as low molecular weight phytoalexins, or some high molecular weight pathogen-related (P-R) proteins (Boller, 1988). Other metabolic pathways, on the contrary, are mainly involved in the mechanisms of insect resistance. The aim of the present chapter is mainly to focus on the latter, trying to briefly summarize their occurrence and the general mechanisms known to be the most important for insect resistance in crop plants.

Plant genes involved in inducible defenses

Wound inducible genes in plants associated with insect resistance or tolerance include representatives in the major groups like proteinase inhibitors (e.g. Graham *et al.*, 1986), lipoxygenases, polifenoloxidases (Hildman *et al.*, 1992), peroxidases, prosystemin (Mc Gurl *et al.*, 1992), chitinases, and myelin basic protein kinases. Studies conducted on these mechanisms revealed their natural presence in many different plant species such as potato, tomato, alfalfa, soybean, pea, tobacco, cabbage, several cucurbits, and in trees such as poplar (e.g. Farmer and Ryan, 1990; Farmer *et al.*, 1992; Bolter, 1993; Kort and Dixon, 1997).

Wound-inducible molecules have been extensively studied in potato where several genes like an anionic peroxidase (Roberts *et al.*, 1988), wun 1 and wun 2 proteins (Logemann *et al.*, 1988), and others (Shirras and Northcote, 1984; Stiekema *et al.*, 1988; Stanford *et al.*, 1989) have been identified. An interesting instance of repression of gene expression as a result of tuber wounding was described for patatin genes (Logemann *et al.*, 1988).

Proteinase inhibitors represent an important family of induced molecules and they are among the best-studied proteins in plant biochemistry and biology (e.g. Brown and Ryan, 1984; Green and Ryan, 1972; Ryan, 1990; Sanchez-Serrano *et al.*, 1986; Peña-Cortes, 1988). Environmental stress, pathogen or insect attack and developmental factors induce proteinase inhibitor genes. The subdivision of the pathways resulting in PI gene transcription is presently a major area of research in signal transduction. The newly synthesized inhibitor proteins are part of the array of defensive chemicals that can protect plants against insect pests (Hilder *et al.*, 1987; Johnson *et al.*, 1989) and pathogens (Pautot *et al.*, 1991) by inhibiting insect digestive enzymes and the extracellular proteinases produced by some pathogens.

There are four known classes of proteinases, which are distinguished on the basis of the central amino acid residue (serine, cysteine, and aspartate) or metal ion involved in catalyzing cleavage of peptide bonds in protein substrates (Hartley, 1970; James, 1976; Barrett, 1986). As the number of different proteinase gene families is strictly limited, therefore, bacteria, fungi, animals and plants have a specific range of inhibitor gene families.

The inhibitory activities of proteinaceous plant proteinase inhibitors (PIs) are effective against a wide range of prokaryotic and eukaryotic proteinases (reviews: Ryan, 1973, 1978, 1990; Richardson, 1977, 1991). In tomatoes (*Lycopersicon esculentum* Mill.) PIs were thought to be the major factors responsible for reducing the leaf damage by beet armyworm feeding activity (Broadway *et al.*, 1986). In addition, two noctuid larvae showed a significant reduction in both growth and development when fed with artificial diets containing soybean trypsin inhibitor or potato PI II (Broadway and Duffay, 1986).

A number of endogenous factors that induce PIs have been identified, including systemin, an 18 amino acid polypeptide derived from plants (Pearce *et al.*, 1991), plant growth regulators such as auxin and ABA (Peña-Cortes *et al.*, 1989), and oligosaccharides from the damaged cell wall (Bishop *et al.*, 1981). Farmer and Ryan (1990) observed that jasmonic acid (JA) and methyl jasmonate (MJ), secondary products of lipoxygenase-catalyzed oxidation of linolenic acid, were also potent inducers of PI in tomato, tobacco and alfalfa leaves when applied exogenously.

Several studies indicate that the induction of Ser proteinase inhibitors may not be responsible for the effective defense mechanism in insects (Wolfson and Murdock, 1990). Benz (1978) demonstrated that although wounding potato leaves led to a doubling of Ser proteinase inhibitor activity, the development of Colorado potato beetle fed with previously injured plants was not significantly different from those individuals reared on intact plants. It was shown that Ser proteinases were not important digestive enzymes in many Coleoptera and that the Kunitz trypsin inhibitor had no effect on the proteinase activity in homogenates of larval bruchid beetles (Kitch and Murdock, 1986) or cowpea weevils (Gatehouse *et al.*, 1985). On the contrary, Murdock *et al.* (1987) demonstrated that Cys (or thiol) proteinases (CPI) were responsible for most of the proteolytic activity in the digestive tract of many Coleoptera. Indeed, the CPI oryzacystatin significantly inhibits the digestive proteinases of two species of stored grain beetles (Liang *et al.*, 1991). In addition, it was shown that the specific CPI, E-64 (L-*trans*-epoxysuccinyl-leucylamido-[4-guanidino]-butane), supplied in insect diet, suppressed the growth and development of larval cowpea weevils (Hines *et al.*, 1990), and Colorado potato beetle (Wolfson and Murdock, 1987).

Wounding by herbivores may simultaneously activate the expression of several genes, involving complex metabolic changes that comprises the induced defensive system of a plant. Stout *et al.* (1994) demonstrated that a group of four components have a primary role in inducing antimetabolic effects on insect feeding on tomato, i.e. proteinase inhibitors, polyphenoloxidases, peroxidases and lipoxigenases. The spatial distribution of such induced compounds suggests that the chemical composition of a plant after wounding is quite heterogeneous and that several putative defensive compounds may be accumulated in different parts of the plant at the same time (Stout *et al.*, 1996). Proteinase inhibitors I and II, for instance, accumulate in tomato and potato leaves not only as a direct consequence of herbivore feeding, but also due to mechanical damage.

By using two isogenic tomato lines, we tried to isolate the contribution of a PI gene among the other activated mechanisms. For this purpose we comparatively measured antimetabolic effects on lepidopteran larvae feeding on wounded plants of the two isolines: the commercial hybrid Money Maker and its isogenic line SIT, that lacks the PI and II gene (seeds were kindly supplied by Prof. M. Koorneef, Wageningen Agricultural University, NL). We performed bioassays on these plants using *Helicoverpa armigera* (Hubner) larvae as test insect.

Four-week old tomato plants (Sit and Money Maker) grown in greenhouse were used for the experiment. We used 24 plants of each line divided in three groups of eight and treated as follows:

- Leaves were mechanically wounded (three holes, 1 cm dia, on a single leaf)
- Insect feeding on plants (5 third instars *H. armigera* maintained for 24 h on each plant before the onset of the experiment);
- Unwounded control plants.

Twenty four hours after treatment, 10 neonate *H. armigera* larvae were placed on each plant to simulate a secondary herbivore attack. Two weeks later, larval mortality and fresh weight of survivors were recorded. The experiment was repeated twice.

The results showed a significant difference in larval mortality on two tomato lines. Percentage mortality was, in fact, significantly different in both cases: (ANOVA 1st experiment: $DF = 1,2$; $F = 11,69$; $p = 0,0017$ and 2nd experiment: $DF = 1,2$; $F = 14,14$; $p = 0,0007$). Also, the difference in larval growth on the two tomato lines was highly significant: (ANOVA 1st experiment: $DF = 1,2$; $F = 9,31$; $p = 0,0026$ and 2nd

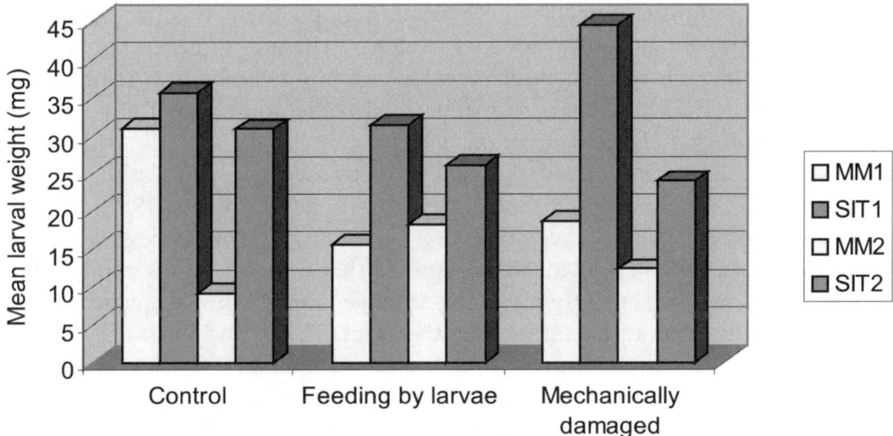

Figure 4.1 Mean body weight of surviving larvae at the end of the experiment. Numbers 1 and 2 refer to two different experiments, repeated with the same tomato isolines.

experiment: DF = 1,2; F = 91,40; p = 0,0001). However, there was no notable difference in the effects due to the wounding mechanism (Figs. 4.1, 4.2).

Our tests showed that a different effect on herbivore development was due to the different genotypes used. The lack of one of the four main defensive compounds in tomato plants, namely the accumulation of a proteinase inhibitor in the leaves, led to very different results. However, quantitative analyses of the different enzymes upon activation should help in elucidating the different role of each gene in plant defense in order to gain more knowledge on the inducible defenses in crop plants.

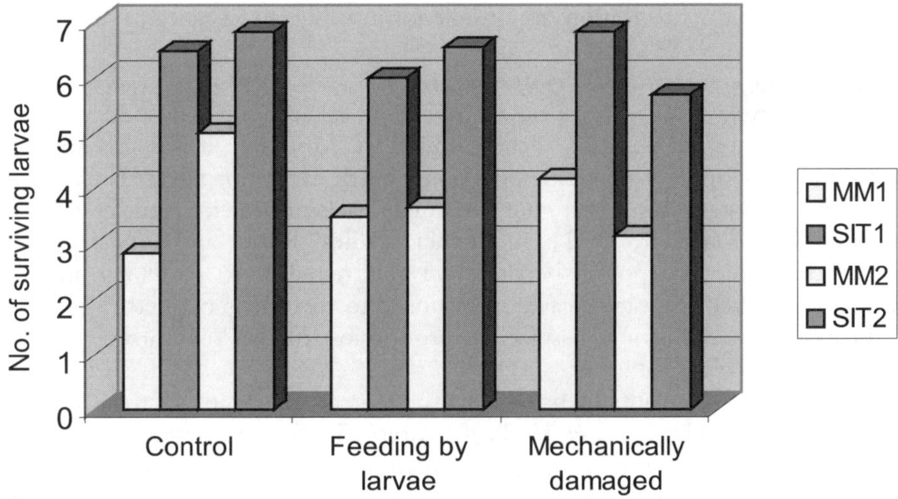

Figure 4.2 Number of larvae survived at the end of the experiment. Numbers 1 and 2 refer to two different experiments, repeated with the same tomato isolines.

Wounding conditions also need to be defined more precisely, because change in the wounding conditions may sometimes lead to drastic changes in gene expression. For instance, sliced tubers incubated under aerobic conditions had a drastic reduction in the patatin gene transcription whereas under anaerobic conditions the patatin gene was normally expressed (Logemann *et al.*, 1988).

Several studies have highlighted the timing of transcript accumulation of wound induced genes and the duration of this activity over time (e.g. Bolter, 1993); less information is instead available on the actual level of feeding activity necessary to induce plant reaction. As a general rule, it can be assumed that gene activation upon herbivory is a fast mechanism. The actual timing is also variable depending upon which genes are involved and the type of injury suffered by the plant (Korth and Dixon, 1997). After the first activation, the mechanism is probably self-maintained if feeding by herbivores continues. The exact relationships between the amount of feeding needed for activating defenses and maintaining transcript accumulations needs to be investigated in different insect-host plant systems in order to apply sound scientific information in breeding programs for insect resistance.

Insect adaptation to inducible defenses

The key feature that enabled insects to successfully become phytophagous, in spite of the large occurrence of plant allelochemicals, is their food specialization. The majority of insect herbivores, in fact, feed only on one or a few plant species and even polyphagous species accept only a few of all plants available to them.

Insect adaptation to plant defenses is made possible by variable mechanisms that sometimes involve the presence of enzymes that specifically detoxify singular compounds, in other cases more general mechanisms are involved such as the microsomal mixed-function oxidases. The latter catalyze diverse reactions and are also involved in the mechanisms of detoxification responsible for insecticide resistance in several insect species (Wilkinson, 1983).

Recently, Jongsma *et al.* (1995) showed that larvae of *Spodoptera exigua* (Hubner) adapt to high levels of serine proteinase inhibitors in tobacco leaves by induction of gut proteinases that are not sensitive to these inhibitors. Moreover, it was also noted that growth and development of Colorado potato beetle were not affected when larvae were reared on potato leaves that contained high levels of proteinaceous cysteine proteinase inhibitors (Bolter, 1992). In further studies, Bolter and Jongsma (1995) reported that Colorado potato beetles were able to adapt to proteinase inhibitors which were induced in potato leaves in response to treatment by methyl jasmonate. The proposed mechanism is a selective production of different proteinases with reduced affinity for the inhibitor.

We transformed eggplants by using a synthetic gene expressing cystatin, a cysteine proteinase inhibitor (La Porta *et al.*, 1998). Colorado potato beetle larvae reared on these plants showed a significant weight reduction compared to the control, only in one experiment out of the eight performed; no difference was evidenced in larval survivorship. In these studies the measured inhibitory activity against the cysteine proteinases was not significantly different among all the analyzed plants. These findings are in agreement with the previous results and support the hypothesis that, a direct response to defense proteins by host plants, larvae produce alternative proteinases with low affinity for the inhibitor and/or very high concentrations of their common proteinases so that only a portion will be inactivated.

Monophagous or oligophagous species may become adapted quite suddenly, when a mutation leads to the presence of a genomic trait responsible for the production of large quantities of a single enzyme that detoxifies the specific allelochemical of host plants. Alternatively, storage mechanisms of these compounds may also evolve.

Polyphagous insects instead, produce large quantities of generic enzymes such as the mixed-function oxidases to cope with a lot of different molecules, usually expressed in lower quantities by host plants. In other cases, insect phenology or behavior is specifically adapted to withstand the antimetabolic effects of defensive compounds. In these cases, insects are capable of avoiding either spatially or temporally the defensive compounds (e.g. Feeny, 1970).

Besides, many leaf feeders eating latex-producing plants, make basal cuts on leaf veins before feeding distally; vein cutting prevents latex flow to the intended feeding site and can be viewed as an insect counteradaptation to the plant defensive system (Dussourd and Eisner, 1987).

Expression of microbial toxins in transgenic plants

Transgenic plants expressing Bt toxins

Bacillus thuringiensis, Gram-positive bacteria in soil, produce crystalline protein inclusions during sporulation that are specifically toxic to lepidopteran, dipteran and coleopteran insects.

Most lepidopteran-active insecticidal crystal proteins (ICPs) are protoxin of MW 130–160 kDa that are, upon ingestion, proteolytically cleaved in the insect midgut into smaller, active forms (MW 60–75 kDa) derived from the N-terminal half of the protein (Chapter 1). Several genes that code for crystal protein production (also called *Cry* genes) were sequenced and several reports were published on the production of transgenic plants, using a number of *Cry1* genes, toxic to lepidopteran, obtained from B. *thuringiensis* var. *kurstaki.* First transgenic plants were obtained by using exclusively constitutive promoters. Adang *et al.* (1987) and Murray *et al.* (1991) reported the introduction of an intact *Cry1Ac* and a truncated *Cry1Ab* gene into tobacco plants; these genes were under the control of the mannopine synthase and the cauliflower mosaic virus (CaMV 35S) promoter, respectively. Barton *et al.* (1987) introduced a toxic truncated *Cry1Aa* gene into tobacco plants under the control of the CaMV 35S promoter.

Transgenic tomato plants containing a truncated *Cry1Ab* gene under the control of the CaMV 35S promoter were analyzed by Fischhoff *et al.* (1987). Vaeck *et al.* (1987) generated transgenic tobacco plants containing intact and truncated *Cry1Ab* gene constructs, as well as truncated *Cry1Ab*-NPTII fusions under the control of the mannopine synthase promoter. Perlak *et al.* (1990) obtained insect-resistant cotton plants containing *Cry1Ab* or *Cry1Ac* truncated structural genes.

In all these cases very low levels of ICPs were detected in plants containing the intact gene; plants containing truncated or fusion gene constructs showed higher levels of protein, however, Northern blot analyses usually could not detect mRNA probably due to its instability.

Transgenic plants obtained by all four research groups were tested for insecticidal activity using feeding assays with *Manduca sexta* (Johannsen) neonate larvae. Although in most cases, ICP expression correlated with toxicity, in some transformed plants ICP levels were below detectable limits enhancing the hypothesis that substances with toxic effects on *M. sexta* naturally occurred in the leaves of *Nicotiana* species (Norris and Kogan, 1980).

Gene sequence modifications enabled to increase the level of *Cry* insecticidal proteins in genetically modified plants; these were obtained either by site-directed mutagenesis to partially modify the gene or by more extensive changes in the DNA to obtain a fully modified synthetic gene. Genes were mutagenized in order to regulate some factors better, such as codon usage in plants, potential secondary structure of mRNA, and potential regulatory sequences.

After the mutagenesis of a *Cry3A* gene, Perlak *et al.* (1993) reported that Russet Burbank potato plants had been genetically improved to resist insect attack and damage by Colorado potato beetle (CPB), *Leptinotarsa decemlineata* (Say) in laboratory and field tests.

Resistance to Lepidoptera was also obtained in transgenic potatoes expressing the Bt *Cry1Ac* gene (Ebora *et al.*, 1994). Leaf disks from transgenic and untransformed potato plants were tested against the potato tuber moth, *Phthorimaea operculella* (Zeller), a major pest of potato, and European corn borer, *Ostrinia nubilalis* (Hubner) which can use potato as an alternative host. Preference tests showed that leaf disks from transgenic plants were less preferred to those from untransformed plants by different instars of the two herbivores.

Other plants were also transformed by modified and/or synthetic Bt gene to introduce insect resistance in maize (Koziel *et al.*, 1993), poplar (Robison *et al.*, 1994), tomato (Rhim *et al.*, 1995), soybean (Stewart *et al.*, 1996), rice (Nayak *et al.*, 1997), peanut (Singsit *et al.*, 1997), broccoli (Metz *et al.*, 1995), alfalfa and tobacco (Strizhov *et al.*, 1996).

We obtained transgenic eggplants by the insertion of a mutagenized *B. thuringiensis* var. *tolworthi* gene (*Cry3B*) into an eggplant commercial F_1 hybrid parent (Arpaia *et al.*, 1997). Such plants showed significant insecticidial activity on neonate larvae of CPB both in laboratory and in the field. The Bt transgene and consequently the toxic effect on CPB larvae were transmitted to progenies derived by selfing; field trials, using transformed lines or hybrids, confirmed the efficacy of these plants for CPB control (Arpaia *et al.*, 1998).

In order to evaluate the expression over time and in different tissues of the *Cry3b* toxin, we used transgenic plants of *Solanum integrifolium*, a wild species related to eggplant, under the control of the constitutive promoter 35S, by using in parallel immunoassays (DAS-ELISA tests) and insect bioassays. Two transgenic lines (4th leaf stage) were firstly used in bioassays with neonate *L. decemlineata* larvae and ELISA. Both tests showed the presence and the toxic activity of the *Cry3b* protein in plants (Table 4.1).

Table 4.1 *Cry3B* expression in young *S. integrifolium* plants measured with insect bioassays and ELISA tests

Line	*Mortality (%)*	*Mean weight of survivors (mg)*	*% of second instars*	*Optical density (405 nm)*
#22 Si Bt	100 a	–	–	0.401a
#24 Si Bt	70 a	0.87 a	0 a	0.437a
Untransformed control	0 b	82 b	100 b	0.015b

When the plants were in blossom, the same experiments were repeated using tissues from different organs of the plants. In particular, bioassays were carried out on tissues from apical leaves, basal leaves and stems, while for the immunoassays, tissues from apical leaves, basal leaves, stems, flowers, anthers and immature berries were used. In Table 4.2 the spectrophotometric values of optical density at 405 nm are reported.

Table 4.2 Results of ELISA tests performed on the same plants in blossom

Line	*Apical leaves*	*Basal leaves*	*Stems*	*Flowers*	*Anthers*	*Berries*
22 Si Bt	0.606	0.228	0.478	0.442	0.446	0.594
24 Si Bt	0.665	0.324	0.538	0.414	0.399	0.718
Control	0.062	0.075	0.006	0.009	0.057	0.050

The analysis of variance showed a statistically significant difference between transgenic lines and untransformed controls. Instead, ELISA tests could not discriminate between protein expression in different tissues of the same plant (data not shown). A lower amount of Cry3B protein was detected only in the basal leaves, probably due to a reduced protein expression in older tissues. The results of insect bioassays, showed that all three types of plant tissues were toxic to *L. decemlineata* larvae (Table 4.3).

These results were confirmed under field conditions when eggplant transgenic lines were assayed for CPB resistance (Arpaia *et al.*, 1998). In Table 4.4, the average OD ratio values between two transgenic lines and the untransformed control are reported. Further studies on hybrids obtained from the transgenic eggplant lines confirmed a

Table 4.3 Results of the insect bioassays on blossoming plants

Line	*Tissue*	*Corrected mortality (%)*	*Mean weight of survivors (mg)*
22 Si Bt	Apical leaves	100	–
24 Si Bt	Apical leaves	100	–
22 Si Bt	Stems	100	–
24 Si Bt	Basal leaves	100	–
24 Si Bt	Stems	94.74	0. 9
22 Si Bt	Basal leaves	78.95	0. 67
Control	Leaves	–	6. 28
Control	Stems	–	1. 90

Table 4.4 Measure of *Cry3B* expression (DAS-ELISA) in different tissues of transgenic eggplant in field conditions

	OD ratio			
Eggplant line	*Young leaves*	*Old leaves*	*Flowers*	*Fruits*
# 9-8	6,867	6,467	5,383	8,517
# 3-2	6,600	6,100	5,650	6,400

constant high expression of *Cry3B* toxins in all the plant tissues along the growing season (unpublished observations).

In other experimental studies, aiming at the evaluation of the expression levels of a modified *Cry3B* gene, the concentration of Bt toxin was determined several times during the vegetative cycle in eight transgenic potato plants resistant to CPB (Arpaia *et al.*, in preparation). The values ranged between 1.27–9.40 μg per gram of leaf fresh weight and they did not vary significantly with the age of the plant. Fig. 4.3 shows the elution profile of a transgenic potato plant in a time range of 16–29 min obtained at 280 nm by Anionic-Exchange High Performance Liquid Chromatography (AE-HPLC).

The brief report of several published studies just outlined, shows how microbial toxins consitutively expressed in higher plants, produced reliable results in several different experimental conditions, these plants have now reached the market place.

Studies on differential expression of genes coding for microbial toxins could be important in order to develop resistant plants with reduced selection pressure against target insects.

Inducible protein expression in transgenic plants

Based on the "systemically acquired resistance" (Collinge and Slusarenko, 1987; Bol and Van Kan, 1988) after the infection of plants with pathogens (with induction of "pathogenesis-related" proteins), several potential gene promoters were studied and tested (e.g., Van de Rhee *et al.*, 1990; Ohshima *et al.*, 1990; Samac and Shah, 1991). The choice, however, was limited since only few suitable gene promoters were available till recently (Roush, 1996).

Induced expression in transgenic plants has been attempted either via chemical induction by spraying of a benign product (Williams *et al.*, 1992) or with a wound induced expression (Johnson *et al.*, 1989).

Several chemicals such as tetracycline, dexamethasone, salicylic acid and copper have been explored as effector molecules to induce/regulate the transgene expression in higher plants. The tetracycline-inducible promoter is considered the most advanced system, and was mainly used for studies on transcription factors *in vivo*. The salicylic acid, which induces genes involved in systemic acquired resistance against pathogens, was used as chemical inducer for protection against insect feeding in transgenic plants (Gatz, 1996; Gatz and Lenk, 1998).

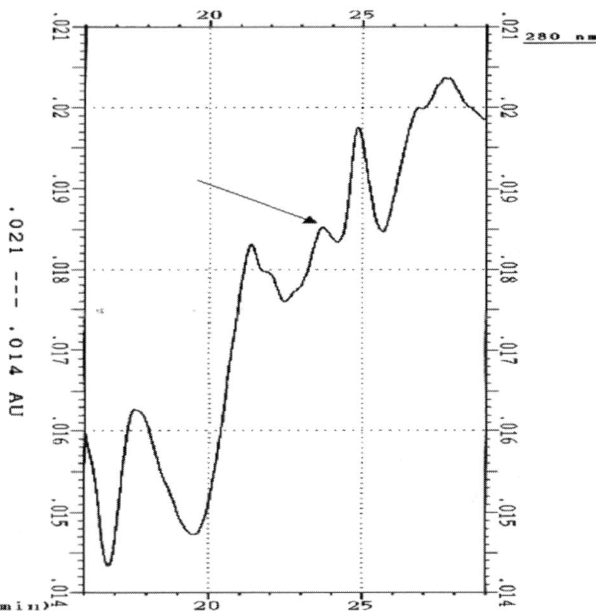

Figure 4.3 AE-HPLC elution profile at 280 nm (time range 16–29 min) of transgenic potato leaf extract. The arrow indicates the peak corresponding to Bt toxin.

Moreover, in the last few years several other examples of targeted expression of transgenes in higher plants have been reported; in particular, auxin-induced (Bellicampi *et al.*, 1996); tissue specific (Rotino *et al.*, 1997); wound-induced (Keinonen Mettala *et al.*, 1998) and ABA-induced (Bommineni *et al.*, 1998) gene expressions were tested and evaluated.

Use of inducible promoters in transgenic plants resistant to pests

Since engineered plants with genes derived from *Bacillus thuringiensis* (Bt) was one of the first projects in plant biotechnology, the evolution of genetic transformation techniques involved the utilization of tissue specific and inducible gene promoters mainly for this class of foreign genes. The first reports on chemical regulation of a native Bt gene under the control of PR-1 promoter (Van de Rhee *et al.*, 1990; Ohshima *et al.*, 1990) showed that transgenic tobacco plants treated with a chemical regulator resulted in accumulation of toxin mRNA, causing the plants to become tolerant to insects (Williams *et al.*, 1992).

To test whether plants treated with the chemical were protected from insects, bioassays with tobacco hornworm larvae were carried out. Both water and chemically treated non-transformed control plants were heavily damaged. However, the transgenic lines sustained less damage when treated with the chemical regulator than when treated with water.

A *Cry1A* anti-lepidopteran gene, under the control of a wound-inducible TR2' promoter, was engineered into potato plants (Jansens *et al.*, 1995). The authors reported on several transformation events that showed a considerable level of gene expression, no data though were presented on possible differences between the TR2' promoter and the constitutive one.

Other microbial toxins

Other resistance genes of microbial origin may be considered as potentially suitable for genetic engineering. The first one transferred in plant, under the control of a wound inducible promoter, is the isopentenyl-transferase gene (ipt) derived from *Agrobacterium tumefaciens* (Smigocki *et al.*, 1993). The gene was already known to be important for fundamental physiological processes in plants (e.g. Li *et al.*, 1992).

Smigocki *et al.* (1993) tested transgenic *Nicotiana* plants for defensive properties against tobacco hornworm, *M. sexta* larvae, and green peach aphid, *Myzus persicae* (Sulzer) nymphs. Wound induction led to a maximum expression of the toxin at the flowering stage, and the effects on both insects were evident in terms of reduction of leaf consumption by the tobacco hornworm (about 70%) and percentage of aphids reaching adulthood (50% compared to control plants). The major obstacle to the successful application of this microbial gene in transgenic plants is obviously the deleterious side-effects that the cytokinin metabolism has on the normal growth and development of transgenic plants (Smigocki *et al.*, 1993; Li *et al.*, 1992; Brzobohaty *et al.*, 1994).

A protein with insecticidal properties, a cholesterol oxidase, was identified from *Streptomyces* spp. fermentation culture filtrates (Purcell *et al.*, 1993). Cholesterol is necessary for cellular membrane functions and the cholesterol oxidase gene (choM) was demonstrated to produce a protein active against boll weevil, *Anthonomus grandis* Boheman, and tobacco budworm, *Heliothis virescens* (Fabricius) larvae at a concentration comparable to the *Bacillus thuringiensis* protein activity (Purcell *et al.*, 1993). The choM gene was isolated and cloned (Corbin *et al.*, 1994), the expression in tobacco protoplasts resulted in the production of an active cholesterol oxidase protein. Successively, the choM gene from the fungus *Streptomyces* spp. was also transferred in tobacco plants (Cho *et al.*, 1995).

A novel insecticidal toxin was more recently isolated from *Photorhabdus luminescens*, a gram-negative bacterium, mutualistic with entomophagous nematodes (Blackburn *et al.*, 1998). The protein was fractionated and one of the toxin complexes (Tca) was shown to cause profound effects on *M. sexta* midgut epithelium (Blackburn *et al.*, 1998).

A different approach was presented by Hanzlik *et al.* (1995) transferring cDNA copies of *Helicoverpa armigera* (Hubner) stunt virus (HaSV), an insect RNA virus, in tobacco. The expression in plant of the RNA1 and the coat protein of HaSV caused severe reduction in growth of *H. armigera* neonate larvae (Hanzlik and Gordon, 1997).

More possibilities are continuously being added to the list of effective genes which could prove useful in developing new transgenic varieties; the way opened by the Bt-expressing transgenic plants now commercialized will certainly speed up the pace at which these new varieties will be developed. Similarly, the experience gained so far in terms of studying insect pest-transgenic variety interactions will certainly enable to predict more accurately the possible value of the resistant germplasm.

Toxin expression and insect population dynamics

Population dynamics of herbivores and plant defenses

As we have learned during several years of use (and abuse) of chemical insecticides, we should always assume as a main objective of insect pest management the control of pest population densities below the level that causes economic damage to the crop (economic injury level, E.I.L.).

Classically, host plant resistance is aimed at the reduction of herbivore damage on crop plants and this strategy directs the efforts in obtaining a high performing plant germplasm, which consequently leads to the study of impact on insect population dynamics. Conversely, biological and chemical control has mainly studied the direct impact of control measures on herbivore populations.

Transgenic plants expressing microbial toxins combine the two strategies since the mechanism of plant resistance is directly derived from that of a microbial insecticide and most of the implications about the use of microbial toxins still hold when using transgenic resistant plants. Some points must then be carefully focused, in order to reach a profitable use of this pest management strategy.

When a chemical/microbial treatment is scheduled, to maximize its efficacy we should properly plan the timing of application of the insecticide; this decision is mainly driven by the characteristics and the status of the target insect and the crop and by environmental conditions. The latter point is not particularly important in the case of transgenic plants, since toxin expression in plants is not hampered under normal field conditions (see above).

Insect seasonal cycles and life cycles are fundamental features to be examined in order to obtain the desired effect when managing insect populations. In this respect, microbial toxins should be present in the field when:

- the maximum number of individuals is exposed to the treatment;
- and the most susceptible stage of the target pest is present.

The physiological status of the crop may also condition the need for insecticide treatments since younger plants at an initial phase of development are usually more susceptible to pest's attack.

By analyzing in parallel insect population dynamics and toxin expression in transgenic plants we can better understand the efficacy of these resistant plants as is outlined in the following examples.

Case studies of the colorado potato beetle and the european corn borer

The Colorado potato beetle's (Coleoptera: Chrysomelidae) life history varies over its geographic range. It is a polivoltine species with largely overlapping generations. The species is oligophagous and can only develop on a few Solanaceae plants. Among crop plants its preferred host is potato, serious damage can also be suffered by eggplant while only in a few areas the pest has adapted to tomato.

The beetles overwinter in the soil as adults, the emergence of post-diapause beetles depends on the latitude, and may start as early as April, but more commonly this happens in late May. Once overwintered beetles have colonized the field, they first feed and then oviposit within 5–6 days depending on the temperature. If overwintered beetles find potato crops, they can complete 1–2 generations on this crop. Depending upon the length of the crop cycle, after potato harvesting, beetles may migrate to seek new hosts. In this case, they may find eggplant where they can continue their development until they complete a third generation in a season.

Eggplants may be primary host plants and in such a situation very young eggplants are attacked. Migration capacity and flexible diapause response are additional biological features of the Colorado potato beetle which render the species capable of minimizing risks by balancing their offspring production between different years and locations in response to adverse climatic and nutritional conditions (Voss and Ferro, 1990). Beetles that emerge under short-day photoperiod do not develop their reproductive system and flight muscles that season. They feed actively for several weeks and then either walk to overwintering sites nearby or burrow into the soil directly in the field (Voss, 1989).

Fig. 4.4 schematically represents the structure of a population of *L. decemlineata* feeding on different crops during 1998 (Metaponto, southern Italy). Data were collected in experimental fields spaced apart less than 100 m. The two peaks of CPB

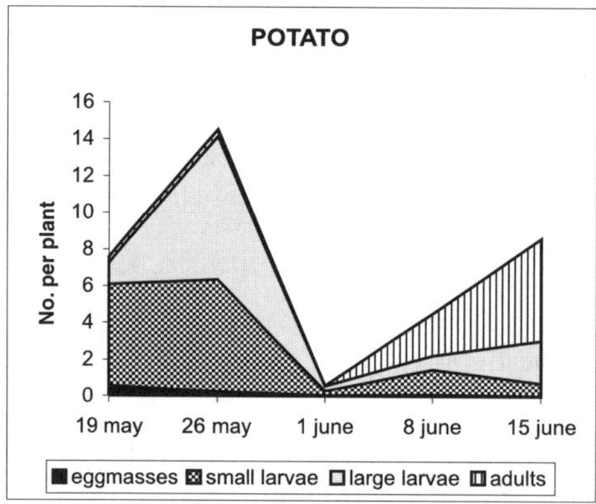

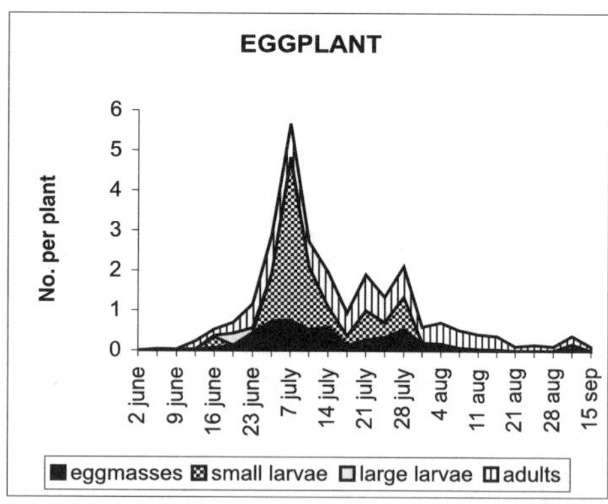

Figure 4.4 Population structure of *L. decemlineata* on different crops during 1998 at Metaponto (Southern Italy). Data can not be pooled to represent population dynamics as a continuous phenomenon since both crops were present in the field during early June. Only part of the beetles after potato harvest moved to the eggplant field.

numbers occurred in late May on potato and early July on eggplant; in both cases the presence of young larvae increased accordingly. Nevertheless, it also occurred during the growing season (June in potato, August in eggplant), that the structure of the population was mainly constituted by adults.

An ideal use of microbial toxins should, therefore, be timed when an even distribution of different insect stages is in the field. Moreover, not all types of feeding damages lead to yield losses, the effect depending on how well a plant can compensate for feeding damage at a given time in a growing season. Both potatoes and eggplants, in fact, are more susceptible at an early stage of development when a relatively limited population of CPB can completely block plant growth. On eggplant, for instance, a level of 8 large larvae was observed to be economically relevant; this was effective even on highly vigorous hybrids (Cotty and Lashomb, 1982). The detrimental effects on plant production though, were compensated by plant growth in the later part of the season. On potato, the presence of high levels of CPB populations late in the season might even be considered useful, because especially large larvae actively feeding prior to harvest may avoid to resort to vine killing, a largely adopted technique aimed at the facilitation of mechanical harvest.

In Fig. 4.5, data of CPB population levels (considering the combined mobile forms) during 1998 season at Metaponto are presented. It also graphically represents the availability of food plants in the area, according to the commonly adopted cultural practices. Early potatoes in most cases do not suffer beetle damage because of the

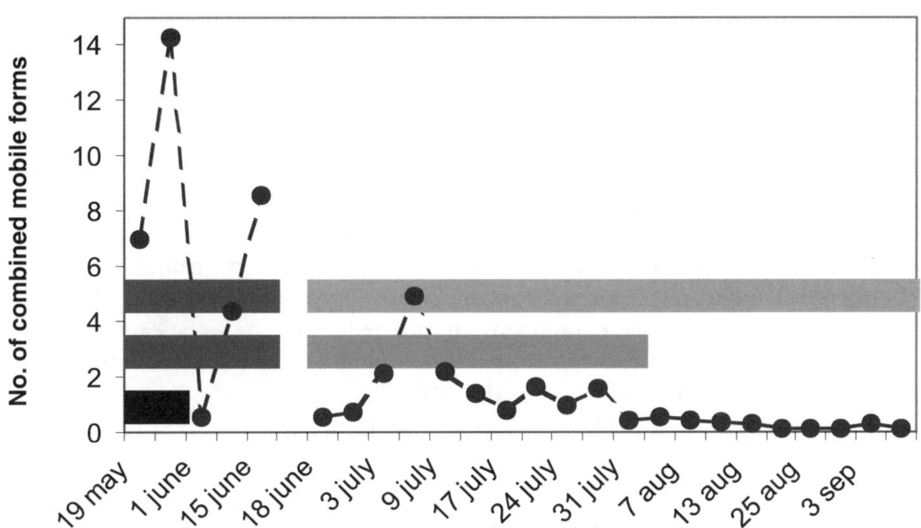

Figure 4.5 Availability of food plants to the Colorado potato beetle during a season. The dotted line represents the number of combined mobile forms (Metaponto, 1998). Bars represent the presence of food crops; lower bar = early potato, middle bar = late potato, upper bar = eggplant. The darker side of the bars indicates higher susceptibility to herbivore damage.

asynchrony between the most susceptible part of the plant cycle and the maximum presence of the herbivore mobile forms. The growing season of late potatoes, on the contrary, is compatible with the presence of fairly large number of beetles in the field. The worst situation (as is apparent in Fig. 4.5) is that of eggplant, where the host plant is available during the whole seasonal cycle of the beetle and it is transplanted in the field when most of the overwintering adults have already resumed activities. It was also proposed (Arpaia *et al.*, 1995) that the long-lasting availability of eggplants during spring and summer has contributed to the shift of the insect pest to this crop.

The overview of pest and host plant's biology furnishes some indications on the strategy that should be used in minimizing the selection pressure on CPB. From this point of view, we should aim at producing transgenic potatoes expressing the highest toxicity in the first part of the cycle; this tactic is not proposable for eggplants where a high expression throughout the cycle, except for the last weeks of cultivation, is a desirable option. Transgenic protection of early potatoes could instead be counterproductive because of continuous expression of Bt-toxins when CPB levels in the field are negligible and only occasionally may reach economic injury levels.

If we hypothesize the use of inducible expression of Bt toxin in transgenic potatoes, it seems difficult to obtain a wound inducible expression tuned in such a way that a turn on/turn off mechanism could match the above mentioned optimal criteria of plant defense. Moreover, a certain amount of feeding by adults might activate the mechanism when the population structure is not ideal for treatments with Bt toxins (e.g. June 15 on potato in Fig. 4.4).

The hypothesis of chemical inducible expression could prove to be more effective in this case, since the turn on mechanism can be activated exactly in space and time during the growing season. Nevertheless, a serious limitation could arise due to the unacceptance of this technique by the public. It may have psychological impact in the sense that a crop protection mechanism aimed at the reduction of the chemicals needs to be replaced by environmentally benign chemical inducible expression for effective control.

Another well-studied case of transgenic resistance is that of maize resistant to the European corn borer (ECB), *Ostrinia nubilalis* (Hubner) (Lepidoptera: Pyralidae), one of the most serious pests of corn in Europe and in North America. While it is generally considered to be linked to maize, this insect represents a very interesting case of disjunctive oligophagy, since it can seriously damage other crop plants belonging to very different families (e.g. potato, pepper, bean). The species has evolved into ecotypes based on the number of generations per year. ECB populations that live at higher altitudes have a typical yearly generation, while in more southern areas two or more generations per year are completed.

In a typical seasonal cycle in the Mediterranean areas with 2 yearly generations (Fig. 4.6), mature larvae overwinter in the residues of past vegetation (stalks, stubble, etc.) and the subsequent pupation occurs inside the tunnels prepared by the larvae. Moths emerge in the spring and only 3–4 days later oviposition begins. Eggmasses are preferentially laid on the lower surface of host leaves. Neonate larvae start feeding on ligules before entering the whorl where they develop. About a month later larvae reach the maturity. The immatures of the next generations may feed on ears, before tunneling into the stalks.

Table 4.6 Percentage yield loss caused by European corn borer for various corn growth stages

Plant stage	*Yield loss/borer/plant (%)*
Early whorl	5.5
Late whorl	4.4
Pre-tassel	6.6
Pollen-shed	4.4
Blister	3.0
Dough	2.0

Source: North Central Regional Extension Publication No. 327

The damage caused by this pest is due to its feeding activity on ears that leads to the production of unmarketable ears and to the mining activity on stalks that causes plant weakening and subsequently the breaking up of the plant. Feeding activity on leaves can be tolerated in most cases, without direct effect on yield.

Usually, the second generation is the most dangerous to the crop. The reasons for this are that no ears are present early in the life cycle of corn plants and secondly the absence of the 2,4-hydroxy-7-methoxy-1,4-benzoxazin-3-one (DIMBOA), known as a potent allelochemical effective against ECB larvae. It may also become very difficult to find access in cornfields for spraying when tall plants are present.

Percentage yield loss differs depending on the plant growth stage at the time the damage occurs. Cavity formation by the first generation borer usually occurs before tassel emergence resulting in approximately 5 per cent yield loss per borer per plant (Witkowski and Wright, 1997). These levels can be sensibly reduced in corn accessions expressing high levels of DIMBOA. Yield losses due to second generation larvae (per borer) vary widely because cavity formation may occur over several weeks, and rapid physiological changes occur as the ear is approaching maximum size and physiological maturity. The pre-tassel stage has been indicated as the most susceptible one, the average value of about 5 per cent yield loss per borer per plant occurs until shortly after pollen shed. As the ear advances from pollen shed to physiological maturity, the yield reduction per borer rapidly decreases (see Table 4.6). The DIMBOA-expressing hybrids maintain this resistance trait up to tassel emergence. Some varieties of corn may also express sheath and collar feeding resistance applicable against second generation borers.

However, the question is that how an inducible activation of Bt defense in plants could help the sustainability of transgenic germplasm? In case of wound inducible genes, the possible turn on mechanism might be afffecting the feeding activity of first generation larvae on young leaves, which can tolerate quite some damage without direct effect on yield. But we still do not have a dependable mechanism of turning off the expression without which our efforts of reducing selection pressure will be jeopardized. Transgenic Bt corns that are presently being cultivated are fundamentally of two types:

- one type whose expression is driven by a constitutive promoter (Ostlie *et al.*, 1997);
- a second type with high levels of toxin in green tissues and lower expression in other plant parts such as silks, kernels and stalks (Koziel *et al.*, 1993).

The first type produces high levels of mortality throughout the growing season, while the second type of corn seems to have a declining activity towards the end of the season; i.e. about 75 percent of mortality (Ostlie *et al.*, 1977). This residual activity might concur in favouring the pest's adaptation to the transgenic plants according to the simulation model of Onstad and Gould (1998). This finding has obvious implications for a strategy of induced expression of toxins in transgenic plants, strongly supporting the idea of the fundamental importance of an accurate targeting of the induced protein expression.

Considering an optimal expression of insect resistance in maize, efforts should be made to genetically engineer corn with high expression of DIMBOA at an early stage of development and then induce expression of Bt toxins (or other toxins active against insects) in a more advanced phase of the growing cycle. Once again, chemically inducible promoters might play an important role to achieve the required goal but we still need to gather scientifically sound information about the mechanisms of wound induction and the relationship between feeding activity and protein expression over time. A corn plant with high expression of DIMBOA, transformed with a wound inducible microbial toxin could inhibit larval feeding at an early stage completely, then let the turn-on mechanism activate when larvae can feed on less-expressing DIMBOA plants, let say, at tassel emergence.

The desired effects of slowing down the spread of resistance alleles in an insect population may be effectively achieved only when a very dependable control of the mechanisms of turn-on and turn-off the gene expression without residual activity over time will be available.

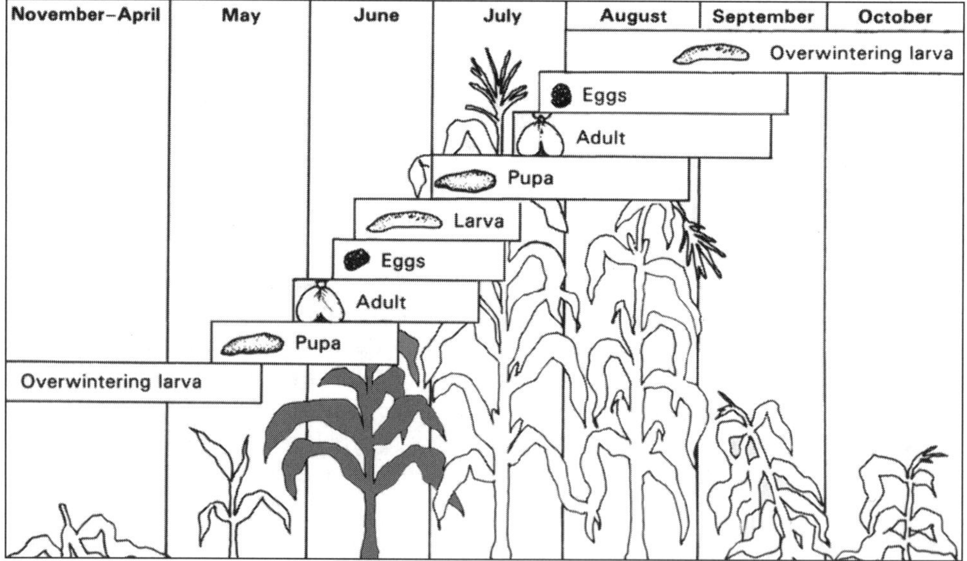

Figure 4.6 Life cycle of *Ostrinia nubilalis* (Hubner) and the corn-growing season in Southern Europe. The darker plant indicates the most susceptible stage during the growing season.

Beyond the 35s: towards a second generation of transgenic plants resistant to insects

Since the first reports about insect adaptation to *Cry* toxins (Tabashnik *et al.*, 1990; Whalon *et al.*, 1993), several simulation models have been used to predict the possible adaptation of target insects to transgenic plants (e.g. Arpaia *et al.*, 1998; Onstad and Gould, 1998; Peck *et al.*, 1999).

It is clear from all the published studies that the onset of resistance to Cry toxins is a likely phenomenon, if transgenic resistant plants will be deployed in the fields at an increasing speed without planning any appropriate resistance management strategy.

As with resistance to conventional insecticides, the development of new toxins or resistance sources will involve costs that must be borne by farmers and consumers (Mc Gaughey *et al.*, 1998). Moreover, in the large areas cultivated with transgenic crops resistant to insects, farmers are experiencing a drastic reduction of use of chemicals with obvious benefits for the consumers, the farm workers and the environment.

On the other hand, there is an increasing concern that Bt-based insecticides, probably the most useful product for crop protection in organic farms and in forests, might lose efficacy since the high selection pressure exerted by transgenic plants will certainly increase over time. This will subsequently increase the resistance to Cry toxins.

Different solutions have been proposed for the improvement of transgenic plants in order to guarantee a long-lasting efficacy of insect resistance characters. The strategy that seems to be adoptable is the high dose/refuge strategy that has become the primary goal of industries and regulatory agencies (Gould, 1998). In perspective, alternative strategies must be carefully studied in order to propose additional possibilities of resistance management.

The first generation of transgenic plants, currently grown on million of hectares has exploited the benefits of the 35S. This constitutive promoter has lead to the genetic transformation of over 50 plant species. A second generation of transgenic plants will certainly include varieties with a more targeted expression; the first example is the ECB-resistant maize, which expresses the transgene via tissue specific promoters (Koziel *et al.*, 1993).

We have briefly outlined above the difficulties still connected to the use of inducible defenses in resistant transgenic plants. Several studies are still in progress on this front, the technical difficulties involved and the costs are still a constraint for their commercial development and thus limit their ready availability for agricultural use.

We believe that the best strategy for each project of genetic engineering for pest resistance might be adopted after a sound consideration of pest and plant's biology, of its damage to the crop and the possible control strategies. Finally, it should be assessed if the transgenic resistant variety could be effectively managed in a contest of integrated pest management (e.g. knowledge of its effects on secondary pests, natural enemies, etc.) of the specific agroecosystem.

It is easy to foresee that each different case will require a different approach, and in this respect the exploitation of all the available techniques should be used to propose a successful, long-lasting and sustainable strategy of pest management with resistant plants.

References

Adang, M.J., Firoozabady, E., Klein, J., DeBoer, D., Sekar, V., Kemp, J.D. *et al.* (1987) Expression of a *Bacillus thuringiensis* insecticidal crystal protein gene in tobacco plants. In C.J. Arntzen and C. Ryan (eds.), *Molecular Strategies for Crop Protection*, UCLA Symposia on Molecular and Cellular Biology, New Series, Vol. 46, Alan R. Liss, New York, pp. 345–353.

Arpaia, S., Lashomb, J.H., and Vail, K. (1995) Valutazione dell' attivita' trofica di Leptinotarsa decemlineata (Say) su melanzana. *Inform. Fitopatol.*, **2**, 55–57.

Arpaia, S., Mennella, G., Onofaro, V., Perri, E., Sunseri, F., and Rotino, G.L. (1997) Production of transgenic eggplant (*Solanum melongena* L.) resistant to Colorado potato beetle (*Leptinotarsa decemlineata* Say). *Theor. Appl. Genet.*, **95**, 329–334

Arpaia, S., Acciarri, N., Di Leo, G.M., Mennella, G., Sabino, G., Sunseri, F. *et al.* (1998) Field performance of Bt-expressing transgenic eggplant lines resistant to Colorado potato beetle. *Proc. Xth EUCARPIA Meeting on Capsicum and Eggplant*, pp. 191–194.

Baldwin, I.T. (1993) Chemical changes rapidly induced by folivory. In E.A. Bernays (ed.), *Insect-Plant Interactions*, Vol. 5, CRC Press Inc., Boca Raton, Fl, pp. 1–23.

Barrett, A.J. (1986) An introduction to the proteinases. In A.J. Barrett and G. Salvesan (eds.), *Proteinase Inhibitors*, Elsevier, Amsterdam, pp. 3–22.

Barton, K.A., Whiteley, H.R., and Yang, Y.S. (1987) *Bacillus thuringiensis* delta-endotoxin expressed in transgenic *Nicotiana tabacum* provides resistance to lepidopteran insects. *Plant Physiol.*, **85**, 1103–1109.

Bellincampi, D., Cardarelli, M., Zaghi, D., Serino, G., Salvi, G., Gatz, C. *et al.* (1996) Oligogalacturonides prevent rhizogenesis in rolB-transformed tobacco explants by inhibiting auxin-induced expression of the rolB gene. *Plant Cell*, **8**, 477–487.

Benz, G. (1978) Insect induced resistance as a means of self-defence of plants. *Bulletin SROP*, **3**, 155–159.

Bishop, P.D., Makus, D.J., Pearce, G., and Ryan, C.A. (1981) Proteinase inhibitor-inducing factor activity in tomato leaves resides in oligosaccharides enzymically released from cell walls. *Proc. Natl. Acad. Sci. USA*, **78**, 3536–3540.

Blackburn, M., Golubeva, E., Bowen, D., and Ffrench-Constant, R.H. (1998) A novel insecticidal toxin from *Photorhabdus luminescens*, toxin complex a (Tca) and its histopathological effects on the midgut of *Manduca sexta. Appl. Environ. Microbiol.*, **64**, 3036–3041.

Bol, J.F., and Van Kan, J.A.L. (1988) The synthesis and possible functions of virus-induced proteins in plants. *Microbiol. Sci.*, **5**, 47–52.

Boller, T. (1988) Ethylene and the regulation of antifungal hydrolases in plants. *Oxf. Surv. Plant Mol. Cell Biol.*, **5**, 145–174.

Bolter, C.J. (1992) Effect of induced resistance mechanisms in potato and tomato plants on the Colorado potato beetle. *Ser. Entomologica*, **49**, 253–254.

Bolter, C.J. (1993) Methyl jasmonate induces papain inhibitor(s) in tomato leaves. *Plant Physiol.*, **103**, 1347–1353

Bolter, C.J., and Jongsma, M.A. (1995) Colorado potato beetles (*Leptinotarsa decemlineata*) adapt to proteinase inhibitors induced in potato leaves by methyl jasmonate. *J. Insect Physiol.*, **41**, 1071–1078.

Bommineni, V.R., Bethune, T.D., Hull, K.L., and Dunstan, D.I. (1998) Expression of uidA under regulation of the ABA-inducible wheat Em promoter in transgenic white spruce somatic embryos. *J. Plant Physiol.*, **152**, 455–462.

Broadway, R.M., and Duffey, S.S. (1986) The effect of dietary protein on the growth and digestive physiology of larval *Heliothis zea* and *Spodoptera exigua. J. Insect Physiol.*, **32**, 673–680.

Broadway, R.M., Duffey, S.S., Pearce, G., and Ryan, C.A. (1986) Plant proteinase inhibitors: a defence against herbivorous insects. *Entomol. exp. appl.*, **41**, 33–38.

Brown, W.E., and Ryan, C.A. (1984) Isolation and characterization of a wound-induced trypsin inhibitor from alfalfa. *Biochemistry*, **23**, 3418–3422.

Brzobohaty, B., Moore, I., and Palme, K. (1994) Cytokinin metabolism: implications for regulation of plant growth and development. *Plant Mol. Biol.*, **26**, 1483–1497.

Carozzi, N., and Koziel, M. (1997) *Advances in Insect Control: The Role of Transgenic Plants*. Taylor & Francis, London, 301 pp.

Cho, H.J., Choi, K.P., Yamashita, M., Morikawa, H., and Murooka, Y. (1995) Introduction and expression of the *Streptomyces* cholesterol oxidase gene (ChoA), a potent insecticidal protein active against boll weevil larvae, into tobacco cells. *Appl. Microbiol. Biotechnol.*, **44**, 133–138.

Collinge, D.B., and Slusarenko, A.J. (1987) Plant gene expression in response to pathogens. *Plant Mol. Biol.*, **9**, 389–410.

Corbin, D.R., Greenplate, J.T., Wong, E.Y., and Purcell, J.P. (1994) Cloning of an insecticidal cholesterol oxidase gene and its expression in bacteria and in plant protoplasts. *Appl. Environ. Microbiol.*, **60**, 4239–4244.

Cotty, S., and Lashomb, J.H. (1982) Vegetative growth and yeld response of eggplant to varying first generation Colorado potato beetle densities. *J. N.Y. Entomol. Soc.*, **XC**, 220–228.

Dussourd, D.E., and Eisner, T. (1987) Vein-cutting behavior: insect counterploy to the latex defense of plants. *Science*, **237**, 898–901.

Ebora, R.V., Ebora, M.M., and Sticklen, M.B. (1994) Transgenic potato expressing the *Bacillus thuringiensis* cry IA(c) gene: effects on the survival and food consumption of *Phthorimea operculella*. *J. Econ. Entomol.*, **87**, 1122–1127.

Farmer, E.E., and Ryan, C.A. (1990) Interplant communication: Airborne methyl jasmonate induces synthesis of proteinase inhibitors in plant leaves. *Proc. Natl. Acad. Sci. USA*, **87**, 7713–7716.

Farmer, E.E., Johnson, R.R., and Ryan, C.A. (1992) Regulation of expression of proteinase inhibitor genes by methyl jasmonate and jasmonic acid. *Plant Physiol.*, **98**, 995–1002.

Feeny, P. (1970) Seasonal changes in oak leaf tannins and nutrients as a cause of spring feeding by winter moth caterpillars. *Ecology*, **51**, 565–581.

Fischhoff, D.A., Bowdish, K.S., Perlak, F.J., Marrone, P.G., McCormick, S.M., Niedermeyer, J.G. *et al.* (1987) Insect tolerant transgenic tomato plants. *Bio/Technology*, **5**, 807–813.

Gatehouse, A.M.R., Butler, K.J., Fenton, K.A., and Gatehouse, J.A. (1985) Presence and partial characterisation of a major protecolytic enzyme in the larval gut of *Callosobruchus maculatus*. *Entomol. exp. appl.*, **39**, 279–286.

Gatz, C. (1996) Chemically inducible promoters in transgenic plants. *Curr. Opinions Biotech.*, **7**, 168–172.

Gatz, C., and Lenk, I. (1998) Promoters that respond to chemical inducers. *Trends Plant Sci.*, **3**, 352–358.

Gould, F. (1988a) Ecological-genetic approaches for the design of genetically engineered crops. In D.W. Roberts and R.R. Granados (eds.), *Biotechnology, Biological Pesticides and Novel Plant-Pest Resistance for Insect Pest Management*, Cornell University, Ithaca, New York, pp 146–151.

Gould, F. (1988b) Genetic engineering, integrated pest management and the evolution of pests. *Biotech.*, **6**, 15–19.

Gould, F. (1991) The evolutionary potential of crop pests. *Am. Scient.*, **79**, 496–507.

Gould, F. (1998) Sustainability of transgenic insecticidal cultivars: integrating pest genetics and ecology. *Annu. Rev. Entomol.*, **43**, 701–726.

Graham, J.S., Hall, G., Pearce, G., and Ryan, C.A. (1986) Regulation of synthesis of proteinase I and II mRNAs in leaves of wounded tomato plants. *Planta*, **169**, 399–405.

Green, T.R., and Ryan, C.A. (1972) Wound-induced proteinase inhibitor in tomato leaves: some effects of light and temperature on the wound response. *Plant Physiol.*, **51**, 19–21.

Hanzlik, T.N., and Gordon, K.H.J. (1997) The Tetraviridae. *Adv. Virus Res.*, **48**, 101–168.

Hanzlik, T.N., Dorrian, S.J., Johnson, K.N., Brooks, E.M., and Gordon, K.H.J. (1995) Sequence of RNA2 of the *Helicoverpa armigera* stunt virus (Tetraviridae) and bacterial expression of its genes. *J. Gen. Virol.*, **4**, 799–811.

Hartley, B.S. (1970) Homologies in serine proteinases. *Phil. Trans. R. Soc. London B*, **257**, 77–87.

Hilder, V.A., Gatehouse, A.M.R., Sheerman, S.E., Barker, R.F., and Boulter, D. (1987) A novel mechanism of insect resistance engineered into tobacco. *Nature*, **330**, 160–163.

Hildman, T., Ebneth, M., Pena-Cortes, H., Sanchez-Serrano, J.J., Willmitzer, L., and Prat, S. (1992) General roles of abscisic and jasmonic acids in gene activation as a result of mechanical wounding. *Plant Cell*, **4**, 1157–1170.

Hines, M.E., Nielsen, S.S., Shade, R.E., and Pomeroy, M.A. (1990) The effect of proteinase inhibitors, E-64 and the Bowman-Birk inhibitor, on the developmental time and mortality of *Acanthoscelides obtectus*. *Entomol. exp. appl.*, **57**, 201–207.

James, M.N.G. (1976) Relationship between the structures and activities of some microbial serine proteinases. II. Comparison of the tertiary structures of microbial and pancreatic serine proteases. In D.W. Ribbons and K. Brew (eds.), *Proteolysis and Physiological Regulations*, Academic Press, New York, pp. 125–142.

Jansens, S., Cornelissen, M., De Clercq, R., Reynaerts, A., and Peferoen, M. (1995) *Phthorimaea operculella* (Lepidoptera: Gelechiidae) resistance in potato by expression of the *Bacillus thuringiensis* cryIA(b) insecticidal crystal protein. *J. Econ. Entomol.*, **88**, 1469–1476.

Jermy, T. (1993) Evolution of insect-plant relationships – a devil's advocate approach. *Entomol. exp. appl.*, **66**, 3–12.

Johnson, R., Narvaez, J., An, G., and Ryan, C.A. (1989) Expression of proteinase inhibitors I and II in transgenic tobacco plants: effects on natural defence against *Manduca sexta* larvae. *Proc. Natl. Acad. Sci. USA*, **86**, 9871–9875.

Jongsma, M.A. (1995) The resistance of insects to plant proteinhase inhibitors. Ph.D. Thesis, Landbouwuiversiteit, Wageningen, 93 pp.

Jongsma, M.A., Bakker, P.L., Peters, J., Bosch, D., and Stiekema, W.J. (1995) Adaptation of *Spodoptera exigua* larvae to plant proteinase inhibitors by induction of gut proteinase activity insensitive to inhibition. *Proc. Natl. Acad. Sci. USA.*, **92**, 8041–8045.

Keinonen Mettala, K., Pappinen, A., Weissenberg, K., and von Weissenberg, K. (1998) Comparisons of the efficiency of some promoters in silver birch (*Betula pendula*). *Plant Cell Rep.*, **17**, 356–361.

Kitch, L.W., and Murdock, L.L. (1986) Partial characterization of a major gut thiol proteinase from larvae of *Callosobruchus maculatus* F. *Arch. Insect Biochem. Physiol.*, **3**, 561–575.

Kort, K.L., and Dixon, R.A. (1997) Evidence for chewing insect-specific molecular events distinct from a general wound response in leaves. *Plant Physiol.*, **115**, 1299–1305.

Koziel, M.G., Beland, G.L., Bowman, C., Carozzi, N.B., Crenshaw, R., Crossland, L. *et al.* (1993) Field performance of elite transgenic maize plants expressing an insecticidal protein derived from *Bacillus thuringiensis*. *Bio/Technology*, **11**, 194–200.

La Porta, N., Belloni, V., Delle Donne, M., Mennella, G., and Rotino, G.L. (1998) *Agrobacterium* mediated genetic transformation and plant regeneration in eggplants with cysteine proteinase inhibitor gene as defense against Colorado potato beetle, *Leptinotarsa decemlineata* Say. *XV EUCARPIA General Congress "Genetics and Breeding for Crop Quality and Resistance*", Viterbo 20–25 Settembre 1998; p 44.

Li, Y., Hagen, G., and Guilfoyle, T.J. (1992) Altered morphology in transgenic tobacco plants that overproduce cytokinins in specific tissues and organs. *Dev. Biol.*, **153**, 386–395.

Liang, C., Brookhart, G., Feng, G.H., Reeck, G.R., and Kramer, K.J. (1991) Inhibition of digestive proteinases of stored grain Coleoptera by oryzacystatin, a cysteine proteinase inhibitor from rice seed. *FEBS Lett.*, **278**, 139–142.

Logemann, J., Mayer, J.E., Schell, J., and Willmitzer, L. (1988) Differential expression of genes in potato tubers after wounding. *Proc. Natl. Acad. Sci. USA.*, **85**, 1136–1140.

McGaughey, W.H., Gould, F., and Gelernter, W. (1998) Bt resistance management. *Nat. Biotechnol.*, **16**, 144–146.

McGurl, B., Pearce, G., Orozco-Cardenas, M., and Ryan, C.A. (1992) Structure, expression and antisense inhibition of the systemin precursor gene. *Science*, **255**, 1570–1573.

Metz, T.D., Roush, R.T., Tang, J.D., Shelton, A.M., and Earle, E.D. (1995) Transgenic broccoli expressing a *Bacillus thuringiensis* insecticidal crystal protein: implications for pest resistance management strategies. *Mol. Breed.*, **1**, 309–317.

Moffat, A.S. (1998) Toting up the early harvest of transgenic plants. *Science*, **282**, 2176–2178.

Murdock, L.L., Brookhart, G., Dunn, P.E., Foard, D.E., Kelley, S. and Kitch, L. (1987) Cysteine digestive proteinases in Coleoptera. *Comp. Biochem. Physiol.*, **87B**, 783–787.

Murray, E.E., Rocheleau, T., Eberle, M., Stock, C., Sekar, V., and Adang, M. (1991) Analysis of unstable RNA transcripts of insecticidal crystal protein genes of *Bacillus thuringiensis* in transgenic plants and electroporeted protoplasts. *Plant Mol. Biol.*, **16**, 1035–1050.

Nayak, P., Basu, D., Das, S., Basu, A., Ghosh, D., Ramakrishnan, N.A. *et al.* (1997) Transgenic elite indica rice plants expressing cry IA (c) delta-endotoxin of *Bacillus thuringiensis* are resistant against yellow stem borer (*Scirpophaga incertulas*). *Proc. Natl. Acad. Sci. USA*, **94**, 2111–2116.

Norris, D.M., and Kogan, N. (1980) Biochemical and morphological bases of resistance. In F.G. Maxwell and P.R. Jennings (eds.), *Breeding Plants Resistant to Insects*, John Wiley & Sons, New York, pp. 29–61.

Ohshima, M., Itoh, H., Matsuoka, M., Murakami, T., and Ohashi, Y. (1990) Analysis of stress-induced or salicylic acid-induced expression of the pathogenesis-related 1a protein gene in transgenic tobacco. *Plant Cell*, **2**, 95–106.

Onstad, D.W., and Gould, F. (1998) Do dynamics of crop maturation and herbivorous insect life cycle influence the risk of adaptation to toxins in transgenic host plants? *Environ. Entomol.*, **27**, 517–522.

Ostlie, K.R., Hutchison, W.D., and Hellmich, R.L. (1997) *Bt corn and european corn borer*. North Central Reg. Ext. Publ., Univ. Minn. Ext. Serv., St. Paul, MN.

Pautot, V., Holzer, F.M., and Walling, L.L. (1991) Differential expression of tomato proteinase inhibitors I and II genes during bacterial pathogen invasion and wounding. *Mol. Plant Microbe Interact.*, **4**, 284–292.

Pearce, G., Strydom, D., Johnson, S., and Ryan, C.A. (1991) A polypeptide from tomato leaves induces wound-inducible proteinase inhibitor proteins. *Science*, **253**, 895–898.
Peck, S.L., Gould, F., and Ellner, S.P. (1999) Spread of resistance in spatially extended regions of transgenic cotton: implications for management of *Heliothis virescens* (Lepidoptera: Noctuidae). *J. Econ. Entomol.*, **92**, 1–16.
Peña-Cortes, H., Sanchez-Serrano, J., Rocha-Sosa, M., and Willmitzer, L. (1988) Systemic induction of proteinase inhibitor-II gene expression in potato plants by wounding. *Planta*, **174**, 84–89.
Peña-Cortes, H., Sanchez-Serrano, J.J., Mertens, R., Willmitzer, L., and Prat, S. (1989) Abscisic acid is involved in the wound-induced expression of the proteinase inhibitor II gene in potato and tomato. *Proc. Natl. Acad. Sci. USA*, **86**, 9851–9855.
Perlak, F.J., Deaton, R.W., Armstrong, T.A., Fuchs, R.L., Sims, S.R., Greenplate, J.T. *et al.* (1990) Insect resistant cotton plants. *Bio/Technology*, **8**, 939–943.
Perlak, F.J., Stone, T.B., Muskopf, Y.M., Petersen, L.J., Parker, G.B., McPherson, S.A. *et al.* (1993) Genetically improved potatoes: protection from damage by Colorado potato beetles. *Plant Mol. Biol.*, **22**, 313–321.
Price, P.W. 1984. *Insect Ecology*, 2nd edition. Wiley, New York, 601 pp.
Purcell, J.P., Greenplate, J.T., Jennings, M.G., Ryerse, J.S., Pershing, J.C., Sims, S.R. *et al.* (1993) Cholesterol oxidase: a potent insecticidal protein active against boll weevil larvae. *Biochem. Biophys. Res. Commun.*, **196**, 1406–1413.
Rhim, S.L., Cho, H.J., Kim, B.D., Schnetter, W., and Geider, K. (1995) Development of insect resistance in tomato plants expressing the δ-endotoxin gene of *Bacillus thuringiensis* subsp. *tenebrionis*. *Mol. Breed.*, **1**, 229–236.
Richardson, M. (1977) The proteinase inhibitors of plants and micro-organisms. *Phytochemistry*, **16**, 159–169.
Richardson, M. (1991) Seed storage proteins: The enzyme inhibitors. In P.M. Dey and J.B. Harborne (eds.), *Methods in Plant Biochemistry*, Vol. 5, Academic Press, New York, pp. 259–306.
Roberts, E., Kutchan, T., and Kolattukudy, P.E. (1988) Cloning and sequencing of cDNA for a highly anionic peroxidase from potato and the induction of its mRNA in suberizing potato tubers and tomato fruits. *Plant Mol. Biol.*, **11**, 15–26.
Robison, D.J., McCown, B.H., and Raffa, K.F. (1994) Responses of gypsy moth (Lepidoptera: Lymantriidae) and forest tent caterpillar (Lepidoptera: Lasiocampidae) to transgenic poplar, *Populus* spp., containing a *Bacillus thuringiensis* δ-endotoxin gene. *Environ. Entomol.*, **23**, 1030–1041.
Rotino, G.L., Zottini, M., Perri, E., Sommer, H., and Spena, A. (1997) Genetic engineering of parthenocarpic plants. *Nature Biotech.*, **15**, 1398–1401.
Roush, R.T. (1996) Can we slow adaptation by pests to insect-resistant transgenic crops? In G. Persley (ed.), *Biotechnology and Integrated Pest Management*, CAB International, Wallingford, UK, pp. 242–263.
Roush, R.T. (1997) Managing resistance to transgenic crops. In N. Carozzi and M. Koziel (eds.), *Advances in Insect Control: the role of transgenic plants*, Taylor & Francis Inc., Bristol, pp. 271–294.
Ryan, C.A. (1973) Proteolytic enzymes and their inhibitors in plants. *Ann. Rev. Plant Physiol.*, **24**, 173–196.
Ryan, C.A. (1978) Proteinase inhibitors in plant leaves: a biochemical model for pest-induced natural plant protection. *Trends Biochem. Sci.*, **5**, 148–150.
Ryan, C.A. (1990) Proteinase inhibitor in plants: genes for improving defenses against insects and pathogens. *Annu. Rev. Phytopathol.*, **28**, 425–449.
Samac, D.A., and Shah, D.M. (1991) Developmental and pathogen-induced activation of the *Arabidopsis* acidic chitinase promoter. *Plant Cell*, **3**, 1063–1072.
Sanchez-Serrano, J., Schmidt, R., Schell, J., and Willmitzer, L. (1986) Nucleotide sequence of proteinase inhibitor II encoding cDNA of potato (*Solanum tuberosum*) and its mode of expression. *Mol. Gen. Genet.*, **203**, 15–20.
Shirras, A.D., and Northcote, D.H. (1984) Molecular cloning and characterisation of cDNAs complementary to mRNAs from wounded potato (*Solanum tuberosum*) tuber tissue. *Planta*, **162**, 353–360.
Singsit, C., Adang, M.J., Lynch, R.E., Anderson, W.F., Aiming, W., Cardineau, G. *et al.* (1997) Expression of a *Bacillus thuringiensis* Cry IA(c) gene in transgenic peanut plants and its efficacy against lesser cornstalk borer. *Transgenic Res.*, **6**, 169–176.
Smigocki, A., Neal, J.W. Jr., McCanna, I., and Douglass, L (1993) Cytokinin-mediated insect resistance in *Nicotiana* plants transformed with the ipt gene. *Plant Mol. Biol.*, **23**, 325–335.
Stanford, A., Bevan, M.W., and Northcote, D.H. (1989) Differential expression within a family of novel wound-induced genes in potato. *Mol. Gen. Genet.*, **215**, 200–208.
Stewart, C.N. Jr., Adang, M.J., All, J.N., Boerma, H.R., Cardineau, G., Tucker, D. *et al.* (1996) Genetic transformation, recovery, and characterization of fertile soybean transgenic for a synthetic *Bacillus thuringiensis* cry IA(c) gene. *Plant Physiol.*, **112**, 121–129.
Stiekema, W.J., Heidekamp, F., Dirkse, W.G., van Beckum, J., de Haan, P., ten Bosch, C. *et al.* (1988) Molecular cloning and analysis of four potato tuber mRNAs. *Plant Mol. Biol.*, **11**, 255–269.

Stout, M.J., Workman, J., and Duffey, S.S. (1994) Differential induction of tomato foliar proteins by arthropod herbivores. *J. Chem. Ecol.*, **20**, 2575–2594.

Stout, M.J., Workman, K.V., and Duffey, S.S. (1996) Identity, spatial distribution, and variability of induced chemical responses in tomato plants. *Entomol. exp. appl.*, **79**, 255–271

Strizhov, N., Keller, M., Mathur, J., Koncz-Kalman, Z., Bosch, D., Prudovsky, E. *et al.* (1996) A synthetic cry IC gene, encoding a *Bacillus thuringiensis* δ-endotoxin, confers *Spodoptera* resistance in alfalfa and tobacco. *Proc. Natl. Acad. Sci. USA*, **93**, 15012–15017.

Tabashnik, B.E., Cushing, N.L., Finson, N., and Johnson, M.W. (1990) Field development of resistance to *Bacillus thuringiensis* in diamondback moth (Lepidoptera: Plutellidae). *J. Econ. Entomol.*, **83**, 1671–1676.

USDA/APHIS Biotechnology permits database. Website: http:\\www.aphis.usda.gov/bbep/bp

Vaeck, M., Reynaerts, A., Hofte, H., Jansens, S., DeBeukleer, M., Dean, C. *et al.* (1987) Transgenic plants protected from insect attack. *Nature*, **328**, 33–37.

Van de Rhee, M.D., Van Kan, J.A.L., Gonzalez-Jaén, M.T., and Bol, J.F. (1990) Analysis of regulatory elements involved in the induction of two tobacco genes by salicylate treatment and virus infection. *Plant Cell*, **2**, 357–366.

Voss, R.H. (1989) Population dynamics of the Colorado potato beetle, (*Leptinotarsa decemlineata Say*) (Coleoptera: Chrysomelidae), in western Massachusetts, with particular emphasis on migration and dispersal processes. Ph.D. Thesis, University of Massachusetts, Amherst.

Voss, R.H., and Ferro, D.N. (1990) Ecology of migrating Colorado Potato beetles (Coleoptera: Chrysomelidae) in western Massachussets. *Environ. Entomol.*, **19**, 123–129.

Whalon, M.E., Miller, D.L., Hollinghworth, R.M., Grafius, E.J., and Miller, J.R. (1993) Selection of a Colorado potato beetle (Coleoptera: Chrysomelidae) strain resistant to *Bacillus thuringiensis*. *J. Econ. Entomol.*, **86**, 226–233

Wilkinson, C.F. (1983) Role of mixed-function oxidases in insecticide resistance. In G.P. Georghiou and T. Saito (eds.), *Pest Resistance to Pesticides*, Plenum Press, New York, pp. 175–205.

Williams, S., Friedrich, L., Dincher, S., Carozzi, N., Kessman, H., Ward, E. *et al.* (1992). Chemical regulation of *Bacillus thuringiensis* delta-endotoxin expression in transgenic plants. *Bio/Technology*, **10**, 540–543.

Witkowski, J., and Wright, R. (1997) The European corn borer: biology & management. 12 pp. HTML document at the web site: http:\\www.ianr.unl.edu.

Wolfson, J.L., and Murdock, L.L. (1987) Suppression of larval Colorado potato beetle growth and development by digestive proteinase inhibitors. *Entomol. exp. appl.*, **44**, 235–240.

Wolfson, J.L., and Murdock, L.L. (1990) Growth of *Manduca sexta* on wounded potato plants: role of induced proteinase inhibitors. *Entomol. exp. appl.*, **54**, 257–264.

Aspects of Nucleopolyhedrovirus Pathogenesis in Lepidopteran Larvae

5

John W. Barrett[1], Mark Primavera[2], Arthur Retnakaran[2], Basil Arif and Subba Reddy Palli[2]

[1]Viral Immunology and Pathology, The John P. Robarts Research Institute, 1400 Western Road, London, Ontario N6G 2V4, Canada, [2]Great Lakes Forestry Center, Canadian Forest Service, Sault Ste. Marie, Ontario P6A 5M7, Canada

Introduction

The environmental advantage of using biological control agents instead of broad spectrum pesticides in controlling insect pests has become increasingly apparent in recent years. While traditional biological control using parasitoids and predators is applicable to specialized situations, the preferred method is the use of microbial pathogens of insects. The bacterial toxin from the parasporal crystals of *Bacillus thuringiensis* (Bt) is widely used in several Bt formulations and the toxin gene has been cloned into many plants to provide protection against insect pests. Entomopathogenic fungi and microsporida are also being tested as potential control agents. Entomopathogenic viruses are of special interest since they are relatively pest specific with little or no impact on the environment and are, therefore, ecologically attractive alternatives to chemical pesticides. One such group, the baculoviruses, has received the attention of many researchers and has been relatively well studied, These viruses contain a circular double stranded DNA inside a protein coat and include two general, the granuloviruses (GVs) and the nucleopolyhedroviruses (NPVs). GVs have one nucleocapsid per envelope and replicate either within the cytoplasm or inside the nucleus of the host cell. NPVs occur as either single nucleocapsids embedded in a protein matrix (SNPV) or multiple nucleocapsids within a protein medium (MNPV) and in both cases they replicate exclusively within the nucleus. The pathogenesis of GVs demonstrated to infect nearly 100 lepidopteran species has been recently reviewed by Federici (1997) and, therefore, this review will be confined to NPV pathogenesis.

The most extensively studied baculovirus that is used for controlling lepidopteran pests is the alfalfa looper virus, *Autographa californica* (Speyer) NPV (AcMNPV). This virus has served as the reference standard for the use of viruses in pest control. The genome of AcMNPV has been completely sequenced and expression systems using this virus have been developed and are being commercially marketed (e.g. Life Technologies Inc., Bac-to-Bac, Invitrogen, BakPak). These large, double-stranded, circular DNA viruses have a host range limited primarily to lepidopteran species. In addition, most NPV species infect a relatively few genera within Lepidoptera. Since NPVs are naturally occurring and have a narrow host range, they are being touted as one of the environmentally attractive biological control agents against many lepidopteran pests of agricultural crops and forests.

General pathology

Although NPVs, in general, infect only the larval stages of Lepidoptera, there are a few reports of low levels of infection occurring in pupae and adults. Infection of the early larval stages, usually with a high dose of the virus often results in killing the host larvae quickly before the virus has had time to complete its life cycle and produce an infectious level of progeny virus. However, if the infection occurs in a later larval instar the virus completes its life cycle including the expression of the entire cascade of viral genes resulting in the production of large quantities of progeny virus. In other words, small early instars succumb rapidly to viral infection because the host cells die before the virus completes its life cycle. The later larger instars survive longer after the infection allowing virus propagation. In addition, massive viral infection of the host cells results in delaying the rate of larval development, the loss of intra-cellular turgor pressure resulting in the larvae becoming flaccid, the induction of feeding inhibition and massive cellular breakdown leading to the mortality of the larva. The dissolution of the virus killed caterpillar results in the release of the newly synthesized progeny virus from the carcass. The liquefied host contaminates the foliage with high levels of NPV. When another larva feeds on the contaminated foliage it acquires the infection and the transmission continues.

The phenotypic effects of the virus are not manifested in the larvae during the initial few days after infection. However, several days following ingestion of the virus the larvae show symptoms of both behavioral and morphological changes. Infected larvae initially feed normally but later exhibit feeding inhibition. Also, the NPV infection prolongs the larval stadium. It has been observed that in some species such as the gypsy moth, *Lymantria dispar* (Linnaeus), infected larvae migrate to the apical parts of the trees and die. In many species including the eastern spruce budworm, *Choristoneura fumiferana* (Clemens), the virus undergoes several generations of replication within the host resulting in the production of large quantities within the larvae. As a result the larva appears swollen, with a glossy cuticle. The hemolymph contains large amounts of virus giving it a cream color. The cuticle of infected larvae of the cabbage looper, *Trichoplusia ni* (Hubner) turns from light green to black during the terminal stage of the infection prior to death. By the time the larva stops feeding completely other signs such as cessation of frass production, loss of turgor pressure and darkening of the cuticle become apparent indicating severe infection, which will result in the mortality of the larvae. In many species, when the infection is advanced, the larva takes on the shape of a loose sack containing a mass of liquefied infected tissues. Most of the tissues such as the fat body, hemocytes and epidermis act as virus factories and the cells become completely filled with polyhedral inclusion bodies (Granados and Lawler, 1981; Barrett *et al.*, 1998 a,b). Eventually, the cells lyse releasing the virus and the cuticular wall of the larva becomes fragile. When the thin body wall of the loose sack containing the liquefied larval tissues bursts the packaged nucleocapsids are released.

NPVs have evolved several strategies that improve their chances of survival; one of which is the production of a proteinaceous coat (polyhedrin) within which the viral nucleocapsids are packaged ensuring protection of the virus from environmental factors such as UV radiation and adverse weather conditions. In addition, the polyhedra have an interesting property of dissolving in an alkaline medium, a feature characteristic of the midgut of most lepidopteran species, which allows the rapid release of the nucleocapsids from the polyhedra.

A second strategy that NPVs have acquired during evolution is the expression of the viral gene that delays molting in the larvae thereby extending the larval stadium. Maintaining the larval stage for a longer time period allows the virus to multiply many times within the host. In lepidopteran larvae, the molting hormone, 20-hydroxyecdysone (20E) controls the molting process by inducing the expression of a sequence of genes, which results in the synthesis and deposition of new cuticle and casting off the old cuticle. The 20E concentrations have to reach threshold levels for gene activation and initiation of molting process. In general, higher concentration of 20E is required to initiate metamorphosis than to initiate molting (Palli *et al.*, 1995). Recent studies on AcMNPV have revealed that baculoviruses have evolved a mechanism to delay the normal molting process by expressing the *ecdysteroid UDP-glucosyltransferase (egt)* gene. The product of *egt* glycosylates the 20E which results in lower levels of functional 20E and as a result the molting process is delayed making the virus infected larvae remain in the larval stage and continue feeding (O'Reilly and Miller, 1989). This delay in molting process and prolongation of larval life is more pronounced at the last instar than at the earlier instars. As explained above more 20E is needed to initiate metamorphosis than to initiate molting. The reduction in 20E levels by NPV results in the 20E concentrations that are below threshold levels for initiation of metamorphosis but not molting.

Internal pathogenesis

Understanding the process of infection and virus transmission within a particular host is by far the most important prerequisite for elucidating the etiology of any viral pathogen. Of all the baculoviruses AcMNPV has been the one that has been extensively studied. The genome of AcMNPV has been completely sequenced (Ayres *et al.*, 1994). The pathogenesis of this virus has been investigated by many researchers in recent years (Granados and Lawler, 1981; Keddie *et al.*, 1989; Flipsen *et al.*, 1993; Engelhard *et al.*, 1994; Kirkpatrick *et al.*, 1994; Flipsen *et al.*, 1995; Barrett *et al.*, 1998a,b).

Role of the midgut

Upon ingestion of the polyhedra, the protective polyhedrin coat is dissolved within the alkaline environment of the larval midgut. The released occlusion derived virions (ODVs) together with the ingested food are contained within the peritrophic membrane, which is a tubular, porous secretion from the anterior end of the midgut that protects the midgut lining. The first barrier for the virions or virus particle to cross in order to reach the midgut cells is the peritrophic membrane and the exact mechanism that allows viral passage is not fully understood (Federici, 1997). The peritrophic membrane, which is made of chitin and protein, varies in detail from species to species but almost all of them contain pores. Estimates of pore sizes based on the passage or exclusion of FITC-labeled dextrans indicate that the pore sizes are smaller than necessary to allow the virion to pass through. For example, it has been shown that the pore size ranges from 21 nm in *Orgyia pseudotsugata* (McDunnough) to 29 nm in *Malacosoma disstria* Hübner (Barbehenn and Martin, 1995). NPVs with nucleocapsid sizes ranging from 30–35 × 250–300 nm, therefore, should be excluded from passing through the peritrophic membrane (Volkman, 1997). Obviously, NPVs have evolved a strategy to overcome this physical barrier since they manage to get through.

Recently, it has been demonstrated that GVs synthesize a protein termed enhancin that facilitates localized digestion of the peritrophic membrane to permit passage of the GV virion (Hashimoto *et al.*, 1991; Wang *et al.*, 1994). Further characterization of enhancin has suggested that it is probably a metalloprotease with a specific affinity for mucins in the peritrophic membrane. Enhancin was able to produce lesions through which the virions could pass (Lepore *et al.*, 1996). Although a homologous enzyme has not been identified in other NPVs, a similar scenario probably exists.

Once the virion has passed the peritrophic membrane it must then enter the epithelial cells of the midgut. It is generally accepted that the released ODVs fuse to the microvilli forming the brush border of the midgut epithelium and pass through into the columnar cells and eventually into the nucleus. Although binding studies suggest that virions might utilize cell surface receptors on the microvilli to pass into the midgut cells, no receptors or specific binding proteins have been identified so far (Horton and Burand, 1993).

Several models have been proposed to explain the NPV infection process within the lepidopteran host upon entry into the midgut. An early model suggests that the parental virus passes directly through the midgut epithelial layer and reaches the hemocytes and infects both the midgut cells as well as the hemocytes concomitantly or sequentially (Granados and Lawler, 1981). Budded virus (BV) released from the hemocytes would spread to other regions of the insect and infect various tissues. Granados and Lawler (1981) observed non-enveloped nucleocapsids (NCs) passing into the midgut epithelial cells from the lumen of the midgut and lining the basal plasma membrane of the epithelial cells within 0.5h post infection (hpi). Subsequently, these NCs had passed through the basal plasma membrane and acquired a viral envelope within 6 hpi. They concluded that both the NCs along the basal plasma membrane and the BV in the hemocoel represented NC of the inoculating virus that had acquired envelopes as they budded through the midgut epithelial cells. Although later studies have questioned these observations, our own work supports aspects of this hypothesis.

Studies on *Trichoplusia ni* (Hubner) infection with a recombinant AcMNPV expressing GFP under the polyhedrin promoter suggested that a proportion of the inoculating virus enters the nuclei of the columnar cells where the virus replicates, synthesizing late and very late proteins, resulting in GFP expression. Subsequently, the virus infected columnar cells are sloughed off from the midgut epithelium into the lumen (Barrett *et al.*, 1998b). Concomitantly, some of the parental NCs pass directly from the midgut lumen through the midgut epithelial layer via the plasma membrane reticular system (PMRS) to gain accesses to the hemocoel. The result of this scenario is that the midgut and hemocytes appear to become infected at approximately the same time (Barrett *et al.*, 1998b). The role of hemocytes in the infection process will be addressed later. Passage directly through the midgut epithelial layer also explains the observation of infection of regenerative cells in the midgut epithelium in the early hours following ingestion of the virus (Flipsen *et al.*, 1995).

Although the exact mechanism of virus transport through the midgut epithelial layer has not been elucidated there is general agreement that NPV infection of the midgut is restricted initially to the columnar epithelial cells of the midgut. Using a modified AcMNPV expressing two biochemical markers in infected second instar *Spodoptera exigua* (Hubner), Flipsen *et al.* (1995) demonstrated that columnar cells were infected within 3 hpi and that virus replication occurred within the columnar cells 12 hpi. Little evidence has been produced to show infection or replication within the goblet cells of the midgut epithelial layer and it is considered a rare event (Flipsen *et al.*, 1995). However, whatever the mechanism, once a productive infection has been established in the columnar cells BVs are produced and released from the columnar cell.

An unusual feature of the infection process of the midgut has been the observation of the presence of the nucleocapsid, virus replication, late gene expression but no production of polyhedra filled with nucleocapsids (Flipsen *et al.*, 1993). This is apparently a feature of AcMNPV pathogenesis and it has been observed in *T. ni*, *Estigmene acrea* and *S. exigua* infected larvae (Granados and Lawler, 1981; Hess and Falcon, 1981; Flipsen *et al.*, 1993). It may or may not be a general feature of NPV pathogenesis, as it has not been observed in other NPV infections in their respective hosts. Nevertheless, the production of small polyhedra that do not contain occluded virions may represent an evolutionary response to the sloughing off of the midgut cells observed in several lepidopteran hosts (Flipsen *et.al.* 1993; Barrett and Palli, unpublished data). AcMNPV infected columnar cells were observed rejected into the lumen 62 hpi in early instars of *S. exigua* (Flipsen *et al.*, 1993). This phenomenon has also been observed 5 days post infection in CfMNPV infected *Choristoneura fumiferana* (Clemens) (Barrett and Palli, unpublished). In contrast, infected regenerative cells were found associated with infected columnar cells in the midgut epithelium (Flipsen *et al.*, 1993, 1995).

Key to internal spread of infection is the tracheal system

The spread of the virus beyond the midgut epithelial layer is not easy to explain because separating the midgut epithelial layer from the rest of the internal organs is the basal lamina, a fibrous matrix of glycoproteins secreted by the epithelial cells, which might serve as an effective barrier against microbial infections. This protective basal lamina surrounds all the internal tissues excepting the hemocytes. As BVs are synthesized and passed out of the midgut epithelial layer, they collect along the basal lamina (Keddie *et al.*, 1989). The question of how the virions pass through the basal lamina has been the subject of investigation by several groups. Earlier researcher suggest that AcMNPV nucleocapsids could bypass the columnar cell nucleus and attach to the basal lamina along the base of the midgut epithelial cells and bud directly into the hemolymph (Granados and Lawler, 1981). They found infectious hemolymph as early as 0.5 hpi suggesting that the inoculating virus had bypassed the midgut and entered the hemolymph. Later studies showing replication within the midgut suggested that this was unlikely (Keddie *et al.*, 1989; Flipsen *et al.*, 1993). We suggested that some of the inoculating virus bypassed the replication step in the midgut, utilizing the PMRS to pass through the midgut. They were then available to infect the hemocytes at the same time that the rest of the inoculating virus was replicating in the columnar cells and producing BV (Barrett *et al.*, 1998a, b).

However, the problem of crossing the basal lamina still existed until a clear study showed that the passage of BV to other tissues occurred via the tracheal matrix (Engelhard *et al.*, 1994). The tracheal system consists of a continuous series of tubes that act to deliver oxygen to the organism's internal organs. These tubes penetrate through the basal lamina of organs and tissues where they supply oxygen (Volkman, 1997). EM studies of NPV infection showed that host tracheolar cells, the terminal tracheal cells entering the tissues (also called tracheoblasts), were the secondary targets of infection after the midgut (Adams *et al.*, 1977). More recently, Engelhard *et al.* (1994) concluded that tracheoles and tracheolar cells were significant contributors to the dissemination of AcMNPV in *T. ni* larvae following infection of the midgut. This was confirmed by Kirkpatrick *et al.* (1994) in *Helicoverpa zea* (Boddie), Washburn *et al.* (1995) in *Heliothis virescens* (Fabricius), Barrett *et al.* (1998b) in *T. ni* and *C. fumiferana* (Barrett and Palli unpublished). By infecting 4th instar *T. ni* with an AcMNPV expressing *β-galactosidase* under a constitutive promoter Engelhard *et al.* (1994) showed that tracheoblasts infiltrating midgut epithelial cells were blue in the presence of X-gal during the early part of the infection process. Tracheoblasts infiltrate all the tissues except the hemocytes, which do not contain a basal lamina allowing free passage. Therefore, the tracheal system offers a conduit for the virus into the entire major organs and tissues. Access to the hemolymph allows BV infection to spread into the hemocytes and other tissues.

Role of hemocytes

Since the tracheal system acts as the major conduit to all tissues for the infection process, the role of hemocytes has to be different. Although we have clear evidence of infection of the hemocytes (Barrett *et al.*, 1998a,b), the role of hemocytes in the transmission and introduction of virus into secondary tissues is unclear. Earlier work suggested that AcMNPV was able to pass directly through the midgut epithelial layer and into the hemolymph where hemocytes would be infected (Granados and Lawler, 1981). These infected hemocytes would then allow introduction of the virus into the fat body and epidermal tissues. Later studies agreed that hemocytes were important, however, their exact role was undefined (Keddie *et al.*, 1989). Nevertheless, by this time the replication within the midgut epithelial layer was identified as being crucial for successful infection. When we used GFP labeled AcMNPV, we observed the simultaneous appearance of GFP both in the hemocytes as well as in the midgut tissue suggesting that, at least some of the inoculating virus was passing through the midgut and infecting the hemocytes (Barrett *et al.*, 1998a). The BV produced from these infected hemocytes can then infect tracheal epithelium at numerous locations within the entire body and act as foci from which infection would radiate. It is also possible that hemocytes spread the virus throughout the tracheal system (Barrett *et al.*, 1998b). More recently, infection of *C. fumiferana* larvae with CfMNPV expressing GFP suggested that infection of hemocytes occurred approximately 12 h later than the infection of the tracheal system (Barrett and Palli, unpublished). This may mean that the role of the hemocytes may be subtly different within different hosts and may be virus specific.

Role of other tissues during infection

Budded virus that was synthesized following secondary infection of the hemocytes and the tracheal matrix initiates further infection of most other tissues in the lepidopteran host. As infection spreads along the tracheal epithelium from the foci of infection, the virus gains access to various other tissues such as the epidermis and the fat body. Contrary to earlier reports that the fat body was infected soon after the hemocytes (Federici, 1997), we observed GFP fluorescence, indicating virus replication and late gene expression in the epidermis which occurred earlier (32 hpi) in the epidermis than in the fat body (55 hpi). In a study of CfMNPV pathogenesis in *C. fumiferana* it was observed that the tracheal epithelial cells associated with the midgut, fat body, epidermis and muscle were observed expressing GFP four days post feeding (dpf). The infection then spread to the fat body, epidermis and muscle cells by 6 dpf (Barrett and Palli, unpublished data). Various other tissues that displayed fluorescence included the silk glands and the tracheae attached to the ovaries, testes, brain, ganglia and Malpighian

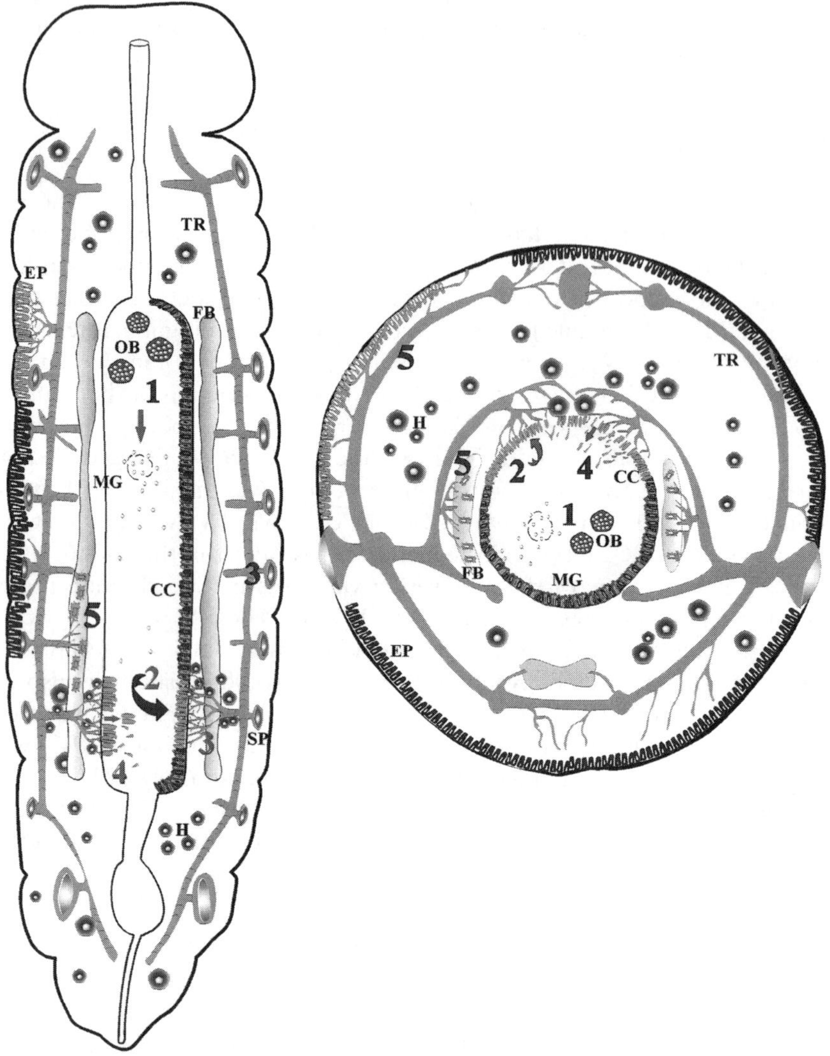

Figure 5.1 Proposed model for infection process of NPVs in lepidopteran larvae. The model is based on our studies on GFP labeled CfMNPV spread *in Choristoneura fumiferana* larvae. The model represents a longitudinal and cross-section through the larvae.

1 The occlusion bodies (OB) are dissolved in the lumen of the midgut (MG) and nucleocapsids are released.
2 Nucleocapsids enter the midgut columnar cells (CC); some of the them reach nucleus and start replicating while rest of them travel to the basal surface of the cell through plasma membrane reticular system.
3 The nucelocapsids infect invading tracheal end cells. The virus replicates in the tracheal epithelial cells and spreads throughout tracheal system (TR).
4 Some of the infected midgut columnar epithelial cells are sloughed off into the lumen and excreted out.
5 The virus replicating in the tracheal system gains access to tissues such as fat body (FB) and epidermis (EP) and infects these tissues at the final stages of infection process. (SP = Spiracles, H = Hemocytes).

tubules but not the tissues themselves (Barrett and Palli, unpublished data). An earlier study utilizing reporter genes under early and late promoters suggested that the Malpighian tubules were infected, however, there was no evidence of virus replication (Knebel-Moersdorf *et al.*, 1996). The appearance of virus within the fat body and epidermis tissue indicates that *in vivo* spread of the virus is almost complete and the larva will soon succumb to the infection.

Conclusions

The speed of virus spread within a host appears to be unique for each virus and its particular host. Regardless of the rate of the infective process the outcome of the infection is similar in most lepidopteran hosts and results in the liquefaction of the host tissues and liberation of masses of polyhedra from within a flaccid carcass. Understanding the mechanism of baculovirus pathogenesis has progressed well during last few years. The model presented in the Fig. 5.1 is based on our current understanding of CfNPV infection process in *C. fumiferana* larvae. Most parts of the model hold true for AcMNV infection process in several lepidopteran larvae. There are still gaps and unanswered questions remain. How do NPV pass through peritrophic membrane and basal lamina? What are the precise roles of hemocytes and tracheal system in spreading virus infection? These and other similar questions should be the focus of baculovirologists' studies in coming years.

References

Adams, J.R., Goodwin, H., and Wilcox, T.A. (1977) Electron microscope investigations on invasion and replication of insect baculoviruses *in vivo* and *in vitro*. *Biol. Cellulaire*, **28**, 261–268.

Ayres, M.D., Howard, S.C., Kuzio, J., Lopez-Ferber, M., and Possee, R.D. (1994) The complete DNA sequence of *Autographa californica* nuclear polyhedrosis virus. *Virology*, **202**, 586–605.

Barbehenn, R.V., and Martin, M.M. (1995) Peritrophic envelope permeability in herbivorous insects. *J. Insect Physiol.*, **41**, 303–311.

Barrett, J.W., Brownwright, A.J., Primavera, M.J., and Palli, S.R. (1998a) Studies of the nucleopolyhedrovirus infection process in insects by using the green fluorescent protein as a reporter. *J. Virol.*, **72**, 3377–3382.

Barrett, J.W., Brownwright, A.J., Primavera, M.J., Retnakaran, A., and Palli, S.R. (1998b) Concomitant primary infection of the midgut epithelial cells and the hemocytes of *Trichoplusia ni* by *Autographa californica* nucleopolyhedrovirus. *Tissue Cell*, **30**, 602–616.

Engelhard, E.K., Kam-Morgan, L.N.W., Washburn, J.O., and Volkman, L.E. (1994) The insect tracheal system: A conduit for the systemic spread of *Autographa californica* M nuclear polyhedrosis virus. *Proc. Natl. Acad. Sci. USA*, **91**, 3224– 3227.

Federici, B.A. (1997) Baculovirus pathogenesis. In L.K. Miller (ed.), *The Baculoviruses*, Plenum Press, New York.

Flipsen, J.T.M., van Lent, J.W.M., Goldbach, R.W., and Vlak, J.M. (1993) Expression of polyhedrin and p10 in the midgut of AcMNPV-infected *Spodoptera exigua* larvae: An immunoelectron microscopic investigation. *J. Invert. Pathol.*, **61**, 17–23.

Flipsen, J.T.M., Martens, J.W.M., Van Oers, M.M., Vlak, J.M., and van Lent, J.W.M. (1995) Passage of *Autographa californica* nuclear polyhedrosis virus through the midgut epithelium of *Spodoptera exigua* larvae. *Virology*, **208**, 328–335.

Granados, R.R., and Lawler, K.A. (1981) *In vivo* pathway of *Autographa californica* baculovirus invasion and infection. *Virology*, **108**, 297–308.

Hashimoto, Y., Corsaro, B.G., and Granados, R.R. (1991) Location and nucleotide sequence of the gene encoding the viral enhancing factor of the *Trichoplusia ni* granulosis virus. *J. Gen. Virol.*, **72**, 2645.

Hess, R.T., and Falcon, L.A. (1981) Electron microscope observations of *Autographa californica* (Noctuidae) nuclear polyhedrosis virus replication in the midgut of the Saltmarsh caterpillar, *Estigmene acrea* (Arctiidae). *J. Invertebr. Pathol.*, **37**, 86–90.

Horton, H.M., and Burand, J.P. (1993) Saturable attachment sites for polyhedron-derived bacuolvirus on insect cells and evidence for entry via direct membrane fusion. *J. Virol.*, **67**, 1860–1868.

Keddie, B.A., Aponte, G.W., and Volkman, L.E. (1989) The pathway of infection of *Autographa californica* nuclear polyhedrosis virus in an insect host. *Science*, **243**, 1728–1730.

Kirkpatrick, B.A., Washburn, J.O., Englehard, E.K., and Volkman, L.E.. (1994) Primary infection of insect tracheae by *Autographa californica* M nuclear polyhedrosis virus. *Virology*, **203**, 184–186.

Knabel-Moersdorf, D., Flipsen, J.T.M., Roncarati, R., Jahnel, F., Kleefsman, A.W.F., and Vlak, J.M. (1996) Baculovirus infection of *Spodoptera exigua* larvae: lacZ expression driven by promoters of early genes pe38 and me53 in larval tissues. *J. Gen. Virol.*, **77**, 815–824.

Lepore, L.S., Roelvink, P.R., and Granados, R.R. (1996) Enhancin, the granulosis virus protein that facilitates nucleopolyhedrosis (NPV) infections, is a metalloprotease. *J. Invertebr. Pathol.*, **68**, 131.

O'Reilly, D.R., and. Miller, L.K. (1989) A baculovirus blocks insect moulting by producing ecdysteroid UDP-glucosyltransferase. *Science*, **245**, 1110–1112.

Palli, S.R., Primavera, M., Lambert, D., and Retnakaran, A. (1995) Age specific effects of RH-5992: a non-steroidal ecdysone agonist, RH-5992, on the spruce budworm, *Choristoneura fumiferana* (Lepidoptera: Tortricidae). *Eur. J. Entomol.*, **92**, 325–332.

Volkman, L.E. (1997) Nucleopolyhedrovirus interactions with their insect hosts. *Adv. Virus Res.*, **48**, 313–348.

Wang, P., Hammer, D.A., and Granados, R.R. (1994) Interaction of *Trichoplusia ni* granulosis virus-encoded enhancin with the midgut epithelium and peritrophic membrane of four lepidopteran insects. *J. Virol.*, **75**, 1961–1967.

Washburn, J.O., Kirkpatrick, B.A., and Volkman, L.E. (1995) Comparative pathogenesis of *Autographa californica* M nuclear polyhedrosis virus in larvae of *Trichoplusia ni* and *Heliothis virescens. Virology*, **209**, 561.

PROSPECTS OF BACULOVIRUSES IN INTEGRATED PEST MANAGEMENT

6

G.S. Battu, Ramesh Arora and G.S. Dhaliwal
Department of Entomology, Punjab Agricultural University, Ludhiana 141 004, India

Introduction

Integrated pest management is a system, which utilizes all suitable methods, in a compatible manner, to maintain pest populations below levels causing economic injury. Integrated pest management programs involve farmers and pest-control techniques in continuing and careful assessment of the populations of pests and their natural enemies in order to decide if and when to intervene with control measures that range from the use of biologicals to resistant varieties and the carefully targeted use of chemical pesticides. The range of IPM options is defined by agroecological, socioeconomic and institutional factors. The key to successful implementation of IPM is the development, by farmers and pest control technicians, of a practical understanding of the ecology of the crops, pests and their natural enemies. The translation of this knowledge into decision tools and practical control tactics to save particular pest problem in a given situation is equally important (Persley, 1996). IPM has been defined as a decision support system for the selection and use of pest control tactics, singly or harmoniously coordinated into a management strategy, based on cost/benefit analyses that take into account the interests of and impacts on producers, society, and the environment (Kogan, 1998).

The world pesticide market from an estimated base of US$ 31 billion is mature with a growth of 1 to 2 per cent per year. Of this market, biologicals contribute only about US$250 million with 60 percent of this being microbial products largely based on *Bacillus thuringiensis* Berliner (Bt). Although biologicals represent a very small market, a variety of drivers are leading to a massive rate of increase of 25 percent per year in biologicals. Softer treatments such as microbials are preferred over broader spectrum synthetic compounds on compelling ecological considerations. In many cases, the regulatory barriers to the development of microbial materials are lower providing an added advantage (Hammock *et al.*, 1999). Among microbials, although baculoviruses have not had the commercial success of Bt, they have significant potential for use in IPM programs. There are number of advantages associated with the use of baculoviruses, which are ideal for IPM because they do not affect predators and parasitoides. They are safe for non-target insects, humans and the environment. The host specificity is, therefore, a useful attribute from an environmental standpoint. Baculoviruses may, in some cases, be the only effective viral-insect control agent available for controlling an insect species (Cunningham, 1988) and provide an avenue available to overcome specific problems, such as pesticide resistance. Even the use of two biologicals like viruses and Bt could lower the possibility for resistance development (Marrone, 1996). In addition to their safety to non-target organisms, the baculoviruses are of particular interest because, (i) they cause mortality in the target insect population, (ii) geographically distinct populations of insects exhibit relatively uniform susceptibility to a virus, (iii) individual populations of insects do not acquire resistance readily upon continual virus pressure, (iv) can be formulated for easier application and long-term storage, (v) can be applied easily using methods similar to those employed in the application of chemical pesticides, and (vi) these are generally compatible with chemical pesticides. Collectively, these properties provide an overwhelming case for pursuing the development of baculoviruses as biological (microbial) alternatives to chemical pest control (Miller, 1998).

Taxonomic status of baculoviruses

Baculoviruses (Family: Baculoviridae) have only been isolated from invertebrates. Most examples have been found in insect species, but there are some reports of baculoviruses, which are pathogenic to Crustacea. Baculovirus infections have been described in over 700 species of invertebrates including Lepidoptera, Hymenoptera, Diptera, Coleoptera, Trichoptera, Thysanura and Neuroptera besides Crustacea (Murphy *et al.*, 1995). Until 1995, family Baculoviridae was subdivided into two subfamilies: Eubaculovirinae, which included the occluded nuclear polyhedrosis virus (NPV) and granulosis virus (GV); and Nudibaculovirinae, encompassing the non-occluded baculoviruses. Currently, Baculoviridae is divided into two genera: *Nucleopolyhedrovirus* and *Granulovirus* (Francki *et al.*, 1991; Murphy *et al.*, 1995). Virions of NPV and GV are occluded in polyhedral and capsular proteinaceous occlusion bodies (OBs), respectively. NPVs have limited host ranges, usually being restricted to one host species or genus, with the exception of the NPVs of *Autographa californica* (Speyer), *Anagrapha falcifera* (Kirby) and *Mamestra brassicae* (Linnaeus). GVs are more specific than NPVs as they have been reported only from Lepidoptera (Battu and Arora, 1996; Moscardi, 1999).

Structure of baculoviruses

Baculoviruses have a large, double-stranded, covalently closed, circular DNA genome of between 88 to 200 kbp. This is associated with a highly basic (arginine-rich) protein of 65 KDa. The DNA-protein complex is contained by a rod-shaped nucleocapsid comprising a 39 KDa or 87 KDa capsid proteins. Other structural components almost certainly remain to be resolved (King *et al.*, 1994). The size of the virus genome determines the length of the nucleocapsid, which may be 200–400 nm. The width remains constant at about 36 nm. One or more nucleocapsids are packaged within a single lipoprotein envelope to form the singly embedded or multiply embedded virus particle or virion, respectively. These structures may be occluded within a crystalline matrix, referred to as a polyhedron or granule. The former range in size from 1 to 15 μm in diameter with an outer envelope, which appears to confer additional strength and protection. Polyhedra consist largely of a single protein (polyhedrin) of about 30 KDa and formed in the nucleus of infected cells. Virions that have been released from polyhedra are called polyhedra-derived virus in the midgut tissues of susceptible insect, whereas virions that are released from cells without occlusion are called extracellular virus (ECV) or budded virus. As opposed with this, GVs contain one virion (singly enveloped nucleocapsid) per virus occlusion body or granule. Granulin, the major granule protein, is similar to polyhedrin in function.

The non-occluded baculoviruses are composed of singly enveloped nucleocapsids but as the name suggests, they are not further packaged into occlusion bodies (King *et al.*, 1994; Battu *et al.*, 1996). The virion lipoprotein envelope fuses with the plasma membrane of the gut wall cells and liberates nucleocapsids into the cytoplasm. The latter serve to transport the virus DNA to the nucleus of the cell. In the infected gut cells, new nucleocapsids are produced by about 8h post-infection (hpi) and begin to bud through the nuclear membrane by 12 hpi thus acquiring the lipid envelope. This membrane appears to be lost in the cytoplasm, but the nucleocapsid gains another as it buds through the plasma membrane. In the course of this latter process, it also acquires a virus-encoded glycoprotein (gp 67) of 67 KDa (Blissard and Rohmann, 1989; Whitford *et al.*, 1989), helping attach the budded virus to another susceptible cell within the insect.

In cell culture, the budded virus is 1000-fold more infectious than virus particles released from polyhedra, which lack gp 67 (Keddie and Wolkman, 1985). The budded virus initiates infection to other tissues in the hemolymph, i.e. fat bodies, nerve cells, hemocytes, etc. The cells infected in the second round of virus replication in the insect larva also produce budded virus, but in addition occlude virus particles within polyhedra in the nucleus. The virus particles occluded within polyhedra, which are genetically identical with the ECV, get their lipid envelope *de novo* within the nucleus and lack the gp 67 found in the budded virus phenotype. The accumulation of polyhedra within the insect proceeds until the host consists almost entirely of a bag of virus. In the terminal stages of infection the insect liquefies and thus releases polyhedra, which can infect other insects if the latter are ingested. Recent evidences indicated that virus encoded chitinase has a role in this process (Hawtin, 1993).

Strategies for utilization of baculoviruses

There are four basic strategies for using baculoviruses in insect pest management.

Introduction and establishment

The introduction and establishment of microbials in an environment is intended to result in permanent suppression of the target pest. Most of the successes of viruses in insect control have been by this method. There have been at least 15 successful introductions of viruses, 5 in crops and 10 in forests (Fuxa, 1990). In the 1930s, an NPV was introduced accidentally into Canada along with parasitoids, which were imported from Scandinavia and released for control of pests. Later this NPV was multiplied and applied in selected locations. The NPV was remarkably successful and no control measures have been required against the pest in Canada for the last 50 years. Its success was attributed to the relative stability of the forest ecosystem and host populations as well as to efficient horizontal and vertical transmission of the virus. Later, the European pine sawfly, *Neodiprion sertifer* (Geoffroy) and the red-headed pine sawfly, *N. lecontei* (Fitch) were also successfully controlled with one or two introductions of respective NPVs into field populations (Cunningham and Entwistle, 1981; Cunningham, 1982). An NPV of *Chrysodeixix includens* (Walker) is possibly the best example of a baculovirus implemented as a classical biological control agent in a crop. This NPV was released on 200–250 ha of soybean in USA and provided control 12–15 years later (Fuxa *et al.*, 1992).

Seasonal colonization

It involves the inoculative release of microbial pathogens to control insect pests for more than one generation, although subsequent releases are required when the pathogen population declines. It requires efficient replication and transmission of the pathogen in host populations. Most important example is the control of velvet bean caterpillar, *Anticarsia gemmatalis* (Hübner) on soybean by application of AgNPV. The virus is applied on 1 million ha annually in Brazil. The AgNPV occurs naturally in Brazil in *A. gemmatalis* populations with pathogenesis similar to other NPVs. Currently, it is produced directly on the farmers' fields. It involves virus application on soybean infested with *A. gemmatalis* larvae, collection of the dead larvae, and storage in large rooms at -4 to -8°C until processed as a formulation. Cost of the formulated product is about US$ 0.7/ha and it reaches the farmer at a mean cost of US$ 1.2–1.5/ha, which is lower than the cost of chemical insecticides (Moscardi, 1999).

Another notable success is the use of non-occluded virus of the rhinoceros beetle, *Oryctes rhinoceros* (Linnaeus) on coconut palms in the Pacific Islands. Releasing infected adults of the beetle in the field spreads the virus. The infected adults do not feed but are attached to breeding and feeding sites. The virus is also transmitted during mating when the uninfected adult ingests virus-laden fecal material from its sexual partner (Zelazny *et al.*, 1992).

The GV of *Plodia interpunctella* (Hübner), a pest of stored grain and grain products has potential for long-term control in storage bins (Kinsinger and McGaughey, 1979). The GV is protected from environmental extremes and persists in the stored grains.

Environmental manipulation

This involves changing the host habitat to favour conservation or augmentation of pathogens in a system where they either occur naturally or have been introduced. Modified cultural practices enhance prevalence of pathogens in insect populations by aiding in persistence or assisting their transport from the soil to the insects feeding substrate (Fuxa, 1991). These practices include changes in cultivation, grazing, sowing and chemical use to increase natural control of *Wilseana* sp. by NPV in New Zealand pastures (Kalmakoff and Crawford, 1982). Movement of cattle similarly enhanced NPV transport and natural control of *Spodoptera frugiperda* (J.E. Smith) in Louisiana pastures (Fuxa and Geaghan, 1983). Environmental manipulation has also been found useful for enhancing the efficacy of non-occluded virus in case of rhinoceros beetle on coconut palms. Viral spread and control of the beetle populations are enhanced if some of the dead palms are left standing and the others are piled and overgrown with crops rather than left lying around the plantation (Pillai *et al.*, 1993).

Microbial insecticides

Most viral pathogens are suitable for use as microbial insecticides. The industry also has maximum interest in this approach, because the multiple applications create the best opportunity for product sales. The NPVs and GVs of lepidopteran caterpillars (Moscardi, 1999) as well as NPVs of several species of sawflies provide short-term control comparable to that with conventional insecticides (Table 6.1).

Field efficacy of baculoviruses

Helicoverpa/Heliothis complex

The genera *Helicoverpa* and *Heliothis* contain some of the most destructive pests of agricultural crops including *Helicoverpa armigera* (Hübner), *H. assulta* (Guenee), *H. zea* (Boddie) and *Heliothis virescens* (Fabricius). Between 1960 and 1980, about 200 field tests were conducted in various countries with NPV, for the control of these pests on different crops. It was evident that crop selection was important and success on cotton was less evident than on maize, control on sorghum was frequently better and control on soybean and chickpea was the best of all (Burges, 1981a). In case of sorghum, a single application of the virus at 11.6×10^{10} OBs per 0.4 ha showed excellent control of the pest. Virus treatment significantly increased yields over untreated checks and it was at least as good as the insecticide in a majority of cases. Four sprays of HzNPV at 100–250 larval

Table 6.1 Examples of commercially available viral pesticides registered for pest control in different countries

Host insect	*Crop(s)*	*Commercial name(s)*	*Country*
	Granulosis viruses		
Adoxophyes orana (Fisher von Roesberstamm)	Apple	Capex 2	Switzerland
Adoxophyes sp.	Tea	–	Japan
Agrotis segetum (Denis & Schiffermuller)	–	Agrovir	Germany, China
Cydia pomonella (Linnaeus)	Apple, pears	Carpovirusine CYD-X Granusal Virin-GyAp Madex	France USA Germany Russia Czech
Erinnyis ello (Linnaeus)	Cassava	–	Brazil
Homona magnanima Diakonoff	Tea	–	Japan
Phthorimaea operculella (Zeller)	Potatoes	PTM baculovirus	Peru, Egypt, Tunisia
Pieris rapae (Linnaeus)	Cabbage	–	China
Plodia interpunctella (Hübner)	Stored almonds, raisins	–	USA
Plutella xylostella (Linnaeus)	Cabbage	–	China
	Nuclear polyhedrosis viruses		
Anagrapha falcifera (Kirby)*	Cotton, vegetables	–	USA
Anticarsia gemmatalis (Hübner)	Soybean	Baculoviron baculovirus nitral, coopervirus, protege	Brazil
Autographa californica (Speyer)*	Cabbage, cotton, ornamentals	VPN 80	Guatemala
Buzura suppressalis (Guenee)	Tea, tung oil tree	–	China
Helicoverpa armigera (Hübner)	Cotton, tomato	Virin-HS	Russia
Helicoverpa zea (Boddie)	Cotton	Elcar, Gemstar	USA
Heliothis virescens (Fabricius)	Cotton	Elcar	USA
Hyphantria cunea (Drury)	Forest, mulberry	Virin-ABB	Russia
Lymantria dispar (Linnaeus)	Forests	Gypchek Disparvirus Virin-ENSH –	USA Canada Russia China
Mamestra brassicae (Linnaeus)	Cabbage	Mamestrin Virin-EKS	France Russia
Neodiprion lecontei (Fitch)	–	Lecont virus	USA, Canada
Neodiprion sertifer (Geoffroy)	–	Monisarmiovirus Virox	Finland UK
Orgyia pseudotsugata (McDummough)	Forests	TB Biocontrol 1 Virtuss	USA Canada

Table 6.1 *(Continued)*

Host insect	*Crop(s)*	*Commercial name(s)*	*Country*
Spodoptera exigua (Hübner)	Ornamentals, vegetables	SPOD-X	USA
	Garden pea, grapes, ornamentals	–	Thailand
Spodoptera frugiperda (J.E.Smith)	Maize	–	Brazil
Spodoptera littoralis (Boisduval)	Cotton	Spodopterin	Africa
Spodoptera litura (Fabricius)	Vegetables, cotton, rice, peanuts	–	China
Spodoptera sunia (Guenee)	Vegetables	VPN82	Guatemala

*These viruses have been developed primarily against lepidopteran species other than original hosts because of their wide host range.
Source: Moscardi (1999), Arora *et al.* (2000)

equivalent (LE, one LE = 6×10^9 OBs) /ha provided effective control (90–96 per cent) on soybean in USA (Ignoffo and Couch, 1981). In India, a large number of field trials have been conducted with NPV for the management of *H. armigera* on tomato, chickpea, sunflower, cotton, groundnut and several other crops (Battu *et al.*, 1993).

Soybean caterpillar

The soybean caterpillar, *A. gemmatalis* is a major pest of soybean in Brazil. As mentioned above the AgNPV is the natural virus in this species with pathogenesis similar to other NPVs. Embrapa (Brazilian Organisation for Agricultural Research) carried out its development and implementation as a microbial insecticide. In pilot scale field trials with AgNPV at 50 LE/ha (1.5×10^{11} OBs/ha) reductions of over 80 percent of *A. gemmatalis* larval populations were obtained and yields were at par with insecticide-treated plots (Moscardi, 1989). Approximately, 20,000 ha were treated in 1983–84 season, which increased to 500,000 ha in 1986–87 and to 1 million ha in 1989–90. In the 1997–98 season, AgNPV use reached approximately 1.2 million ha (Moscardi, 1999).

Spodoptera complex

Several species in the genus *Spodoptera*, including *S. litura* (Fabricius), *S. frugiperda*, *S. littoralis* (Boisduval) and *S. exigua* (Hübner), are important pests on many crops. In Brazil, an indigenous isolate of SfNPV at 2.5×10^{11} OBs/ha provides effective control of early larval instars of *S. frugiperda* on maize (Valicente and Cruz, 1991). The virus is applied in approximately 2000 ha annually. In USA, Crop Genetics International (CGI) commercially produces SPOD-X (SeNPV) for control of *S. exigua* on cotton and vegetable crops. NPVs have also been developed and used against *S. litura* (China, India,

Taiwan), *S. littoralis* (Egypt), *S. exigua* (USA, Guatemala, Thailand) and *S. sunia* (Guenee) (Guatemala) (Moscardi, 1999). Battu *et al.* (1998a) reported that three weekly sprays of SlNPV at 375 LE/ha caused more than 95 per cent mortality of *S. litura* larvae on cauliflower crop in large scale field trials. This treatment also conserved the braconid and ichneumonid parasitoids, which were killed in the insecticide-treated plots.

Codling moth

The codling moth, *Cydia pomonella* (Linnaeus) is a worldwide key pest of apples, pears and walnuts. A granulosis virus isolated from *C. pomonella* was highly virulent to the pest and killed it rapidly, protecting fruit from economic damage in numerous field trials in several countries (Hübner, 1990). Currently, commercial formulations of CpGV are available in France, Switzerland, Germany, Russia and several other European countries.

Forest pests

The NPVs of several species of sawflies in the forest ecosystem seem particularly suitable for long-term control. *N. sertifer*, *N. lecontei*, *N. swainei* (Middleton) and *Gilpinia hercyniae* (Hartig) have been successfully controlled with one or two introductions of their respective NPVs into field populations (Cunningham, 1982). Other species of sawflies, such as *N. taedai linearis* Ross and *N. pratti banksianae* (Dyar) are also effectively controlled (short-term control) when the respective NPVs are applied against early instars. Forest Lepidoptera subjected to applications of baculoviruses include the gypsy moth, *Lymantria dispar* (Linnaeus); the Douglas-fir tussock moth, *Orgyia pseudotsugata* (McDunnough); the spruce budworm, *Choristoneura fumiferana* (Clemens); the Western spruce budworm, *C. occidentalis* Freeman; the jackpine budworm, *C. pinus* Freeman; the pine beauty moth, *Panolis flammea* (Denis & Schiffermuller); and the fall webworm, *Hyphantria cunea* (Drury) (Hübner, 1986). Several of these baculoviruses are now used commercially in USA, Canada and other countries (Table 6.1).

Pests of plantation crops

A non-occluded virus of the rhinoceros beetle, *O. rhinoceros*, a major pest of coconut palms, has been successfully used in pest control in South-east Asia and the Pacific. Adults from the field or reared in the laboratory are infected by immersing them in a viral suspension and then allowed to crawl through sawdust mixed with the virus. The infected adults do not feed but are attracted to breeding and feeding sites. The infected adult midgut cells produce large numbers of viral particles that are disseminated with the feces. When an infected adult beetle visits the breeding site for oviposition, it defecates and the larvae already present in the breeding site become infected upon ingestion of the contaminated fecal matter (Tanada and Kaya, 1993).

An NPV of the hornworm, *Perigonia lusca*, an important pest of Paraguay tea, has been found highly effective against the pest in pilot scale field trials in Argentina (Sosa-Gomez *et al.*, 1994). In Japan, two pests of tea, the smaller tea tortrix, *Adoxophyes* sp. and the Oriental tea tortrix, *Homona magnanima* Diakonoff are being controlled by their GVs. The treated area increased from 460 ha in 1991 to 5850 ha in 1995 (Nishi and Nonaka, 1996). An NPV of *Buzura suppressaria* (Guenee) has been used in China to control this insect on tea and tung oil trees in over 2000 ha (Yi and Li, 1989).

Pests of stored products

Current methods of pest control in stored grains and other products rely on fumigation and insecticide application. With the ban on the use of ethylene bromide and concerns with other chemicals, attention has been focussed on baculoviruses and other microbial agents for the control of insect pests in stored products. The lepidopteran storage pests like *Ephestia cautella* (Walker), *Galleria mellonella* (Linnaeus) and *Ephestia kuehniella* Zeller are known to suffer from epizootic due to baculovirus infections (FAO, 1973; Dales, 1994). The granulosis virus (GV) of the Indian meal moth, *P. interpunctella* has the potential for long-term control in storage bins (Kinsinger and McGaughey, 1976). A freeze-dried, powdered formulation of PiGV with a relatively long shelf life was developed. This formulation could be applied as a spray or dust (Cowan *et al.*, 1986). It was estimated that an application rate of 14 μg/g, a 200 g production was sufficient to treat 14000 kg of commodity. The cost of GV application at US $ 2.45–4.08 compared favourably with the cost of phosphine fumigation at US $ 2.58–4.98 (Vail and Tebbets, 1991). The virus is now recommended for the control of *P. interpunctella* in stored almonds and raisins and has been patented by the USDA (Moscardi, 1999).

A broad-spectrum baculovirus, the *A. falcifera* NPV has also been demonstrated to be effective against the raisin moth, *Ephestia figulilella* (Gregson) in addition to the Indian meal moth (Vail *et al.*, 1993). A GV isolated from the potato tuber moth, *Phthorimaea operculella* (Zeller), has shown high efficacy in protecting the potato crop in the field as well as potato tubers under storage. The GV has been developed as a microbial insecticide by the International Potato Center in Peru and is widely used in Colombia, Ecuador, Peru, and Bolivia (Raman *et al.*, 1992).

Field stability and persistence

The persistence, accumulation and denaturation of baculoviruses in the environment are critical factors in determining the successful use of these agents. Entomopathogens are highly susceptible to damage by desiccation, and by exposure to sunlight, or to ultraviolet (UV) radiations (Ignoffo and Batzar, 1971; Battu and Ramakrishnan, 1989). Formulations of entomopathogens need to be modified to minimize such effects in overall achievements for their better persistence over crop foliage so that pest larvae at various times, get an opportunity to ingest their lethal inocula. Angus and Luthy (1971) listed various additives/adjuvants (like charcoal, India ink, egg-albumin, molasses, optical brightener, etc.) to be used along with formulations of various entomopathogens including baculoviruses.

According to Young and Yearian (1974), the persistence of *Heliothis* NPV was significantly better on tomato (up to 96 h) than on soybean and cotton. Further, they observed that persistence was 10 times more on the calyx, inner surface of mature and terminal leaves. The half-life at unprotected sites was 24 h, at protected leaf sites 24–48 h and at protected floral sites 96 h. Exposure of 0 to 24 h could not inactivate the viral potency. It, however, declined drastically with relatively higher subsequent sunshine exposure of 36, 48, 60, 72, 84 and 96 h as was evident from respective 96.7, 80.2, 66.5, 55.5, 30.0 and 10.0 per cent observed larval mortality of *H. armigera* in bioassays of residual viral (HaNPV) deposit study (Kaushik, 1991). On soybean foliage, Ignoffo *et al.* (1974) observed half-life of *Heliothis* NPV as 2–3 days, while its persistence was detected even after 14 days exposure. Half life values for the NPV alone and the virus when used with soybean and cotton seed adjuvants, were 1.8, 3.5–4.3 and 6.0 days, respectively against *H. zea* on soybean foliage (Smith and Hostetter, 1982). Tuan *et al.* (1989) reported that weak alkaline dew (pH 8.1) inactivated HaNPV collected from soybean leaves. However, it remained active on the dew from maize, tomato and asparagus (pH 7.2-7.3).

Heliothis NPV-bait formulations when used on cotton remained active for at least 6 days during hot, dry and sunny weather (McLaughlin *et al.*, 1971). *Heliothis* NPV was known to lose its activity more rapidly on cotton foliage of which some activity was also lost at night. Young and Yearian (1974) reported most rapid inactivation of *Heliothis* NPV on cotton, with little activity remaining after 24 h. Dhandapani *et al.* (1990) reported that addition of crude sugar (15 per cent) to the HaNPV spray fluid, increased the persistence of the virus both under natural sunlight and shade. Only low levels of HaNPV remained on sorghum heads at 4 days after application (Young and McNew, 1994). Huger *et al.* (1996) also concluded that loss of effectiveness of an NPV of *Mythimna separata* (Walker) on sorghum foliage occurred mainly due to its rapid inactivation by sunlight exposure.

In North Indian conditions, an NPV of *Spilosoma obliqua* Walker lost a total of 33.3 to 50 per cent of its original activity on the sunflower foliage within a comparatively shorter exposure period between 4 to 12 h. However, upon exposure to sunlight up to a period of 72 h, the virus could still persist with 25 per cent activity. At maximum exposure period of 4 days, 75 per cent of the activity was lost (Battu and Sidhu, 1992). In case of groundnut foliage, on the other hand, the same virus under similar exposure conditions lost 70 per cent of the original activity within 4 days (Battu and Bakhetia, 1992). The SINPV persisted on the sunflower foliage for a period of 6 days with 6.6 per cent of its original activity intact (Kaler, 1996). An evening spraying of NPV of *S. litura* helped significantly to minimize the photoinactivation of this virus on cotton foliage besides allowing its greater ingestion by *S. litura* larvae. The same virus, however, has been reported to persist on banana crop for one day in Southern India (Santharam *et al.*, 1978) although it could tolerate sunshine exposure with a severe loss in its virulence up to 8 days (Narayanan *et al.*, 1977).

Certain substrates such as boric acid (Morales *et al.*, 1997), chitinase (Shapiro *et al.*, 1987), extracts of neem tree (Cook *et al.*, 1996), and optical brightners of the stilbene group (Shapiro, 1995) have enhanced baculovirus activity. Mixtures of baculoviruses with optical brightners of the stilbene group seem to have excellent potential for use in formulated products because they can enhance viral activity at concentrations as low as 0.01 per cent, reduce time to kill the host, and provide protection against UV solar radiation. These substances have enhanced the activity of NPVs of *A. californica*, *A. falicifera*, *A. gemmatalis*, *H. virescens*, *H. zea*, *L. dispar*, *S. exigua* and *T. ni* (Shapiro and Argauer, 1997). Argauer and Shapiro (1997) evaluated 8 optical brightners (Blankophor HRS, P167, BBH, RKH, BSU, DML, LPG and Tinopal LPW) of the stilbene group for their activity as virus enhancers. Five of the eight compounds acted as enhancers and the most active brightners (BBH, RKH and LPW) reduced LC_{50} of Gypsy moth, *L. dispar* NPV by 800 to 1300-fold. The most effective compounds were those exhibiting the greatest fluorescence. The brightner acts on the insect midgut and has no effect on the virus *per se.* The virus and the optical brightner must be ingested. Within 48 h the insects stop feeding, midguts are clear and the gut pH is greatly reduced. The brightner allows the virus to replicate in a non-permissive tissue (Columnar cells of the midgut). More importantly, the host spectrum of the baculovirus can be expanded by use of these compounds. None of the components or derivatives of Tinopal LPW was found to be as active as the parent compound (Shapiro and Argauer, 1997).

Commercialization of baculoviruses

The first baculovirus to be developed for commercial use was Elcar (Sandoz Inc.), an NPV of *H. zea*, primarily developed for use on cotton and registered by Environmental Protection Agency in USA in 1975 (Ignoffo, 1981). Elcar was active against all the major *Helicoverpa/Heliothis* species and provided efficient control in soybean, sorghum, maize, tomato, chickpea and navy beans (Ignoffo and Couch, 1981; Teakle, 1994). The advent of synthetic pyrethroids in late 1970s resulted in reduced interest in Elcar and production was stopped in 1982. However, during the last two decades several GVs, and NPVs were registered in Europe and other parts of the world for use in insect pest control (Table 6.1). In 1996, Biosys introduced GemStar LC, a liquid concentrated formulation of HzNPV for control of *H. zea* and *H. virescens* (Fabricius) in US cotton. The NPV of soybean caterpillar, *A. gemmatalis* is the most widely used viral pesticide and is applied annually on approximately 1 million ha of soybean crop in Brazil. The virus is produced directly in the farmers' fields to lower rearing costs (Moscardi, 1999).

Production through cell/tissue culture

More than 200 cell lines have been established from approximately 70 species of insects. The majority of these cell lines have been described from Lepidoptera, Diptera, Orthoptera, Hemiptera, Coleoptera, and Hymenoptera. Many established cell lines from lepidopteran species have proved to be invaluable tool for the *in vitro* propagation of insect-pathogenic viruses. During the past decade and a half significant progress has been made in understanding the replication and molecular biology of baculoviruses in cell culture, and these basic studies are providing the basis for understanding the nature of virus-host interactions including pathogenicity, host range, virulence and latency (Granados *et al.*, 1987).

More than 14 different multinucleocapsid NPVs (MNPVs) including that of *A. californica* have been grown in different cell lines. In addition to AcMNPV, NPVs from *Bombyx mori* (Linnaeus), *L. dispar*, and *S. frugiperda* grow readily in cell cultures, and are easily plaqued and should be amenable to genetic and molecular biological analysis (Miller, 1987). Until recently, the *H. zea* single nucleocapsid NPV (HzSNPV) was the only SNPV to have been grown in an established cell line. Many insect pathologists believed earlier that SNPVs might be more difficult to grow in cell cultures than MNPVs. However, at least three new SNPVs from *H. armigera* (SNPV) (Zhu and Zhang, 1985), *Orgyia leucostigma* (J.E.Smith) (SNPV) (Sohi *et al.*, 1984) and *T. ni* (SNPV) (Granados *et al.*, 1986) have been propagated *in vitro.* Granados *et al.* (1986) established 36 new *T. ni* cell lines from embryonic tissues and 29 such lines had supported replication of *T. ni* SNPV, and it appeared that susceptibility of these lines to this virus was stable. All of the new cell lines were highly susceptible (> 95 per cent of cells infected) to AcMNPV infection and several were susceptible to *T. ni* granulosis virus (TnGV). The ability of many of these new cell lines to support growth of different baculoviruses may be related to the tissues used to initiate the cultures.

Prior to 1984, attempts to replicate GVs in primary organ cultures or established cell lines had met with minimal or no success. It was primarily due to the lack of cell viral receptors or missing host enzymes needed for replication. Miltenburger *et al.* (1984) in Germany reported the first successful *in vitro* replication of *C. pomonella* GV (CpGV) in primary cell lines from *C. pomonella*, 81 were screened for CpGV replication and 9 were susceptible based on light and electron microscopy, and dot immuno assays with monoclonal antibodies. Another development was the successful establishment of several new *T. ni* cell lines, which were susceptible to TnGV (Granados *et al.*, 1986). Even a total of 26 new *T. ni* embryonic cell lines, 15 different cell lines and 3 sublines were susceptible to TnGV as determined by the peroxidase-antiperoxidase (PAP) assay. Further, Granados *et al.* (1987) reported that none of the floating cells was susceptible to TnGV confirming the finding of Miltenburger *et al.* (1984). This implies that other new cell lines from different insect species could be developed for the growth of new GVs and their subsequent commercial exploitation to produce viral pesticides.

If cell culture methods could be developed for the production of insect viruses, several advantages can be envisaged. Selected cell strains with desired production qualities and free from contaminating microorganisms could be stored as frozen materials to provide a stable supply of cells without the risk of genetic alteration. During production, the conditions for cell growth and virus replication could be controlled to assure a uniform final product.

The ultimate goal of research in insect cell culture is the production of viral pathogens in large volume on a commercial scale. A number of satisfactory culture media have been developed for the growth of insect cells. Two methods of large volume cell culturing, i.e. attached cell culture and suspension cell culture from *S. frugiperda* in roller bottles and production of AcMNPV are well known. The principal advantage with these is the economy of the space and labour compared to flask cultures (Battu *et al.*, 1993, 1994; Battu and Arora, 1997). The significant achievement in the development of low cost protein-free media is bound to enable the production of viral pesticides (Rabindra and Rajasekaran, 1996). Finally, relatively simple and inexpensive procedures would be required to harvest the viral occlusion bodies or the infectious entities in case of non-occluded baculoviruses.

Interaction with insecticides

One of the major limitations of baculoviruses is the slow speed of kill, which is unacceptable in most agricultural crops. To overcome this problem, baculoviruses may be combined with conventional pesticides (Yearian and Young, 1982). The NPVs of a number of important insect pests like *H. armigera, A. gemmatalis, H. zea, T. ni* and *S. litura* have been reported to be compatible with chemical insecticides (Battu *et al.*, 1993; Tanada and Kaya, 1993; Moscardi, 1999). However, an antagonistic response between NPVs of *S. litura* and *S. obliqua* with selected dosages of some insecticides (endosulfan, fenitrothion) has also been reported (Chaudhari and Ramakrishnan, 1980; Battu *et al.*, 1992; Kaler *et al.*, 1999). The interactions of baculoviruses with each chemical insecticide should, therefore, be studied critically before recommending such combinations for pest control.

Host-parasitoid-baculovirus interactions

Host-parasitoid-baculovirus interactions occur with the development of larval parasitoids on or in infected host individuals from eggs laid on, in, or near the host by the adult female parasitoid. Interactions may also originate when a parasitized host subsequently is infected by a pathogen. As a result, parasitoids may be directly affected at the organismic level by their development in infected hosts or indirectly at the population level through the influence of pathogenic microbes on the populations of their hosts. Although host-parasitoid and host-pathogen relationships have been extensively studied for over 100 years, yet there have been few detailed studies on their multitrophic interactions (Brooks, 1993).

Harmful effects

Premature host death is the most common consequence of a host-pathogen (baculovirus) interaction, and many examples are known where the premature deaths of infected hosts have resulted in deaths of the parasitoids. Deleterious interactions have been especially well documented in cases of baculovirus infections of insects involving both circumstantial evidence from field studies (Teakle *et al.*, 1985) as well as numerous direct observations of the parasitoids unable to survive the baculovirus-induced deaths of their hosts. About seven braconids, one encyrtid, four ichneumonids, one pteromalid and six tachinids have been observed to be unable to complete their development due to the baculovirus-induced deaths of their hosts (Brooks, 1993). As there is no evidence of the direct susceptibility of various parasitoids to insect viruses, the outcome of parasitoid development depended primarily on the timing of parasitoid emergence from the host and host death due to virus.

Preferred development in baculovirus-infected hosts

A few parasitoids have been reported to develop normally or to even prefer development in virus-infected hosts. Three species of braconid parasitoids, *Apanteles* sp., *Apanteles prodeniae* Vier. and *Chelonus* sp. and ichneumonid wasps, *Campoletis* sp. at times parasitized *S. litura* successfully irrespective of the fact whether or not the host species carried its own NPV infection in the cabbage and cauliflower/castor ecosystems (Battu *et al.*, 1998b).

Battu *et al.* (1997) studied the interaction of two important biotic agents, viz. nucleopolyhedrosis alone and in combination with a braconid parasitoid, *Apanteles glomeratus* (Linnaeus) responsible for natural mortality of the cabbage butterfly, *Pieris brassicae* (Linnaeus) attacking the cauliflower, cabbage, *taramira (Eruca sativa)* and *sarson (Brassica campestris)* crops. The percentage mortality due to NPV alone, *A. glomeratus* alone and a combination of NPV + *A. glomeratus* was 10 to 25, 5.7 to 8.0 and 10.0 to 20.7, respectively. It was observed that coexistence of the nucleopolyhedrosis in the same host larva did not deter the incipient parasitoid to feed, develop and mature as grown up grubs, which were subsequently observed to emerge successfully from the same larva. NPV infection did not affect the emergence of parasitoid grubs and their subsequent development. The caterpillars showing abnormal behaviour such as sluggish movements, reduced feeding and stunted growth indicated the incipient infestation by the parasitoid. The presence of fairly good amount of the virus load (1.13 to 2.0×10^8 OBs/larva) within the parasitized host larvae was quite interesting and indicated the innocuous nature of this virus towards *A. glomeratus* of *P. brassicae* larvae.

An integrated pest management system involving release of either or both the agents, along with other components may be utilized for managing this important pest.

Quality control and safety testing

Quality control of production of the given baculovirus batches is essential to ensure manufacture of pre-determined standards at which a product has been safety tested. There are two aspects, i.e. identity of the viral agent and recognition of contaminants or contaminating organisms. While commercializing viral insecticides, the data for registration should describe the complete production methods, various checks required and maintenance details to ensure contaminant free quality production. Usually, by careful maintenance of seed stocks as well as by individual batch checking during production and harvesting phases, one can very easily ensure the correct identity of a given virus-isolate being produced. Holding the viral seed-inoculation as inert either cold-stored and/or freeze-dried materials help in maintaining the virulence of a given virus isolate and necessarily the freeze drying of baculoviruses helps in retention of purity. Reports with some baculoviruses suggest that freeze drying might greatly reduce their (viral) viability (Burges, 1981b).

Quality control standards

Podgwite and Bruen (1978) have detailed exclusively the quality control standards given by the USDA Forest Service (as requirements) for the production of NPV of the gypsy moth (TM Bio-Control I: NPV of the Douglas-fir tussock moth). The activity is based on the LC_{50} determined for *O. pseudotsugata* larvae (strain GL-1), not exceeding 13.066 ng of final product giving a titre ≥76.534 million units GL/g. Purity, permitted only for baculovirus morphotypes, was determined by darkfield microscopy and electron microscopy. Cytoplasmic polyhedrosis or polyhedrosis caused by unicapsid virus are not permitted, nor are the deaths by insecticidal chemicals detected by per oral and intrahemocoelic injection of a group of larvae at the LC_{50} level. *B. thuringiensis* and its toxins must be absent. The aerobic bacterial count must not exceed 10^9 colonies/g by plate count on trypticase soy agar. Brewer agar was used to detect the anaerobic and microacerophilic bacteria.

Regarding the physical features, the virus preparation must pass a screen of 100 meshes/inch (Tyler-scale) with no more than 1 per cent residue. A total residue was determined at 104°C. For safety to vertebrates, no fecal coliform bacteria, or any other bacteria or agents pathogenic for warm blooded vertebrates are permitted as detected by intraperitoneal injection in mice, oral administration to mice and by plate tests. Coliform bacteria (lactose fermentors), typhoid, paratyphoid and dysentery bacteria are detected on formula II Endoagar. Fecal coliforms were detected in EC (*Escherichia coli*) fermentation tubes (44.5°C) after enrichment in lauryl tryptose broth. Pathogenic Enterobacteriaceae, e.g. *Shigella, Salmonella* (= SS) were detected by spread plates of SS agar. The sensitivity of detection on agar plates depended upon the amount of the preparation plated

(0.001 g for the tussock moth NPV). The fungal contaminants with similar temperature requirements to those of the pesticidal fungus raise the problem of deciding how many spores to plate to get adequate sensitivity (Burges, 1981b). At the registration of Elcar (*Heliothis* NPV), a very vigorous limit of 10^7 total bacteria per gram was set. This is less per acre of crop than the number of coliform bacteria permitted in 2 gallons of unpasteurized milk (Rogoff, 1973). Non-occluded viruses may infect insects, even simultaneously with NPV. It is desirable that the healthy and infected stock be examined for non-occluded viruses and other contaminants.

Quality of baculovirus preparations

Following the regulation of NPV products like "Spod-X" and "Gemstar" for *S. exigua* and *Heliothis* spp., respectively during the year 1994 in USA, many other Asian and European countries like Thailand and Holland allowed the registration of these products (Kolodny-Hirsch and Dimock, 1996). In India too interest in commercialization of baculovirus based insecticides has developed recently and NPV products involving respective baculovirus species from *H. armigera* and *S. litura* are available. However, the wide-spread use of these products is still not achieved though the market is huge for *H. armigera* and *S.litura* crop protection products (HaNPV: 4.26 $\times 10^{23}$ and SlNPV: 1.59 $\times 10^{23}$ viral occlusion bodies) to fulfill the needs of at least 10 per cent of the crop area under cotton, chickpea, oilseeds, vegetables, etc.(Sathiah and Jayaraj, 1996).

Unfortunately, the quality of many NPV preparations is extremely poor and is totally ineffective in killing target pests especially when field-evaluation reports are evaluated analytically. Grzywacz *et al.* (1997) conducted investigations on such problematic aspects in Egypt, where NPV of *S. littoralis* was mass produced and distributed for use against *S. littoralis* management in Egyptian crop ecosystem. It was astonishing to note that in a survey of NPV being either sold or distributed to farmers in 1994, none of the 17 samples examined showed the declared level of NPV. In 5 of the samples, no NPV was recorded at all and in another 7, there was insufficient viral DNA present to get an identifying restriction enzyme profile. Of those containing NPV, none contained more than 10 per cent of the expected NPV and most of them had less than one per cent.

Similarly, some NPV products produced in India were ineffective under laboratory conditions when fed via contaminated foliage and contained actual NPV content far below the required quantum (i.e. 6×10^9 OBs making one LE/ml of the product) (Battu, 1999). One of the South Indian situation was particularly bad where a survey of commercial supply of NPV of *H. armigera* in 1996-99 (Kennedy *et al.*, 1999) indicated that all the eleven samples examined had low levels of viral occlusion bodies to be effective. These samples contained no active material at least in two of the lots examined while in rest of the nine samples the OBs level varied between 1.50×10^6 to 4.305×10^5 per ml which means 4000 to 196 times lower than the actual requirement of the NPV.

It is thus expected that future field applications of baculoviruses will be strictly defined and separated in terms of infectivity of the products and regulatory agencies will have to ensure that the standardization of products is strictly inscribed on product labels.

There could be several reasons for the poor quality but the main drawbacks are related to deficiencies in production techniques and quality control procedures. A problem for producers, customers and regulators is that the standard technique for assessing chemical pesticides through chemical analysis is not appropriate for an infective biological agent such as NPV and GV. Even the standard toxicity assessment methodologies applied for rapid action contact pesticides are often inappropriate for the relatively slow acting biopesticides such as NPV and GV, which have to be ingested as viral occlusion bodies applied to pest's food material (Kennedy *et al.*, 1999). The use of LE as a standard measure of NPV activity must be based on actual counts of occlusion bodies that can easily be done reliably and efficiently using a hemocytometer on aqueous suspensions through optical microscopy (Battu *et al.*, 1993). Many producers and research workers enthusiastically engaged in entrepreneurships involving propagation of baculoviruses as cottage industries in developing countries, are lacking the basic technical training in discriminating between NPV occlusions and artifacts involving cellular debris and developmental stages of many saprophytic/infectious microorganisms (Battu *et al.*, 1994). Thus their training will go a long way in avoidance of this commonly observable mistaken identification of artifacts in such propagation.

Safety testing

In November 1980, the WHO held an important informal consultation of experts to propose mandatory safety tests for microbial agents intended for use in pest control. The group benefited from the reports of various meetings, the draft guidelines and the policy statement. It emphasized that microbial pesticides pose inherently different hazards to mammals than do chemicals, that test methods for microbials should reflect these differences and that negative as opposed to positive results are to be expected. A system of tests in three tiers was proposed.

The *first tier* is designed to expose animals to very severe acute tests with the microbial agents, such that the agents can be considered safe if all the results are negative. Any positive results lead to either rejection of the agent or to quantification of the effect in the *second tier* tests. These results, in turn, lead either to a hazard analysis resulting in a decision that the agent is safe without any limitations, to label restrictions, to long term tests in a *third tier*, or to rejection. For viruses, emphasis is placed on infectivity. Some viral occlusion bodies are not known to be normally associated with toxins. Mutagenicity screens and 90-day subacute and long-term toxicity tests are regarded as irrelevant. In tier 1, a series of tests with tissue cultures is included and considered as more sensitive than tests with whole mammals. For viruses, any evidence of insertion of viral DNA into the host genome leads to tier 2 tests for transformation of host cells. The value of long term carcinogenicity tests for viruses is regarded as dubious. Recommendations were also made about the tests necessary at different stages in the development of the use of a microbial agent and for research into safety testing methodology. These proposals (Burges, 1981b) provide tests that will generate data from which confident conclusions about the degree of safety of microbial agent can be made, while not embarking on unnecessary tests or tests that produce data of dubious value.

Microbial biodiversity and intellectual property rights

The existing genetic potential of bioagents needs to be conserved by the preservation of natural and man-made systems and sites, and, where there are difficulties in recovering such diversified microbes from the environment, by maintaining regional as well as centralized culture collections of organisms of current and potential value. In recent years the TRIPS (Trade Related Aspects of Intellectual Property) agreement is one of the most important steps toward world wide harmonized laws in intellectual property. Article 27 provides the patentability of inventions in all fields of technology, including biotechnology. Living organisms as well as genes must be patentable, if the normal criteria for the patentability (novelty, inventive step and industrial applicability) are met. However, the patenting of plants or animals stays optional according to article 27(3)(b) (Stauffer, 1999). Patents are also available for products consisting of, or containing biological materials. The latter means any material containing genetic information and is capable of reproducing itself or can be reproduced in a biological system.

European biotechnology industry as well as universities have waited for a long time to get clear guidelines on how they can protect their inventions. The latest legal instrument which determines the biotechnology patent law (The Biotechnology Patenting Directive of the European Union) in European countries finally provides a clear basis for what can be patented and what not (Stauffer, 1999). An ordinance was promulgated on January 1, 1995 by the Government of India to give effect to the provisions under the TRIPS Agreement, for granting product patents and their subsequent enforcement, yet it could not get the shape of a law, due to various reasons. Accordingly, again the Central Government promulgated the Patents (Amendment) Ordinance 1999 on January 8, 1999, for meeting the requirements under TRIPS Agreement and shall be deemed to have come into force on January 1, 1995.

It is significant to note that now a system for granting product patents in respect of drugs and agrochemicals has been made available through stipulation of the TRIPS Agreement which did not have such a system earlier. It must introduce two provisions in its law. Firstly, a system for accepting product patent applications in the above two areas which must be put in place on the date of entry into force of the WTO Agreement, i.e. January 1, 1995. However, such applications may not be considered for examination until product patent laws are introduced. India has a transition time up to December 31, 2004 for introducing a system for granting the product patents. Secondly, exclusive marketing rights (EMRs) shall be granted in respect of drugs and agrochemicals for a period of five years or until a product patent law is put in place, whichever period is shorter. The Controller General will apply various criteria for assessing the eligibility of an application for EMR (Anonymous, 1999). Section 39 of the Patent Act 1970, which prohibited residents of India from applying for patents outside India without permission, has been dropped. Thus, any process for the medicinal, surgical, curative, prophylactic or other treatment of human beings or any process for a similar treatment of animals or plants to render them free of disease or to increase their economic value or that of their products, will be the subject for their patentability in India.

Perhaps, a naturally occurring virus that kills more than a dozen different kinds of crop eating caterpillars may be marketed as a future biopesticide (Battu and Arora, 1996). Biosys, Incorporation of Columbia, M.D., has registered a naturally occurring virus (celery looper virus-an NPV) with the US Environmental Protection Agency. The firm has signed an agreement with Zeneca Agrochemical, Surrey, England for worldwide commercialization of new insecticide products based on the microorganisms. The virus quickly kills caterpillars that accidentally eat virus particles while munching on the plant. The microorganism does not target people, pests or wildlife. It fells its namesake, the celery looper caterpillar, and other cotton and vegetable pests including the cabbage looper, tomato and tobacco hornworm, cotton bollworm (also called corn earworm and tomato fruitworm), beet armyworm, diamondback moth and pink bollworm. USDA has granted licenses for the celery looper virus to Biosys and to Sandoz Agro, Incorporation, Des Plains, IL. Both companies are continuing outdoor tests of the virus (Dureja, 1999).

Significantly, this development elsewhere, is bound to encourage scientists working in the developing world to go ahead with researches on such baculovirus strains known to have relatively wider host range for the class Insecta (particularly the pestiferous species) either from the indigenously available baculoviruses or through negotiations with those possessing them elsewhere in the world with the overall aim of development and commercialization of ecofriendly microbial products. That *Condica* species caterpillars have been mass produced at Rahuri in Maharashtra State of India for the propagation of one of the local strains of NPV of *H. armigera* is a noteworthy recent development supported by the Department of Biotechnology, Government of India, New Delhi (Wahab, 1997). The Department of Biotechnology has provided similar financial support to the National Institute of Immunology, New Delhi for the development of an NPV of *A. californica*. This is known for its wide host range (Battu and Arora, 1996), therefore, an ecofriendly viral insecticide may be in the offing for the IPM programme against *H. armigera* and other susceptible pestiferous species. It is thus important for the scientists in developing countries to evolve codes and ethics necessary for getting their indigenous discoveries patented locally and abroad for the protection of the microbial wealth serving as a resource base for the future development and commercialization of microbial biopesticides.

Conclusions

The selective toxicity of baculoviruses to major insect pests and their safety to non-target organisms makes them ideal tools for use in IPM programmes. Nearly thirty nuclear polyhedrosis and granulosis viruses are already being commercially utilized for the management of important pests. These positive trends, however, need to be accompanied by strengthening of research efforts to overcome some of the major limitations in production, use and efficacy of baculoviruses.

The relatively slow speed with which baculoviruses kill their hosts has hampered their effectiveness as well as acceptance by potential users. Genetic improvement, using traditional methods as well as genetic engineering may produce strains of baculoviruses with improved pathogenesis and virulence. Until *in vitro* techniques are developed for mass production of viruses, the production process will remain laborious and costly. Recent advances in virus production using insect cell lines offer a way out of this situation. Quality control in commercially produced microbial pesticides is another area requiring urgent attention. It is necessary to maintain the viability and virulence of the pathogens till use. The allelochemical interactions among plants, herbivores, entomopathogens and entomophages need further study. Efforts should be made to minimize the loss in infectivity of baculoviruses due to photo-inactivation. Finally, interaction of baculoviruses with other methods of pest control should be thoroughly studied to develop stronger IPM strategies. The self-perpetuating nature and selectivity of baculoviruses would certainly prove to be an asset in sustainable agriculture.

References

Anonymous (1999) The Patents (Amendment) Ordinance 1999. *IPR Bull.*, **5**, 6–7.

Angus, F.A., and Luthy, P. (1971) Formulations of microbial insecticides. In H.D. Burges, and N.W. Hussey (eds.), *Microbial Control of Insects and Mites*, Academic Press, New York, pp. 623–638.

Argauer, R., and Shapiro, M. (1997) Fluorescence and relative activities of stilbene optical brightness as enhancers for the Gypsy moth (Lepdioptera: Lymantriidae) baculovirus. *J. Econ. Entomol.*, **90**, 416–420.

Arora, R., Battu, G.S., and Ramakrishnan, N. (2000) Microbial pesticides: current status and future outlook. In G.S. Dhaliwal and B. Singh (eds.), *Pesticides and Environment*, Commonwealth Publishers, New Delhi, pp. 344–395.

Battu, G.S. (1999) Viral pathogenesis mediated development of microbial pesticides. In *Abstr.Natn.Symp.Emerging Trends in Biotechnological Application for Integrated Pest Management*, Chennai, pp. 3–4.

Battu, G.S., and Arora, R. (1996) Genetic diversity of baculoviruses: Implications in insect pest management. In T.N. Ananthakrishnan (ed.), *Biotechnological Perspectives in Chemical Ecology of Insects*, Oxford & IBH Publishing Co. Pvt.Ltd., New Delhi, pp. 179–200.

Battu, G.S., and Arora, R. (1997) Insect pest management through microorganisms. In K.R. Dadarwal (ed.), *Biotechnological Approaches in Soil Microorganisms for Sustainable Crop Production*, Scientific Publishers, Jodhpur, India, pp. 222–246.

Battu, G.S., and Bakhetia, D.R.C. (1992) Foliar persistence of the nuclear polyhedrosis virus of *Spilosoma obliqua* (Walker) on groundnut. In *Proc. Nat. Symp. Recent Advances in Integrated Pest Management*, Ludhiana, pp.98.

Battu, G.S., and Ramakrishnan, N. (1989) Comparative role of various mortality factors in the natural control of *Spilosoma obliqua* (Walker) in Northern India. *J. Ent. Res.*, **13**, 38–42.

Battu, G.S., and Sidhu, R.S. (1992) Persistence of the nuclear polyhedrosis virus of *Spilosoma obliqua* on the sunflower foliage. In *Proc Natn Symp Recent Advances in Integrated Pest Management*, Ludhiana, pp. 98–99.

Battu, G.S., Arora, R., and Bath, D.S. (1997) Inocuity of the nucleopolyhedrosis towards *Apanteles glomeratus* (Linnaeus) attacking *Pieris brassicae* (Linnaeus) caterpillars around Ludhiana. *Indian J. Ecol.*, **24**, 90–93.

Battu, G.S., Arora, R., and Bath, D.S. (1998a) Evaluation of native nuclear polyhedrosis virus for ecosafe management of *Spodoptera litura* (Fabricius) on cauliflower crop. In G.S. Dhaliwal, N.S. Randhawa, R. Arora and A.K. Dhawan (eds), *Ecological Agriculture and Sustainable Development* Vol. 2, Indian Ecological Society and Centre for Research in Rural and Industrial Development, Chandigarh, India, pp. 315–322.

Battu, G.S., Arora, R., and Bath, D.S. (1998b) Natural enemy complex of some lepidopteran pests of agricultural crops in the Punjab, India. In G.S. Dhaliwal, N.S. Randhawa, R. Arora and A.K. Dhawan (eds),

Ecological Agriculture and Sustainable Development, Vol. 2, Indian Ecological Society and Centre for Research in Rural and Industrial Development, Chandigarh, pp. 231–248.

Battu, G.S., Dhaliwal, G.S., and Raheja, A.K. (1994) Biotechnology: Perspectives in insect pest management. In G.S., Dhaliwal and R. Arora (eds.), *Trends in Agricultural Insect Pest Management*, Commonwealth Publishers, New Delhi, pp. 417–468.

Battu, G.S., Ramakrishnan, N., and Dhaliwal, G.S. (1993) Microbial pesticides in developing countries: Current status and future potential. In G.S.Dhaliwal and B. Singh (eds.), *Pesticides: Their Ecological Impact in Developing Countries*, Commonwealth Publishers, New Delhi, pp. 270–334.

Battu, G.S., Ramakrishnan, N., and Prakash, N. (1996) Electron microscopy of the nuclear polyhedrosis virus of *Spilosoma obliqua* Walker-alkali dissolution of occlusion bodies. *Shashpa*, **3**, 37–45.

Battu, G.S., Sidhu, R.S., and Dhaliwal, G.S. (1992) Interaction between chemical insecticides and NPV of *Spilosoma obliqua* (Walker). *Indian J. Appl. Ent.*, **6**, 49–53.

Blissard, G.W., and Rohrmann, G.F. (1989) Location, sequence, transcriptional mapping and temporal expression of the gp 64 envelope glyptoprotein gene of the *Orgyia pseudotsugata* multicapsid nuclear polyhedrosis virus. *Virology*, **170**, 537–555.

Brooks, W.M. (1993) Host-parasitoid-pathogen interactions. In N.E. Beckage, S.N. Thompson and B.A. Federici (eds.), *Parasites and Pathogens of Insects,Vol.2, Pathogens*, Academic Press, San Diego, USA, pp. 231–272.

Burges, H.D. (1981a) *Microbial Control of Pests and Plant Diseases 1970–1980*. Academic Press, New York.

Burges, H.D. (1981b) Safety, safety testing and quality control of microbial pesticides. In H.D. Burges (ed.), *Microbial Control of Pests and Plant Diseases 1970–1980*, Academic Press, London, pp. 737–767.

Chaudhari, S., and Ramakrishnan, N. (1980) Field efficacy of baculovirus and its combination with sublethal dose of DDT and endosulfan on cauliflower against tobacco caterpillar, *Spodoptera litura* (Fabricius). *Indian J. Ent.*, **42**, 596–599.

Cook, S.P., Webb, R.E., and Thorpe, K.W. (1996) Potential enhancement of the gypsy moth (Lepidoptra:Lymantriidae) nuclear polyhedrosis virus with the triterpene azadirachtin. *Environ. Entomol.*, **25**, 1209–1214.

Cowan, D.K., Vail, P.V., Kok-Yokomi, M.L., and Schreiber, F.C. (1986) Formulation of granulosis virus of *Plodia interpunctella* (Hubner) (Lepidoptera: Pyralidae): efficacy, persistence and influence on oviposition and larval survival. *J. Econ. Entomol.*, **79**, 1085–1090.

Cunningham, J.C. (1982) Field trials with baculoviruses: control of forest pests.In E. Kurstak (ed.), *Microbial and Viral Pesticides*, Marcel Dekker, New York, pp. 335–386.

Cunningham, J.C. (1988) Baculoviruses: their status compared to *Bacillus thuringiensis* as microbial insecticides. *Outlook Agric.*, **17**, 10–17.

Cunningham, J.C., and Entwistle, P.F. (1981) Control of sawflies by baculovirus. In H.D. Burges (ed.), *Microbial Control of Pests and Plant Diseases 1970–1981*, Academic Press, New York, pp. 379–407.

Dales, M.J. (1994) *Controlling Insect Pests of Stored Products Using Insect Growth Regulators and Insecticides of Microbial Origin*, NRI Bulletin 64, Natural Resource Institute, Chatam, UK.

Dhandapani, N., Jayaraj, S., and Rabindra, R.J. (1990) Influence of sunlight and crop improvement on the efficacy of nuclear polyhedrosis virus against *Heliothis armigera* (Hubner) on groundnut (*Arachis hypogaea* L.). *J Appl. Ent.*, **116**, 523–526.

Dureja, P. (1999) PRJ-News and views. *Pestic. Res. J.*, **11**, 229– 236.

FAO (1973) *The Use of Viruses for the Control of Insect Pests and Disease Vectors*. Report Joint Meeting FAO/WHO on Insect Viruses, Nov.22–27, 1972, FAO Studies No.91, Geneva.

Francki, R.I.B., Fauquet, C.M., Knudson, D.L., and Brown, F. (1991) Classification and nomenclature of viruses: Fifth Report of the International Committee on Taxonomy of Viruses. *Arch. Virol.*, **2**, 117–123.

Fuxa, J.R. (1990) New directions for insect control with baculoviruses. In R.R. Baker and P.E. Dunn (eds.), *New Directions in Biological Control*, Wiley-Liss, New York, pp. 97–113.

Fuxa, J.R. (1991) Insect control with baculoviruses. *Biotech. Adv.*, **9**, 425–442.

Fuxa, J.R., and Geaghan, J.P. (1983) Multiple regression analysis of factors affecting prevalence of nuclear polyhedrosis virus in *Spodoptera furgiperda* (Lepidoptera: Noctuidae) population. *Environ. Entomol.*, **12**, 311–316.

Fuxa, J.R., Richter, A.R., and McLeod, J. (1992) Virus kills soybean looper years after its introduction into Louisiana. *La Agric.*, **35**, 20–23.

Granados, R.R., Derksen, A.C.G., and Dwyer, K.G. (1986) Replication of the *Trichoplusia ni* granulosis and nuclear polyhedrosis viruses in cell cultures. *Virology*, **152**, 472–476.

Granados, R.R., Kathleen, G.D., and Auja, C.G.D. (1987) Production of viral agents in invertebrate cell cultures. In K. Maramorosch (ed.), *Biotechnology in Invertebrate Pathology and Cell Culture*, Academic Press, San Diego, USA, pp. 167–181.

Grzywacz, D., McKinley, D., Jone, K.A., and Moawad, G. (1997) Microbial contamination in *Spodoptera littoralis* nuclear polyhedrosis virus produced in insects in Egypt. *J. Invertebr. Pathol.*, **69**, 151–156.

Hammock, B.D., Inceoglu, A.B., Rajendra, W., Fuxa, J.R., Chejanovsky, N., Jarvis, D. *et al.* (1999) Impact of biotechnology on pesticide delivery. In G.T. Brooks and T.R. Roberts (eds.), *Pesticide Chemistry and Bioscience:The Food-Environment Challenge*, Royal Society of Chemistry, Cambridge, pp. 73–99.

Hawtin, R.E. (1993) The chitinase of *Autographa californica* nuclear polyhedrosis virus. Ph.D. Thesis, Oxford Brookes University, Oxford, U.K.

Hubner, J. (1990) Viral insecticides: profits, problems and prospects. In J.E. Casida (ed.), *Pesticides and Alternatives*, Elsevier Scientific Publishers, Amsterdam, pp. 117–122.

Hubner, J. (1986) Use of baculovirus in pest management programmes. In R.R. Granadoes and B.A. Federici (eds.), *The Biology of Baculoviruses*. Vol I, *Biological Properties and Molecular Biology*. CRC Press, Boca Raton, pp. 181– 202.

Hugar, H., Kulkarni K.A., Lingappa, S., and Hugar, P. (1996) Persistence of NPV of *Mythimna separata* (Walker) on sorghum foliage. *Karnatka J. Agric. Sci.*, **9**, 51–55.

Ignoffo, C.M. (1981) Living microbial insecticides. In J.R. Norris and M.H. Richmond (eds.), *Essays in Applied Microbiology*, John Wiley & Sons, New York, pp. 2–31.

Ignoffo, C.M., and Batzar, O.F. (1971) Microencapsulation and ultraviolet protectants to increase sunlight stability of an insect virus. *J. Econ Entomol.*, **64**, 850–853.

Ignoffo, C.M., and Couch, T.L. (1981) The nucleopolyhedrosis virus of *Heliothis* species as a microbial insecticide. In H.D. Burges (ed.), *Microbial Control of Pests and Plant Diseases*, Academic Press, New York, pp. 329–362.

Ignoffo, C.M., Hostetter, D.L., and Pinell, R.E. (1974) Stability of *Bacillus thuringlensis* and *Baculovirus heliothis* on soybean foliage. *Environ. Entomol.*, **3**, 117–19.

Kaler, D. (1996) Utilization of nuclear polyhedrosis virus and *Bacillus thuringiensis* Berliner against *Spodoptera litura* (Fabricius). M.Sc.Thesis, Punjab Agricultural University, Ludhiana

Kaler, D., Battu, G.S., and Arora, R. (1999) Interaction between three chemical insecticides and nuclear polyhedrosis virus of *Spodoptera litura* (Fabricius). *Indian J. Ecol.*, **26**, 177–181.

Kalmakoff, J., and Crawford, A.M. (1982) Enzootic virus control of *Wiseana* spp. in the pasture environment. In E. Kurstak (ed.), *Microbial and Viral Pesticides*, Marcel Dekker, New York, pp. 435–438.

Kaushik, H.D. (1991) Studies on nuclear polyhedrosis virus of *Heliothis armigera* (Hubner) on tomato (*Lycopersicon esculentum* Miller), Ph.D. Dissertation, Haryana Agricultural University, Hisar.

Keddie, B.A., and Volkman, L.E. (1985) Infectivity difference between the two phenotypes of *Autographa californica* nuclear polyhedrosis virus: importance of the 64K envelope glycoprotein. *J. Gen. Virol.*, **66**, 1195–1200.

Kennedy, J.S., Rabindra, R.J., Sathiah, N., and Gryzwacz, D. (1999) The role of standardisation and quality control in the successful promotion of NPV insecticides. In S. Ignacimuthu and A. Sen (eds.), *Biopesticides and Insect Pest Management*, Phoenix Publishing House Pvt. Ltd., New Delhi, pp. 170–174.

King, L.A., Possee, R.D., Hughes, D.S., Atkinson, A.E., Palmer, C.P., Marlow, S.A. *et al.* (1994) Advances in insect virology. *Adv. Insect Physiol.*, **25**, 1–73.

Kinsinger, R.A., and McGaughey, W.H. (1976) Stability of *Bacillus thuringiensis* and a granulosis virus of *Plodia interpunctella* on stored wheat. *J. Econ. Entomol.*, **69**, 149–154.

Kinsinger, R.A., and McGaughey, W.H. (1979) Susceptibility of populations of Indian meal moth and almond moth to *Bacillus thuringiensis. J. Econ. Entomol.*, **69**, 149–154.

Kogan, M. (1998) Integrated pest management: Historical perspectives and contemporary developments. *Annu. Rev. Entomol.*, **43**, 243–270.

Kolodny-Hirsch, D., and Dimock, M. (1996) Commercial development and use of Spod-X, a wild type baculovirus insecticide for beet armyworm. Paper presented at *29th Annual Meeting of Society of Invertebrate Pathology and III Colloquim* on *Bacillus thuringiensis*, Cardoba, Spain.

Marrone, P. (1996) Biological products for integrated pest management. In G.J. Persley (ed.), *Biochemistry and Integrated Pest Management*, CAB International, Wallingford, U.K., pp.150–163.

McLaughlin, R.E., Andrews, G.L., and Bell, M.R. (1971) Field tests for control of *Heliothis* spp. with a nuclear polyhedrosis virus included in a boll weevil bait. *J. Invertebr. Pathol.*, **18**, 304.

Miller, L.K. (1987) Expression of foreign genes in insect cells.In K. Maramorosch (ed.), *Biotechnology in Invertebrate Pathology and Cell Culture*, Academic Press, San Diego, USA, pp. 295–304.

Miller, L.K. (1998) Genetically improved baculoviruses for insect pest management. In V.L. Chopra, R.B. Singh and Anupam Varma (eds). *Crop Productivity and Sustainability-Shaping the Future*, Oxford & IBH Publishing Co.Pvt. Ltd., New Delhi, pp. 269–276.

Miltenburger, H.G., Naser, W.L., and Harvey, J.P. (1984) The cellular substrate: a very important requirement for baculovirus *in vitro* replication. *Z. Naturforsch. Biosci.*, **39C**, 993–1002.

Morales, L., Moscardi, F., Sosa-Gamez, D.R., Paro, F.E., and Soldorio, I.L. (1997) Enhanced activity of *Anticarsia gemmatalis* Hub (Lepidoptera: Noctuidae) nuclear polyhedrosis virus by boric acid in the laboratory. *Ann. Soc. Ent. Bros.*, **26**, 115–120.

Moscardi, F. (1989) Use of viruses for pest control in Brazil: the case of nuclear polyhedrosis virus of the soybean caterpillar, *Anticarsia gemmatalis. Mem. Inst. Oswaldo Fruz*, **84**, 51–56.
Moscardi, F. (1999) Assessment of the application of baculoviruses for control of Lepidoptera. *Annu. Rev. Entomol.*, **44**, 257–289.
Murphy, F.A., Faquet, C.M., Bishop, D.H.L., Gabrial, S.A., Jarvis, A.W., Martelli, G.P. *et al.* (eds) (1995), *Classification and Nomenclature of Viruses: Sixth Report of the International Committee on Taxonomy of Viruses*, Springer-Verlag, Berlin.
Narayanan, K. Govindarajan, R., and Jayaraj, S. (1977) Preliminary observations on the persistence of nuclear polyhedrosis virus of *Spodoptera litura* (F.). *Madras Agric. J.*, **64**, 487–488.
Nishi, Y., and Nonaka, T. (1996) Biological control of the tea tortrix-using granulosis virus in the field. *Agrochem. Japan*, **69**, 7–10.
Persley, G.J. (1996) *Biotechnology and Integrated Pest Management*, CAB International, Oxon, U.K.
Pillai, G.B., Sathiamma, B., and Dangar, T.K. (1993) Integrated control of rhinoceros beetle. In M.K. Nair, H.H. Khan, P. Gopalsundaram and E.V.V. Bhaskara Rao (eds.), *Advances in Coconut Research and Development*, Oxford & IBH Publishing Co. Pvt. Ltd., New Delhi, pp. 455–463.
Podgwaite, J.D., and Bruen, R.B. (1978) *U.S. Dep. Agric. Forest Serv.Gen.Tech. Rep.NE-38*. Northeast Far Expt. St., Broomall, PA.
Rabindra, R.J., and Rajasekaran, B. (1996) Insect cell cultures: A tool in basic research, biotechnology and pest control. In T.N.Ananthakrishnan (ed.), *Biotechnological Perspectives in Chemical Ecology of Insects*, Oxford & IBH Publishing Co.Pvt. Ltd., New Delhi, pp. 223–239.
Raman, K.V., Alcazar, J., and Valdez, A. (1992) *Biological Control of Potato Tuber Moth using Phthorimaea* Baculovirus. CIP Training Bull. 2, International Potato Centre, Lima, Peru.
Rogoff, M.H. (1973) Industrialization. *Ann. N.Y.Acad. Sci.*, **217**, 200–210.
Santharam, G., Regupathy, A., Easwaramoorthy, S., and Jayaraj, S. (1978) Effectiveness of nuclear polyhedrosis virus against field populations of *Spodoptera litura* (F) on banana *Musca paradisica* L. *Indian J. Agric. Sci.*, **48**, 676–678.
Sathiah, N., and Jayaraj, S. (1996) Technology for mass production of biopesticides. *Silver Jubilee Seminar on Employment Opportunities for Biologists*, Chennai, India.
Shapiro, M. (1995) Radiation protection and activity enhancement of viruses. In F.R. Hall and J.W. Barray (eds), *Biorational Pest Control Agents: Formulation and Delivery*, American Chemical Society, Washington, DC, pp. 153–164.
Shapiro, M., and Argauer, R. (1997) Components of the stilbene brightner Tinopal LPW as enhancers for the Gypsy moth (Lepidoptera: Lymantriidae) baculovirus. *J. Econ. Entomol.*, **90**, 899–904.
Shapiro, M., Preisler, H.K., and Robertson, J.L. (1987) Enhancement of baculovirus activity on gypsy moth (Lepidoptera: Lymantiidae) by chitinase. *J. Econ. Entomol.*, **85**, 1120–1124.
Smith, D.B., and Hostetter, D.L. (1982) Laboratory and field evaluations of pathogen-adjuvant treatments. *J. Econ. Entomol.*, **75**, 472–476.
Sohi, S.S., Percy, J., Arif, B.M., and Gunningham, J.C. (1984) Replication and serial passage of a singly enveloped baculovirus of *Orgyia leucostigma* in homologus cell lines. *Intervirology*, **21**, 50–60.
Sosa-Gomez, D.R., Kitajima, E.W., and Rolon, M.E. (1994) First records of entomopathogenic diseases in the Paraguay tea agroecosystem in Argentina. *Fla. Entomol.*, **77**, 378–382.
Stauffer, T.P. (1999) Patenting of biological material in European countries. *IPR Bull.*, **5**, 3–5.
Tanada, Y., and Kaya, H.K. (1993) *Insect Pathology*, Academic Press, Inc., Harcourt Brace Jovanovich Publishers, San Diego.
Teakle, R.E. (1994) Virus control of *Heliothis* and other key pests: potential and use, and the local scene. In C.J. Monsour, S. Reid and R.E. Teakle (eds.), *Proc Ist Symp. on Biopesticides: Opportunities for Australian Industry*, University of Australia, Brisbane, pp. 51–56.
Teakle, R.E., Jensen, J.M., and Mulder, J.C. (1985) Susceptibility of *Heliothis armigera* (Lepidoptera: Noctudiae) on sorghum to nuclear polyhedrosis virus. *J. Econ. Entomol.*, **78**, 1373–1378.
Tuan, S.J., Tang, J.C., and Hov, R.F. (1989) Factors effecting pathogenicity of NPV preparations to the corn earworm, *Heliothis armigera. Entomophaga*, **34**, 541–549.
Vail, P.V., and Tebbets, J.S. (1991) The granulosis virus of the Indian meal moth: recent developments. In *Proc Fifth International Working Conference on Stored Product Protection.*, Bordeaux, France, pp. 1247–1254.
Vail, P.V., Hoffman, D.F., Street, D.A., Manning, J.S., and Tebbets, J.S. (1993) Infectivity of a nuclear polyhedrosis virus isolated from *Anagrapha falcifera* (Lepidoptera: Noctuidae) against production and post-harvest pests and homologous cell lines. *Environ. Entomol.*, **22**, 110–145.
Valicente, F.H., and Cruz, I. (1991) Control biologico da lagasta-do-cortucho, *Spodoptera frugiperda*, com o baculovirus. *Embrapa, Sete Lagoas Circ. Tec.*,15, 23 pp.
Wahab, S. (1997) Biotechnological approaches for the management of pests, diseases and weeds in agriculture-country status. *Pestic. Wld.* (September-October): 13–20.

Whitford, M., Stewart, S., Kugio, J., and Faulkner, P. (1989) Identification and sequence analysis of a gene encoding gp 67, an abundant envelope glycoprotein of the baculovirus *Autographa californica* polyhedrosis virus. *J. Virol.*, **63**, 1393–1399.
Yearian, W.C., and Young, S.Y. (1982) Control of insect pests of agricultural importance by viral insecticides. In E. Kurstak (ed.), *Microbial and Viral Pesticides*, Marcel Dekker, New York, pp. 387–423.
Yi, P., and Li, Z. (1989) Microbial pest control in China: viruses and fungi. *Symp. Sino-American Biocontrol Research*, San Antonio.
Young, S.Y., and McNew, R.W. (1994) Persistence and efficacy of four nuclear polyhedrosis viruses for corn earworm (Lepidoptera: Noctuidae) on heading grain sorghum. *J. Entomol. Sci.*, **29**, 370–380.
Young, S.Y., and Yearian, W.C. (1974) Persistence of *Heliothis* NPV on foliage of cotton, soybean and tomato. *Environ. Entomol.*, **3**, 253–255.
Zelazny, B., Lolong, A., and Patang, B. (1992) *Oryctes rhinoceros* (Coleoptera: Scarabaeidae) populations suppressed by a baculovirus. *J. Invertebr. Pathol.*, **59**, 61–68.
Zhu, G., and Zhang, H. (1985) The multiplication characteristics of *Heliothis armigera* in the established cell lines. *Abstr. 3rd Int. Cell. Cult. Congr.*, p. 66.

Beauveria Bassiana and other Entomopathogenic Fungi in the Management of Insect Pests

7

G.G. Khachatourians, E.P. Valencia[1] and G.S. Miranpuri[2]
Bioinsecticide Research Laboratory/Microbial Biotechnology Laboratory, Department of Applied Microbiology and Food Science, College of Agriculture, University of Saskatchewan, Saskatoon S7N 5A8, Canada [1]Aventis Crop Science, Carrera 77A # 45–61, Bogota, Columbia [2]Department of Medical Microbiology and Immunology, 436 Service Memorial Institute, 1300 University Avenue, University of Wisconsin-Madison, Madison, WI 53706, USA

Introduction

After almost twenty years of progress in discovery science and development of technology for insect pest control agents, it is possible to critically examine outcomes of the four classes of microorganisms, viruses, bacteria, fungi and protozoa in meeting the challenge (Khachatourians, 1986; Koul and Dhaliwal, 2001). With the advent of biotechnology there was even a euphoric suggestion not only for the success of bacterial and viral agents, but as well that of the transgenic plants containing, e.g. the *Bacillus thuringiensis* (Bt) insecticidal gene(s). However, history often reminds that many predictions are premature as in this context, where sufficient successes or failures were observed with insect pest management.

Plant injurious insects can cause diseases of crops, damage to fruits and seeds. Irrespective of the human perspectives on the definition of a pest, insects are components of our agroecosystems. Many insects depend on vegetation for survival and will compete with other herbivores and human food demand. It is human experience that decides the relationship between insects that are beneficial and those that are not (pests). If considered pests, modern agriculture needs periodic reduction of pest population size or control of their spread. However, absolute control, annihilation of an insect species is not possible through the use of microbial pest control practice and its intensified application at least for Bt, in the long term has yielded resistant races.

For many years, the focus of application of entomopathogenic fungi (EPF) for their role in insect pest diseases has been increasing. Research on applied mycology and biotechnology of Hyphomycetous EPF has provided a better understanding of pathogen-host interactions. The genetics and applied mycology of several EPF especially that of *Beauveria bassiana* (Bb) with prerequisite industrial R & D is in place. New reports in the use of these fungi from both temperate and tropical regions confirm the pivotal role of Bb in pest management practices. The EPF and more importantly Bb are poised to generate essential knowledge based technologies for insect pest management.

Taxonomy of hyphomycetous EPF

The distinguishing characteristic of true fungi is the presence of chitin in their cell walls. True fungi are placed in four phyla (Ascomycota, Basidiomycota, Chytridiomycota, Zygomycota) and the phylum, Deuteromycota, or fungi imperfecti (due to a lack of known sexual forms, anamorphs). These 5 phyla contain many of the EPF. In terms of ecological niche, EPF are found in both aquatic and terrestrial (forest and agricultural) habitats and usually found to be associated with insects, i.e. they are entomogenous. In some instances, for example, Bb isolates can be pathogens of insects of both terrestrial and aquatic habitats (Miranpuri and Khachatourians, 1990, 1991; Agarwala *et al.*, 1998). Additionally, Reithinger and coworkers (1997) made a valuation of Bb as a potential biological control agent against phlebotomine sand flies in Colombian coffee plantations. These workers show that Bb is pathogenic to *Phlebotomus papatasi* (Scopoli) and *Lutzomyia longipalpis* (Lutz & Neiva) (Diptera: Psychodidae), important vectors of tropical-endemic diseases. While histopathologic examination indicates that Bb is unable to infect sand flies under natural conditions, dead sand flies were shown to be readily infected and laboratory bioassays where flies were exposed to the fungus applied onto coffee plants showed lower mean survival times than the control. Luz *et al.* (1998) tested Bb on 3rd instars of nine *Triatoma* spp., four *Rhodonius* spp., two *Panstrongylus* spp. and *Dipetalogaster maxima* (Uhler). These insects can be transmission vehicles for Chagas disease agent, *Trypanosoma cruzi.* Most susceptible vectors to Bb infection were *Panstrongylus herreir* Wygodzinsky, *D. maxima, Triatoma picturata* Usinger, *Rhodnius robustus* Larrousse, *Rhodnius prolixus* Stal, *Triatoma infestans* Klug, and *Triatoma brasiliensis* Neiva.

The majority of EPF identified to date belong to three phyla: Ascomycota (classes: Laboulbeniales and Pyrenomycetes), Deuteromycota (class Hyphomycetes), and Zygomycota (class Zygomycetes).

Members of the phylum Deuteromycotina also called Fungi Imperfecti, lack a sexual cycle and form asexual conidia. There are a few reports suggesting that some members of the genera *Beauveria, Fusarium, Paecilomyces* and *Verticillium* through anastomasis of somatic hyphae exchange nuclei and can provide karyogamy or diploid cells (Khachatourians, 1991). Members of the Deuteromycete classes, Coelomycetes (Sphaeropsidales) and Hyphomycetes (Moniliales), are considered to be insect pathogens. Two genera, *Aschersonia* and *Tetranacrium*, of the class Coelomycetes are important pathogens of whiteflies and scale insects. The class Hyphomycetes contains over 40 entomopathogenic genera distributed worldwide. Interesting members include *Verticillium lecanii*, which can sporulate on live, infected aphids that continue to produce viviparous young, aiding with the dispersal of spores to new hosts. *Culicinomyces clavisporus* and *Tolyplocadium cylindrosporum* infect mosquito larvae *per os*.

The taxonomic consideration of 12 classes of EPF can be based on traditional systems of morphology and ultrastructure or modern biochemistry and molecular genetics (Samson *et al.* 1988; Khachatourians, 1991, 1996). In general, taxonomic criteria as indicated in Table 7.1, can have specific range of applications, such as identification of a genus and species, or that of the population. The newer tools of modern biochemistry and molecular genetics such as use of RAPDs in conjunction with protein and isoenzyme polymorphism is a powerful system for both study of pathogenicity and species dissemination and genetic drift (Khachatourians, 1996).

Table 7.1 Taxonomic tests for identification of entomopathogenic fungi

Taxonomic criteria/test	*Taxonomic groups*					*Population*
	Species	*Genus*	*Family*	*Order*	*Class*	
Morphological						
Mycelial forms			+	+	+	
Cytoplasmic				+	+	
Spore		+	+	+	+	
Biochemical						
2D-PAGE	+	+	+	+	+	+
Isozymes	+	+				
Enzymes	+	+				
Fatty acids	+	+	+	+	+	
1′ & 2″ metabolites	+	+	+	+	+	
Toxins	+	+				
Genetical						
Chromosome no's.	+					
Karyotype (OFAGE)	+					+
G+C content	+					
RAPDs	+	+	+	+	+	+
ITS spacers	+	+				
RRNA genes	+	+	+			
Protoplast fusion	+	+	+			
Mt DNA	+	+				+
RFLPs	+	+				+
Protein genes	+	+	+	+		

Interactions of *Beauveria bassiana* with host insects

In spite of many genera and species of Hyphomycetous fungi that have the potential for commercialization, the large varieties of agriculturally important target insects (e.g. see Bidochka *et al.* 1998; Miranpuri and Khachatourians, 1994a,b, 1995a–d; Miranpuri *et al.* 1993) and a clearer understanding of host pathogen interactions at the molecular level (Khachatourians, 1991, 1996; Hajek and St. Leger, 1994) makes Bb an ideal mycoinsecticide. This is because in addition to the knowledge of associative events governing the early phases of interaction, the subsequent events contributing to insect disease are both complex and challenging, as its overall expression represents the interplay of multiple genetic traits. It is for certain that growth with the concomitant production of extracellular hydrolytic enzymes, adhesive mucilaginous substance(s) and appressoria followed by penetration into the pest, spread of hyphal bodies and production of toxic metabolites are important elements in interaction (Khachatourians, 1991, 1996; Hajek and St. Leger, 1994).

Variations in many aspects of fungal interaction with host can, in fact, be found amongst a collection of natural isolates of Bb. Therefore, the study of a collection of EPF within an institution beyond the obvious has a two-fold importance, (i) understanding of quantitative aspects of the interaction with the target insect, and (ii) ultimate production and commercialization.

Given the availability of commercial kits for determination and quantification of enzymes, nucleic acids, lipid and other biochemicals, several investigators have made important contributions to the field (Lecuona, 1999). Rivera *et al.* (1997) studied a total of 36 isolates of Bb, for the presence of 9 extracellular enzymes involved in the degradation of proteins, chitin and polysaccharides. While differences in enzyme profiles were seen, all isolates showed glucosidase, N-acetylglucosaminidase and elastase activity, but none had fucosidase activity. With respect to two enzymes required in the onset of infection, protease and chitinase respectively two broad groupings high/low and low/high were detected (Rivera *et al.*, 1997). Polymerase chain reaction (PCR) amplification with 4 different primers showed no simple correlation with host, geographic origin or pathogenicity.

Duque and Arango (1998) made a qualitative biochemical characterization of isolates of Bb in the collection held at CENICAFE Colombia. The goal was to find the best strain for the biocontrol of the coffee berry borer (CBB), *Hypothenemus hampei* (Ferrari). This was based on three criteria: (i) Enzymatic activity and capacity for uptake of substrates, (ii) Factors, which determine the process of virulence of Bb to the insect, and (iii) Survival in natural conditions. Among a collection of 93 isolates, those isolates from their original host (OH) and passed through the CBB, or reactivated Bb (RB), were the basis for characterization and selection. In the enzymatic evaluation with the solid substrate method it was found that some isolates showed variable response to the chitinase and lipase production. However, after reactivation, i.e. passage through the insect against which the control is directed, i.e. CBB, a positive response was recorded. According to de-Marcano *et al.* (1999), Bb and *P. fumosoroseus* treated adults of *Cylas formicarius elegantulus* (Summers) showed reduction of movement and feeding after 48 h and death of some after 72 h. Total mortality was higher with Bb in its commercial and reactivated forms (98.34 and 96.6%, respectively), while with *P. fumosoroseus*, these were 63.33 and 98.33 per cent, respectively. Although reactivation of some EPF by passage through the host insect is often shown to increase insecticidal potency, its molecular basis remains unknown (Hayden *et al.*, 1992).

Other elements of EPF interaction with target pests are toxins. Toxins of EPF can be organic non-proteinaceous chemicals and peptides/proteins (Khachatourians, 1991, 1996) and other secondary metabolites many of which are essential in the pest control process. Further, the success of EPF in a winning strategy against the host relies on insect defenses, the physiology of EPF-pest interaction and coupling with environmental conditions. Bb appears to possess a multifaceted capability for both suppressing and eluding the cellular defense response (Hung and Boucias, 1992). The insect immunity (Gillespie *et al.*, 1997) and the molecular mimicry of Bb to escape detection by insect's cellular immune surveillance is important to the outcome of the host-pathogen relationship (Gillespie and Khachatourians, 1992; Hung and Boucias, 1992; Boucias *et al.*, 1994; Pendland *et al.*, 1993).

Time course of the infection is a key parameter in the insect pest interaction with EPF. For the farmers, it is the speed or time with which they kill target pests, that counts. The existing paradigm in the use of synthetic chemicals is that they act quickly and noticeably, resulting in a killing within a span of hours or a day. In spite of accumulating evidence, the perception that has been maintained by both the scientific and industrial community is that the performance of the microbial agents falls short of the chemicals. This being the reality, a negative perception has been generated, i.e. EPF as biocontrol agents take much longer to control insects.

To understand and appropriately communicate the unique prerequisites of insect pathogenesis and death inflicted by EPF is one of the solutions here. The rate limiting step for chemical pest control agents is adsorption, whereas that of the EPF is a multistep and complex process of physical interactions, germination, and penetration, each of which have physiological, physical and chemical specificities (Jeffs *et al.*, 1999). Bioassay of a collection of the mycoinsecticide, Bb, isolated from around the world has produced LT_{50} values ranging from 2 to 12 days (Khachatourians, 1992). It appears as though some "critical milestone events" (CME) determine the minimum time required for the insect biocontrol process. The importance of CME in relation to the search for or engineering of isolates with a faster kill remains to be seen. In a given insect the concept of a CME suggests that there is a threshold of time which is required to reach the point where the insect host demise is irreversible and death is unavoidable (see Hegedus and Khachatourians, 1996b).

The cumulative effects of EPF-pest interaction result in changes in normal physiology and development. For example, insects infected with certain fungal pathogens show stimulation of feeding, followed by a sequence of reduction to stoppage of feeding, reduced mobility, behavioral fever response, and changes in the horizontal and vertical migration patterns (Khachatourians, 1996). These findings can be used in an agronomic sense to provide better control of insect pest activity or reduction of damage to a crop.

Preference of EPF for insect biocontrol

In pest control situations there are times when insects impose limitations on the choice of biological agents. For example, insecticidal protein toxins or viruses can not be ingested by plant sucking insects or those going through metamorphosis and developmental stages, e.g. eggs, pupae and occasionally adults, where ingestion of food is absent or impaired. This situation presents opportunities for the preferential use of EPF as biocontrol agents (Valencia and Khachatourians, 1998).

Besides the contact mode of action, EPF also have some other important advantages as microbial agents for pest control. More than 95 per cent of the bacterial insecticides now available for control of agricultural pests and insect vectors, are based on formulations of Bt. However, both Bt and baculoviruses have a relatively narrow host specificity, i.e. their major commercial target is lepidopteran larvae. Therefore certain fraction or many major pest species from Coleoptera, Diptera, Orthoptera, Hemiptera, Homoptera and Thysanoptera, become obvious candidates for biocontrol via EPF (Valencia and Khachatourians, 1998). Furthermore, a large group of tetranichydae, tarsonemiidae and eryophyidae pest mites that are important agricultural pests that do not belong to the Insecta but to the Arachnida class are not susceptible to Bt or baculoviruses. This group of insects has EPF, such as *Hirsutella* spp and *Neozygites* spp, as their natural enemies.

Due to the single-site of action of Bt toxins, certain lepidopteran insects have become resistant to this insecticide (Chaufaux *et al.*, 1997). Although the possibility of resistance development against virtually any agent making selective pressure has to be accepted, it is apparent that complex mechanisms of action pose a lower risk of resistance development. In this respect, EPF have the comparative advantage of having a multiple-site/multiple-step mechanism of action (Khachatourians, 1991), which further minimizes chances of occurrence of resistance in the target pests. Inspite of this, one publication suggests the selection of Bb resistant insects. Junior and Alves (1998) studied the effects of two isolates of Bb on the development of the maize weevil, *Sitophilus zeamais* Matschulsky. However, a number of adults emerged at 40, 50 and 60 days after infestation of rice. The length of life cycle and weight of emerged adults were less affected by fungal inoculation, which Junior and Alves (1998) attribute to a selection of more vigorous insects that survived from the action of Bb. In spite of this suggestion, the infective process of EPF depends on the success of a series of physiological events, from the attachment of conidiospores to the insect cuticle, to the interaction with the host immune system at the level of the hemolymph (Khachatourians, 1996). To be "selected as a more vigorous insect" several concurrent adaptive mutations at different action sites would be required in order to render the target host survive a fungal pathogen.

On the other hand, the preference for the use of EPF in insect control has a context not really shared by other microbial biocontrol agents. This is because EPF are ubiquitous to many environments where insects occur. This situation clearly represents a second opportunity in that there is a tremendous potential for the discovery of native-indigenous isolates for the control of indigenous pests. Likewise, there is also a good opportunity for the isolation of EPF strains in the original habitats of exotic pests, thus making possible to use these pathogens within classical biological control programs. In either case, basic studies on the impact on non-target and beneficial organisms should be ideally completed before the utilization of EPF products in a wider scale (Valencia, 1995a,b). Also, EPF can be highly virulent against phytophagus-pest insects while still presenting a significantly lesser impact on their natural enemies because of a certain level of physiological specialization on the pest insect host, as a result of the repetitive selection (Stich and Jackson, 1997). We are beginning to understand that the interaction of EPF-host depends on the occurrence of non-specific and specific events between the conidiospores and the insect cuticle (Jeffs and Khachatourians, 1997; Jeffs *et al.*, 1999). As a result, conidia with their particular biophysical properties can be readily selected and even modified, for a further enhancement of preferential action of EPF.

Considering all the evidences and scientific information currently gathered on EPF, it is sensible to suggest that they can be preferentially selected as a major and important component of insect control and management programs. Needless to say, such a preference is urgently needed for the implementation of sustainable and competitive agriculture in developing countries worldwide.

General and molecular genetics of EPF

Because of their position in the fungi imperfecti until mid 1990s, Hyphomycetous EPF have been difficult subjects for genetic studies. Therefore, an understanding of the molecular basis for growth, development and pathogenicity of EPF is seriously lacking. Some genetic studies on *B. brongniartii, M. anisopliae* and *V. lecanii* are available. A complete genetic study is a difficult proposition. It requires collection of biochemical- and developmental- and conditional lethal-mutants of several types, built into isogenic background. We set out to achieve such a collection for Bb and have shown mutational stability, parasexual crosses, karyogamy, recombinant stability, transposition, and segregation of paired markers (Khachatourians, 1996 and unpublished results). Bello and Paccola-Meirelles (1998) also reported parasexual crosses among strains of Bb. These authors located auxotrophic markers and resistance to benomyl and analyzed gene transfer leading to formation of heterocaryons. Segregants were selected and genetic markers were assigned to four linkage groups. Viaud *et al.* (1998) performed protoplast fusion of diauxotrophic mutants of a Bb and a toxinogenic strain of *B. sulfurescens* (Bs2). Hybrids, which were significantly different from the parents in pathogenicity, resulted.

Traditional general genetic studies that compliment newer information about molecular genetics have been gathered by electrophoretic karyotyping of the chromosome(s), for the development of physical genomic maps and localization of genes involved in EPF-pest relationships (Pfeifer and Khachatourians, 1992, 1993; Shimizu *et al.*, 1993). Other molecular genetic tools or studies such as restriction fragment length polymorphisms (RFLP), DNA probes for Bb or other EP fungal tracking and proprietary strain fingerprinting should be of immediate use (Khachatourians, 1996; Hegedus and Khachatourians, 1993a,b). The multigenic nature of pathogenicity and the EPF-pest relationship are being examined through isolation of several catabolic genes through homologous gene sequence probing of genomic DNA or libraries. Equally important is the study of extrachromosomal DNA elements (Maurer *et al.*, 1997) and mitochondrial genetics of EPF, as they lend to both basic and practical applications. To our knowledge other than two groups (Rodriguez *et al.*, 1999; Kouvelis *et al.*, 1999; Mavridou and Typas, 1998; Hegedus and Khachatourians, 1993b; Hegedus *et al.*, 1991, 1998; Pfeifer *et al.*, 1993) which demostrated mtc DNA variations e.g. tropical-subtropical vs isolates from temperate regions, there is not much in the literature.

Through the use of technique of differential display, Bb genes expressed during growth on insect cuticle indicate particular genes involved either for substrate utilization or pathogenesis (Berretta *et al.*, 1998). Such studies in conjunction with the cloning and sequencing work on cuticle degrading enzyme-encoding genes should help the development of "improved strains" via recombinant DNA technologies. However, such constructs will face increased requirements for registration because the "genetic engineered" component would remove the "natural" isolate status of the EPF and create need for more rigorous test of environmental impact.

The other benefit of molecular genetics of EPF is determination of DNA characteristics, which represents an enormous advantage not only for taxonomy and biosystematics, but also for commercial product R & D. DNA probes can identify genera, species, varieties and even mutants (Khachatourians, 1996). The DNA based technology systems include chromosomal DNA probes or differentiate between isolates rDNA and rRNA or mtDNA sequence comparisons, PCR amplification RAPD, RFLPs, single stranded conformational polymorphism (which can identify base change per 100 bases, Hegedus and Khachatourians 1996b; Urtz and Rice, 1997).

Applied R & D in mycoinsecticides

Applied research and development of mycoinsecticides with its three elements; production, formulation and application, culminate the realization of discovery and bench research and make the transfer of insect pest control/management in the real world possible. Eight major issues towards the quality of a fully formulated mycoinsecticide can be introduced, as follows: viability of the fungal propagules, virulence, humectability, dispersibility, suspendibility, solubility, particle size and when necessary, compatibility with other agricultural products required for crop protection (Morales, 1995).

Production of mycoinsecticides

Production of EPF at the industrial scale has several requirements most important of which is that pathogenecity and virulence of the isolates must be maintained through the whole production process. Draganova (1997) reported that Bb (9 strains) and

B. tenella [Bb] (9 strains), grew on media containing various carbon sources with a concentration of 1 per cent. Most Bb strains metabolised trehalose, saccharose (sucrose), maltose, glucose, adonitol and sorbitol and few strains utilised amino acids. The *B. tenella* strains metabolised trehalose, saccharose, maltose and glucose. Arabinose, rhamnose, inulin and L-lysine were not used by half of the Bb strains or by most of the *B. tenella* strains.

As reviewed in Feng *et al.* (1994), mass production of EPF can occur by either diphasic or submerged fermentation techniques. Diphasic fermentation has the advantage of combining liquid and solid media, to allow the mass production of mycelia and conidiospores, respectively. Because the conidiospores are the ready-to-use infective propagules of EPF, special emphasis has been given to the production of conidia, either in solid or in liquid media. Solid media production ensures that typical aerial conidia will be obtained, but faces serious limitations in terms of volumetric productivity. On the other hand, liquid media production optimizes volumetric productivity, but the yields of submerged conidia can be limited by several factors, including a diminished aeration due to high fermentation volumes (Feng *et al.*, 1994). The first scientific study of microcycle conidiation of Bb was in 1987 by Thomas, Khachatourians and Ingledew. This study showed the physiological requirements of production characteristics, infectivity and others (Thomas *et al.*, 1987; Hegedus *et al.*, 1990, 1992). Recently, Bosch and Yantorno (1999) further elaborated on Bb microcycle conidiation process in terms of different C (glucose and starch) and N sources (KNO_3, NH_4Cl, glutamate and peptone). In addition, they showed that absence of the C source, at every level of the N source, produced daughter conidia at the tip of very short germ tubes. The secondary conidia produced by microcycle conidiation were viable and capable of repeating the cycle for at least two generations.

Bradley *et al.* (1992) has also reviewed solid culture of the Bb by means of packed bed solid culture system. In this case, a liquid phase is absorbed in a substrate containing starch and the fungus grows on the substrate particles. Because the gaseous phase remains available for aeration, this factor is no longer limiting the fungal growth and conidia productivity. Yields of 10^{13} conidia were obtained in a fermenter smaller than 1 litre and the pilot plant rendered a dry powder containing 2.6×10^{11} conidia g^{-1} (Bradley *et al.*, 1992). Desgranges *et al.* (1993) used a solid state fermentation system to produce Bb for the biological control of European corn borer. The product in this case showed a field efficiency of 80 per cent and persisted for 3 weeks. Suresh and Chandrasekaran (1998) have shown that substrates such as prawn waste can be utilized for growth and high levels of chitinase production by a marine isolated Bb under solid state fermentation. Puzari and coworkers (1997) have designed an inexpensive medium with rice hull, sawdust and rice bran at a ratio of 75:25:100, respectively, to mass culture Bb for the biocontrol of rice hispa, *Dicladispa armigera* (Olivier). This type of production is not only convenient in terms of supply of ingredients, but also capital inexpensive when compared to most schemes presented by Bradley *et al.* (1992). The growth medium used by Puzari *et al.* (1997) produced 39.33×10^7 conidia/ml of water 24 days after inoculation and 78.6 per cent mortality of the adult insects was achieved at spray concentration of 10^6 spores/ml of water.

The relationship between mycoinsecticidal activity and biophysical, physiological and biochemical characteristics of conidia must be considered because of their importance to infectivity as much as to their survival in the environment. Conidia should be compatible with the auxiliary ingredients required to formulate the commercial product. In this regard, the nutrients and overall components of the growth media can affect the minimum water activity (a_w) required for germination of conidia and fungal growth (Pitt and Miscamble, 1994). Furthermore, tolerance of the conidia of EPF to low moisture environments can be enhanced by the addition of specific compatible solutes to the media. As shown recently by Hallsworth and Magan (1999), the effects of temperature (5–50°C), water availability (0.998–0.880 water activity, $a(_w)$, and $a(_w)$ × temperature interactions (15–45°C) on growth of Bb, *M. anisopliae*, and *P. farinosus* were significant, with interspecies variations in growth rates on media modified with each of the three a_w-modifying solutes. These growth relationships could equally impact on environmental limits that determine efficacy of entomogenous fungi as biocontrol agents in nature. For example, polyalcohols play a major role to regulate the water relations between EPF and their environment.

Formulation of mycoinsecticides

Following the assessment of developmental and pathogenecity parameters of the active (non-formulated) ingredient using the best performing strain, the process moves forward to the formulation steps. Five major criteria: high virulence, high productivity in liquid or solid media, high genetic stability, high physiological stability under storage and high compatibility with the auxiliary ingredients must be considered as absolutely met. As expected, the assessment of these criteria requires extensive characterization and quantitative studies. As a practical example, the characterization of the biophysical properties of conidiospores based on determinations of spore hydrophobicity and lectin binding properties, can be of significant importance in relation to the type and amounts of auxiliary ingredients required in the final formulated product (Jeffs and Khachatourians, 1997; Jeffs *et al.*, 1999; Valencia and Khachatourians, in preparation). Simple preparation and application of conidia and dry mycelium preparations of two isolates of Bb against the sugarcane borer, *Diatraea saccharalis* (Fabricius) was reported to be successful by Arcas and coworkers (1999). On the other hand, auxiliary ingredient requirement is one of the major factors affecting the final costs of the formulation.

Phytochemicals such as neem are natural products that have for centuries been used as biopesticides (Isman, 2001). Bajan *et al.* (1998) studied the reaction of various ecotypes of Bb to a preparation of the pyrethroid Fastak [alpha-cypermethrin] or BioNEEM™ (extract of *Azadirachta indica*). Depending on the Bb isolated, Fastak either considerably inhibited, did not affect, or stimulated growth of a fungal isolate. However, it significantly reduced the pathogenicity of the isolate from the Fastak heavily concentrated zone. Bajan *et al.* (1998) suggested that preparations of BioNEEM or Fastak and appropriate Bb isolates can be applied simultaneously in pest control.

In general Bb conidiospores are hydrophobic (Jeffs and Khachatourians, 1997) and therefore, better suited for suspension and dispersion in non- or partially- aqueous formulation. Thus, it should be of no surprise that evaluation by Vandenberg *et al.* (1998) of Bb for control of the diamondback moth on crucifers (grown in growth chambers, the greenhouse, and the field) provided significant reductions in larval counts.

They used conidiospores formulated as (i) wettable powder in aqueous and oil formulae and applied at 2 rates, and (ii) as an emulsifiable suspension at a high rate. Carballo (1998) showed that with Bb formulations of 10, 15 or 20 per cent oil and 5×10^8 fungal conidia/ml, 100 per cent mortality of *Cosmopolites sordidus* (Germar) was observed. Masuka and Manjonjo (1996) used *M. flavoviride* and Bb spores formulated in commercial soya oil for the control of *Mecostibus pinivorus.* However, some oils can be phytotoxic and, therefore, unpractical (Smart and Wright, 1992).

In other examples, where mixtures of EPF and other chemicals were used, water or oil can be the basis for the formulation of final product. Delgado-Francisco *et al.* (1999) compared Bb conidia formulated in a mineral oil carrier, formulated diflubenzuron, a combination of Bb plus formulated diflubenzuron, and fenitrothion. These large-scale field trials which occurred in Mali, tested mycoinsecticidal potential against unconfined grasshopper populations in field plots of 10 ha each. All treatments after 14 days post treatment showed a grasshopper populations decrease by 38.1 per cent in plots treated with Bb alone, 29.4 per cent in plots treated with diflubenzuron alone, and 55.6 per cent in plots treated with the Bb plus diflubenzuron. Effects of the diflubenzuron-Bb mixture were additive and not synergistic.

Smith *et al.* (1999) have made a unique use of formulation of pheromone and spores. They made use of hydrogenated rapeseed oil as a carrier for conidia of Bb for the control of *Prostephanus truncatus* (Horn). Melting the oil, which is solid at temperatures below 32°C, allows the incorporation of materials such as aggregation pheromones and conidia; sudden cooling produces solid fat pellets. Pellets containing *P. truncatus* aggregation pheromone attracted significantly higher numbers of beetles which were retained over a period of storage in glass bottles. Pheromones and conidia incorporated into the same pellets and stored in a freezer or refrigerator retaining over 80 per cent viability after 51 weeks; those stored in an incubator at 27°C showed significantly lower germination at 20.7-27.2 per cent after the same time. Insects exposed to pellets for 24 h showed 96-100 per cent mortality within six days of exposure.

As indicated in Frankland *et al.* (1996), fungi and certainly EPF face a continuously changing environment, from solar radiation to dry or humid situations. Humectants are important in determining the retention of moisture and the speed of the final product to become wettable, thus affecting the ease for the preparation of the product mix. Alves *et al.* (1996) prepared powder formulations of a Bb, originally isolated in Brazil, in talc (hydrous magnesium silicate), silica gel, powdered rice and cornstarch which were stored under: (a) ambient temperature (15-38°C); (b) refrigerator (+6 to -2°C); and (c) freezer (-10 to -7°C). Unlike formulations stored under the first condition, which completely lost viability after 1-8 months, unformulated conidia were totally non-viable after just 2 months. All formulations stored under refrigerator and freezer conditions maintained 100 per cent viability for 7 years. After 30 months of storage, the unformulated conidia and the formulations stored in the refrigerator showed slow germination and had low virulence. Likewise, Sandhu *et al.* (1993) showed that by choosing favorable RH, and temperature they could prolong storage of Bb conidia and virulence for 24 months for the use of the Bb product for third instar larvae of chickpea borer, *Helicoverpa armigera* (Hubner).

The effect of temperature and humidity, on Bb mycosis in *Hippodamia conivergens* Guerin-Meneville (Coleoptera: Coccinellidae) using standard EPA bioassay method (exposure by immersion) was tested by James *et al.* (1998). The result of immersion assays in a per-insect dose was 5 times greater than that from spray applications. Differences in environmental conditions between the laboratory and greenhouse, i.e. the range of temperatures had a significant effect on both germination rate and vegetative growth and insect mycosis. Contrary to these results, levels of mycosis in lady beetles decreased as temperatures increased over this same range. Although high levels of humidity are required for conidial germination, James and co workers found no well-defined threshold period of high humidity exposure required for mycosis in this insect. Therefore, temperature and humidity also affect physiological interactions between the host and pathogen.

As it is known that a significant portion of the early stage of pathogenesis depends on the onset of spore germination, any attempt in speeding this event should shorten the time needed for insecticidal action (Khachatourians, 1996). Judicious inclusion of some nutritional aids for spore germination could also be in the formulation. Agarwala *et al.* (1998) confirmed earlier findings of Miranpuri and Khachatourians (1990, 1991) that Bb was insecticidal to *Aedes aegypti* (Linnaeus). What is unique in their study was that conidia were mixed with 1 per cent dextrose as a spore germinating agent and resulted in 100 per cent mortality directly proportional to the concentration of conidia applied.

Similarly, formulation suspendibility is conditioning the time during which, the mycoinsecticide particles remain suspended in the liquid matrix. If this time is shorter than required to get the application equipment empty, the dose applied will not be uniform, as the product will settle out and will be more concentrated at the bottom of the application tank. Additionally, the particle size and dispersibility also determines the ease of the application, otherwise the nozzles can be clogged and the product droplets on the plant surfaces will be non-homogeneous. Overall, there is no single formula-fit all as formulation requirements of different EPF species and isolates may vary.

Application of mycoinsecticides

Provided that production and formulation of a mycoinsecticide have been optimized, the next steps to be considered are those of product application; timing to fit particulars of the location and application techniques. Timing should relate to the optimum period of the day for exposure of the insect pest population in consideration of wind, rain, sun, etc., and the other impacting conditions. A strategic timing should be mindful of the natural enemies (NEs), so that they are less affected by application process. As well, EPF-based products can be applied in the evening or at night in particularly dry or hot regions, to protect the fungal agent from solar radiation and heat.

The topic of efficacy of the EPF application under exposure to sunlight or shade remains ambiguous. Arango (1997) tested the viability of Bb suspended in 10 ml oil/20 litres water or in 10 per cent tarsol oil plus emulsifier at 500 ml/100 litres under sunlight and artificial shade in Colombia. Low-volume sprayings were carried out on 30 branches with 120 coffee berries per branch. The spraying volume per branch was 17.5 ml with 4.02×10^9 conidia. The spraying was directed to the upper and lower sides of branches. The viability was evaluated after 0, 2, 4, 5, 24, 48 and 336 h. Results showed no difference between sunny and shaded plots. There is, on the other hand, the expectation that the opposite results should also be possible.

The utility of Bb conidia formulated as a dustable powder, as oil suspensions and as a novel hydrogenated rapeseed oil pellet for the control of stored grain pest, *S. zeamais* was reported by Hidalgo *et al.* (1998). Mineral oil and a mineral and maize oil mixture, with Bb at a concentration of 10^9 conidia/ml (both at 20 ml/kg grain) was shown to have the highest level of net control in maize grains. The fat pellet formulation resulted in low levels of mortality (21–31%) when used in maize grains. The authors indicate that this approach may be recommendable in case of small farms. Likewise, the application of Bb for the control of the coffee berry borer in Colombia is recommended to be done preferentially during the wet-rainy seasons. During these time periods, the pest exhibits a lower population dynamics and the higher relative humidity facilitates the activity of the pathogen (Valencia, 1994, 1995b). It is important to mention that, if applications of EPF are performed during a wet season, it is necessary to add a surfactant to the application mix, with the purpose of reducing the washing off effect of rains.

The place of application of EPF in the agroecosystem is also relevant. Most crops present specific microhabitats that should be considered for the optimal utilization of these mycoinsecticides. In all cases, the application should be ideally directed (focused) to the cropping area or plant structure, where the target pest is located. For instance, many sucking insects, such as aphids, scales and whiteflies, reside under the leaves of the crop plant, therefore, particular application equipment and devices and techniques (e.g. amount of atomization or turbulence to cover under the plant canopy) will be required. For example, equipment having its nozzles oriented upwards (instead of downwards), to directly impact under leaf surfaces or sprays with high pressure nozzles, will produce very small droplets, that can easily create a turbulence able to reach all plant surfaces available.

In some cases, two or more developmental stages of the same pest can occur in different locales. The more contrasting cases are those where one stage occurs in the plant and another in the soil. The cotton boll weevil, *Anthonomus grandis* Boheman, is a good study case. We have previously proposed, that applications of Bb directed to the fallen buds on the ground, are likely to affect the population dynamics of this pest (Valencia and Khachatourians, 1998). Although oil or water-based formulations can be appropriate for the aerial parts of the cotton plants, applications to the soil would probably required different formulations. Granular, dust powder or pelletized forms of the EPF would be more convenient than liquids to perform soil applications (Morales, 1995). Furthermore, because the soil can be considered as a more favorable environment for EPF than plant surfaces, other fungal propagules, i.e. mycelia and blastospores, can be used. Application of these formulations between the crop rows at the base of the cotton plants will provide an extra-protection to the fungus from sun radiation.

Incorporation of mycelia and blastospores into alginate pellets has been applied to formulate Bb (Pereira and Roberts, 1991). Corn starch and corn starch-oil formulations of Bb mycelia as well as non-formulated dry mycelial preparations have been developed (Pereira and Roberts, 1990). One of the major advantages of mycelia as a propagule of EPF, is its capacity to rapidly colonize the soil under favorable conditions. Mycelia have the potential to start a massive production of conidia, even after several months of storage (Feng *et al.*, 1994). These characteristics suggest that formulations of EPF based on mycelia, should be suitable to offer an effective pest control, as they can attain abundant sporulation after re-hydration (Feng *et al.*, 1994).

Bioinsecticidal activity of conidia and dry mycelium preparations of two isolates of Bb against the sugarcane borer, *D. saccharalis*, was studied by Arcas *et al.* (1999). Fungi were grown in solid- state and submerged cultures in order to obtain conidia and dry mycelium preparation. Concerning the samples obtained in submerged conidia, in which a culture medium based on glucose and yeast extract was employed, it is interesting to point out that although both strains showed similar behaviors with yields of approximately 1.50×10^{10} conidia per gram of dry mycelium, both preparations failed in their bioinsecticidal activities. Despite satisfactory yields, bioinsecticidal activity of the Bb strains dropped drastically showing a larval mortality below 2 per cent. Clearly, control of soil pests using mycelial formulations, would be an excellent subject of IPM modeling aimed to sustainable crop protection in tropical as well as in temperate agricultural zones.

Mycoinsecticide registration

Efficiency of action for EPF should be a specific requirement of successful mycoinsecticides. Even then, EPF could be made more efficient through various biotechnological modifications of the organisms or their formulation into final product. Provided that the safety and efficacy of a mycoinsecticide is accounted, registration of EPF today is not as difficult or an impediment to the commercialization. The review of our rationale for replacement of chemical pest control agents with biological alternatives is of current public and global interest. With consideration to important issues of biodiversity, sustainability of the agricultural eco-systems and health of the environment, we must move forward to the commercialization of EPF.

Safety concerns

The efficiency of commercial scale production of EPF, however, paints a picture different from that needed for the synthetic chemicals. Further, production schemes have minimum chance of contamination with other microorganisms and are not labor intensive. Their bulk storage or transportation does not pose much danger either. Product efficiency can be expanded by formulations to permit control of different target pests with distinct geographical niches. The user of EPF must know that they are dealing with a microbiological agent and due care must be taken in translating the instructions from handling to spraying. Although safe at final spray dosage, accidental human exposure to concentrates of EPF can be prevented through special label warnings and education.

Regulatory concerns

Many EPF are naturally occurring microorganisms. The regulatory agencies in many jurisdictions must acknowledge this and become flexible and coordinated with respect to the kinds of test data required. Regulatory bodies, whether national or international, deal with ever changing state of the knowledge and public perception. Tightening of the rules on the regulation for biologicals and biotechnologically derived products, especially when they are to be consumed or released into the environment should be based on *a priori* scientific and validated proof. Coordination of regulatory requirements currently occurs through the Organization of Economic Cooperation and Development treaty nations. What is included in this rationale is the fact that it will be another 7–10 years before newer EPF presented, as alternatives in pest management, could become available.

Field trials with a number of EPF are currently under way on every continent. As indicated earlier, reduction of temperature sensitive strains (Hegedus and Khachatourians, 1994, 1996c; Chelico and Khachatourians, 1999) for applications in unique time environmental windows should eliminate some of the concerns associated with the deliberate release of fungi and destruction of insect biodiversity. The use of species specific DNA probes and molecular genetic tracking techniques for taxonomic identification and environmental release/monitoring will make significant changes in our perceptions of risk and persistence of EPF.

Pathways for mycoinsecticides development

Estimated time lapse between discovery and applied research and development work needed for an ultimate registration of a naturally occurring microbial insecticide may vary. Bt based patents either as formulations obtained from batch fermentation or the incorporation of the delta endotoxin gene into transgenic plants occupy a prominent position (Schnepf *et al.*, 1998). However, in terms of number of EPF registered and in the process of commercialization, as shown in Table 7.2, there is a significant rise (Khachatourians, 1986; Leathers *et al.*, 1993; Feng *et al.*, 1994; Miranpuri and Khachatourians, 1993, 1994a).

Based on a careful examination of the present situation, two different pathways for bioinsecticide development can be suggested. With any newly isolated microbial agent (viruses, bacteria, fungi or protozoa) three distinct pathways leading to production of three distinct generations of commercial pest control products are possible. With newly isolated fungus, about which little research or development exists, as shown in Fig. 7.1 (I), a number of required R & D activities in 7–10 years will lead to its use as a generation I product commercialization. After this, any additional application of the same agent but for a different insect, jurisdiction or geographic loci will need entomological data to prove efficacy for specific registration as a new commercial product as shown in pathway described in Fig. 7.1 (II). The demands in the last two pathways for discovery, research and development are greater, and years and large expenditures are required.

Table 7.2 Commercial mycoinsecticide products in use

Fungus	*Product*	*Manufacturer*	*Target insect*
B. bassiana	Naturalis-L®	Fermone Corp.	Cotton boll weevil, sweet potato whitefly, *Bemisia aregentifolii*
147	Ostrinil®	Calliope	European corn borer
	Conidia WG®	AgrEvo S.A.	Coffee berry borer, *Hypothenemus hampei*
GHA	Mycotrol®	Mycotech Corp.	Grasshoppers and locusts
	Gh-Of		
	Boverin®	Fermone Corp.	
	Naturalis®225 and L	Fermone Corp.	Aphids, thrips and whiteflies
	Brocaril®	Laverlam	Coffee berry borer
V. lecanii	Verticillin®	NPO Vector	Plant-eating and sucking insects
	Mycotal®	Koppert	*Trialeurodes vaporariorum*, *Bemisia tabaci* and thrips
	Vertalecv	Kipper	Complex of aphids, *Myzus persicae, Aphis gossypii* and *Macrosiphum euphorbiae*
M. anisopliae	Bio 1020	Bayer	Black vine weevil, *Otiorhynchus sulcatus*
	Bio-Path®	EcoScience	Cockroaches

It is conceivable that new combinations of various insect pest specific feeding deterrents and plant encoded proteins, enzyme inhibitors and lectins that have a protective role within storage tissues and seeds of plants and insecticidal toxic proteins or toxins of entomopathogenic fungi could be genetic material for strategies to generate insect resistant transgenic plants (Hegedus *et al.*, 2000). These approaches are now described as "pyramiding genes" (Gatehouse, 1999). This concept relies on the ability of plant geneticists and breeders to "pyramid" agriculturally desirable traits in the form of packages of different genes. The traits built into a given plant may provide a more durable resistance to insect pests as they act through a multiple mechanistic form of insect control, which is a rather recent concept. Secondly, they may increase the protective efficacy spectrum of activity for different crops and particular insect pests at any one place. Another avenue for changing the pathways for new mycoinsecticide development that is awaiting contextual exploitation is the perspective that engineering of the pest (Pfeifer and Grigliatti, 1996) has a place in the practice of insect pest management. Overall, such strategies should lead to newer pathways for mycoinsecticide development. However, it is difficult to forecast how such systems can mature from theory to commercial reality.

New opportunities for mycoinsecticide use

Winds of change blow over all agricultural lands around the world. The globalization process demands that all countries have to be competitive. However, to reach this goal, most of these countries will have to attain a balance between traditional, and the

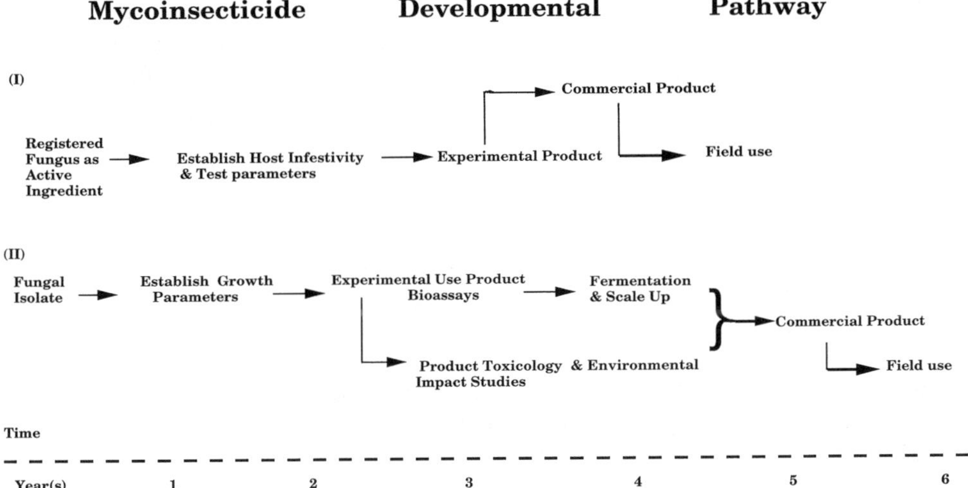

Figure 7.1 Mycoinsecticide development pathways. The development of a new mycoinsecticide for field application from an existing registered fungus for a different insect than that registered for, upper pathway marked (I), and for a newly isolated fungus pathway marked (II) is shown. The duration of research, development, impact studies and commercialization are approximated from the time axis (in years) from the start to field use time.

most innovative technologies of crop protection and production. In order to produce a significant impact on the provision of agricultural products, a special emphasis has to be done on the management of major-exportation crops, which usually correspond to the most extensive areas involving farming activities.

Various regions of Asia, Central and South America, North America and Europe represent a rich complex of cropping systems with a great diversity of pests and, therefore, diversity of pest control methods and practices. Besides the emphasis given to exportation crops, the control of major-key pests in these agroecosystems, should be the first priority. The fact that many of the major pests occurring in these continents are not lepidopterans suggests the need and the opportunity to use EPF species for insect biocontrol in these regions. To illustrate market opportunities, several practical examples of the utilization of EPF within IPM models including cases of mycoinsecticides already used as commercial products in some countries will be given.

Opportunities in Asia

Asia is a continent where agriculture represents a very important sector of the societal sustainability including the economy. Among the different crops cultivated, rice is probably the most important one, not only because of the grown area but also mainly because of the number of people who depend on this crop. Indeed, most people in Asia depend on rice as a primary source of food, as a major source of employment in country areas, as well as for exportation activities. Rice exhibits several sucking insect pests belonging to Homoptera and Hemiptera orders, which are a serious limiting factor for the production of this crop. As explained in previous sections of this chapter, most sucking insects are not likely to be controlled by means of bioinsecticides having an ingestion mode of action. For this reason we believe, there is an opportunity for the development and commercialization of EPF for the biocontrol of sucking insects in rice. However, it is also important to make clear that other types of NE of sucking insects have been successfully utilized in some countries as biocontrol agents. In Philippines, for example, it has been found, that by maintaining a healthy population of beneficial spiders, it is possible to achieve good control of the sucking insects complex, thus making sometimes unnecessary the utilization of chemical insecticides. Hazarika and co-workers (1998) have shown the presence of seasonal and host-correlated variation in the susceptibility of rice hispa to Bb in the field. Their survey of rice ecosystems in Assam, India, revealed the virulence of Bb on eggs, pupae and adults of *D. armigera*. Infection was more prevalent in the egg stage and fluctuated during different months.

Masarrat *et al.* (1998) showed the effectiveness of Bb on 1st-instar nymphs of the mango pest, *Drosicha mangiferae* (Green). Mortality was noticed after 3 days, and cumulative mortality was found to increase with time for all concentrations. The highest percentage mortality (100% 14 days after treatment) was obtained with the highest spore concentration (6×10^9). Hem *et al.* (1998) performed a field evaluation of Bb against *Helicoverpa armigera* (Hubner), infecting chickpeas, under field conditions in India for two crop seasons. At a concentration of 2.68×10^7 spores/ml, the average pod damage was 6 per cent and the yield was 2,377 tones/ha. The untreated control recorded 16.3 per cent pod damage with a yield level of 1,844 tons/ha.

In the case of one or several species of sucking insects requiring biological control, EPF represent a good alternative to be used alone or in combination with chemical insecticides within an IPM system. Among major pests in rice crops in Asia, the Homopterans, *Nilapavata lugens* (Stal) and *Nephotettix* spp, are a serious threat to rice production. *Nephotettix* spp. and *N. lugens* show good susceptibility to some species of EPF, among which several isolates of Bb have been found virulent against these pests (Feng *et al.*, 1994; Xu 1988). We suggest that an IPM model could be elaborated for these insects in rice crops, considering the agroecological and micro-enviromental conditions of rice fields.

As rice is extensively cultivated in flooded lands in Asia, the crop is under permanent conditions of humidity and moisture. We hypothesize that these conditions can be very favorable to the survival and biological activity of EPF, against the sucking insect complex. As indicated in other cases, the IPM model for *N. lugens* in rice would require the determination of the specific action sub-threshold doses for the application of EPF. Likewise, the model should provide the information on the optimal doses, timing and application techniques needed to obtain the best results. Importantly, timing of the application of EPF in rice has to be carefully managed in relation to the application of chemical fungicides. Considering the predominant modes of action of fungicides as protectants or curatives, experimental applications of EPF can be performed few days before or after the chemical treatments, to determine the time frame within which they can be applied. Also, EPF can be safely rotated with these fungicides. Todorova *et al.* (1998) examined the compatibility of Bb with six fungicides (chlorothalonil, maneb, thiophanate-methyl, mancozeb, metalaxyl+mancozeb and zineb) and two herbicides (diquat and glufosinate-ammonium) used commercially in potato fields. All six fungicides tested under controlled conditions, along with the herbicide glufosinate- ammonium inhibited Bb mycelial growth and sporulation. However, the second herbicide tested, diquat, had no noticeable effect on mycelial growth and sporulation. What was remarkable here is that diquat synergized the insecticidal activity of Bb in the simultaneous treatments on Colorado potato beetle adults and caused 50–76.6 per cent mortality. Thus, serious concerns about certain interactions can be minimal.

Regarding commercial uses of EPF in Asia, the species Bb has been extensively used in China for the control of several forest and crop insects. The annual production of Bb conidial powder in China (10,000 tons approximately) has made possible to treat between 0.8 to 1.3 million hectares (Xu, 1988; Zhang, 1992; Ying, 1992). More recently, studies by Feng and coworkers (1998) on the survival and infection of Bb on *Dendrolimus punctatus* (Walker) in natural environments by using scanning electron microscopy showed that more than 85 per cent RH was required for the development of white muscardine disease. Also, RH was a restricting factor on *D. punctatus* control in natural environments in China. Finally, the fungus could not be released in early winter in the region where the pest produces 2 to 3 generations per year, but in late June provided more effective control. Lin *et al.* (1998) studied pathogenic effect of Bb infected on *D. punctatus* under different temperature and humidity. Conidium germination, the infection process and pathogenic effects of Bb on larvae of *D. punctatus* within RH range of 10–95 per cent were observed. At 33 combinations of temperature and humidity host pathogen relationships were studied. The relationship between germination rate (V) of conidia and temperature (T) was described by the equation V = (T4.25)/27.2, where 4.25 was the initial temperature, and 27.2 was the effective accumulated temperature in a day for conidium germination. Humidity was the main restrictive factor for the germination and infection of conidia, and temperature and humidity had compensatory and comprehensive effects. The relationship between environmental conditions and insect control opportunity is not only very important, but at times could be somewhat restricted to limited options.

One final cautionary note is that in regions where indigenous fungal isolates may exist, insect control by deliberately released fungal agent must be cautiously interpreted. The reason is that in the case of such co-inoculation of the insects, various types of synergistic, neutral and antagonistic interactions must be considered in explaining the outcomes. Lin (1998) tested strain types of Bb by electrophoretic patterns of esterase isoenzyme extracted from the mycelia of 40 isolates before or after the release of the fungus into forests. Zymograms of isoenzymes were related to the geographical distribution and the hosts. A total of 18 isolates from different sites in the same region belonged to the same strain. However, the author shows that a fungal disease of *D. punctatus* was caused by an indigenous strain of Bb in the spring. This type of observation always raises the question of the role of such indigenous fungi in creating co-inoculation and indeed synergistic, additive or antagonistic interactions, which could result. If such an analytical methodology (isozyme studies or DNA

RAPDs) were unavailable, most often an erroneous conclusion due to no one's fault is one of the outcomes. We have also observed such results in a field cage trial of Bb on local migratory grasshoppers.

Central and South American opportunities

Crop systems in Central and South America present unique characteristics for agricultural production. This continental region holds a long tradition of practices on agroecology and IPM, which facilitate the introduction of new strategies of crop protection. However, the implementation of novel technologies exclusively, will initially be expensive (Valencia and Khachatourians, 1998). Because of the limited economic resources of most Latin American countries, the challenge of sustainable agriculture will require a balance between traditional and the most innovative technologies, particularly in case of small farmers. Many countries in this region have successfully implemented practices for the management of key pests (Valencia and Khachatourians, 1998).

Due to the biodiversity of arthropods in these crops and their peri-agricultural zones, outbreaks of potential and secondary pests are always risks (Valencia, 1994). This risk is particularly high when methods such as the non-selective use of chemical insecticides, the clearing out and destruction of natural flora around crops and the total eradication of weeds are widely adopted by farmers (Altieri, 1995).

One of the outstanding characteristics of most crops in Central and South America, is the abundant fauna of beneficial insects present in their crops (Valencia, 1995a). The maintenance of this fauna offers the farmer a complementary pest control in combination with other measures including bioinsecticides (Rodriguez del Bosque, 1994). However, the generalized adoption of bioinsecticides by farmers will not occur, unless they can receive an economic benefit for such adoption. Few studies document an increase in net profits as a result of the implementation of IPM practices, in comparison to conventional pesticide programs under field conditions (Trumble *et al.*, 1997). Because the use of EPF and other microbial biocontrol agents (MBAs) may not be sufficient in today's economic environment, it is IPM modeling which can enhance EPF to meet the needs of traditional or highly technical agroecosystems (Valencia, 1995a).

Among the major crops in Central and South America, coffee and cotton are very important, from the economic and employment point of view. Results of De La Rosa and coworkers (1997) show the potential of Bb integrated control directed against *H. hampei.* When ranked in terms of the three most aggressive strains Bb-4 (Ecuador), Bb-25 (Mexico), and Bb-26 (Mexico) the LC_{50s} of 0.003, 0.004, and 0.006 per cent, respectively have been obtained, which are equivalent to 2.2×10^6, 4.1×10^6 and 5.9×10^6 conidia per ml of suspension. The cotton budworm, *Heliothis virescens* (Fabricius), is one of the key pests of this crop and distributed in the whole region. The insect is one of the most limiting pests in cotton, presenting a tremendous capacity to develop resistance against a wide variety of chemical insecticides (Sparks, 1981). Further, physiological resistance due to high levels of activity of esterases, carboxyl-esterases and the mixed function oxidases system has been confirmed (Valencia, 1993a,b). *Heliothis virescens* is not particularly susceptible to *B. thuringiensis* or baculoviruses when occurs as a pest in cotton, as often these MBA offered erratic controls under field conditions (Cardona, 1995). This fact suggests the need to look for other type of MBA for the control of this pest in cotton.

The Zygomycete EPF *Zoophthora radicans* has been reported as pathogenic to lepidopteran larvae belonging to the *Heliothis* complex (Glare *et al.*, 1987). This fungus has several characteristics making promising its R & D for *Heliothis* control in cotton. The Zygomycete is adapted to a range of temperatures (0–36°C) wider than most EPF (Glare *et al.*, 1987). Furthermore, its infective and transmission capacities increase during dew periods. Most cotton growing zones are located in very hot zones and usually have abundant dew on the crop, especially at night and during the early morning. These conditions are favorable to the activity of *Z. radicans.* Here, the IPM model for *H. virescens* using *Z. radicans* is aimed to manage the early infestations of the insect, as it occurs as small larvae (first to third instars) at the upper and middle thirds of the cotton plants. The central goal of the IPM model would be to diminish the population dynamics of the pest during the period of its establishment in the cotton field and to minimize the potential damage to the crop. At the same time the immigrant fauna of beneficial insects is allowed to increase (Valencia and Khachatourians, 1998). Normally the action threshold for chemical control of the tobacco budworm in cotton is around 15 per cent of infestation of small larvae in young leaves and buds. We have recently proposed that in case of applications of *Z. radicans* for the control of this pest, several action subthresholds (i.e. 7 and 10 per cent of infestation) should also be tested under experimental conditions (Valencia and Khachatourians, 1998).

Although the control of *H. virescens* with *Z. radicans* appears as a market opportunity, several biotechnological improvements will be required to facilitate the commercialization of this Zygomycete in cotton. First of all, it is necessary to optimize the mass production and development of adequate formulations of this EPF. The induction of the protoplast stage of this fungus should be a priority to attain a cost-effective mass production. Special formulations such as wettable granules or liquids should be preferred to fine powder or dust formulations, considering that spores of most Entomoththorales are fragile and very sensitive to physical and chemical agents. A more intensive research than in case of some Deuteromycetes would be probably necessary, as the fundamental knowledge on genetics, physiology and mass production of *Z. radicans* is currently being gathered.

Another examples of successful commercial uses of EPF in Latin America can be introduced. The EPF Bb has been extensively used in Colombia for the IPM of the coffee berry borer (CBB) in combination with cultural practices and a chemical insecticide. A good quality formulation under the trademark Conidia WG® has been commercialized during several years by the multinational company AgrEvo S.A. (now Aventis Crop Science). This company and the local firms Live Systems Technology (LST) and Laverlam have undertaken intensive R&D efforts for the introduction and commercialization of mycoinsecticides in Colombia (Valencia and Khachatourians, 1998). *Metarhizium anisopliae* has been commercially used for several years in Brazil for the control of major pests in sugarcane and cotton (Cardona, 1995). Other Latin

American countries such as Mexico, Costa Rica, Honduras, Guatemala, El Salvador, Nicaragua, Chile, Ecuador, Peru, Venezuela and Panama, have implemented IPM programs including an extensive utilization of EPF and other agents (Andrews and Quezada, 1989; Thompson, 1998).

North American opportunities

Temperate and sub-tropical regions in Europe and North America present several particularities regarding pests and pest management. We have mentioned that the occurrence of a single or relatively few key pests per cropping season in some of the major crops in these regions, represent an excellent opportunity to develop and successfully implement fungal biocontrol programs. Among the major pests occurring in several crops in Europe and North America are the whiteflies, *Bemisia tabaci* (Gennadius), *B. argentifolii* and *Trialeurodes vaporariorum* (Westwood). The whiteflies, *B. tabaci* and *B. argentifolii* are particularly limiting due to their good capacity for the transmission of plant pathogenic viruses (Duffus, 1987). Another factor, which makes EPF attractive for the IPM modeling of whiteflies, is the significant capacity of these insects to develop resistance against chemical insecticides. On the basis of exact measurement in cotton crop reported resistance of *T. vaporariorum* as early as 1971, several workers (Prabhaker *et al.*, 1985; Dittrich and Ernst, 1983; Satpute and Subramanian, 1983) have reported resistance of *B. tabaci* to many chemical compounds. Interestingly, most species of whiteflies have a wide range of natural enemies (NE = parasitoids, predators and fungi) which maintain their populations under control in natural ecosystems. However, few species of NE can be efficiently used as biocontrol agents in the agroecosystems, because their efficacy depends on several parameters. Among these parameters, crop-pest interactions, pest-beneficial interactions and interference of wide spectrum chemical pesticides, are the most important (Onillon, 1990).

With respect to NE, most if not all MBAs reported on Aleyrodidae have been only fungi, which through the cuticle infect these sucking insects (Fransen *et al.*, 1987). From all fungal pathogens of whiteflies, the genus *Aschersonia* (Coelomycetes: Deuteromycotina) is characterized by its pathogenecity and specificity to infect whiteflies. Several species of *Aschersonia* have been found infecting Aleyrodidae under field and greenhouse conditions (Mains, 1959). Because of its specificity, *Aschersonia* spp. are excellent candidates to be used in combination with parasitoids, among which, the genus *Encarsia* spp. offers excellent efficiency for controlling whiteflies (Gerling, 1986). However, limitations in the mass production of *Aschersonia* spp. hindered its commercialization.

Isolates of *M. anisopliae* (CNPSo-Ma12) and Bb (CNPSo-Bb56) were tested under field conditions as biological control agents of soybean stink bugs, *Nezara viridula* (Linnaeus), *Piezodorus guildinii* (Westwood) and *Euschistus heros* (Fabricius), which usually occur as a complex in soybean in Brazil (Sosa-Gomez and Moscardi, 1998). Kaolin-based powder formulations of the two fungi were applied to soybean plots at a rate of 1.5×10^{13} conidia per ha and depending on field parameters gave variable control. These are important informations toward the development of entomopathogenic fungi as microbial insecticides of these pests in Brazil. Other EPF species such as Bb, *Paecilomyces* spp. and *V. lecanii*, also have been reported as pathogenic to *Bemisia* and *Trialeurodes* whiteflies (Fransen *et al.*, 1987). The results of Wraight *et al.* (1998) indicate that highly virulent strains of *P. fumosoroseus* and Bb with considerable whitefly control potential of diverse origins are widespread and numerous in nature. Akey *et al.* (1998) in a field experiment conducted in 1993–94 in Arizona with cotton cv. Deltapine 5415, showed that treatments with *P. fumosoroseus* or Bb were as effective in controlling *B. argentifolii* as best chemical control regimes.

The elaboration of an IPM model for the control of *B. tabaci* with the EPF species *A. aleyrodis*, Bb or *Paecilomyces* spp. represents an ideal opportunity for the commercial development of EPF in Europe and North America. Due to its specificity, *A. aleyrodis* could be utilized to manage the early-first infestations of *B. tabaci*, as this EPF will allow the establishment and increase of other NE of whiteflies. For this reason, some (less susceptible) species of parasitoids could be released in the same time period, to reinforce the biocontrol strategy. The application of systemic chemical insecticides to the soil in the early season, is another component of IPM, recommendable in zones of especially high pressure of whitefly infestation. As the dynamics of the pest increases with the augmentation of the crop biomass, other EPF presenting a higher virulence and a faster killing effect than the ones offered by *Aschersonia*, would be required. We have found several isolates of Bb and *P. fumosoroseus* to be highly virulent against adults and immature stages of *B. tabaci*. The species *P. fumosoroseus* is specially promising, as recently researchers have obtained conidia in submerged culture (De la Torre and Cardenas-Cota, 1996). In this case, production of conidia directly from blastospores through microcyclic sporulation was observed.

Narvaez and co-workers (1997) studied spore production of 10 Bb and 12 *M. anisopliae* isolates obtained from Coleoptera, Homoptera, Lepidoptera and Orthoptera which caused pathogenicity levels of 80 per cent in coffee berry borer, *H. hampei*. Isolates were cultured on a rice substrate and *H. hampei*, 30 individuals per isolate being inoculated and incubated in a humid chamber for 10 days. Each colonizing fungus was transferred to a test tube containing 4.5 ml distilled water for counting. In the rice substrate, isolates Bb9207, Bb 9301 and Ma 9236 produced the greatest number of spores (3.57×10^{10}, 1.53×10^{10} and 1.31×10^{10} spores g^{-1}, respectively). Ma 9303 produced the fewest spores (1.65×10^{8} spores g^{-1}). Using *H. hampei* as substrate, Bb 9218 produced the greatest number of spores (1.36×10^{7} spores per insect). Pathogenicity and sporulation were observed as independent characteristics. The results showed that spore production depends on the isolate and interaction with the culture media in which it develops.

The above results are encouraging because their optimization towards high spore yields would open a good opportunity for the industrial scale up and commercial development of mycoinsecticides. Continuing with the IPM model proposed for *Bemisia* whitefly, during the final period of the cropping season, particularly high populations of adult whiteflies can occur. In these cases EPF alone may not be enough to reduce the population dynamics of the pest. Consequently, contact chemical insecticides can be used to drastically reduce populations and as a part of a rotation program. One such report comes from Alves *et al.* (1998) who examined the utilization of pesticides with Bb to control coffee berry borer and coffee rust, *Hemileia vastatrix* in coffee (cv. Catuai) during 1995–96 in Sao Paulo, Brazil.

Treatments included combinations of disulfoton + triadimenol (as Baysiston GR), Bb, copper oxychloride (as Cupravit) and endosulfan (as Thiodan). The greatest CBB infestations occurred in disulfoton+triadimenol-treated and control plots. The greatest CBB mortality due to Bb occurred after 2 applications of Bb, Cupravit+Bb (one and two applications), two applications of disulfoton+triadimenol+Bb. The greatest rust infestations were found in plots not treated with disulfoton+triadimenol, and the greatest leaf loss occurred in the control plot.

Relationships to integrated pest management and climates

The distribution and insecticidal activity of EPF in nature can be found in several contexts of the ecosystem. Foremost in the ecosystem is the changing environment that EPF must confront, survive and carry out their interactions with the plant and insect world (Frankland *et al.*, 1996). Depending on broad environmental conditions, whether in tropical or temperate zones, EPF must interact with their hosts, create infections loci and after the death of insects sporulate and aid spore transmission and dispersal with some degree of specificity (i.e. non-pathogenicity to non-target invertebrates). In the ecosystem, the success of EPF depends on how it approaches the host insect, grows as a population (density, behavior and distribution) and helps with its own persistence. Questions of epizootics and enzootics in particular seasons, pest generations, weather conditions and regional distribution in abiotic environments, are determinant to ensure species preservation, augmentation or management. The occurrence of *M. anisopliae* and Bb in soils from temperate and near-northern habitats of Canada were shown by Bidochka *et al.* (1998) indicating that geographical and spatial-distribution of these fungi is independent of soil temperature. After the death of pests infected with EPF, fungal outgrowth, the production and dispersal of spores to new hosts and the environment occurred. Lin *et al.* (1999) performed RAPD tests on 28 Bb strains from different sources. While polymorphism was observed among the different strains, cluster analysis showed that it was not related to the collection points or hosts. Differences among the strains from different collection sites or hosts and also among different isolates collected from the same location and same host were observed. These results indicate the dilemma of uses of the RAPD analysis in environmental context, which is sometimes unclear.

The diversity of characteristics of the agricultural climates and their particular insect pests around the world, represent a challenge for the implementation of EPF in terms of biological control and IPM programs. In South American ecosystems, many NEs of insect pests, are unable to adapt to the whole range of macro-environmental conditions where the pest occurs (Valencia, 1994). In addition, the agroecosystem itself presents a variety of microclimates and microhabitats, which interact in a very dynamic fashion during the phenological development of the crop. While following the original principles of IPM (Smith, 1974) the climatic situation requires that the insect pest control strategy should be designed with utmost consideration of specifics for a particular region, crop, insect pest and its key NEs. We have recently proposed the development and implementation of IPM modeling systems, which can enhance the efficiency of NEs and particularly of EPF, to meet the needs of traditional and highly technology intensive crops in tropical and temperate regions (Valencia and Khachatourians, 1998). We have defined the IPM model, as the selection of factors and conditions, giving the highest possibility of success to each one of the methods of pest control. Because the IPM model has to be designed for a specific situation, it can be considerably flexible and, therefore, adapted to a diversity of climates, crops and pests in different climates in world regions (Valencia and Khachatourians, 1998).

There are significant differences between tropical and temperate countries, which represent the starting line for the use of EPF towards an implementation of global strategies of sustainable agriculture. The first major difference is the occurrence of four seasons in temperate regions. This factor not only limits agricultural activities to a single cropping season in the year, but also the survival, persistence and dissemination of EPF and the number of generations of insect host with which they interact, whether pest- or beneficial- insects. Extremely low temperatures during winter exert a tremendous selective pressure on poikilothermic organisms such as insects, so that only those species that have evolved adaptations to over winter will be able to survive and reproduce. The overall consequence of the abiotic-climatic selective pressure, is a limited biodiversity of plants and insects in temperate regions in comparison to tropical habitats. Not surprisingly, pest complexes are less likely to occur in these regions, since often a single-major pest is required to be controlled per cropping season. This situation represents an extraordinary opportunity for the use of EPF and other biocontrol agents for successful implementation of classical and non-classical biological control. Indeed, in these cases, all the pest control efforts can be focused on a single target pest, thus taking advantage of the generally high specificity of NE (DeBach and Rosen, 1991). In this respect, the utilization of parasitoids or highly pathogenic and host insect specific EPF is particularly promising in temperate regions, given the possibility of isolation of indigenous strains, which can be much more adapted to the seasonal changes and overall climatic conditions existing in these zones (Valencia and Khachatourians, 1998).

Temperate climates

Considering that temperate regions mostly correspond to developed countries, specific and rigorous regulations of the release of microbiological insecticides by government agencies are and will remain the rule. Besides, the eventual impact of entomopathogens on NTOs and their persistence in the agroecosystem, probably are the major environmental concerns in these countries. In this regard, unlike any other system, EPF offer a very practical solution: the development of conditional lethal mutants. More specifically, the technologies for the selection and management of heat-sensitive and cold-sensitive mutants are now in place (Hegedus and Khachatourians, 1994; Chelico and Khachatourians, 1999). Although isolation of such mutants requires time and trained human resources, they offer significant environmental benefits in both tropical and temperate regions. These mutants die off in the environment, when a certain threshold of high- or low- temperatures is reached. Cold-sensitive mutants in particular can be applied between spring and summer times (when the crop sensitivity to damage as well as the pressure of insect pests are usually high), with the significant additional advantage that these strains will not be able to over-winter. As a result, these EPF can only persist for one cropping season. Needless to say we selected our Bb mutants on the basis that their virulence on the intended target pest remained unaffected. Additionally, it is also necessary to ensure the genetic stability of these mutants, which is normally assessed based on their molecular characterization (Khachatourians, 1996). It is also possible to consider the need of developing cold-tolerant mutants of indigenous fungal isolates. Considering that these fungi are already a part of our ecosystem, their impact on NTOs rather than their persistence, is the more relevant environmental issue.

The development and implementation of EPF to be used within the IPM models requires a functional approach to incorporation of biocontrol agents in cropping systems. An IPM model proposed for the flea beetle, *Phyllotreta cruciferae* (Goeze), in canola crop using Bb in combination with brand name chemical insecticides, is a good example of a biorational pest control approach that could be implemented in a temperate region. Flea beetles appear in the canola crop early in the season, when a new generation of adult forms emerges from over-wintering pupae. Population peaks of this pest are generally coincident with a highly susceptible phenological state of the crop (seedlings and young plants) which can further aggravate the economic impact on the crop. Traditionally, flea beetles have been controlled by means of seed treatment with a residual chemical insecticide, usually followed by the application of a systemic compound at sowing, which offers protection until the stage of the sixth leaf of the crop. Despite these treatments, farmers often have to perform an additional application of an insecticide with knock-down effect between 30 and 40 days after sowing, as the pressure of the emerging new adults can be too high at that moment. This situation suggests that the incorporation of an entomopathogen in the management of *P. cruciferae* may contribute to improve the efficacy and cost- effectiveness of the control of this pest in canola. Some isolates of Bb have been proven to be highly virulent to this coleopteran pest under laboratory conditions (Khachatourians, unpublished data).

A theoretical model proposed for IPM of flea beetles in canola, considers two possible scenarios: first, the rotation of chemical insecticides with Bb and second, the mixture of this EPF with one of the chemical compounds. The idea is to take advantage of the wide protection period generally offered by the EPF to complement the fast killing effect and high mortality produced by chemical insecticides. In the case of rotation, the bioinsecticide could be applied in a window between the seed treatment and/or the application at sowing, and the foliage spray of the knock down (synthetic chemical) insecticide product. We suggest that the application of the EPF alone should be done between germination and the stage of the third or fourth leaf of the canola crop, thus, earlier than the moment when the application of the foliage chemical treatment is recommended. These early applications of the EPF, can eventually make possible to eliminate either the seed treatment or the application at sowing, although it is unlikely that both chemicals could be simultaneously removed from the system, because of the dynamics of emergence and crop damage of the over-wintering flea beetles. As well, early applications of Bb may allow delaying the treatment with the knock down product, depending on the population dynamics of the pest. As a result, a significantly wider protection period to the crop should be attained, based on the combined efficacy of the sequentially sprayed EPF and the chemical insecticides. The second scenario previously mentioned is the mixture of Bb with a chemical insecticide. In this case, the EPF is utilized to prolong the control of the pest, once the chemical compound (Valencia, 1995b) has drastically diminished its population.

This approach is particularly functional at times of maximal pressure of the pest on the crop, and when additional applications of insecticides are not recommendable. The application of Bb in mixture with a knock down insecticide at the peak of infestation of the flea beetles, can provide a very fast control and yet, an ongoing regulation of the remaining or new emerging insects, due to the activity of the fungal pathogen. Additional benefits could be related to the possibility to reduce the dose of either the chemical or the biological insecticides in the mixture and to diminish the costs of the treatment, as compared to the application of both products separately. As well, according to the existing knowledge on insect resistance management, the eventual occurrence of resistance of flea beetles against chemical insecticides, can be successfully delayed or even prevented based on this strategy (Miranpuri and Khachatourians, 1993). As mentioned before, the theoretical IPM model must be experimentally validated in different agroecological zones before its commercial implementation.

Tropical climates

Tropical agroecosystems also present a set of unique characteristics, which have to be considered for a successful pest management. The biodiversity of the insects occurring in these regions implies a higher complexity of phytophagous and even disease vector insects such as phlebotomine species, often associated with some crops or cropping areas. Whenever such an agroecological balance of the crop is disrupted, potential and secondary pests become a threat. The outbreak of potential pests is often related to the reduction or elimination of the fauna of beneficial insects, frequently as a result of the non-selective use of broad-spectrum chemical pesticides.

This beneficial fauna in turn is represented by a complex of predators, parasitoids and antagonists, which usually maintain and regulate the populations of phytophagous and pest arthropods (Rodriguez del Bosque, 1994). Haseeb and Murad (1997) conducted a laboratory trial to evaluate the effect of the Bb on insect predators at 25°C. Although most predatory species were susceptible in the laboratory, Bb and other EPF are likely to be less detrimental under field conditions, where the differential behavior and susceptibility of NEs, make possible the survival of a big proportion of the existing beneficial fauna. As an example, *Coccinella septempunctata* Linnaeus and *Coccinella* spp. were highly susceptible, while the coccinellid, *Brumoides suturalis* (Fabricius) and *B. syrphids* were less susceptible.

Remarkably enough, it has been demonstrated that in tropical agroecosystems, the additive action of several species of predatory insects can have a better regulatory effect on the pest population, than a single, host specific parasitoid (Valencia, 1995a).

Importantly, the application of an EPF for the control of a pest species must be done in such a way, that its impact on the overall population of NTOs can be minimized. In other words, we are suggesting that similarly to the use of chemical pesticides, biological insecticides also have to be used rationally. Following this thought, it is necessary to discuss now, the importance of the delicate balance between pathogenecity-virulence and the specificity of MBAs to selectively affect the host pest. Feng (1998) suggests the reliability of protease as an indicator of virulence index. Further, he suggests that this can be a criterion for early-stage selection of candidate isolates of EPF. Feng (1998) admits that protease could not entirely replace conventional virulence assay methods. Such a scheme is an oversimplification of the screening criteria as many EPF have multiple proteases.

Bioinsecticides having an ingestion mode of action, generally exhibit a good level of specificity. However, their utilization is very limited, if two or more pest insects occur as a complex. Entomopathogenic fungi naturally have a wider spectrum of pest control and the possibility of developing co-formulations of two or more compatible isolates affecting different species of pest, is a very promising approach for the management of pest complexes in tropical ecosystems. Additionally, there are evidences that the repetitive selection of EPF fungal strains on a specific pest host is reducing their comparative impact on NEs and NTOs (Stich and Jackson, 1997). For us, this would be a logical consequence of the enzyme-substrate specificity, since hydrolytic enzymes of a given species of EPF, will certainly find different substrates, inducers and inhibitors in the cuticle of a beneficial insect as compared to the pest.

We believe that as a result of a well designed and implemented IPM program, each pest control measure could be used at a minimum input (including biopesticides) with a maximum of efficacy, thus optimizing the overall efficiency of the crop protection strategy. Similarly to temperate regions, agricultural systems in tropical countries can be greatly benefited from the implementation of IPM models. An important advantage of the use of EPF to manage pest complexes in these agroecosystems is that the utilization of these products at the beginning of the season favors the establishment and increase of beneficial species. Accordingly, the first generalized applications of wide spectrum chemical insecticides can be delayed (Valencia, 1995a). This aspect is particularly important in case of biannual crops, which because of their occurrence in the nature usually do not support a permanent population of beneficial insects. We would like to conclude this section, by introducing an IPM model proposed for the management of the cotton boll weevil, *A. grandis* using Bb. Efficacy of Bb species against *A. grandis* has been previously established (Wright, 1993). The goal is to design an IPM program giving the fungus and else control measures, the better opportunity to achieve a cost-effective and efficient control of the boll weevil.

Four different populations of *A. grandis* usually occur in cotton fields: immigrant, establishing, resident and emigrant adult boll weevils. The IPM model is aimed to manage the population dynamics of the pest, based on the control of immigrant and establishing populations through early applications of Bb. This objective can be achieved by optimizing the opportunities for contact between the pest and the pathogen. Immigrant *A. grandis* may appear in cotton crops before or soon after cotton buds are formed. The insects occur initially in focus and their first activities are walking around and inside the squares and feeding on these structures.

This behavior offers an excellent opportunity to initiate the IPM program, by ground level applications of Bb directed to the squares. At this time the application of Bb should be done at an average infestation level, lower than the threshold recommended to apply chemical insecticides. We have previously described this as an action sub-threshold (Valencia and Khachatourians, 1998). Terrestrial applications of EPF directed to the squares and other plant structures offer an additional protection to the pathogen, because the cotton bud sepals can reduce the exposure of the fungal spores to direct sun light. If the number of foci in the total area increases or the boll weevil infestation appears to be generalized, aerial applications of Bb should be made, provided that the general average infestation is still lower than the threshold for chemical control. The application of Bb in these instances not only can reduce the population of immigrant boll weevils, but more importantly, can also diminish the potential of adult females' oviposition in the cotton buds. If this oviposition takes place, the buds will fall and subsequently will release the first generation of weevils in the cotton field. Therefore, the manual or mechanical collection of the infested fallen buds, is one of the most important cultural practices to control *A. grandis* in several tropical countries (Murillo and Cifuentes, 1995).

The emergence of the new adults from the plant structures on the ground, is another special opportunity to significantly affect the population dynamics of the pest, by means of Bb applications directed to the fallen buds. The new adults will walk around on the soil and the fallen structures, before climbing to the cotton flowers. Consequently, these insects will come in contact with the pathogen, prior to the initiation of feeding and oviposition (Valencia and Khachatourians, 1998). Once the infestation of the establishing boll weevils increases (beyond the action sub-threshold) and the pest is generalized in the field, applications of Bb alone, may not be enough to offer a good protection to the crop. At this point in time, the initiation of chemical applications may be necessary, based on the action threshold already defined previously (Andrews and Quezada, 1989). If the infestation pressure of the boll weevil is particularly high or there are overlapping generations in the field, the chemical insecticide can be applied in mixture with Bb to attain a combined effect, whether synergistic, additive or complementary (Valencia, 1995a,b; Valencia and Khachatourians, 1998).

The final considerations about temperate and tropical agroecosystems are related to the economics of crop production. Because the philosophy of IPM modeling is the optimization of the pest management system by minimizing the inputs, an important reduction of costs of pest control can be expected. This proposal is compatible with the principles of low input-sustainable agriculture (LISA) and good agricultural practices (Nene, 1996). The pest management optimization under the scheme of IPM modeling can go further. As not only action thresholds but also action sub-thresholds are introduced into the system, the average accumulated damage to the crop during the season is likely to be reduced or at least disallowed to go any higher than that, which would have been in the case of the utilization of chemical control exclusively (Valencia, 1994).

Consequently, if pest control costs are lowered and the harvest quality and yields are maintained at previous levels or increased, there is an opportunity for farmers to attain higher profits. Recent experiences in Latin America show that this can be the case. The

IPM of the coffee berry borer *Hypothenemus hampei*, in Colombia, (Brustillio and Posda, 1996; Bustillio *et al.*, 1999; Sosa-Gomez and Lanteri, 1999), Brazil (Gravena and Yamamoto, 1994), India (Balakrishna *et al.*, 1994, 1995; Reddy and Rao, 1999), Mexico (Galavez *et al.*, 1999), and Togo and Cote d'Ivoire (Vega *et al.*, 1999) using Bb can be taken as a practical example. The pest arrived in Colombia in 1988. During several years farmers have gained experience in the management of the insect, based on an IPM, via combination of cultural practices, chemical control and applications of Bb. As a consequence of the implementation of demonstrative farms on IPM modeling, entrepreneurial coffee growers have attained an average reduction of pest control costs of 30 per cent in comparison to conventional programs. However, most of these farmers were able to obtain exportation quality coffee and good yields at the harvest (Murillo, 1998). Beyond the economic benefits for growers, the coffee agroecosystem itself can be also benefited. Integrated pest management modeling facilitates the reduction and rational use of chemical insecticides in coffee plantations. As a result, a rich complex of beneficial insects is better maintained and a lower impact on NTOs in the agroecosystem can be expected (Valencia, 1995b).

Finally, independent of the region under consideration (either temperate or tropical), a different IPM approach has to be undertaken for perennial crops (plantations, orchards, etc.) when compared to temporary-short cycled crops, such as wheat, canola, corn, vegetables and others. This is because unlike temporary crops, perennials usually have a significant population of resident beneficial insects, which account for a permanent regulation of the phytophagus complex. Additionally, several pest species present a relatively long life cycle, adjusted to the phenology of the crop. Therefore, strategies for classical and nonclassical biological control in perennial crops can be facilitated by the complementary effect of the resident beneficial fauna.

In general and in particular cases, IPM models using EPF can be implemented in diversity of agricultural and agro-forestry systems, with the purpose of attaining cultural efficiency and cost-effectiveness of crop protection. Clearly, this approach will favor sustainability and competitiveness of agriculture, with the subsequent social, economic and environmental benefits in temperate as well as in tropical regions around the world.

Conclusions

Trends in fungal biotechnology research and developments project a significant change in adoption and utilization of EPF in insect pest control and management schemes. These schemes are going to offer new paradigms in IPM and the management of pests resistant to synthetic chemical pesticides and even the first generation *B. thuringiensis* resistant (Hegedus *et al.*, 2001).

Practice of IPM in its relationship to various EPF is complex both in its nature and in its implementation (Valencia and Khachatourians, 1998; Khachatourians and Valencia, 1999) specially when it comes to insect pest "control" versus "management." Fungi such as Bb have been used in traditional insect pest control, but given the current discovery and applied research findings, management by reduction of pest numbers and hence their effect on cropping can be achieved. Present day pest management research and the transfer of the information has shifted in terms of boundaries. The public vs. private investment and the notion of return on the investment, extension services and sharing of the new information should play increased role, either for growers who are older and nearing retirement or for the new generation of farmers. The sources and manners of adoption of new technology or information such as the use of EPF can help or hinder the ultimate goal of advancing of biopesticide use and research. But definitely, integrating farmers' experiential-or-experimental knowledge and understanding of the agricultural business and the general public environmental concerns, to the design and use of innovative technologies such as biopesticides and IPM modeling, will be an important strategic and communication asset for agriculture in the new millennium.

Altieri (1987) makes the point that there is a lack of training in "holistic thinking," so far as crop protection is concerned and those in the profession are trained too narrowly and with a focus only within certain disciplines. Plant protection research tends to depend on the use of component technologies, such as plant sciences, entomology, biological control, and microbiology, etc. but relatively little research on linking these components together. Biopesticide research programs require extensive crosscutting knowledge and research skills. The powers and promises of a select few fungi in insect control strategies should be congruent with the powerful but still unrealized IPM strategies.

Sometime soon there will be greater global demand for biocontrol agents and consequently the focus on EPF will increase. We have recently introduced the concept of rational design of biopesticides (RADBIO) as one biotechnological approach to improve the pest control performance of EPF (Valencia and Khachatourians, 1998). RADBIO can be accomplished by four major avenues: conventional selection of EPF fungal strains, mutation and selection, hybridization of compatible isolates and genetic engineering (*in vitro* recombinant DNA techniques), to render genetically improved EPF. We want to bring the attention to the point, that of all these alternatives, only the last one implies the genetic alteration of EPF with foreign genes. We believe that RADBIO can expedite the R & D process and facilitate the registration of more reliable and cost-effective mycoinsecticides worldwide. Should action follow and a confirmed market demand be secured, EPF are poised to meet both the challenges and the opportunities that lie ahead.

References

Agarwala, S.P., Sagar, S.K., and Sehgal, S.S. (1998) A laboratory evaluation of conidia of *Beauveria bassiana* (Hyphomycetes) as a biological control agent against *Aedes aegypti* larvae. *Shashpa*, **5**, 217–220.

Akey, D.H., Henneberry, T.J., Dugger, P., and Richter, D. (1998) *Proceedings Beltwide Cotton Conference*, San Diego, pp. 1073–1077.

Altieri, M.A. (1987) *Agroecology: The Scientific Basis of Alternative Agriculture*. IT Publications, London.

Altieri, M.A. (1995) *Agroecology: The Science of Sustainable Agriculture*, Second Edition, IT Publications, London.

Alves, S.B., Almeida, J.E.M., Salvo. S.de, and De-Salvo, S. (1998) Utilization of pesticides with *Beauveria bassiana* to control coffee berry borer and coffee rust. *Manejo Integrado de Plagas*, **48**, 19–24.

Alves, S.B., Pereira, R.M., Stimac, J.L., and Vieira, S.A. (1996) Delayed germination of *Beauveria bassiana* conidia after prolonged storage at low, above- freezing temperatures. *Biocontrol Sci. Technol.*, **6**, 575–581.

Andrews, K.L., and Quezada, J.R. (1989) Manejo integrado de plagas insectiles en la agricultura: estado actual y futuro. *Departamento de Proteccion Vegetal. Escuela Agricola Panamericana, El Zamorano, Honduras, Centroamerica*, 623 pp.

Arango, P.E.V. (1997) Evalucion de formulaciones en aceite y en agua el hongo *Beauveria bassiana* (Balsamo) Vuillemin en campo. *Rev. Colomb. Entomol.*, **23**, 59–64.

Arcas, J.A., Diaz, B.M., and Lecuona, R.E. (1999) Bioinsecticidal activity of conidia and dry mycelium preparations of two isolates of *Beauveria bassiana* against the sugarcane borer *Diatraea saccharalis. J. Biotechnol*, **67**, 151–158.

Bajan, C., Kmitowa, K., and Popowska-Nowak, E.T.I. (1998) Reaction of various ecotypes of entomopathogenic fungus *Beauveria bassiana* to the botanical preparation NEEM and pyrethroid Fastak. *Arch. Phyto. Plant Protect.*, **31**, 369–375.

Balakrishnan, M.M., Sreedharan, K., and Bhat, P.K. (1994) Occurrence of the entomopathogenic fungus *Beauveria bassiana* on certain coffee pests in India. *J. Coffee Res.* **24**, 33–35.

Balakrishnan, M.M., Vijayan, V.A., Sreedharan, K., and Bhat, P.K. (1995) New fungal associates of the coffee berry borer *Hypothenemus hampei. J. Coffee Res.* **25**, 52–54.

Bello, V.A., and Paccola–Meirelles, L.D. (1998) Localization of auxotrophic and benomyl resistance markers through the parasexual cycle in the *Beauveria bassiana* (Bals.) Vuill entomopathogen. *J. Invertebr. Pathol.*, **72**, 119–125.

Berretta, M.F., Lecuona, R.E., Zandomeni, R.O., and Grau, O. (1998) Genotyping isolates of the entomopathogenic fungus *Beauveria bassiana* by RAPD with fluorescent labels. *J. Invertebr. Pathol.*, **71**, 145–150.

Bidochka, M.J., Kasperski, J.E., and Wild, G.A.M. (1998) Occurrence of the entomopathogenic fungi *Metarhizium anisopliae* and *Beauveria bassiana* in soils from temperate and near-northern habitats. *Can. J. Bot.*, **76**, 1198–1204.

Bosch, A., and Yantorno, O. (1999) Microcycle conidiation in the entomopathogenic fungus *Beauveria bassiana* Bals. (Vuill.) *Proc. Biochem.*, **34**, 707–716.

Boucias, D.G., Mazet, I., Pendland, J., and Hung, S.Y. (1995) Comparative analysis of the *in vivo* and *in vitro* metabolites produced by the entomopathogen *Beauveria bassiana. Can. J. Bot.*, **73**, S1092–S1099.

Bustillio, A.E., Bernal, M.G., Benavides, P., and Chaves, B. (1999) Dynamics of *Beauveria bassiana and Metarhizium anisopliae* infecting *Hypothenemus hampei.* (Coleoptera: Scolytidiae) populations emerging from fallen coffee berries. *Florida Entomol.* 82, 491–498.

Bustillio, P.A.E., and Posda, F.E.J. (1996) El uso de entomopathogenos en el control de la broca del cafe in Colombia. *Manejo Integrad. Plagas.* **42**, 1–13.

Bradley, C., Kearns, R., and Woods, P. (1992). The role of production technology in mycoinsecticide development. In *Frontiers in Industrial Mycology*, Mycological Society of America, Symposium on Industrial Mycology, Madison, WI, June 1990. Chapman and Hall, New York.

Carballo, V.M. (1998) Mortalidad de Cosmopolites sordidus con diferentes formulaciones de *Beauveria bassiana. Manejo Integ.Plagas.*, **48**, 45–48.

Cardona, C. (1995) Entomologia y control de plagas: analisis y perspectives para el futuro. MIP: manejo-integrado de plagas en Cultivos y medio ambiente. *Capitulo 1: ciencia y tecnica para el MIP en cultivos*, **1**, 1–6.

Chaufaux, J., Muller-Cohn, J., Buisson, C., Sanchiz, V., Lereclus, D., and Pasture, N. (1997) Inheritance of resistance of *Bacillus thuringiensis* Cry I toxin in *Spodoptera littoralis* (Lepidoptera: Noctuidae). Insecticide resistance and management. *J. Econ. Entomol.*, **90**, 873–878.

Chelico, L., and Khachatourians, G.G. (1999) Initial characterization of cold-sensitive mutants of the entomopathogen, *Beauveria bassiana. Programme Joint Ann. Meet.*, Canad. Entomol. Soc. and Entomol. Soc. Sask., September 25–29, Abstracts, p. 37.

De La Rosa, W., Alatorre, R., Trujillo, J., and Barrera, J.F. (1997) Virulence of *Beauveria bassiana* (Deuteromycetes) strains against the coffee berry borer (Coleoptera: Scolytidae). *J. Econ. Entomol.*, **90**, 1534–1538.

De la Torre, M., and Cardenas-Cota, H.M. (1996) Production of *Paecilomyces fumosoroseus* conidia in submerged culture. *Entomophaga*, **41**, 443–453.

de-Marcano, D.A., Marcano, A.J., and Morales, M. (1999) Patogenicidad de *Beauveria bassiana* y *Paecilomyces fumosoroseus* sobre adultos del picudo de la batata *Cylas formicarius elegantulus* Summers (Curculionidae). *Rev. Facul. Agronom., Universidad del Zulia.*, **16**, 52–63.

DeBach, P., and Rosen, D. (1991) *Biological Control by Natural Enemies*. Second edition, Cambridge University Press, Cambridge, UK.

Delgado-Francisco X., Britton, J.H., Onsager, J.A., and Swearingen, W. (1999) Field assessment of *Beauveria bassiana* (Balsamo) Vuillemin and potential synergism with diflubenzuron for control of savanna grasshopper complex (Orthoptera) in Mali. *J. Invertebr. Pathol.*, **73**, 34–39.

Desgranges, C., Vergoignan, C., Lereec, A., Riba, G., and Durand, A. (1993) Use of solid state fermentation to produce *Beauveria bassiana* for the biological control of European corn borer. *Biotechnol. Adv.*, **11**, 577–587.

Dittrich V., and Ernst, G.H. (1983) Resistance pattern in whiteflies of Sudanese cotton Mittelungen der Dutschen Gesellschaft fur allgemaine und angewandte. *Entomologie*, **4**, 96–97.

Draganova D. (1997) Utilization of carbon sources by the strains of the *Beauveria* Vuill. genus (Moniliales, Deuteromycetes). *Bulgar. J. Agric. Sci.*, **3**, 557–562.

Duffus, J.E. (1987) Whitefly transmission of plant viruses. In K.F. Harris (ed), *Current Topics in Vector Research*, Spring-Verlag, New York, pp. 73–91.

Duque, B.E.V., and Arango, P.E.V. (1998) Caracterizacion bioquimica cualitativa de aislamientos de *Beauveria bassiana* de una coleccion de hongos entomopatogenos de Cenicafe. *Revis. Colomb. Entomol.*, **24**, 61–66.

Feng, L.H., Huang, B., Li, Z.Z., and Hu, C. (1998) Electron microscopic scanning observation of the survival and infection of *Beauveria bassiana* on *Dendrolimus punctatus. Mycosystema.*, **17**, 344–348.

Feng, M.G. (1998) Reliability of extracellular protease and lipase activities of *Beauveria bassiana* isolates used as their virulence indices. *Weishengwu Xuebao.*, **38**, 461–467.

Feng, M.G., Poprawski, T.J., and Khachatourians, G.G. (1994) Production formulation application of the entomopathogenic fungus *Beauveria bassiana* for insect control worldwide. *Biocontrol Sci. Technol.*, **4**, 3–34.

Frankland J.C., Magan , N., and Gadd, G.M. (1996) *Fungi and Environmental Change*. Cambridge Universtiy Press, Cambridge.

Fransen, J.J., Winkelman, C., and Lenteren, J.C. van. (1987) The differential mortality of various life stages of the greenhouse whitefly, *Trialeurodes vaporariorum* (Homoptera: Aleyrodidae) by infection with the fungus *Aschersonia aleyrodis* (Deuteromicotina: Coelomycetes). *J. Invertebr. Pathol.*, **50**, 158–165.

Galavez, R.J., Berrera, J., Nelson, K.C., and Martinez-Quezada, A. (1999) Metodos no quimicos contra la broca del cafe y su transferencia tecnologia en los altos de Chiapas, Mexico. *Agrociencia*, **33**, 431–438.

Gatehouse, A.M.R. (1999) Biotechnological applications of plant genes in production of Insect resistant crops. In S.L. Clement and S.R. Quisenberry (eds.), *Global Resources for Insect-Resistant Crops*, CRC Press, Boca Raton, Florida, pp. 263–280.

Gerling, D. (1986) Natural enemies of *Bemisia tabaci*, biological characteristics and potential as biological control agents: a review. *Agriculture, Ecosyst. Environ.*, **17**, 99–110.

Gillespie, J.P., and Khachatourians, G.G. (1992) Characterization of the *Melanoplus sanguinipes* hemolymph after infection with *Beauveria bassiana* or wounding. *Comp. Biochem. Physiol.*, **103B**, 455–463.

Gillespie, J.P., Kanost, M.R., and Trenczek, T. (1997) Biological mediators of insect immunity. *Annu. Rev. Entomol.*, **42**, 611–643.

Glare, T.R., Milner, R.J., Chilvers, G.A., Mahon, R.J., and Brown, W.V. (1987) Taxonomic implications of intraspecific variation amongst isolates of the aphid pathogen fungi *Zoophthora radicans* Brefeld and *Zoophthora phalloides* Batko (Zygomycetes: Entomophthoraceae). *Aust. J. Bot.*, **35**, 49–67.

Gravena, S.S., and Yamamoto, P.T. (1994) Manajo de pragas se cafeerio com endosulfan na regiao de Espirito Sanato do Pinhal, S.P. *Cien. Jabotica.* **22**, 123–131.

Hajek, A.E., and St. Leger, R.J. (1994) Interactions between fungal pahogens and insect hosts. *Ann. Rev. Entomol.*, **39**, 293–322.

Hallsworth, J.E., and Magan, N. (1999) Water and temperature relations of growth of the entomogenous fungi *Beauveria bassiana, Metarhizium anisopliae*, and *Paecilomyces farinosus. J. Invertebr. Pathol.*, **74**, 261–266.

Haseeb, M., and Murad, H. (1997) Pathogenicity of the entomogenous fungus, *Beauveria bassiana* (Bals.) Vuill. to insect predators. *Int. Pest Contr.*, **40**, 50–51.

Hayden, T.P., Bidochka, M.J., and Khachatourians, G.G. (1992) Entomopathogenecity of several fungi towards the English grain aphid (Homoptera: Aphididae) and enhancement of virulence with host passage of *Paecilomyces farinosus. J. Econ. Entomol.*, **85**, 58–64.

Hazarika, L.K., Puzari, K.C., and Saikia, D.K. (1998) Seasonal and host correlated variation in the susceptibility of rice hispa to *Beauveria bassiana* in the field. *Indian J. agric. Sci.*, **68**, 361–363.

Hegedus, D.D., and Khachatourians, G.G. (1993a) Construction of cloned DNA probes for the specific detection of the entomopathogenic fungus *Beauveria bassiana* in grasshoppers. *J. Invertebr. Pathol.*, **62**, 233–240.

Hegedus, D.D., and Khachatourians, G.G. (1993b) Identification of molecular variants in mitochondrial DNAs of members of the genera *Beauveria, Verticillium, Paecilomyces, Tolypocladium*, and *Metarhizium. Appl. Environ. Microbiol.*, **59**, 4283–4288.

Hegedus, D.D., and Khachatourians, G.G. (1994) Isolation and characterization of conditional lethal mutants of *Beauveria bassiana. Can. J. Microbiol.*, **47**, 766–776.

Hegedus, D.D., and Khachatourians, G.G. (1996a) Detection of the entomopathogenic fungus *Beauveria bassiana* within infected *Melanoplus sanguinipes* using PCR DNA probe. *J. Invertebr. Pathol.*, **67**, 21–27.

Hegedus, D.D., and Khachatourians, G.G. (1996b) Identification differentiation of the *Beauveria bassiana* using PCR single-stranded conformation polymorphism analysis. *J. Invertebr. Pathol.*, **67**, 289–299.

Hegedus, D.D., and Khachatourians, G.G. (1996c) The effects of temperature on the pathogenicity of hs- mutants of the entomopathogenic fungus *Beauveria bassiana* toward *Melanoplus sanguinipes. J. Invertebr. Pathol.*, **68**, 160–165.

Hegedus, D.D., Bidochka, M.J., and Khachatourians, G.G. (1990) *Beauveria bassiana* submerged conidia production in a defined medium containing chitin, two hexosamines or glucose. *Appl. Microbiol. Biotechnol.*, **33**, 641–647.

Hegedus, D.D., Bidochka, M.J., Miranpuri, G.S., and Khachatourians, G.G. (1992) A comparison of the virulence, stability and cell-wall surface characteristics of three spore types produced by the entomopathogenic fungus *Beauveria bassiana. Appl. Microbiol. Biotechnol.*, **36**, 785–789.

Hegedus, D.D., Gruber, M., Braun, L., and Khachatourians, G.G. (2001) Strategies for genetic engineering of plants for resistance to insects. In G.G. Khachatourians, W.- K. Nip, A. McHughen, R. Scorza and Y.-H. Hui (eds.), *Handbook of Transgenic Food Plants*, Marcel Dekker Inc., New York, (in press).

Hegedus, D.D., Pfeifer, T.A., MacPherson, J.M., and Khachatourians, G.G. (1991) Cloning and analysis of five mitochondrial tRNA-encoding genes from the fungus *Beauveria bassiana. Gene*, **109**, 149–154.

Hegedus, D.D., Pfeifer, T.A., Mulyk, D.S., and Khachatourians, G.G. (1998) Characterization and structure of the mitochondrial small rRNA gene of the entomopathogenic fungus *Beauveria bassiana. Genome*, **41**, 471–476.

Hem, S., Ahmad, R., and Saxena, H. (1998) Field evaluation of *Beauveria bassiana* (Balsamo) Vuillemin against *Helicoverpa armigera* (Hubner) infecting chickpea. *J. Biol. Contr.*, **11**, 93–96.

Hidalgo, E., Moore, D., and Le Patourel, G. (1998) The effect of different formulations of *Beauveria bassiana* on *Sitophilus zeamais* in stored maize. *J. Stored Prod. Res.*, **34**, 171–179.

Hung, S.Y., and Boucias, D.G. (1992) Influence of *Beauveria bassiana* on the cellular defense response of the beet armyworm, *Spodoptera exigua. J. Invertebr. Pathol.*, **60**, 152–158.

Isman, M.B. (2001) Biopesticides based on phytochemicals. In O. Koul and G.S. Dhaliwal (eds.), *Phytochemical Biopesticides*, Harwood Academic Publishers, Amsterdam, The Netherlands, pp. 1–12.

James, R.R., Croft, B.A., Shaffer, B.T., and Lighthart, B. (1998) Impact of temperature and humidity on host-pathogen interactions between *Beauveria bassiana* and a coccinellid. *Environ. Entomol.*, **27**, 1506–1513.

Jeffs, L.B., and Khachatourians, G.G. (1997) Estimation of spore hydrophobicity for members of the genera *Beauveria, Metarhizium* and *Tolypocladium* by salt-mediated aggregation and sedimentation. *Can. J. Microbiol.*, **43**, 23–28.

Jeffs, L.B., Xavier, I.J., Mattai, R., and Khachatourians, G.G. (1999) Relationships between fungal spore morphologies and surface properties of the genera *Beauveria, Metarhizium, Paecilomyces, Tolypocladium* and *Verticillium. Can. J. Microbiol.*, **45**, 936–948.

Junior, M.A., and Alves, B.S. (1998) Efeito de *Beauveria bassiana* sobre o desenvolvimento de *Sitophilus zeamais. Manejo Integr. Plagas.*, **50**, 51–54.

Khachatourians, G.G. (1986) Production and use of biological pest control agents. *Trends Biotechnol.*, **4**, 20–124.

Khachatourians, G.G. (1991) Physiology and genetics of entomopathogenic fungi. In D.K. Arora, L. Ajello and K.G. Mukerji (eds.), *Handbook of Applied Mycology*, Volume 2, *Humans, Animals, Insects*, Marcel Decker, New York, pp. 613–661.

Khachatourians, G.G. (1992) Virulence of five *Beauveria* strains, *Paecilomyces farinosus* and *Verticillium lecanii* against the migratory grasshopper *Melanoplus sanguinipes. J. Invertebr. Pathol.*, **59**, 212–214.

Khachatourians, G.G. (1996) The relationship between biochemistry molecular biology of entomopathogenic fungi insect diseases. In *THE MYCOTA*, K. Esser and P.A. Lemke (Series eds.) and D.H. Howard and J.D. Miller (eds.), Vol. VI, *Animal Human Relationships*, Springer Verlag, Berlin, pp. 331–362.

Khachatourians, G.G., and Valencia, E. (1999) Integrated pest management and entomopathogenic fungal biotechnology in the Latin Americas. II. Key Research and Development Prerequisites. *Rev. Acad. Colomb. Cienc.*, **23** (in press)

Koul, O. (1996) Advances in neem research and development- Present and future scenario. In S.S. Handa and M.K. Koul (eds.), *Supplement to Cultivation and Utilization of Medicinal Plants*, National Institute of Science Communication, New Delhi, pp. 583–611.

Koul, O and Dhaliwal, G.S. (2001) Microbial biopesticides: An introduction. In O. Koul and G.S. Dhaliwal (eds.), *Microbial Biopesticides*, Harwood Academic Publishers, Amsterdam, The Netherlands, pp. 1–11.

Kouvelis, V.N., Zare, R., Bridge, P.D., and Typas-M-A. (1998) Differentiation of mitochondrial subgroups in the Verticillium lecanii species complex. *Let. Appl. Microbiol.* **28**, 263–268.

Leathers, T.D., Gupta, S.C., and Alexer, N.J. (1993) Mycopesticides: status, challenges and potential. *J. Ind. Microbiol.*, **12**, 69–75.

Lin, H.F. (1998) Testing strain types of *Beauveria bassiana* by means of electrophoretic patterns of esterase isoenzyme. *Acta Phytophyl. Sinica.*, **25**, 315–320.

Lin, H.F., Fan, M.Z., Li, Z.Z., and Hu, C. (1998) Pathogenic effect of *Beauveria bassiana* infected on *Dendrolimus punctatus* under different temperature and humidity. *Chinese J. Appl. Ecol.*, **9**, 195–200.

Lin, H.F., Huang, B., Li, Z.Z., and Hu, C. (1999) Analysis of the relationship between DNA amplified fingerprints and source of *Beauveria bassiana* strains. *Mycosystema*, **18**, 73–78.

Luz, C., Silva, I.G., Cordeiro, C.M.T., and Tigano, M.S (1998) *Beauveria bassiana* (Hyphomycetes) as a possible agent for biological control of Chagas disease vectors. *J. Med. Entomol.*, **35**, 977–979.

Mains, E.B. (1959) North American species of *Aschersonia* parasitic on Aleyrodidae. *J. Invertebr. Pathol.*, **1**, 43–47.

Masarrat, H., Srivastava, R.P., and Haseeb, M. (1998) Dose-mortality relationship of entomogenous fungus, *Beauveria bassiana* (Bals.) Vuill. against mango mealy bug, *Drosicha mangiferae* (Green). *Insect Environ.*, **4**, 74–75.

Masuka, A.J., and Manjonjo, V. (1996) Laboratory screening of *Metarhizium flavoviride* and *Beauveria bassiana* for the control of *Mecostibus pinivorus*, a *Pinus patula* defoliator in Zimbabwe. *J. Agric. Sci. South Afric.*, **2**, 91–95.

Maurer, P., Rejasse, A., Capy, P., Langin, T., and Riba, G. (1997) Isolation of the transposable element hupfer from the entomopathogenic fungus *Beauveria bassiana* and insertion mutagenesis of the nitrate reductase structural gene. *Mol. Gen. Genet.*, **256**, 195–202.

Mavridou, A.,Typas, M.A. (1998) Intraspecific polymorphism in Metarhizium anisopliae var. anisoplia revealed by analysis of rRNA gene complex and mtDNA RFLPs. *Mycol. Res.* **102**, 1233–1241.

Miranpuri, G.S., and Khachatourians, G.G. (1990) Larvicidal activity of blastospores and conidiospores of *Beauveria bassiana* (strain GK 2016) against age groups of *Aedes aegypti. Vet. Parasitol.*, **37**, 155–162.

Miranpuri, G.S., and Khachatourians, G.G. (1991) Infection sites of the entomopathogenic fungus *Beauveria bassiana* in larvae of the mosquito, *Aedes aegypti. Entomol. exp. appl.*, **59**, 19–27.

Miranpuri, G.S., and Khachatourians, G.G. (1993) Role of bioinsecticides in integrated pest management (IPM) and insect resistance management (IRM) practices. *J. Insect Sci.*, **6**, 161–172.

Miranpuri, G.S., and Khachatourians, G.G. (1994a) Fungal and bacterial control of hematophagous vectors: A review. *J. Insect Sci.*, **7**, 150–168.

Miranpuri, G.S., and Khachatourians, G.G. (1994b) Pathogenicity of *Beauveria bassiana* (Bals) Vuill and *Verticillium lecanii* (Zimm) toward blister beetle *Lytta nuttali* Say. *J. Appl. Entomol.*, **118**, 103–110.

Miranpuri, G.S., and Khachatourians, G.G. (1995a) Application of *Beauveria bassiana* and *Verticillium lecanii* against Saskatoon berry leaf aphid *Acyrthosiphon macrosiphium* (Wilson). *J. Insect Sci.*, **8**, 93–95.

Miranpuri, G.S., and Khachatourians, G.G. (1995b) Comparative virulence of different isolates of *Beauveria bassiana* and *Verticillium lecanii* against Colorado potato beetle. *J. Insect Sci.*, **8**, 160–166.

Miranpuri, G.S., and Khachatourians, G.G. (1995c) Entomopathogenicity of *Beauveria bassiana* (Balsamo) and *Verticillium lecanii* (Zimmerman) toward English grain aphid *Sitobion avenae. J. Insect Sci.*, **8**, 34–39.

Miranpuri, G.S., and Khachatourians, G.G. (1995d) Entomopathogenicity of *Beauveria bassiana* toward *Phyllotreta cruciferae* Goeze. *J. Appl. Entomol.*, **118**, 167–170.

Miranpuri, G.S., Bidochka, M.J., and Khachatourians, G.G. (1993) Pathogenicity of *Beauveria bassiana* toward lygus bug (Hemiptera:Miridae). *J. Appl. Entomol.*, **115**, 313–317.

Morales, E. (1995) Propuesta de Proyecto Marco Agrobiologicos. *Publicacion Interna, Hoechst Schering Agro*, AgrEvo S.A. Santafe de Bogota, pp. 1–53.

Murillo, A. (1998) Microbial control of coffee berry borer: Colombian coffee. In A. Sagenmueller (ed.), *AgrEvo's Contribution to Integrated Crop Management*, Hoechst Schering AgrEvo GmbH, pp. 25–26.

Murillo, A., and Cifuentes, F. (1995) Una propuesta de modelo MIP en algodon. MIP: Manejo Integrado de Plagas en cultivos y medio ambiente. Capitulo 3: *Implementacion del MIP en Colombia en diferentes cultivos*. pp. 66–70.

Narvaez, G.M.P.-del, Gonzalez, G.M.T., Bustillo, P.A.E., Chaves, C.B., Montoya, R.E.C., and del-Narvaez, G.M.P. (1997) Produccion de esporas de aislamientos de *Beauveria bassiana* y *Metarhizium anisopliae* en diferentes sustratos. *Rev. Colomb. Entomol.*, **23**, 125–131.

Nene, Y.L. (1996) Sustainable agriculture: future hope for developing countries. *Can. J. Plant Pathol.*, **18**, 133–140.

Onillon, J.C. (1990) The use of natural enemies for the biological control of whiteflies. In D. Gerling (ed.), *Whiteflies: Their Bionomics, Pest Status and Management*, Intercept Ltd., Andover, pp. 287–313.

Pendland, J.C., Hung, S.Y., and Boucias, D.G. (1993) Evasion of host defense by *in vivo* produced protoplast-like cells of the insect mycopathogen *Beauveria bassiana*. *J. Bacteriol.*, **175**, 5962–5969.

Perira, R.M., and Roberts, D.W. (1990) Dry mycelium preparations of the entomopathic fungi *Metarhizium anisopliae* and *Beauveria bassiana*. *J. Invertebr. Pathol.*, **56**, 39–46.

Perira, R.M., and Roberts, D.W. (1991) Alginate and cornstarch mycelial formulations of the entomopathogenic fungus *Beauveria bassiana* and *Metarhizium anisopliae*. *J. Econ. Entomol.*, **84**, 1657–1661.

Pfeifer, T.A., and Grigliatti, T.A. (1996) Future perspectives on insect pest management: Engineering of the pest. *J. Invertebr. Pathol.*, **67**, 109–119.

Pfeifer, T.A., and Khachatourians, G.G. (1992) *Beauveria bassiana* protoplast regeneration and transformation using electroporation. *Appl. Microbiol. Biotechnol.*, **38**, 376–381.

Pfeifer, T.A., and Khachatourians, G.G. (1993) Electrophoretic karyotyping of the entomopathogenic Deuteromycete *Beauveria bassiana*. *J. Invertebr. Pathol.*, **61**, 231–235.

Pfeifer, T.A., Hegedus, D.D., and Khachatourians, G.G. (1993) The mitochondrial genome of the entomopathogenic fungus *Beauveria bassiana*: Analysis of the ribosomal RNA region. *Can. J. Microbiol.*, **39**, 25–31.

Pitt, J.I., and Miscamble, B.F. (1994) Water relations of *Aspergillus flavus* and closely related species. *J. Crop Prot.*, **58**, 86–90.

Prabhaker, N., Coudriet, D.L., and Meyerdrik D.E. (1985) Insecticide resistance in the sweet potato whitefly *Bemisia tabaci* (Homoptera: Aleyrodidae). *J. Econ. Entomol.*, **78**, 748–752.

Puzari, K.C., Sarmah, D.K., and Hazarika, L.K. (1997) Medium for mass production of *Beauveria bassiana* (Balsamo) Vuillemin. *J. Biol. Contr.*, **11**, 97–100.

Reddy, A.G.S.M., and Rao, L.V.A. (1999) Incidence of coffee berry borer in non-conventional coffee area of Karnataka. *Indian Coffee*, **63**, 15–16.

Reithinger, R., Davies, C.R., Cadena, H., and Alexander, B. (1997) Evaluation of the fungus *Beauveria bassiana* as a potential biological control agent against phlebotomine sand flies in Colombian coffee plantations. *J. Invertebr. Pathol.*, **70**, 131–135.

Rivera, M.A., Bridge, P.D., and Bustillo, P.A.E. (1997) Caracterizacion bioquimica y molecular de aislamientos de *Beauveria bassiana* procedentes de la broca del cafe, *Hypothenemus hampei*. *Rev. Colomb. Entomol.*, **23**, 1–2, 51–57.

Rodriguez, R., Cullen, D., Kurtsman, C., and Khachatourians, G. (1999) Estimation of fungal diversity via molecular methods. In P. Cannell. (ed). *Measuring and Monitoring Biological Diversity: Standard Methods for Fungi*, Section 5, Chapter 3. Smithsonian Inst. Press, Washington DC, 48 pp.

Rodriguez del Bosque, L.A. (1994) Teoria y bases ecologicas del control biologico. V Congreso y curso de control biologico de plagas. *Memorias. Inst. Tecnol. Agro.*, Oaxaca, Mejico, pp. 6–19.

Samson, R.A., Evans, H.C, and Latge, J-P. (1988) *Atlas of Entomopathogenic Fungi*. Springer Verlag, Berlin.

Sandhu, S.S., Rajak, R.C., and Agarwal, G.P. (1993) Studies of prolonged storage of *Beauveria bassiana* conidia: effects of temperature and relative humidity on conidial viability and virulence against chickpea borer, *Helicoverpa armigera*. *Biocontr. Sci. Technol.*, **3**, 47–53.

Satpute, U.S., and Subramanian, T.R. (1983) A note on the secondary outbreak of whitefly (*Bemisia tabaci*) on cotton with phosalone treatment. *Pestology*, **7**, 4.

Schnepf, E., Crickmore, N., Van Rie, J., Lereclus, D., Baum, J., Feitelson, J. *et al.* (1998) *Bacillus thuringiensis* and its pesticidal crystal proteins. *Microbiol. Molec. Biol. Rev.*, **62**, 775–806.

Shimizu, S., Higashiyama, R., and Matsumoto, T. (1993) Chromosome length polymorphism in *Beauveria bassiana*. *J. Seric. Sci. Japan*, **62**, 45–49.

Sitch, J.C., and Jackson, C.W. (1997) Pre-penetration events affecting host specificity of *Verticillium lecanii*. *Mycol. Res.*, **101**, 535–541.

Smart, J.R., and Wright, J.E. (1992) Phytotoxicity of *Beauveria bassiana* oil carriers to selected crops. *Subtrop. Plant Sci.* [Weslaco, Texas]: Rio Grande Valley Hort. Soc., **45**, 27–31.

Smith, R.F. (1974) The origins of the integrated pest control in California: an account of the contributions of Charles W. Woodworth. *Pan Pacific Entomol.*, **50**, 426–440.

Smith, S.M., Moore, D., Karanja, L.W., and Chandi, E.A. (1999) Formulation of vegetable fat pellets with pheromone and *Beauveria bassiana* to control the larger grain borer, *Prostephanus truncatus* (Horn). *Pestic. Sci.*, **55**, 711–718.

Sosa-Gomez, D.R., and Lanteri, A.A. (1999) Estado actual de control biologico de plagas agricolas con hongos entomopathogenos. *Revis. Socied. Entomol. Argentina.* **58**, 295–300.

Sosa Gomez, D.R., and Moscardi, F. (1998) Laboratory and field studies on the infection of stink bugs, *Nezara viridula, Piezodorus guildinii*, and *Euschistus heros* (Hemiptera: Pentatomidae) with *Metarhizium anisopliae* and *Beauveria bassiana* in Brazil. *J. Invertebr. Pathol.*, **71**, 115–120.

Sparks, T.C. (1981) Development of insecticide resistance in *Heliothis* spp. in the Americas. *Bull. Entomol. Soc. Am.*, 27.

Stich, J.C., and Jackson, C.W. (1997) Pre-penetration events affecting host specificity of *Verticillium lecanii. Mycol. Res.*, **101**, 535–541.

Suresh, P.V., and Chandrasekaran, M. (1998) Utilization of prawn waste for chitinase production by the marine fungus *Beauveria bassiana* by solid state fermentation. *World J. Microbiol. Biotechnol.*, **14**, 655–660.

Thomas, K.C., Khachatourians, G.G., and Ingledew, W.M. (1987) Production and properties of *Beauveria bassiana* conidia cultivated in submerged culture. *Can. J. Microbiol.*, **33**, 12–20.

Thompson, A. (1998) Colombian coffee. *Latin Trade*, **6**, 74–75.

Todorova, S.I., Coderre, D., Duchesne, R.M., and Cote, J.C. (1998) Compatibility of *Beauveria bassiana* with selected fungicides and herbicides. *Environ. Entomol.*, **27**, 427–433.

Trumble, J.T., Craosn, W.G., and Kund, G.S. (1997) Economics and environmental impact of a sustainable integrated pest management program in celery. *J. Econ. Entomol.*, **90**, 136–146.

Urtz, B.E., and Rice, W.C. (1997) RAPD-PCR characterization of *Beauveria bassiana* isolates from the rice water weevil *Lissorhoptrus oryzophilus. Lett. Appl. Microbiol.*, **25**, 405–409.

Valencia, P.E. (1993a) Resistencia enzimatica a insecticidas en larvas de *Heliothis virescens* (Lepidoptera: Noctuidae). *Rev. Colomb. Entomol.*, **19**, 131–138.

Valencia, P.E. (1993b) Posibles modelos para el manejo integrado de la broca. *Agric. Am.*, **217**, 20–22.

Valencia, P.E. (1994) Modelos para el manejo integrado de la broca del cafe. *Agric. Am.*, **218**, 26–28.

Valencia, P.E. (1995a) El manejo integrado de plagas (MIP) y la regulacion biologica: verdaderas alternatives para el desarrollo agricola sostenible. *MIP: manejo integrado de plagas en cultivos y medio ambiente. Capitulo I: ciencia y tecnica para el MIP en cultivos* **1**, 57–60.

Valencia, P.E. (1995b) Manejo integrado de plagas (MIP): alternative para el desarrollo agricola sostenible. *Agric. Am.*, **233**, 19–21.

Valencia, P.E., and Khachatourians, G.G. (1998) Integrated pest management and entomopathogenic fungal biotechnology in the Latin Americas I. Opportunities in a global agriculture. *Rev. Acad. Colomb. Cinc.*, **22**, 193–202.

Vandenberg, J.D., Shelton, A.M., Wilsey, W.T., and Ramos, M. (1998) Assessment of *Beauveria bassiana* sprays for control of diamondback moth (Lepidoptera: Plutellidae) on crucifers. *J. Econ. Entomol.*, **91**, 624–630.

Vega, F.E., Mercadier, G., Damon, A., and Kirk, A. (1999) Natural enemies of the coffee berry borer, *Hypothenemus hampei.* (Ferrarri) (Coleoptera: Scolytidae) in Togo and Cote d'Ivorie, and other insects associated with coffee beans. *Afric. Entomol.* **7**, 243–248.

Viaud, M., Couteaudier, Y., and Riba, G. (1998) Molecular analysis of hypervirulent somatic hybrids of the entomopathogenic fungi *Beauveria bassiana* and *Beauveria sulfurescens. Appl. Environ. Microbiol.*, **64**, 88–93.

Wraight, S.P., Carruthers, R.I., Bradley, C.A., Jaronski, S.T., Lacey, L.A., Wood, P. *et al.* (1998) Pathogenicity of the entomopathogenic fungi *Paecilomyces* spp. and *Beauveria bassiana* against the silverleaf whitefly, *Bemisia argentifolii. J. Invertebr. Pathol.*, **71**, 217–226.

Wright, J.E. (1993) Control of the boll weevil (Coleoptera: Curculionidae) with Naturalis-L: a mycoinsecticide. *J. Econ. Entomol.*, **86**, 1355–1358.

Xu, Q.F. (1988) Some problems about study and application of *Beauveria bassiana* against agricultural and forest pests in China. In Y.W. Li, Z.Z. Li, J.W. Wu, Z.K. Wu, and Q.F. Xu (eds.), *The Study and Application of Entomogenous Fungi in China*, Academic Periodical Press, Beijing, China, pp. 1–9.

Ying, F.W. (1992) Current situation of *Beauveria bassiana* for the control of the pine caterpillar and its prospects in China. *Proc. 19th Int. Cong. Entomol.* Beijing, China, pp. 300.

Zhang, A.I. (1992) The recent developments on the study of *Beauveria bassiana* in China. *Proc. 19th Int. Cong. Entomol.*, Beijing, China, pp. 268.

Entomopathogenic Nematodes and Insect Pest Management

8

Albrecht M. Koppenhöfer[1] and Harry K. Kaya[2]
[1]Department of Entomology, Rutgers University, New Brunswick, NJ 08901, USA
[2]Department of Nematology, University of California, Davis, CA 95616, USA

Introduction

Associations between nematodes and insects are very common with the relationships ranging from phoresis to parasitism and pathogenesis. The term "entomophilic nematodes" includes all these associations. "Entomogenous nematodes" are only those that have a facultative or obligate parasitic associations with insects. Parasitism by entomogenous nematodes can have various deleterious effects on their hosts including sterility, reduced fecundity, reduced longevity, reduced flight activity, delayed development, or other behavioral, physiological and morphological aberrations, and in some cases, rapid mortality. Parasitic associations with insects have been described from 23 nematode families. Seven of these families contain species that have potential for biological control of insects: Mermithidae and Tetradonematidae (Order: Stichosomida); Allantonematidae, Phaenopsitylenchidae, and Sphaerulariidae (Order: Tylenchida); Heterorhabditidae and Steinernematidae (Order: Rhabditida). With the exception of the latter two families and the tylenchid, *Deladenus siricidicola*, which has been successfully used for classical and inoculative control of woodwasps (Bedding, 1993), the microbial control potential of these nematodes is rather limited because of problems with their culture and/or limited virulence. This chapter will concentrate on the Heterorhabditidae and Steinernematidae. Detailed information on the other groups can be obtained from Poinar (1979), Nickle (1984, 1991), and Kaya and Stock (1997).

Nematodes in the families Heterorhabditidae and Steinernematidae are called entomopathogenic nematodes in reference to their ability to quickly kill hosts (1–4 d dependingon nematode and host species) thanks to their mutualistic association with bacteria in the genus *Xenorhabdus* for Steinernematidae and *Photorhabdus* for Heterorhabditidae. Because of their similarity in most aspects to be discussed in this chapter, they will be considered together. However, although both families belong to the same order, they are not closely related and have distinctly different reproductive strategies (Blaxter *et al.*, 1998). Several species from both genera are presently used as microbial insecticides and are produced commercially by various companies around the world. This chapter will provide a general background on entomopathogenic nematodes and their symbiotic bacteria but will focus on recent advances and current critical issues and their use in insect control. For more details about research until 1993, we refer the reader to books edited by Gaugler and Kaya (1990) and Bedding *et al.* (1993), a review by Kaya and Gaugler (1993), and an extensive bibliography compiled by Smith *et al.* (1992). For descriptions of techniques used with entomopathogenic nematodes, we refer to Kaya and Stock (1997), Koppenhöfer (2000), and Lacey and Kaya (2000).

Taxonomy and biology of the nematode/bacterium complex

Nematode taxonomy

Entomopathogenic nematodes have been recovered from soils throughout the world and are very common soil organisms (Hominick *et al.*, 1996). Numerous surveys continuously recover new isolates and many of these constitute new species. Presently, 33 species of entomopathogenic nematodes in two families and 3 genera are recognized (Table 8.1). Several taxonomic changes at the generic and species levels have been made, especially in the last 10 years, resulting in confusion in the literature. A recent publication from a meeting of most of the specialists in the area of entomopathogenic nematode taxonomy and systematics gives a synopsis of the current status, protocols, and definitions for the biosystematics of entomopathogenic nematodes (Hominick *et al.*, 1997).

Associated bacteria

Xenorhabdus and *Photorhabdus* spp. are motile, Gram-negative, facultatively anaerobic rods in the family Enterobacteriaceae. Presently, 8 species are recognized but most isolates from recently described entomopathogenic nematode species have yet to be determined (Table 8.1). Until recently all symbionts isolated from *Heterorhabditis* spp. were assigned to *P. luminescens*, but because of the high phenotypic, genetic, and physiological diversity, more recent research supported the division of *Photorhabdus* into relatedness groups (Akhurst *et al.*, 1996) or different species (Liu *et al.*, 1997; Han and Ehlers, 1998; Ehlers and Niemann, 1998) and Fischer-LeSaux *et al.* (1999) suggested 2 new *Photorhabdus* species.

Primary differences between the two genera include that most *Photorhabdus* isolates are able to luminesce while *Xenorhabdus* spp. do not luminesce and that *Photorhabdus* spp. are catalase positive whereas *Xenorhabdus* spp. are catalase negative. Both genera produce phenotypic variant forms. Phase I is the form naturally associated with the nematodes whereas phase II can arise spontaneously when the bacterial cultures are in the stationary non-growth stage. Differences occur between the two phases; phase I produces antibiotics, adsorbs certain dyes, and develops large intracellular inclusions composed of crystal proteins, whereas phase II does not adsorb the dyes, does not produce antibiotics, and forms intracellular inclusions inefficiently. Phase I and II have distinctly different colony morphologies. Differences in pathogenicity between the phases have been observed in some hosts (e.g. Volgyi *et al.*, 1998) but not other (e.g. Jackson *et al.*, 1995; Nishimura *et al.*, 1995). Phase I is claimed to be superior to phase II in its ability to support nematode propagation *in vitro*, although recent evidence suggests that this is not always the case (Ehlers *et al.*, 1990; Volgyi *et al.*, 1998). Reversion from phase II to phase I has only been documented with *Xenorhabdus* spp. For detailed information on the associated bacteria, see Akhurst and Boemare (1990), Boemare *et al.* (1996), Forst and Nealson (1996), and Forst *et al.* (1997).

Biology of nematode/bacterium complex

Steinernematidae and Heterorhabditidae are obligate pathogens in nature. The only stage that survives outside of a host is the non-feeding third stage infective juvenile (IJ) or dauer juvenile. The IJs carry cells of their bacterial symbiont in their intestinal tract. After locating a suitable host, the IJs invade it through natural openings (mouth, spiracles, anus) or thin areas of the host's cuticle (common only in Heterorhabditidae, see Wang and Gaugler, 1998) and penetrate into the host hemocoel. The IJs recover from their developmental arrestment, release the symbionts, and bacteria and nematodes cooperate to overcome the host's immune response. The bacteria propagate and produce substances that rapidly kill the host and protect the cadaver from colonization by other microorganisms. The nematodes start developing, feed on the bacteria and host tissues metabolized by the bacteria, and go through 1–3 generations. Depleting food resources in the host cadaver lead to the development of a new generation of IJs that emerges from the host cadaver in search of a new host. A major difference between Steinernematidae and Heterorhabditidae is that *Heterorhabditis* adults are hermaphrodites in the first generation but amphimictic in following generations whereas *Steinernema* adults are always amphimictic (Fig. 8.1).

Each nematode species is specifically associated with one symbiont species, although a symbiont species may be associated with more than one nematode species (Table 8.1). This specificity has been demonstrated to operate at 2 levels (Akhurst and Boemare, 1990). First, although the nematode can develop on other bacteria, best reproduction occurs on their natural symbiont. Second, natural symbiont cells are

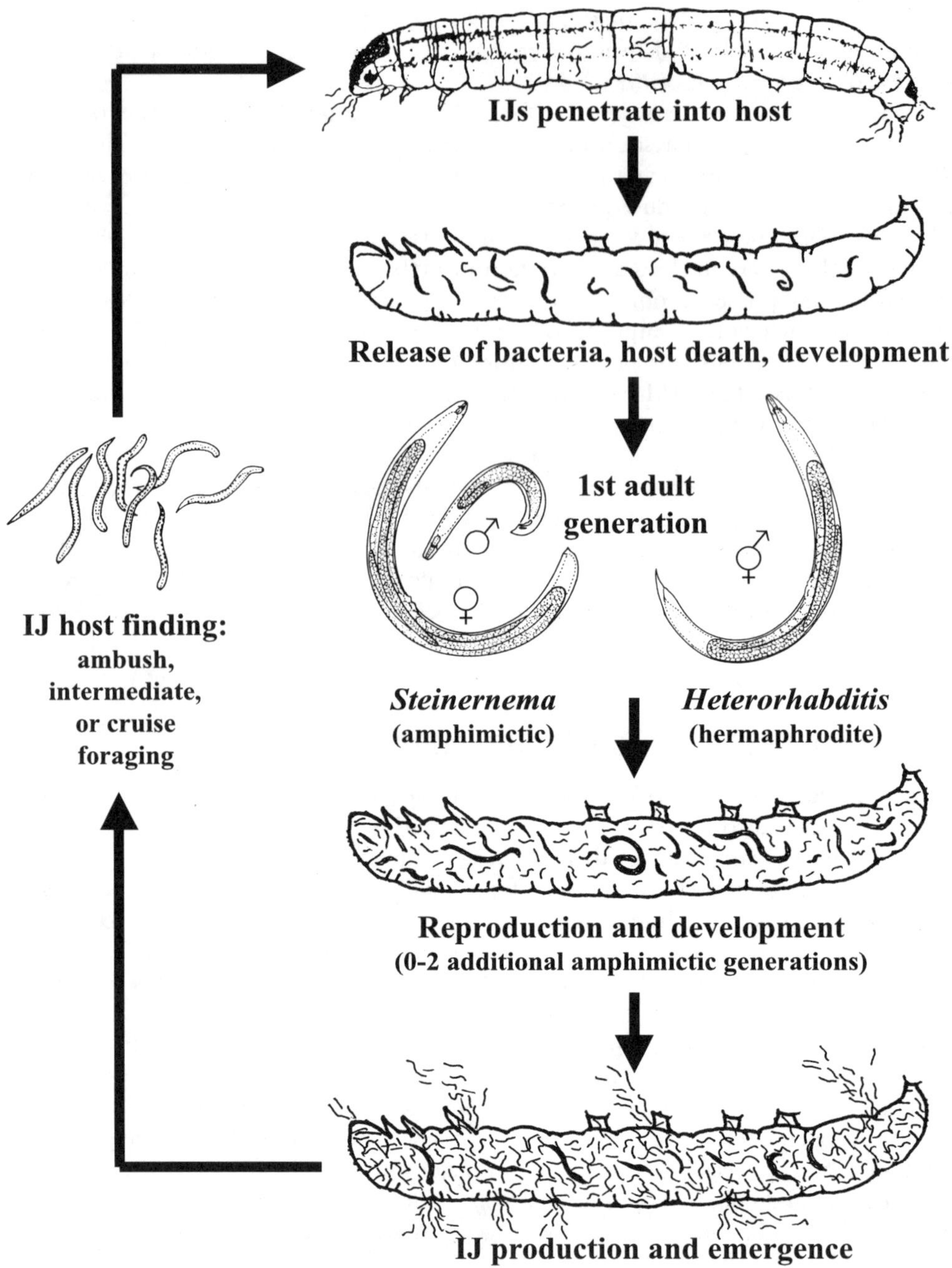

Figure 8.1 Life cycle of entomopathogenic nematodes (see text for detail).

Table 8.1 Described species of entomopathogenic nematode species and their respective symbiotic bacterial species

Nematode genus	*Nematode species*	*Symbiont species*
Steinernema[1]	*abbasi*	Undescribed
"	*affine*[2]	*Xenorhabdus bovienii*
"	*arenarium (=anomali)*[3]	*Xenorhabdus* sp.
"	*bicornutum*	Undescribed
"	*carpocapsae*	*X. nematophilus*
"	*caudatum*	Undescribed
"	*ceratophorum*	Undescribed
"	*cubanum*	*X. poinarii*
"	*feltiae (= bibionis)*	*X. bovienii*
"	*glaseri*	*X. poinarii*
"	*intermedium*	X. bovienii
"	*karii*	Undescribed
"	*kraussei*	*X. bovienii*
"	*kushidai*	*X. japonicus*
"	*longicaudum*	Undescribed
"	*monticolum*	Undescribed
"	*neocurtillae*	Undescribed
"	*oregonense*	Undescribed
"	*puertoricense*	Undescribed
"	*rarum*	*Xenorhabdus* sp.
"	*riobrave*	*Xenorhabdus* sp.
"	*ritteri*	*Xenorhabdus* sp.
"	*scapterisci*	*Xenorhabdus* sp.
"	*siamkayai*	*Xenorhabdus* sp.
"	Undescribed	*X. beddingii*
Neosteinernema	*longicurvicauda*	Undescribed
Heterorhabditis	*argentinensis*	*Photorhabdus luminescens*[5]
"	*bacteriophora (= heliothidis)*	*P. luminescens / P. temperata*[6]
"	*brevicaudis*	*P. luminescens*
"	*hawaiiensis*[4]	*P. luminescens*
"	*indica*	*P. luminescens*
"	*marelatus (= hepialus)*	*P. luminescens*
"	*megidis*	*P. temperata*
"	*zealandica*	*P. temperata*

[1]The species "*carpocapsae*" has been referred to as "*feltiae*" between 1983 and 1989. The name "*feltiae*" is valid and takes precedent over "*bibionis*". [2]Endings of some specific epithets in *Steinernema* were corrected to reflect that *–nema* is neuter gender. [3]In brackets previously used names. [4]Appears to be conspecific with *H. indica* (Adams *et al.*, 1998; Hashmi and Gaugler, 1998). [5]*P. luminescens* has been divided into several species (Fischer-LeSaux *et al.*, 1999); species not included in this study may be affected in the future. [6]Some strains of *H. bacteriophora* are associated with *P. luminescens* (Brecon, HP88), others with *P. temperata* (NC1).

retained better than cells of other bacteria. In this association, the nematode is dependent upon the bacterium for (a) quickly killing its insect host, (b) creating a suitable environment for its development by producing antibiotics that suppress competing secondary microorganisms, (c) and transforming the host tissues into a food source. The bacterium requires the nematode for (a) protection from the external environment, (b) penetration into the host's hemocoel, and (c) inhibition of the host's antibacterial proteins.

Host range

Many of the well-studied entomopathogenic nematode species (e.g. *S. carpocapsae*, *S. feltiae*, and *H. bacteriophora*) attack a wide spectrum of insects in the laboratory where host contact is assured, environmental conditions are optimal, and no ecological or behavioral barriers to infection exist. The range of insects infected by entomopathogenic nematodes after inundative field releases is considerably smaller, as evident from a long list of control failures. The natural host range of most species is even more restricted due to the ecology of the nematodes and their potential hosts as well as environmental factors (Peters, 1996). The effect of inundative applications of entomopathogenic nematodes on non-target organisms is negligible (Akhurst, 1990; Bathon, 1996; Boemare *et al.*, 1996), adding to their safety as biological control agents. Because isolation of new nematode strains/species is usually done using larvae of the greater wax moth, *Galleria mellonella* (Linnaeus), as bait insects, the host range of known species is likely to be biased towards generalists or species adapted to Lepidoptera. Some of those species that have been isolated from host cadavers in the field have a rather restricted host range. For example, *S. scapterisci*, is adapted to Orthoptera, especially mole crickets, and poorly performs in other insects (Grewal *et al.*, 1993; Parkman and Smart, 1996). *S. kushidai* appears to be so specialized to scarabaeid larvae (Mamiya, 1989) that it hardly infects and cannot reproduce in other hosts.

Nematode/bacterium interactions with hosts

The efficacy of entomopathogenic nematodes can vary with many biological factors, including nematode species and strain, as well as insect species and developmental stage (Eidt and Thurston, 1995; Simões and Rosa, 1996). Many soil-dwelling insects have developed behaviors that reduce host-finding, attachment, or penetration of IJs [e.g., high defecation rate to reduce infection via the anus (scarabaeid grubs), low CO_2 output or CO_2 release in bursts to minimize chemical cues (lepidopteran pupae, scarabaeid grubs), formation of impenetrable cocoons or soil cells before pupation (many Lepidoptera and Scarabaeidae), walling-off infected individuals to avoid contamination (termites), or aggressive grooming or evasion behavior (scarabaeid grubs) (Gaugler *et al.*, 1994)]. IJs can penetrate into insects using several different routes, depending on which routes are accessible and the specific stage of the insect (Eidt and Thurston, 1995). In some insects, however, the usual routes may be inaccessible. For example, the mouth may be blocked by oral filters (wireworms) or too narrow (insect with sucking/piercing mouthparts or small insects with chewing mouthparts), the anus may be normally constricted by muscles or other structures (wireworms), the spiracles may be covered with septa (wireworms) or sieve plates (scarabaeid grubs) or simply too narrow for passage (some Diptera and Lepidoptera).

To enter the body cavity, the IJs have to penetrate through natural barriers at some point, whether thin cuticle, tissues, cells, mucus, etc. To achieve this, the nematode may use physical force (e.g., rupture thin tracheae by body thrusting or, as in

Heterorhabditis, use a tooth situated terminally on their mouth). The IJs may also digest holes into the tissue with proteolytic secretions (AbuHatab *et al.*, 1993; Peters and Ehlers, 1994). In many insects, intersegmental membranes, fore- and hindgut cuticular linings, or the peritrophic membrane may be too thick or dense to allow for penetration into the hemocoel.

Once inside the body cavity, the nematodes and bacteria have to overcome the host's immune response (Simões and Rosa, 1996; Forst *et al.*, 1997). The insect non-self response system consists of interacting humoral and cellular factors. To eliminate the bacteria, the insects may use antibacterial proteins and/or phagocytosis followed by nodule formation; to eliminate the nematodes, the insects may encapsulate them followed by melanization. The nematodes are evidently capable of evading recognition as non-self as has been observed for *S. carpocapsae* in some insects including *G. mellonella* larvae. The mechanism of this avoidance is not well established but may involve the structure/chemistry of the nematode cuticle. *S. glaseri*, although initially recognized and encapsulated in larvae of the Japanese beetle *Popillia japonica* Newman, escapes from the capsules (Wang *et al.*, 1995). *Heterorhabditis* IJs are capable of avoiding encapsulation in tipulid larvae by exsheathing from the J2 cuticle during penetration (Peters *et al.*, 1997).

One mechanism that *Xenorhabdus* bacteria use to tolerate or evade the insects' humoral response is to inhibit the activation of the insect enzyme, prophenoloxidase, by excreting a lipopolysaccharide. The nematodes produce immuno-inhibiting factors that destroy the antibacterial factors produced by the insect, allowing the bacteria to produce insecticidal toxins that rapidly kill the host (Bowen *et al.*, 1998). Surface coat proteins of *S. glaseri* suppress the host immune response in *P. japonica* larvae and destroy hemocytes (Wang and Gaugler, 1998) but are ineffective in the house cricket *Acheta domesticus* (Linnaeus) (Wang *et al.*, 1994). Nematodes can also produce paralyzing exotoxins and cytotoxic and proteolytic extracellular enzymes. To protect the host resource from colonization by other microorganisms, the symbiotic bacteria produce antibiotic and antimycotic substances (Akhurst, 1993; Boemare *et al.*, 1996; Georgis and Kelly, 1997). The degree to which any of the above reactions by insect and/or nematode/bacterium occurs depends on the insect host and nematode/bacterium complex involved, thus contributing to the variable efficacy of entomopathogenic nematodes against different insects.

Ecology

Behavior

Matching the nematode's and the pest's biology and ecology are essential for successful pest control with entomopathogenic nematodes. This approach requires understanding the nematodes' infection process as it relates to host selection, search, attachment, and recognition. Although the infection process of entomopathogenic nematodes contains aspects of both predation and parasitism, nematodes require only one single prey which is usually orders of magnitude larger than the nematode. Concepts essential to predation theory including functional response, learning, search image, switching, encounter rate, attack rate, handling time, are not relevant.

One of the factors restricting nematode host range is the type of foraging behavior exhibited by the IJs. Entomopathogenic nematodes employ different foraging strategies and behaviors to locate and infect hosts (Gaugler *et al.*, 1997). Some species are widely searching foragers or cruisers that are characterized by a high motility and active distribution throughout the soil profile, ability to orientate to volatile host cues and switch to localized search after host contact. These species are well adapted to infecting sessile hosts (e.g., *S. glaseri, H. bacteriophora*) (Fig. 8.2). Other species are sit-and-wait strategists or ambushers characterized by low motility and tendency to stay near the soil surface, and lack of response to volatile and contact host cues unless presented in an appropriate sequence. Ambushers efficiently infect mobile host species near the soil surface (e.g., *S. carpocapsae, S. scapterisci*) (Fig. 8.2). Most known entomopathogenic nematode species appear to be situated somewhere along a continuum between these two extremes using an intermediate type of foraging strategy (e.g., *S. riobrave, S. feltiae*) (Campbell and Gaugler, 1997).

All entomopathogenic nematode species studied exhibit a behavior termed body-waving where 30–95 per cent of the body is raised off the substrate for a few seconds. Some nematode species can raise >95 per cent of their body off the substrate, standing on a bend in their tail and assuming a straight posture that can be maintained for extended periods of time with alternating periods of motionlessness and waving. This behavior is termed nictation. Typical ambushers nictate >70 per cent of their foraging time. Species with intermediate foraging strategies nictate less frequently and for shorter periods (e.g. *S. riobrave*) or cannot nictate (e.g. *S. feltiae*). All cruisers cannot nictate. IJs of nictating species are also capable of jumping. Directed jumping appears to be used for host attachment (only in typical ambushers); non-directed jumping may play a role in dispersal (Campbell and Kaya, 1999a,b).

Dispersal

Active IJ dispersal, although rather limited with up to 90 cm in both horizontal and vertical direction within 30 days (Kaya, 1990), gives entomopathogenic nematodes the ability to actively seek out hosts. Passive nematode dispersal by water, wind, phoresis, infected hosts, human activity, etc. can cover much greater distances and may account for their widespread distribution. Important factors that influence the motility of IJs are moisture, temperature, and soil texture. The most important factor is moisture because nematodes need a water film for effective propulsion. In soil, IJs move through the water film that coats the interstitial spaces. If this film becomes too thin (in dry soil) or the interstitial spaces are completely filled with water (in saturated soil), nematode movement can be restricted (Koppenhöfer *et al.*, 1995). Different nematode species/strains have different temperature optima and ranges (Griffin, 1993; Grewal *et al.*, 1994), but generally nematodes will become sluggish at low tempera-

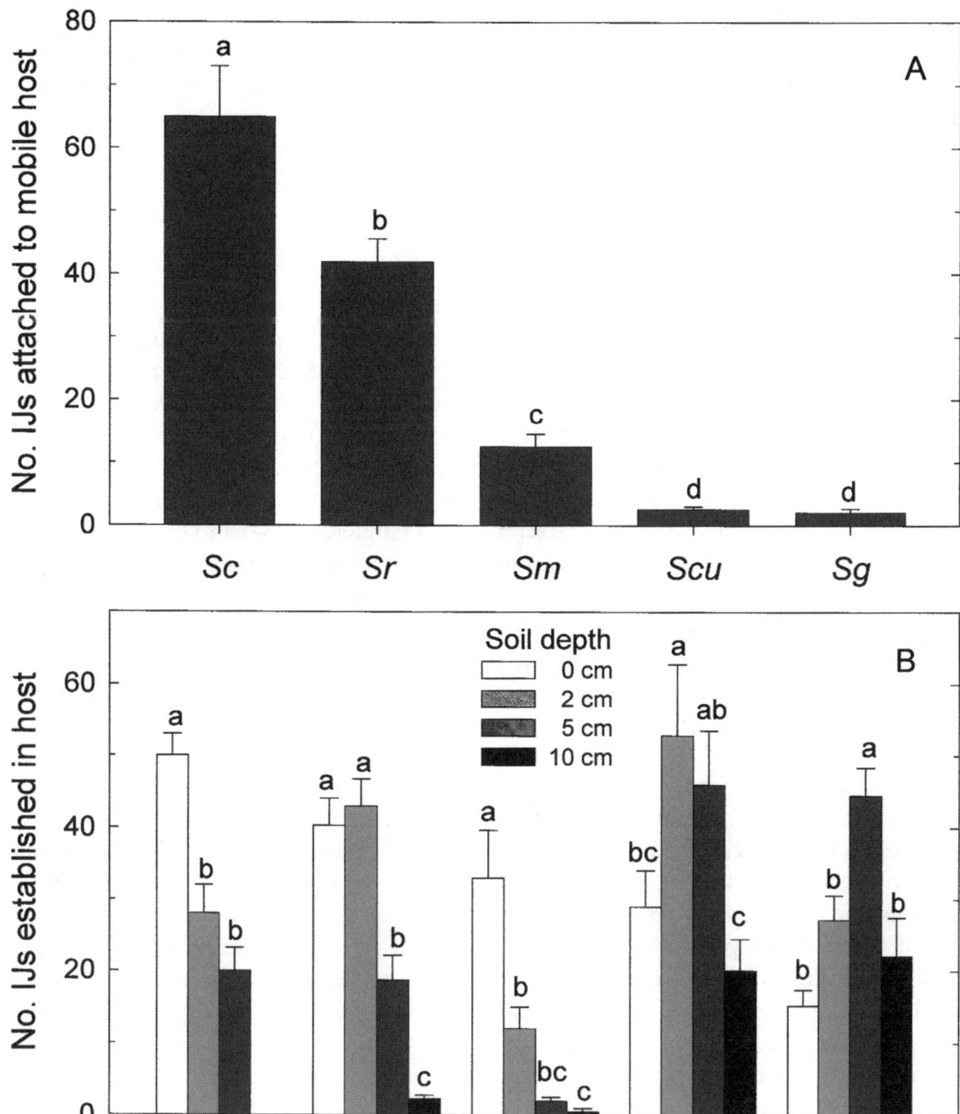

Figure 8.2 Ability of entomopathogenic nematode species with different foraging strategies (ambusher: *Sc* = *Steinernema carpocapsae*; intermediate: *Sr* = *S. riobrave*, *Sm* = *S. monticolum*; cruiser: *Scu* = *S. cubanum*, *Sg* = *S. glaseri*) to attach to a mobile host (A) and to infect hosts at different soil depths (B). Attachment to a mobile host was measured in a petri dish lined with sand-sprinkled filter paper treated with 1,000 infective juvenile nematodes (IJs). One wax moth larva was kept moving in the dish for 10 min by disturbance with a prod before IJs were rinsed off. Infection at different soil depths was measured in soil columns with one wax moth larva placed at 0, 2, 5, or 10 cm depth. The soil surface was treated with 1,000 IJs and the larvae were recovered and the number of nematodes established in them counted after 3 days. Means sharing the same letter are not significantly different among species (A) or within species (B).

tures (<10–15°C) and will be inactivated at higher temperatures (>30–40°C). Porosity or texture of soil affects nematode dispersal, with less dispersal occurring, as soil pores become smaller (Kaya, 1990). Soil salinity only affects IJ dispersal at extremely high levels (Thurston *et al.*, 1994).

Survival

Persistence of IJs is rather limited. When applied onto the soil surface, losses can reach 50 per cent within hours of application due primarily to UV radiation and desiccation (Smits, 1996). Even those IJs that settle in the soil suffer 5–10 per cent losses per day until usually only around 1 per cent of the original inoculum survive after 1–6 weeks. Because of this limited persistence, entomopathogenic nematodes should only be applied when susceptible stages of the target pest are present. Persistence of applied IJs is influenced by various intrinsic [e.g. behavioral, physiological (Womersley, 1993; Wright *et al.*, 1998), and genetic characteristics (Gaugler, 1993)] and extrinsic factors. This section will concentrate on the extrinsic factors including abiotic factors [extreme temperatures, soil moisture, osmotic stress, soil texture, RH, UV radiation (Kaya, 1990; Glazer, 1996; Smits, 1996)] and biotic factors [antibiosis, competition and natural enemies (Kaya and Koppenhöfer, 1996)].

Moisture is a central factor in IJ survival. IJs can survive desiccation to relatively low moisture levels if water removal is gradual giving them time to adapt to an inactive stage (Womersley, 1990). This is the case in natural soils. Due to the induced inactivity, IJs may actually persist longer in dry soil. On foliage and in other exposed habitats, nematode survival is generally a matter of minutes to hours unless the RH is close to 100 per cent. Nematodes may also avoid desiccation by remaining inside the host cadaver until moisture conditions improve (Brown and Gaugler, 1997; Koppenhöfer *et al.*, 1997).

The effect of temperature on nematode survival varies with nematode species and strains (Griffin, 1993; Grewal *et al.*, 1994). Extended exposure to temperatures below 0°C and above 40°C is lethal to most species of entomopathogenic nematodes but the effect depends on exposure time (Brown and Gaugler, 1996). In the soil environment, IJs are normally buffered from temperature extremes or have enough time to disperse into deeper soil layers where the buffering effect is stronger. For most species, the best longevity of IJs has been observed between 5 and 15°C. Higher temperatures will increase metabolic activity and depletion of energy reserves, and shorten lifespan.

UV light can inactivate and kill nematodes within minutes. Direct exposure to UV light (i.e., sunlight) has to be minimized by applying IJs early in the morning or evening, or using sufficient amounts of water to rinse the IJs into the soil.

Soil texture has an effect on IJ survival. Generally, nematode survival is lower in fine textured soils with the lowest survival in clay soil. The lower survival rate is probably related to lower oxygen levels in the smaller soil pores. Similarly, oxygen may become a limiting factor in water-saturated soils and soils with high contents of organic matter. The pH value of the soil does not have a strong effect on IJ survival. Thus, pH values between 4 and 8 do not vary in their effect on IJs, but at pH 10, IJ survival declines rapidly.

Soil salinity has only limited negative effects on entomopathogenic nematode survival even at salinities well above the tolerance levels of most crop plants (Thurston *et al.*, 1994). High NaCl has a negative effect on *H. bacteriophora* survival but not on *S. glaseri* survival. $CaCl_2$ and KCl have no effects on either nematode species. Seawater has no negative effects on survival of several *Heterorhabditis* species/strains (Griffin *et al.*, 1994) and, because of the frequent isolation of *Heterorhabditis* from sites near the ocean, Griffin *et al.* (1994) argue that these species might have been spread between landmasses by ocean currents.

The effects of various biotic factors on nematode survival have been extensively reviewed by Kaya and Koppenhöfer (1996). Antibiosis can occur when chemicals with adverse effects on the nematodes are released from roots in the soil affecting IJ host finding, or when such chemicals are present in an infected host and affect nematode infection and reproduction. Intraspecific competition may reduce nematode fitness when too many IJs infect one host. Although more than one species of *Steinernema* may be able to reproduce in one host cadaver, the prevalence of one nematode's bacterial symbiont will dramatically reduce the fitness of the other nematode's progeny and eventually lead to its local extinction. By having different foraging strategies and/or host specificities, different nematode species can coexist in the same habitat. Interspecific competition may also occur with other insect pathogens, especially if they are applied in the same location as the nematodes. The outcome of the competition will depend on the kind of competitor (e.g. entomopathogenic fungi, bacteria, or viruses), the timing of infection, and environmental factors such as temperature or soil moisture. Among the natural enemies of nematodes, nematophagous fungi are the best studied. Other natural enemies include invertebrate predators such as collembolans, mites, tardigrades and predatory nematodes. These natural enemies reduce IJ populations in soil in laboratory experiments, but their impact under field conditions is poorly understood. Finally, entomopathogenic nematodes are also susceptible to "predation" of nematode-killed insects by scavengers (Baur *et al.*, 1998).

Infectivity

The infectivity of IJs is influenced by many factors including those that affect their dispersal and survival. Obviously, IJs that cannot disperse because of unfavorable moisture or temperature conditions will also not be able to infect a host (Fig. 8.3). However, even under apparently optimal laboratory conditions, only a portion of inoculated nematodes (usually <40 per cent) can be recovered from susceptible hosts exposed to them. Because this phenomenon has been observed for many different species under various experimental conditions, it led to the wide acceptance of the "phased infectivity hypothesis" in the literature (e.g. Kaya and Gaugler, 1993; Griffin, 1996). Specifically, Hominick and Reid (1990) stated that ". . . an effective survival strategy might be for infectivity to be phased over time. Thus, upon emergence from a host, some individuals may be immediately infectious, while others become dormant for a time."

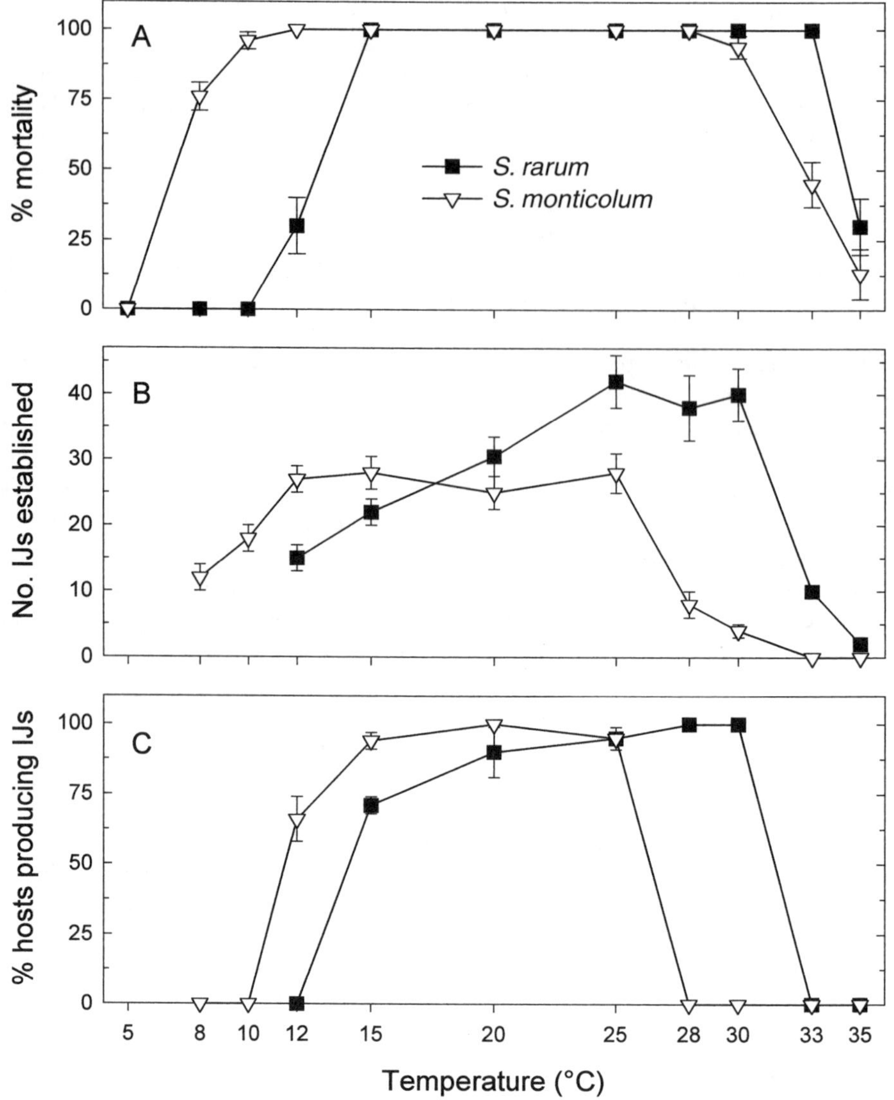

Figure 8.3 Effect of temperature on mortality (A), time until death (B), and percentage of hosts producing nematode progeny (C) in wax moth larvae exposed to 50 infective juveniles of *Steinernema rarum* and *S. monticolum*.

A population of IJs emerging from a host cadaver can be divided into an infectious portion and a non-infectious portion. If the "phased-infectivity-hypothesis" is correct, the non-infectious portion can be further divided into a permanently non-infectious and a temporarily non-infectious portion. While there is considerable evidence that temporary non-infectiousness can be induced in already emerged IJs by environmental factors such as low moisture (Kung *et al.*, 1991) and low temperature (e.g. Griffin,

1996; Brown and Gaugler, 1996), experimental evidence for an innately non-infectious proportion is limited to one strain of *S. feltiae* (Bohan and Hominick, 1996, 1997a). Recent research suggests that "phased infectivity" may only occur under specific conditions and/or in a small proportion of the IJ population of some species (Campbell *et al.*, 1999).

The infection process may be influenced by many different factors. Hay and Fenlon (1995) observed that initial *S. feltiae* infection facilitated secondary infection by additional IJs in sciarid larvae and hypothesized that a part of the IJ population would only penetrate into hosts that had already been infected. In *S. feltiae*, females appear to infect hosts faster than males (Bohan and Hominick, 1997b; Renn, 1998a). Male IJs in *S. arenarium*, *S. carpocapsae*, *S. scapterisci*, and *S. glaseri*, but not in *S. feltiae*, tend to disperse faster and respond stronger to volatile host cues than female IJs, and infect hosts faster than females in sand columns (Grewal *et al.*, 1993). Stuart *et al.* (1998), however, found no differences between male and female invasion timing for *S. glaseri* using wider sand columns. The tendency of *S. glaseri* males to emerge earlier from hosts may still allow them to find and infect hosts before females do (Lewis and Gaugler, 1994). Other findings suggest that infected hosts release a substance within hours of infection that reduces subsequent infection by conspecific IJs. This was observed in different lepidopteran hosts and for 4 *Steinernema* species (Glazer, 1997; Ishibashi and Wang, 1998).

Recycling of nematodes

As obligate pathogens, natural populations of entomopathogenic nematodes have to recycle in their hosts, but only few studies have examined the dynamics of nematode populations and the factors affecting them. Within-site distribution of nematode populations is patchy (Stuart and Gaugler, 1994; Campbell *et al.*, 1997, 1998; Strong *et al.*, 1996) and may depend on various biotic and abiotic factors including seasonal fluctuations, foraging strategy of the IJs, host population dynamics, alternate hosts, etc. It is likely that entomopathogenic nematode populations persist as metapopulations which exhibit a "shifting mosaic" type of dynamics with asynchronous fluctuations and little migration between patches (Levins, 1970). Patches are highly vulnerable to extinction, and in order to persist as a metapopulation, the founding rate of local populations has to be the same as the extinction rate (Lewis *et al.*, 1998).

Recycling is also very desirable after inundative applications of entomopathogenic nematodes because it can provide additional and prolonged control of a pest. Numerous studies have provided evidence for nematode recycling in the soil environment after inundative releases (Kaya, 1990; Klein, 1993). Although recycling probably is quite common, it is not clear what factors influence its occurrence. Most of the abiotic and biotic factors that influence persistence, infectivity, and motility of individual IJs, also influence nematode recycling. Some of these factors may even be more stringent for recycling than for persistence and infection. For example, the temperature range for successful reproduction inside a host cadaver is narrower than that for infection (Grewal *et al.*, 1994; Fig. 8.3).

Genetics

Genetic improvement

The growing knowledge of the biology and ecology of entomopathogenic nematodes and their symbiotic bacteria has increased interest in improving their beneficial traits or eliminating weaknesses by means of genetic manipulation. The bases for genetic improvement is understanding the genetic traits relevant to efficacy. While several genes of the symbiotic bacteria have already been cloned and characterized, genetic characterization of the nematode is lagging behind. The necessary genetic foundation has to be acquired by isolating, characterizing, and mapping morphological and movement mutants, before proceeding to genetic dissection to understand the biological basis of traits. The sequencing of the genome of the nematode *Caenorhabditis elegans* has recently been completed (Blaxter, 1998), and this closely related nematode could serve as a useful model for improving the understanding of entomopathogenic nematode genetics.

The main targets for genetic improvement in entomopathogenic nematodes are increased efficacy, resistance to environmental extremes, development of anhydrobiotic strains, and increased suitability of *Heterorhabditis* for culture in the liquid fermentation process (Burnell and Dowds, 1996). A starting point for genetic research should be the screening of natural isolates that may contain many desirable attributes that can be transferred into commercial strains by cross hybridization as has been demonstrated for heat tolerance (Shapiro *et al.*, 1997a). Because many of the complex behavioral and physiological traits are likely to result from the interaction of many genes, an effective means of genetically improving such traits is by selective breeding. Examples of successful selective breeding include the selection for cold tolerance (Griffin and Downes, 1994; Grewal *et al.*, 1996), improved control efficacy (Tomalak, 1994), and nematicide resistance (Glazer *et al.*, 1997). In situations where key regulatory genes may control the expression of several genes, mutagenesis may be appropriate. Mutagenesis has been used to isolate desiccation tolerant strains (O'Leary and Burnell, 1997).

Recently, genetic engineering has been adopted for the improvement of beneficial traits of entomopathogenic nematodes. Conferring commercial rather than an ecological advantages to the nematodes may be the best way to improve nematode performance while limiting regulatory problems for field releases and commercialization. Using microinjection and microprobes (Hashmi *et al.*, 1995a,b), plasmid containing heat-shock protein genes from *C. elegans* have been introduced into *H. bacteriophora*. The resulting transgenic strain has a higher tolerance to short temperature spikes (a deleterious condition often encountered during transportation). Field trials showed no increased persistence of the transgenic strain compared to the wild-type (Wilson *et al.*, 1999).

The main targets for genetic improvement of the bacterial symbionts are pathogenicity, host specificity, symbiont specificity, resistance to environmental extremes, and control of phase variation (Burnell and Dowds, 1996). Genes of these bacteria such as outer membrane protein genes, low-temperature induced genes, maltose metabolism genes, lux genes, extracellular enzyme genes, and crystalline protein genes have already been cloned (Forst *et al.*, 1997). Cloned genes can be easily identified in *Escherischia coli* where they are expressed from their own promoters, followed by the production of active gene products and recognizable secretion signals. DNA transfer in the symbiotic bacteria has been achieved by transformation, conjugation, and transduction, using *E. coli* plasmids, mobilizable plasmids, and phageλ, respectively; transposon banks have been constructed (Burnell and Dowds, 1996; Forst *et al.*, 1997). However, methodologies used with one strain are often ineffective with other strains, and DNA uptake by both bacterial genera is poor.

Laboratory colonization

A central issue in biological control is that genetic variation for attributes affecting natural enemy success is either missed during collection or lost during importation and rearing (Roush, 1990a,b). The three topics pertaining to preservation of genetic variation are founder effect, inbreeding, and inadvertent selection (Stuart and Gaugler, 1996). Founder effect is a serious problem for entomopathogenic nematode rearing because many nematode strains are isolated from a limited number of insect cadavers at single geographical sites resulting in loss of genetic variance. A way of maintaining genetic diversity is to collect from as many geographical sites as possible and hybridize the isolates. Repositories for entomopathogenic nematodes as proposed by Hominick and Reid (1990), wherein hundreds of identified isolates are stored and available, would facilitate the development of hybrid strains. Caution has been expressed that hybridization could break up co-adapted gene complexes, resulting in reduced host adaptation. This seems more likely for nematodes with narrow host ranges.

Inbreeding depression during the rearing phase tends to be a problem in small populations and should not be a major issue for entomopathogenic nematodes which are reared on large scales even under laboratory conditions (Stuart and Gaugler, 1996). The relatively low dispersal capability of nematodes suggests that inbreeding is not uncommon in natural populations, and any deleterious recessive alleles may have already been eliminated by natural selection.

Inadvertent selection or laboratory adaptation can lead to the loss of field-adapted alleles important for biological control. The potential for inadvertent selection is very significant for entomopathogenic nematodes because they are reared in large populations, especially when reared *in vitro* (Stuart and Gaugler, 1996). Inadvertent selection can be reduced by minimizing generation turnover. The development of liquid nitrogen procedures allows for rapid storage from field isolation to a stable stock inoculum in a few generations, without risk of subsequent laboratory adaptation. However, many of the widely used strains are already likely to be laboratory adapted, and strong consideration should be given to correcting any laboratory adaptation by outcrossing with field material. Researchers should also give consideration to inadvertent selection, inasmuch as they may be studying laboratory rather than field-adapted behaviors.

Commercialization

Mass production

Entomopathogenic nematodes can be mass produced *in vivo* and *in vitro* in three-dimensional solid media or liquid fermentation (Ehlers, 1996; Grewal and Georgis, 1998). Advantages of the solid media process are that capital costs are low, limited expertise is required, and the logistics of production are flexible. Because of limited economics of scale, this method is mostly feasible for countries with low labor costs. Conversely, for the liquid fermentation process the proportion of labor and capital costs decreases in upscale, while operating cost increase. Liquid fermentation allows the lowest cost mass-production of entomopathogenic nematodes and is the method of choice for industrialized countries. Nematodes that have been successfully produced in 7500–80,000 liter fermenters include *S. carpocapsae, S. riobrave, S. feltiae, S. glaseri, S. scapterisci, H. bacteriophora*, and *H. megidis* with yield capacity as high as 250,000 IJs/ml depending on the nematode species. However, *Heterorhabditis* liquid culture has still unstable yields and prolonged process time due to the variable recovery of the IJs inoculated into the cultures and the inability of the amphimictic adults to mate under liquid culture conditions (Ehlers *et al.*, 1998; Strauch and Ehlers, 1998).

During the past few years, a distinct cottage industry has emerged that produces entomopathogenic nematodes mostly *in vivo*, especially for the home lawn and garden markets. Although the *in vivo* process lacks any economy of scale and is increasingly sensitive to problems with increasing scale of production, it requires minimal expertise and capital investments and may be an important future sector in nematode commercialization for specific niche markets.

Formulations/storage

IJs can be stored in aqueous suspension at 4–15°C (depending on nematode species) without much loss of activity for 6–12 months for *Steinernema* species and 3–6 months for *Heterorhabditis* species. At higher temperatures, storage life is considerably shorter. Many commercial nematode products are still formulated on moist substrates (e.g. sponge, vermiculite, aqueous suspensions) and require continuous refrigeration to maintain nematode quality for extended periods. To improve IJ shelf-life and resistance to temperature extremes, a large number of formulations have been developed including alginate, clays, activated charcoals, polyacrylamide, and water dispersible granules (Georgis and Kaya, 1998; Grewal and Georgis, 1998). These formulations reduce IJ metabolism by immobilization or partial desiccation. Optimal formulations differ for the various nematode species because of their specific requirements for moisture and oxygen. Presently, the most promising formulations are water dispersible granules that combine long nematode shelf-life without refrigeration (6 months at 4–25°C; 2 months at 30°C) with ease of handling. The partially desiccated IJs rehydrate after application to a moist environment such as soil. However, to achieve optimal infectivity the IJs need to rehydrate for up to 3 d in soil (Baur *et al.*, 1997b). In desiccating environments like foliage, rehydration has to occur before application making this formulation impractical for these situations.

Regulations

Regulations concerning the use of entomopathogenic nematodes for insect control vary among countries (Richardson, 1996; Bedding *et al.*, 1996; Rizvi *et al.*, 1996). The nematodes are exempted from registration in many countries, e.g. Australia, Denmark, Germany, the Netherlands, Spain, UK, or USA, but in other countries they are subject to similar registration procedures as a chemical pesticide, e.g. Austria, Belgium, Hungary, Japan, Switzerland, and New Zealand. The importation and use of non-indigenous or exotic nematode species are subject to strict regulations in most countries, as is the case for transgenic nematodes and their symbiotic bacteria. However, some countries already consider foreign strains of endemic species to be exotic. This can be a major obstacle for the commercialization of entomopathogenic nematodes that, due to the enormous developmental costs involved, usually concentrates on few species and strains thereof.

Regulatory bodies have to be aware of their responsibility for the future development of safe control agents such as entomopathogenic nematodes by developing realistic and coordinated guidelines. The current volume of markets does not justify the cost of registration procedures currently required for chemical or microbial insecticides. A workshop with 15 expert participants from 10 countries came to the following conclusions and recommendations (Ehlers and Hokkanen, 1996) that should form a strong base for future regulatory decisions concerning the use of entomopathogenic nematodes. First, because of biological and ecological features that make entomopathogenic nematodes exceptionally safe for use in biological control, they should not be subject to any kind of registration. Second, the introduction of non-indigenous nematode species should be regulated at the species level.

Efficacy

Key target pests

Entomopathogenic nematodes have been tested against a large number of insect pest species with results varying from no effect to excellent control (Klein, 1990; Begley, 1990; Bedding *et al.*, 1993). Many factors can influence the success of nematode applications. Numerous failures can be attributed to poor understanding of the nematodes' (and pests') ecology. Matching biology and ecology of both the nematode and the target pest are of great importance if nematode applications are to result in significant pest reductions. Among the factors that need to be considered are the foraging behavior and temperature requirements of a nematode species, as well as the pest's accessibility in its habitat to nematodes and its suitability as a host.

Table 8.2 Target pests for commercially available entomopathogenic nematodes[1]

Pest group	*Common name*	*Life-stage*[2]	*Application site*	*Commodity*	*Nematode sp.*[3]
BLATTODEA					
Blattellidae	German cockroach	A/N	Baits	Apartments/structures	Sc
COLEOPTERA					
Chrysomelidae	Flea beetles	L	Soil	Mint, potato, sweet potato, sugar beets, vegetables	Sc
	Rootworms	L	Soil	Corn, peanuts, vegetables	Sc, Sr
Curculionidae	Billbugs	L	Turf	Turf	Sc, Hb
	Root weevils	L	Soil	Berries, citrus, forest seedlings, hops, mint, Ornamentals, sweet potato,sugar beet, banana[4]	Sc, Hb, Hm, Sr
Scarabaeidae	White grubs	L	Soil, turf	Berries, field crops, ornamentals, turf	Hb, Sg, Hm
DIPTERA					
Agromyzidae	Leaf miners	L	Foliage	Ornamentals, vegetables	Sc
Ephydridae	Shore flies	L	Soil	Ornamentals, vegetables	Sf
Sciaridae	Sciarid flies	L	Soil	Mushrooms, ornamentals, vegetables	Sf
Muscidae	Filth flies	A	Baits	Animal rearing facilities	Sf, Hb
Tipulidae	Crane flies	L	Turf, soil	Turf, ornamentals	Sc, Hm
LEPIDOPTERA					
Carposinidae	Peach borer moth	L	Soil	Apple	Sc
Noctuidae	Cutworms, armyworms	L/P	Soil, turf	Corn, cotton, peanuts, turf, vegetables	Sc
Cossidae	Carpenter worms	L	Cryptic	Ornamentals, shrubs	Sc
	Leopard moth	L	Cryptic	Apple, pear	Sc
Pyralidae	Webworms	L	Soil, turf	Cranberries, ornamentals, turf	Sc
Sesiidae	Crown borers	L	Cryptic	Berries	Sc
	Stem borers	L	Cryptic	Cucurbits, ornamentals, fruit trees	Sc
ORTHOPTERA					
Gryllotalpidae	Mole crickets	A/N	Turf, soil	Turf, vegetables	Ss, Sr
SIPHONAPTERA					
Pulcidae	Cat fleas	L/P	Soil, turf	Pet/vet	Sc
NEMATODA	Plant-parasitic nematodes		Turf	Turf	Sr

[1] After Georgis and Manweiler (1994), Georgis *et al.* (1995), Klein (1990), Begley (1990).
[2] L = larva; P = pupa; N = nymph; A = adult.
[3] Sc = *S. carpocapsae*; Sf = *S. feltiae*; Sr = *S. riobrave*; Ss = *S. scapterisci*; Hb = *H. bacteriophora*; Hm = *H. megidis*.
[4] Applied to residual rhizomes in a cryptic habitat (see text).

Entomopathogenic nematodes have been applied most successfully in habitats that provide protection from environmental extremes, especially in soil, their natural habitat, and in cryptic habitats. Excellent control has been achieved against insects that bore into plants, probably because their cryptic habitats contain less limiting factors (e.g. natural enemies of the nematodes) than soil and can more easily be manipulated to the nematodes' advantage. Low or highly variable efficacy has been achieved in manure because of high temperatures in animal rearing facilities and toxic effects of manure contents (ammonia) on IJs. Control attempts in aquatic habitats have been unsuccessful because the nematodes are not adapted to directed motility in this environment. Finally, on foliage and similar habitats the IJs are exposed to detrimental conditions that can be only marginally remedied by adjuvants. Table 8.2 provides a list of insect pests along with the commodities in which they have been successfully controlled with entomopathogenic nematodes.

Entomopathogenic nematodes can also have indirect effects on other soil organisms that can be exploited for pest control. Numerous laboratory and greenhouse studies have shown that inundative applications of several entomopathogenic nematodes have a suppressive effect on populations of various plant-parasitic nematodes (e.g. Bird and Bird, 1986; Gouge *et al.*, 1994). Grewal *et al.* (1997) showed that *S. carpocapsae* and especially *S. riobrave* can be as effective as chemical nematicides in the suppression of root-knot, sting, and ring nematodes in turfgrass at economic application rates. *S. riobrave* is now marketed for control of plant-parasitic nematodes in turfgrass. Originally, it was hypothesized that the suppressive effect of the entomopathogenic nematodes on the plant-parasitic nematodes was because of competition for spaces at the plant roots or due to an increase in nematode antagonists after the inundative application of the entomopathogenic nematodes. Recent studies indicate that allelopathic interactions occur between the plant-parasitic and the entomopathogenic nematodes (Grewal *et al.*, 1999). Root-penetration of root-knot nematode, *Meloidogyne incognita*, juveniles is temporarily reduced due to the repellent effect of metabolites of the entomopathogenic nematodes' symbiotic bacteria; living entomopathogenic nematodes IJs have no suppressive effects. Long-term effects are reduced egg production and egg hatch in *M. incognita* females exposed as juveniles to soil treated with entomopathogenic nematodes. *M. incognita* females inside the plant roots are not affected.

Application strategies

Releases of entomopathogenic nematodes have almost exclusively used the inundative approach where high numbers of IJs are released in a uniform distribution and control of pest populations is expected to be fast and thorough. However, their limited shelf life, susceptibility to environmental extremes, high price, etc. often make nematodes, as most other biologicals, poorly fit for an approach following the chemical pesticide paradigm. Other approaches including inoculative and augmentative releases, and conservation and management of endemic nematode populations need to receive considerably more attention in the future, as they may be more promising and feasible in many pest situations.

Inoculative release of entomopathogenic nematodes, i.e. the release of relatively small numbers of IJs with the expectation that they establish new populations for long-term pest suppression, has only been attempted a few times and little is known about the optimal approach to this strategy. *Steinernema glaseri*, isolated originally from scarabaeid larvae, was released in a massive inoculative control program in New Jersey from 1939 to 1942 against the Japanese beetle, an introduced pest. Although reisolated from southern New Jersey (Gaugler *et al.*, 1992), the elimination of bacterial symbionts by the use of antimicrobials in the *in vitro* rearing procedure, and possibly poor climatic adaptation of this neotropical nematode limited the success of this program. More recently, *S. scapterisci* originally isolated from southeast South America was successfully introduced into Florida for the classical biological control of mole cricket pests (Parkman and Smart, 1996). The nematode established successfully after treatment of 50 m^2 plots in pastures with either IJs (2×10^9/ha) or nematode-infected mole crickets (4 cadavers/m^2) or after release of mole crickets that had been exposed to nematodes in mole cricket sound traps.

For successful inoculative releases of entomopathogenic nematodes, long-term, multi-generational survival and recycling of the nematode populations are essential. To achieve this goal, several conditions are important including (i) presence of moderately susceptible insect hosts throughout most of the year, (ii) high economic threshold level of the target insect pests, and (iii) soil conditions favorable for nematode survival (Kaya, 1990). The optimal release method for inoculative releases of entomopathogenic nematodes may depend on the systems into which they are released, i.e. spatial and temporal distribution and susceptibility of target hosts and potential alternative hosts, seasonal fluctuations in other biotic and abiotic factors, etc.

Periodic augmentative releases into established nematode populations, and/or management of the susceptibility of the host/pest populations (for example using stressors such as other control agents) are two other approaches that may be used to boost or manage established nematode populations and warrant more attention. The possibilities and requirements for using entomopathogenic nematodes in a conservation approach of biological control have been extensively discussed by Lewis *et al.* (1998).

Application methods

The most commonly used application method for entomopathogenic nematodes is spraying directly onto the soil (or other) surface. Nematodes can be applied with most commercially available spray equipment including hand or ground sprayers, mist blowers, and aerial equipment on helicopters (Georgis *et al.*, 1995). The IJs can withstand pressures up to 1068 kPa and pass through all common nozzle type sprayers with openings as small as 100 μm in diameter. However, the screens in the nozzles should be removed to minimize damage to the IJs. Nematodes can also be delivered via irrigation systems including drip, microjet, sprinkler, and furrow irrigation (Georgis *et al.*, 1995; Cabanillas and Raulston, 1996). Post-application (in the case of dry soil also pre-application) irrigation as well as continued moderate soil moisture are essential for good nematode performance. When water is limited, subsurface injection of nematodes appears to be an efficient delivery way (Klein, 1993).

Boring insects have been successfully controlled by injecting nematode suspensions directly into the borer holes or blocking the holes with sponges soaked with nematode suspensions (tree borers: Huaiwen *et al.*, 1993), or adding nematode suspension to insect-attracting cuts in residual rhizomes (adult banana weevils: Treverrow and Bedding, 1993). Baits containing IJs can offer a cost-effective way of controlling mobile insects when trap stations are used that ensured intimate IJ-pest contact and protected the IJs from light and desiccation (housefly adults: Renn, 1998b; German cockroach: Appel *et al.*, 1993).

The detrimental effects of environmental extremes often can be alleviated by the addition of adjuvants to the nematode formulation/suspension. Because nematodes are especially affected by desiccation and UV radiation after foliar applications, adjuvants have been used to improve nematode performance against foliage-feeding pests. Solar radiation can be filtered with stilbene brighteners, especially Blankophor BBH (Nickle and Shapiro, 1994; Baur *et al.*, 1997a). Effective antidesiccants can be TX7719, Rodspray oil, and Nufilm P (Baur *et al.*, 1997a), Folicote (Glazer *et al.*, 1992), or glycerin (Broadbent and Olthof, 1995). Surfactants such as Silwett L-77, Kinetic, or dish detergents may also improve nematode speed of penetration into soil (Schroeder and Sieburth, 1997), but further studies are necessary to determine the mechanism of this interaction and whether these combinations are feasible under field conditions.

Effects of Agrochemicals and Other IPM Components

Entomopathogenic nematodes are usually applied to systems/substrates that are regularly treated with many other agents, including chemical or botanical pesticides, bioinsecticides, soil amendments, and fertilizers. Depending on the agents, application timing, physico-chemical characters of the system, etc., the nematodes may or may not interact with these other agents, with interactions ranging from antagonistic to synergistic.

Entomopathogenic nematodes appear to be compatible with many herbicides, fungicides, acaricides, insecticides, nematicides (e.g. Rovesti and Deseö, 1990; Ishibashi, 1993; Georgis and Kaya, 1998), azadirachtin (Stark, 1996), *Bacillus thuringiensis* products (Kaya *et al.*, 1995), and pesticidal soap (Kaya *et al.*, 1995). However, many other pesticides have limited to strong toxic effects on IJs (e.g. Patel and Wright, 1996; Rovesti and Deseö, 1990). On the other hand, synergistic interactions between entomopathogenic nematodes and other control agents has been observed for various insecticides (e.g. Koppenhöfer and Kaya, 1998; Nishimatsu and Jackson, 1998) and pathogens (Thurston *et al.*, 1994; Koppenhöfer *et al.*, 1999) (Fig. 8.4). In view of the diversity of available chemical and biorational insecticides, a generalization on pesticide-nematode compatibility cannot be made.

Inorganic fertilizer may be compatible with nematodes for short-term inundative pest control (Bednarek and Gaugler, 1997). Similarly, composted manure or urea does not have negative effects on nematode virulence but fresh manure does (Shapiro *et al.*, 1997b). Natural nematode populations, on the other hand, have been negatively affected by inorganic fertilizers, but positively affected by organic manure (Bednarek and Gaugler, 1997).

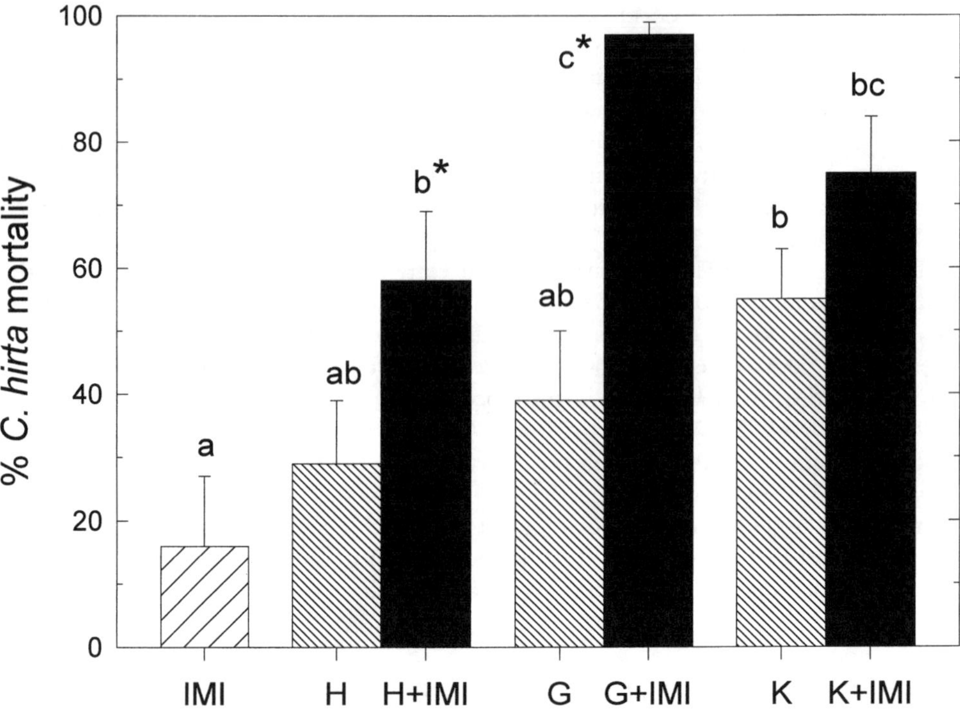

Figure 8.4 Effect of treatment with the chloronicotinyl insecticide imidacloprid (IMI) (200 g AI/ha), the entomopathogenic nematodes *Heterorhabditis bacteriophora* (H), *Steinernema glaseri* (G), or *S. kushidai* (K) (each at 0.4×10^9 infective juveniles/ha), or their combination with imidacloprid on mortality (mean of 7 replicates + SE) of third-instar *Cyclocephala hirta* (Coleoptera: Scarabaeidae) (6 per pot) in pots with grass. Pots were destructively sampled 14 days after application. Means sharing the same letter are not significantly different ($P < 0.05$). Significant synergistic interactions between nematode and imidacloprid are indicated by *.

Conclusions

The history of entomopathogenic nematode use in insect control dates to the pioneering research of R.W. Glaser and his coworkers in the 1930s and early 1940s (Gaugler and Kaya, 1990). With the emergence of inexpensive chemical insecticides after World War II, research and use of these nematodes were curtailed. In the 1960s, the documentation of problems associated with chemical pesticides resulted in a resurgence of biological control alternatives for pest management; and by the early 1980s, entomopathogenic nematodes became a major force in microbial control, being second only to *B. thuringiensis*. However, the efficacy of nematodes against insect pests was mixed. The reason for success or lack of success of controlling insect pests, particularly in the soil environment, were often unknown, emphasizing the need to obtain basic information on the biology, behaviour, ecology and genetics of these nematodes. Indeed, the 1990s have resulted in a surge of more basic information and the discovery and description of many more species. Recent research in behavioural ecology has clearly demonstrated that these nematodes are not generalist pathogens and are adapted to hosts in a particular environment depending on their foraging strategy. That is, their behaviour restricts much of their activity to a certain soil stratum eliminating many insects from infection. Understanding their behavioural patterns and genetics will allow us to match the right nematode species with the insect pests and enhance their use and production for insect control in the field. Another major advancement has been the molecular engineering of a heat-shock protein into the nematodes that offers the possibility of extended shelf life. Insertion of other genes, for example desiccation tolerance, may also extend shelf-life and/or survival in the field. Although the transgenic nematodes with the heat-shock protein gene have been field tested with no adverse effect, environmental risk assessments are needed for each new transgenic nematode.

These nematodes are highly successful for they are ubiquitous in nature. Yet, their populations usually are not sufficiently high to cause epizootics and reduce pest populations. Further research in the field is needed to better understand the factors that regulate their populations. The survival of the infective juveniles, for example, can be affected by abiotic (UV light, temperature extremes) and biotic (predators and pathogens) and "predation" nematode-killed insects by scavengers, but the effects of these mortality factors on long-term insect control have not been studied in detail in the field. As we learn more about nematode survival, their use as more effective and selective inundative agents against a number of insect pests and as inoculative agents for classical biological control can be fully realized. In some instances, stressors can be used as a pest management strategy against an insect pest to increase its susceptibility to the nematode. This approach has been highly successful against white grubs in turfgrass (Fig. 8.4). Finally, conservation and augmentation of natural nematode populations through proper management practices and periodic nematode releases offer possibilities for insect pest suppression.

Significant advances are also being made with the bacteria associated with entomopathogenic nematodes. Insecticidal compounds have been isolated and some of the genes show potential to be incorporated into plants for insect control (Bowen *et al.*, 1998). Basic information on how the mutualistic bacteria are retained in the infective juvenile's intestine is being unraveled.

Entomopathogenic nematodes are fascinating animals that have contributed to suppression of soil insect pests and insects in cryptic habitats. Besides this applied area, they will contribute to science as useful tools in understanding the evolution of parasitism and symbiosis and the the mechanism of insect resistance to infection.

Acknowledgments

We thank Patricia Stock for assistance in the preparation of the life cycle drawing.

References

AbuHatab, M.A., Selvan, S., and Gaugler, R. (1993) Role of proteases in penetration of insect gut by the entomopathogenic nematode *Steinernema glaseri* (Nematoda: Steinernematidae). *J. Invertebr. Pathol.*, **66**, 125–130.

Adams, B.J., Burnell, A.M., and Powers, T.O. (1998) A phylogenetic analysis of *Heterorhabditis* (Nemata: Rhabditidae) based on internal spacer 1 DNA sequence data. *J. Nematol.*, **30**, 22–39.

Akhurst, R.J. (1990) Safety to nontarget invertebrates of nematodes of economically important pests. In M. Laird, L.A. Lacey and E.W. Davidson (eds.), *Safety of Microbial Insecticides*, CRC Press, Boca Raton, pp. 233–240.

Akhurst, R.J. (1993) Bacterial symbionts of entomopathogenic nematodes – the power behind the throne. In R. Bedding, R. Akhurst and H. Kaya (eds.), *Nematodes and the Biological Control of Insects*. CSIRO, East Melbourne, pp. 127–135.

Akhurst, R.J., and Boemare, N.E. (1990) Biology and taxonomy of *Xenorhabdus*. In R. Gaugler and H.K. Kaya (eds.), *Entomopathogenic Nematodes in Biological Control*. CRC Press, Boca Raton, pp. 75–90.

Akhurst, R.J., Mourant, R.G., Baud, L., and Boemare, N.E. (1996) Phenotypic and DNA relatedness between nematode symbionts and clinical strains of the genus *Photorhabdus* (Enterobacteriaceae). *Intern. J. System. Bacteriol.*, **46**, 1034–1041.

Appel, A.G., Benson, E.P., Ellenberger, J.M., and Manweiler, S.A. (1993) Laboratory and field evaluations of an entomogenous nematode (Nematoda: Steinernematidae) for German cockroach (Dictyoptera: Blatellidae) control. *J. Econ. Entomol.*, **86**, 777–784.

Bathon, H. (1996) Impact of entomopathogenic nematodes on non-target hosts. *Biocontr. Sci. Technol.*, **6**, 421–434.

Baur, M.E., Kaya, H.K., Gaugler, R., and Tabashnik, B. (1997a) Effects of adjuvants on entomopathogenic nematode persistence and efficacy against *Plutella xylostella. Biocontr. Sci. Technol.*, **7**, 513–525.

Baur, M.E., Kaya, H.K., and Strong, D.R. (1998) Foraging ants as scavengers on entomopathogenic nematode-killed insects. *Biol. Contr.*, **12**, 231–236.

Baur, M.E., Kaya, H.K., and Tabashnik, B. (1997b) Efficacy of a dehydrated steinernematid nematode against black cutworm (Lepidoptera: Noctuidae) and diamondback moth (Lepidoptera: Plutellidae). *J. Econ. Entomol.*, **90**, 1200–1206.

Bedding, R.A. (1993) Biological control of *Sirex noctilio* using the nematode *Deladenus siricidicola*. In R. Bedding, R. Akhurst and H. Kaya (eds.), *Nematodes and the Biological Control of Insects*, CSIRO, East Melbourne, pp. 11–20.

Bedding, R., Akhurst, R., and Kaya, H. (1993) *Nematodes and the Biological Control of Insect Pests*, East Melbourne: CSIRO, 178 pp.

Bedding, R.A., Tyler, S., and Rochester, N. (1996) Legislation on the introduction of exotic entomopathogenic nematodes into Australia and New Zealand. *Biocontr. Sci. Technol.*, **6**, 465–475.

Bednarek, A., and Gaugler, R. (1997) Compatibility of soil amendments with entomopathogenic nematodes. *J. Nematol.*, **29**, 220–227.

Begley, J.W. (1990) Efficacy against insects in habitats other than soil. In R. Gaugler and H.K. Kaya (eds.), *Entomopathogenic Nematodes in Biological Control*, CRC Press, Boca Raton, pp. 215–321.

Bird, A.F., and Bird, J. (1986) Observations on the use of insect parasitic nematodes as a means of biological control of root-knot nematodes. *Int. J. Parasitol.*, **16**, 511–516.

Blaxter, M.L., De Ley, P., Garey, J.R., Liu, L.X., Scheldeman, P., Vierstraete, A. *et al.* (1998) A molecular evolutionary framework for the phylum Nematoda. *Nature*, **392**, 71–75.

Blaxter, M. (1998) *Caenorhabitis elegans* is a nematode. *Science* **282**, 2041–2046.

Boemare, N., Laumond, C., and Mauleon, H. (1996) The entomopathogenic nematode- bacterium complex; biology, life cycle and vertebrate safety. *Biocontr. Sci. Technol.*, **6**, 333–345.

Bohan, D.A., and Hominick, W.R. (1996) Investigations on the presence of an infectious proportion amongst populations of *Steinernema feltiae* (Site 76 strain) infective stages. *Parasitology*, **112**, 113–118.

Bohan, D.A., and Hominick, W.R. (1997a) Long-term dynamics of infectiousness in the infective-stage pool of the entomopathogenic nematode *Steinernema feltiae* (Site 76 strain) Filipjev. *Parasitology*, **114**, 301–308.

Bohan, D.A., and Hominick, W.R. (1997b) Sex and the dynamics of infection in the entomopathogenic nematode *Steinernema feltiae*. *J. Helminthol.*, **71**, 197–201.

Bowen, D., Rocheleau, T.A., Blackburn, M., Andreev, O., Golubeva, E., Bhartia, R. *et al.* (1998) Insecticidal toxins from the bacterium *Photorhabdus luminescens*. *Science*, **80**, 2129–2132.

Broadbent, A.B., and Olthof, Th. H.A. (1995) Foliar application of *Steinernema carpocapsae* (Rhabditida: Steinernematidae) to control *Liriomyza trifolii* (Diptera: Agromyzidae) larvae in chrysanthemums. *Environ. Entomol.*, **24**, 431–435.

Brown, I.M., and Gaugler, R. (1996) Cold tolerance of steinernematid and heterorhabditid nematodes. *J. therm. Biol.*, **21**, 115–121.

Brown, I.M., and Gaugler, R. (1997) Temperature and humidity influence emergence and survival of entomopathogenic nematodes. *Nematologica*, **43**, 363–375.

Burnell, A.M., and Dowds, B.C.A. (1996) The genetic improvement of entomopathogenic nematodes and their symbiont bacteria: phenotypic targets, genetic limitations and an assessment of possible hazards. *Biocontr. Sci. Technol.*, **6**, 435–447.

Cabanillas, H.E., and Raulston, J.R. (1996) Effects of furrow irrigation on the distribution and infectivity of *Steinernema riobravis* against corn earworm in corn. *Fundam. appl. Nematol.*, **19**, 273–281.

Campbell, J.F., and Gaugler, R. (1997) Inter-specific variation in entomopathogenic nematode foraging strategy: dichotomy or variation along a continuum? *Fundam. appl. Nematol.*, **20**, 393–398.

Campbell, J.F., and Kaya, H.K. (1999a) How and why a parasitic nematode jumps. *Nature,* **397**, 485–486.

Campbell, J.F., and Kaya, H.K. (1999b) Mechanism, kinematic performance, and fitness consequences of jumping behavior in entomopathogenic nematodes (*Steinernema* spp.). *Can. Entomol.,* **77**, 1947–1955.

Campbell, J.F., Koppenhöfer, A.M., Kaya, H.K., and Chinnasri, B. (1999) Are there temporarily non-infectious dauer stages in entomopathogenic nematode populations: a test of the phased infectivity hypothesis. *Parasitology,* **118**, 499–508.

Campbell, J.F., Lewis, E.E., Yoder, F., and Gaugler, R. (1997) Entomopathogenic nematode (Heterorhabditidae and Steinernematidae) spatial distribution in turfgrass. *Parasitology,* **113**, 473–482.

Campbell, J.F., Orza, G., Yoder, F., Lewis, E., and Gaugler, R. (1998) Spatial and temporal distribution of endemic and released entomopathogenic nematode populations in turfgrass. *Entomol. exp. appl.,* **86**, 1–11.

Ehlers, R.-U. (1996) Current and future use of nematodes in biocontrol: practice and commercial aspects with regard to regulatory policy issues. *Biocontr. Sci. Technol.,* **6**, 303–316.

Ehlers, R.-U., and Hokkanen, H.M.T. (1996) Insect biocontrol with non-endemic entomopathogenic nematodes (*Steinernema* and *Heterorhabditis* spp.): Conclusions and recommendations of a combined OECD and COST workshop on scientific and regulatory policy issues. *Biocontr. Sci. Technol.*, **6**, 295–302.

Ehlers, R.-U., Lunau, S., Krasomil-Osterfeld, K., and Osterfeld, K.H. (1998) Liquid culture of the entomopathogenic nematode-bacterium complex *Heterorhabditis megidis/Photorhabdus luminescens. BioControl*, **43**, 77–86.

Ehlers, R.-U., and Niemann, I. (1998). Molecular identification of *Photorhabdus luminescens* strains by amplification of specific fragments of the 16S ribosomal DNA. *System. Appl. Microbiol.*, **21**, 509–519.

Ehlers, R.-U., Stoessel, S., and Wyss, U. (1990). The influence of the phase variants of *Xenorhabdus* spp. and *Escherischia coli* (Enterobacteriaceae) on the propagation of entomopathogenic nematodes of the genera *Steinernema* and *Heterorhabditis. Rev. Nematol.,* **13**, 417–424.

Eidt, D.C., and Thurston, G.S. (1995) Physical deterrents to infection by entomopathogenic nematodes in wire worms (Coleoptera: Elateridae) and other soil insects. *Can. Entomol.*, **127**, 423–429.

Fischer-LeSaux, M., Viallard, V., Brunel, B., Normand, P., and Boemare, N.E. (1999). Polyphasic classification of the genus *Photorhabdus* and proposal of new taxa: *P. luminescens* subsp. *luminescens* subsp. nov., *P. luminescens* subsp. *akhurstii* subsp. nov., *P. luminescens* subsp. *laumondii* subsp. nov., *P. temperata* sp. nov. *P. temperata* subsp. sp. nov., *P. asymbiotica* sp. nov. *Int. J. System. Bacteriol.*, **49**, 1645–1656.

Forst, S., Dowds, B., Boemare, N., and Stackebrandt, E. (1997) *Xenorhabdus* and *Photorhabdus* spp.: bugs that kill bugs. *Annu. Rev. Microbiol.*, **51**, 47–72.

Forst, S., and Nealson, K. (1996) Molecular biology of the symbiotic-pathogenic bacteria *Xenorhabdus* spp. and *Photorhabdus* spp. *Microbiol. Rev.*, **60**, 21–43.

Gaugler, R. (1993) Ecological genetics of entomopathogenic nematodes. In R. Bedding, R. Akhurst and H. Kaya (eds.), *Nematodes and the Biological Control of Insects*, CSIRO, East Melbourne, pp. 89–95.

Gaugler, R., Campbell, J.F., Selvan, S., and Lewis, E.E. (1992) Large-scale inoculative releases of the entomopathogen *Steinernema glaseri*: assessment 50 years later. *Biol. Contr.*, **2**, 181–187.

Gaugler, R., and Kaya, H.K. (1990). *Entomopathogenic Nematodes in Biological Control*, CRC, Boca Raton 365 pp.

Gaugler, R., Lewis, E., and Stuart, R.J. (1997) Ecology in the service of biological control: the case of entomopathogenic nematodes. *Oecologia*, **109**, 483–489.

Gaugler, R., Wang, Y., and Campbell, J.F. (1994) Aggressive and evasive behaviors in *Popillia japonica* (Coleoptera: Scarabaeidae) larvae: defenses against entomopathogenic nematode attack. *J. Invertebr. Pathol.*, **64**, 193–199.

Georgis, R., Dunlop, D.B., and Grewal, P.S. (1995) Formulation of entomopathogenic nematodes. In F. R., Hall and J.W. Barry (eds.), *Biorational Pest Control Agents: Formulation and Delivery*, ACS Symposium Series No. 595, Amer. Chem. Soc., Washington, DC, pp. 197–205.

Georgis, R., and Kaya, H.K. (1998) Formulation of entomopathogenic nematodes. In H.D. Burges. (eds.), *Formulation of Microbial Biopesticides: Beneficial Microorganisms, Nematodes and Seed Treatments*. Kluwer, Dordrecht, The Netherlands, pp. 289–308.

Georgis, R., and Kelly, J. (1997) Novel pesticidal substances from the entomopathogenic nematode-bacterium complex. In P.A. Hedin, R.M. Hollingworth, E.P. Masler, J. Miyamoto and D.G. Thompson (eds.), *Phytochemicals for Pest Control*, ACS Symposium Series No. 658, Am. Chem. Soc., Washington, DC, pp. 134–143.

Georgis, R., and Manweiler, S.A. (1994) Entomopathogenic nematodes: A developing biological control technology. *Agric. Zool. Rev.*, **6**, 63–94.

Glazer, I. (1996) Survival mechanisms of entomopathogenic nematodes. *Biocontr. Sci. Technol.*, **6**, 373–378.

Glazer, I. (1997) Effects of infected insect on secondary invasion of steinernematid entomopathogenic nematodes. *Parasitology*, **114**, 597–604.

Glazer, I., Klein, M., Navon, A., and Nakache, Y. (1992) Comparison of efficacy of entomopathogenic nematodes combined with antidesiccants applied by canopy sprays against 3 cotton pests (Lepidoptera: Noctuidae). *J. Econ. Entomol.*, **85**, 1636–1641.

Glazer, I., Salame, L., and Segal, D. (1997) Genetic enhancement of nematicide resistance in entomopathogenic nematodes. *Biocontr. Sci. Technol.*, **6**, 499–512.

Gouge, D.H., Otto, A.A., Schirocki, A., and Hague, N.G.M. (1994) Effects of steinernematids on the root-knot nematode, *Meloidogyne javanica. Tests Agrochem. Cultiv. No. 15*, [*Ann. Appl. Biol.*, **124**, (Suppl.)], 134–135.

Grewal, P.S., Gaugler, R., Kaya, H.K., and Wusaty, M. (1993) Infectivity of the entomopathogenic nematode *Steinernema scapterisci* (Nematoda; Steinernematidae). *J. Invertebr. Pathol.*, **62**, 22–28.

Grewal, P.S., Gaugler, R., and Wang, Y. (1996) Enhanced cold tolerance of the entomopathogenic nematode *Steinernema feltiae* through genetic selection. *Ann. Appl. Biol.*, **129**, 335–341.

Grewal, P.S., and Georgis, R. (1998) Entomopathogenic nematodes. In F.R. Hall and J. Menn (eds.), *Methods in Biotechnology, Vol. 5: Biopesticides: Use and Delivery*, Humana Press, Totowa, NJ, pp. 271–299.

Grewal, P.S., Lewis, E.E., and Venkatchari, S. (1999) Allelopathy: A possible mechanism of suppression of plant-parasitic nematodes by entomopathogenic nematodes. *Nematology*, **1**, 734–743.

Grewal, P.S., Martin, W.R., Miller, R.W., and Lewis, E.E. (1997) Suppression of plant parasitic nematode populations in turfgrass by application of entomopathogenic nematodies. *Biocontr. Sci. Technol.*, **7**, 393–399.

Grewal, P.S., Matsuura, M., and Converse, V. (1997) Mechanism of specificity of association between the nematode *Steinernema scapterisci* and its symbiotic bacterium. *Parasitology*, **114**, 483–488.

Grewal, P.S., Selvan, S., and Gaugler, R. (1994) Thermal adaptation of entomopathogenic nematodes: niche breadth for infection, establishment, and reproduction. *J. therm. Biol.*, **19**, 245–253.

Grewal, P.S., Selvan, S., Lewis, E.E., and Gaugler, R. (1993) Males as the colonizing sex in insect parasitic nematodes. *Experientia*, **49**, 605–608.

Griffin, C.T. (1993) Temperature responses of entomopathogenic nematodes: implications for the success of biological control programmes. In R. Bedding, R. Akhurst and H. Kaya (eds.), *Nematodes and the Biological Control of Insects*, CSIRO, East Melbourne, pp. 115–126.

Griffin, C.T. (1996) Effects of prior storage conditions on the infectivity of *Heterorhabditis* sp. (Nematoda: Heterorhabditidae). *Fundam. appl. Nematol.*, **19**, 95–102.

Griffin, C.T., Finnegan, M.M., and Downes, M.J. (1994) Environmental tolerances and the dispersal of *Heterorhabditis*: Survival and infectivity of European *Heterorhabditis* following prolonged immersion in seawater. *Fundam. appl. Nematol.*, **17**, 415–421.

Griffin, C.T., and Downes, M.J. (1994) Recognition of low temperature active isolates of the entomopathogenic nematode *Heterorhabdits* spp. (Rhabditida: Heterorhabditidae). *Nematologica*, **40**, 106–115.

Han, R.C., and Ehlers, R.-U. (1998) Cultivation of axenic *Heterorhabditis* spp. dauer juveniles and their response to non-specific *Photorhabdus luminescens* food signals. *Nematologica*, **44**, 425–435.

Hashmi, G., and Gaugler, R. (1998) Genetic diversity in insect-parasitic nematodes (Rhabditida: Heterorhabditidae). *J. Invertebr. Pathol.*, **72**, 185–189.

Hashmi, S., Hashmi, G., and Gaugler, R. (1995a) Genetic transformation of an entomopathogenic nematode by microinjection. *J. Invertebr. Pathol.*, **66**, 293–296.

Hashmi, S., Ling, P., Hashmi, G., Reed, M., Gaugler, R., and Trimmer, W. (1995b) Genetic transformation of nematodes using arrays of micromechanical piercing structures. *Biotechnique*, **19**, 166–770.

Hay, D.B., and Fenlon, J.S. (1995) A modified binomial model that describes the infection dynamics of the entomopathogenic nematode *Steinernema feltiae* (Steinernematidae, Nematoda). *Parasitology*, **11**, 627–633.

Hominick, W.R., Briscoe, B.R., del Pino, F.G., Heng, J., Hunt, D.J., Kozodoy, E. *et al.* (1997) Biosystematics of entomopathogenic nematodes: current status, protocols and definitions. *J. Helminthol.*, **71**, 271–298.

Hominick, W.M., and Reid, A.P. (1990) Perspectives on Entomopathogenic Nematology. In R. Gaugler and H.K. Kaya (eds.), *Entomopathogenic Nematodes in Biological Control*, CRC Press, Boca Raton, pp. 327–345.

Hominick, W.M., Reid, A.P., Bohan, D.A., and Briscoe, B.R. (1996) Entomopathogenic nematodes: Biodiversity, geographical distribution and the convention on biological diversity. *Biocontr. Sci. Technol.*, **6**, 317–331.

Huaiwen, Y., Gangying, Z., Shangao, Z., and Heng, J. (1993) Biological control of tree borers (Lepidoptera: Cossidae) in China with the nematode *Steinernema carpocapsae*. In R. Bedding, R. Akhurst and H. Kaya (eds.), *Nematodes and the Biological Control of Insects*, CSIRO, East Melbourne, pp. 33–40.

Ishibashi, N. (1993) Integrated control of insect pests by *Steinernema carpocapsae*. In R. Bedding, R. Akhurst and H. Kaya (eds.), *Nematodes and the Biological Control of Insects*, CSIRO, East Melbourne, pp. 105–113.

Ishibashi, N., and Wang, X.D. (1998) Infection behavior for species maintenance in steinernematid nematodes. *Proc. VIIth Int. Coll. Invertebr. Pathol. Microbial conf., Sapporo, Aug 23–28 1998*, pp. 28–32.

Jackson, T.J., Wang, H., Nugent, M.T., Griffin, C.T., Burnell, A.M., and Dowds, B.C.A. (1995) Isolation of insect pathogenic bacteria, *Providencia rettgeri*, from *Heterorhabditis* spp. *J. Appl. Bacteriol.*, **78**, 237–244.

Kaya, H.K. (1990) Soil ecology. In R. Gaugler and H.K. Kaya (eds.), *Entomopathogenic Nematodes in Biological Control*, CRC Press, Boca Raton, pp. 93–115.

Kaya, H.K., Burlando, T.M., Choo, H.Y., and Thurston, G.S. (1995) Integration of entomopathogenic nematodes with *Bacillus thuringiensis* or pesticidal soap for control of insect pests. *Biol. Contr.*, **5**, 432–441.

Kaya, H.K., and Gaugler, R. (1993) Entomopathogenic nematodes. *Annu. Rev. Entomol.*, **38**, 181–206.

Kaya, H.K., and Koppenhöfer, A.M. (1996) Effects of microbial and other antagonistic organisms and competition on entomopathpogenic nematodes. *Biocontr. Sci. Technol.*, **6**, 333–345.

Kaya, H.K., and Stock, S.P. (1997) Techniques in insect nematology. In *Manual of Techniques in Insect Pathology*, Academic Press, San Diego, pp. 281–324.

Klein, M.G. (1990) Efficacy against soil-inhabiting insect pests. In R. Gaugler and H.K. Kaya (eds.), *Entomopathogenic Nematodes in Biological Control*, CRC Press, Boca Raton, pp. 195–214.

Klein, M.G. (1993) Biological control of scarabs with entomopathogenic nematodes. In R. Bedding, R. Akhurst and H. Kaya (eds.), *Nematodes and the Biological Control of Insects*, CSIRO, East Melbourne, pp. 49–57.

Koppenhöfer, A.M. (2000). Nematodes. In L. Lacey and H.K. Kaya (eds.), *Field Manual of Techniques for the Application and Evaluation of Entomopathogens*, Kluwer, Dordrecht, pp. 283–301.

Koppenhöfer, A.M., Baur, M.E., Stock, S.P., Choo, H.Y., Chinnasri, B., and Kaya, H.K. (1997) Survival of entomopathogenic nematodes within host cadavers in dry soil. *Appl. Soil. Ecol.*, **6**, 231–240.

Koppenhöfer, A.M., Choo, H.Y., Kaya, H.K., Lee, D.W., and Gelernter, W. (1999) Improved field and greenhouse efficacy with a combination of entomopathogenic nematode and *Bacillus thuringiensis* against scarab grubs. *Biol. Contr.*, **14**, 37–44.

Koppenhöfer, A.M., and Kaya, H.K. (1998) Synergism of imidacloprid and entomopathogenic nematodes: A novel approach to white grub control in turfgrass. *J. Econ. Entomol.*, **91**, 618–623.

Koppenhöfer, A.M., Kaya, H.K., and Taormino, S. (1995) Infectivity of entomopathogenic nematodes (Rhabditida: Steinernematidae) at different soil depths and moistures. *J. Invertebr. Pathol.*, **65**, 193–199.

Kung, S.P., Gaugler, R., and Kaya, H.K. (1991) Effects of soil temperature, moisture, and relative humidity on entomopathogenic nematode persistence. *J. invertebr. Pathol.*, **57**, 242–249.

Lacey, L., and Kaya, H.K. (2000) *Field Manual of Techniques for the Application and Evaluation of Entomopathogens*, Dordrecht: Kluwer, 911 pp.

Levins, R. (1970) Extinction. In M. Gerstenhaber (ed.), *Lectures on Mathematics in the Life Sciences, Vol. 2*, Amer. Math. Soc., Providence, pp. 77–107.

Lewis, E.E., Campbell, J.F., and Gaugler, R. (1998) A conservation approach to using entomopathogenic nematodes in turf and landscapes. In P. Barbosa (ed.), *Conservation Biological Control*, Academic Press, San Diego, pp. 235–254.

Lewis, E.E., and Gaugler, R. (1994) Entomopathogenic nematode (Rhabditida: Steinernematidae) sex ratio relates to foraging strategy. *J. Invertebr. Pathol.*, **64**, 238–242.

Liu, J., Berry, R., Poinar, G., Jr., and Moldenke, A. (1997) Phylogeny of *Photorhabdus* and *Xenorhabdus* species and strains as determined by comparison of partial 16S rRNA gene sequences. *Int. J. System. Bacteriol.*, **47**, 948–951.

Mamiya, Y. (1989) Comparison of infectivity of *Steinernema kushidai* (Nematoda: Steinernematidae) and other steinernematid and heterorhabditid nematodes for three different insects. *Appl. Ent. Zool.*, **24**, 302–308.

Nickle, W.R. (1984) *Plant and Insect Nematodes*, Marcel Dekker, New York, pp. 925.

Nickle, W.R. (1991) *Manual of Agricultural Nematology*, Marcel Dekker, New York, pp. 1035.

Nickle, W.R., and Shapiro, M. (1994) Effects of eight brighteners as solar radiation protectants for *Steinernema carpocapsae*, All strain. *Suppl. J. Nematol.*, **26**, 782–784.

Nishimatsu, T., and Jackson, J.J. (1998) Interaction of insecticides, entomopathogenic nematodes, and larvae of the Western corn rootworm (Coleoptera: Chrysomelidae). *J. Econ. Entomol.*, **91**, 410–418.

Nishimura, Y., Hagiwara, A., Suzuki, T., and Yamanaka, S. (1994) *Xenorhabdus japonicus* sp. Nov. associated with the nematode *Steinernema kushidai. World. J. Microbiol. Biotech.*, **10**, 207–210.

O'Leary, S.A., and Burnell, A.M. (1997) The isolation of mutants of *Heterorhabditis megidis* (Strain UK211) with increased desiccation tolerance. *Fundam. Appl. Nematol.*, **20**, 197–205.

Parkman, J.P., and Smart, G.C., Jr. (1996) Entomopathogenic nematodes, a case study: Introduction of *Steinernema scapterisci* in Florida. *Biocontr. Sci. Technol.*, **6**, 413–419.

Patel, M.N., and Wright, D.J. (1996) The influence of neuroactive pesticides on the behaviour of entomopathogenic nematodes. *J. Helminthol.*, **70**, 53–61.

Peters, A. (1996) The natural host range of *Steinernema* and *Heterorhabditis* spp., and their impact on insect populations. *Biocontr. Sci. Technol.*, **6**, 389–402.

Peters, A., and Ehlers, R.-U. (1994) Susceptibility of leatherjackets (*Tipula paludosa* and *T. oleracea*; Tipulidae: Nematocera) to the entomopathogenic nematode *Steinernema feltiae*. *J. Invertebr. Pathol.*, **63**, 163–171.

Peters, A., Gouge, D.H., Ehlers, R.-U., and Hague, N.G.M. (1997) Avoidance of encapsulation by *Heterorhabditis* spp. infecting larvae of *Tipula oleracea*. *J. Invertebr. Pathol.*, **70**, 161–164.

Poinar, G.O., Jr. (1979) *Nematodes for Biological Control of Insects*, CRC Press, Boca Raton, pp. 277.

Renn, N. (1998a) Routes of penetration of the entomopathogenic nematode *Steinernema feltiae* attacking larval and adult houseflies (*Musca domestica*.). *J. Invertebr. Pathol.*, **72**, 281–287.

Renn, N. (1998b) The efficacy of entomopathogenic nematodes for controlling housefly infestations of intensive pig units. *Med. Vet. Entomol.*, **12**, 46–51.

Richardson, P.N. (1996) British and European legislation regulating rhabditid nematodes. *Biocontr. Sci. Technol.*, **6**, 449–463.

Rizvi, S.A., Hennessey, R., and Knott, D. (1996) Legislation on the introduction of exotic nematodes in the US. *Biocontr. Sci. Technol.*, **6**, 477–480.

Roush, R.T. (1990a) Genetic variation in natural enemies: critical issues for colonization in biological control. In M. Mackauer, L.E. Ehler and J. Roland (eds.), *Critical Issues in Biological Control*, Intercept, Andover, pp. 263–88.

Roush, R.T. (1990b) Genetic considerations in the propagation of entomophagous species. In R.P. Baker and P.E. Dunn (eds.), *New Directions in Biological Control*, Alan R. Liss, New York, pp. 373–387.

Rovesti, L., and Deseö, K.V. (1990) Compatibility of chemical pesticides with the entomopathogenic nematodes, *Steinernema carpocapsae* Weiser and *S. feltiae* Filipjev (Nematoda: Steinernematidae). *Nematologica*, **36**, 237–245.

Schroeder, W.J., and Sieburth, P.J. (1997) Impact of surfactants on control of the root weevil *Diaprepes abbreviatus* larvae with *Steinernema riobravis*. *J. Nematol.*, **29**, 216–219.

Shapiro, D.I., Glazer, I., and Segal, D. (1997a) Genetic improvement of heat tolerance in *Hetrorhabditids bacteriophora* through hybridization. *Biol. Contr.*, **8**, 153–159.

Shapiro, D.I., Tylka, G.L., and Lewis, L.C. (1997b) Effects of fertilizers on virulence of *Steinernema carpocapsae*. *Appl. Soil Ecol.*, **3**, 27–34.

Simões, N., and Rosa, J.S. (1996) Pathogenicity and host specificity of entomopathogenic nematodes. *Biocont. Sci. Technol.*, **6**, 403–411.

Smith, K.A., Miller, R.W., and Simser, D.H. (1992) Entomopathogenic nematode bibliography – heterorhabditid and steinernematid nematodes, *Southern Coop. Ser. Bull. 370*, Ark. Agric. Exp. Sta. Fayetteville, 81pp.

Smits, P. (1996) Post-application persistence of entomopathogenic nematodes. *Biocontr. Sci. Technol.*, **6**, 379–387.

Stark, J.D. (1996) Entomopathogenic nematodes (Rhabditida: Steinernematidae): toxicity of neem. *J. Econ. Entomol.*, **89**, 68–73.

Strauch, O., and Ehlers, R.-U. (1998) Food signal production of *Photorhabdus luminescens* inducing the recovery of entomopathogenic nematodes *Heterorhabditis* spp. in liquid culture. *Appl. Microbiol. Biotechnol.*, **50**, 369–374.

Strong, D.R., Kaya, H.K., Whipple, A.V., Child, A.L., Kraig, S., Bondonno, M. *et al.* (1996) Entomopathogenic nematodes: natural enemies of root-feeding caterpillars on bush lupine. *Oecologia*, **108**, 167–173.

Stuart, R.J., and Gaugler, R. (1994) Patchiness in populations of entomopathogenic nematodes. *J. Invertebr. Pathol.*, **64**, 39–45.

Stuart, R.J., and Gaugler, R. (1996) Genetic adaptation and founder effect in laboratory populations of the entomopathogenic nematode *Steinernema glaseri. Can. J. Zool.*, **74**, 164–170.

Stuart, R.J., AbuHatab, M., and Gaugler, R. (1998) Sex ratio and the infection process in entomopathogenic nematodes: are males the colonizing sex? *J. Invertebr. Pathol.*, **72**, 288–295.

Tomalak, M. (1994) Selective breeding of *Steinernema feltiae* (Filipjev) (Nematoda: Steinernematidae) for improved efficacy in control of a mushroom fly, *Lycoriella solani* Winnertz (Diptera: Sciaridae). *Biocontr. Sci. Techn.*, **4**, 187–198.

Thurston, G.S., Ni, Y., and Kaya, H.K. (1994) Influence of salinity on survival and infectivity of entomopathogenic nematodes. *J. Nematol.*, **26**, 345–351.

Thurston, G.S., Kaya, H.K., and Gaugler, R. (1994) Characterizing the enhanced susceptibility of milky disease-infected scarabaeid grubs to entomopathogenic nematodes. *Biol. Contr.*, **4**, 67–73.

Treverrow, N.L., and Bedding, R.A. (1993) Development of a system for the control of the banana weevil borer, *Cosmopolites sordidus* with entomopathogenic nematodes. In R. Bedding, R. Akhurst and H. Kaya (eds.), *Nematodes and the Biological Control of Insects*, CSIRO, East Melbourne, pp. 41–47.

Volgyi, A., Fodor, A., Szentirmai, A., and Forst, S. (1998) Phase variation in *Xenorhabdus nematophilus. Appl. Environ. Microbiol.*, **64**, 1188–1193.

Wang, Y., Campbell, J.F., and Gaugler, R. (1995) Infection of entomopathogenic nematodes *Steinernema glaseri* and *Heterorhabditis bacteriophora* against *Popillia japonica* (Coleoptera: Scarabaeidae) larvae. *J. Invertebr. Pathol.*, **66**, 178–184.

Wang, Y., Gaugler, R., and Cui, L. (1998) Variations in immune response of *Popillia japonica* and *Acheta domesticus* to *Heterorhabditis bacteriophora* and *Steinernema* species. *J. Nematol.*, **25**, 11–18.

Wang, Y., and Gaugler, R. (1998) Host and penetration site location by entomopathogenic nematodes against Japanese beetle larvae. *J. Invertebr. Pathol.*, **72**, 313–318.

Wang, Y., and Gaugler, R. (1998) *Steinernema glaseri* surface coat protein suppresses the immune response of *Popillia japonica* (Coleoptera; Scarabaeidae) larvae. *Biol. Contr.*, **14**, 45–50.

Wilson, M., Xin, W., Hashmi, S., and Gaugler, R. (1999) Risk assessment and fitness of a transgenic entomopathogenic nematodes. *Biol. Contr.* **15**, 81–87.

Womersley, C.Z. (1990) Dehydration survival and anhydrobiotic potential. In R. Gaugler and H.K. Kaya (eds.), *Entomopathogenic Nematodes in Biological Control*, CRC Press, Boca Raton, pp. 117–137.

Womersley, C.Z. (1993) Factors affecting physiological fitness and modes of survival employed by dauer juveniles and their relationship to pathogenicity. In R. Bedding, R. Akhurst and H. Kaya (eds.), *Nematodes and the Biological Control of Insects*, CSRIO, East Melbourne, pp. 79–88.

Wright, D.J., Piggott, S., and Patel, M.N. (1998) Physiological aspects of survival by entomopathogenic nematodes. *Proc. VIIth Int. Coll. Invertebr. Pathol. Microbiol. Conf.*, Aug 23–28 1998, Sapporo, pp. 20–23.

Bioherbicides: Potential Successful Strategies for Weed Control

9

Robert J. Kremer
US Department of Agriculture, Agricultural Research Service, Cropping Systems and Water Quality Unit and Department of Soil and Atmospheric Sciences, University of Missouri, Columbia MS 65211, USA

Introduction

Recognition of the potential for using natural enemies in weed control developed from observations by early naturalists and agriculturists, and was the basis for the first attempted use in 1902 of natural enemies in controlling the troublesome weed lantana (*Lantana camara*) in Hawaii (Harley and Forno, 1992). From these early beginnings, the concept of biological control was developed. By definition, biological control of weeds is the intentional use of living organisms (biotic agents) to reduce the vigor, reproductive capacity, density, or impact of weeds (Quimby and Birdsall, 1995). The strategies of biological control of weeds can be classified in two broad categories: (i) classical or inoculative, and (ii) inundative or mass exposure. The classical strategy is based on introduction of host-specific organisms (insects, pathogens, nematodes, etc.) from the weed's native range into regions where the weed has established and become a widespread problem. The biotic agents, after quarantine to assure host specificity, are released into weed-infested sites and are allowed to adapt and flourish in their new habitat over time eventually establishing a self-perpetuating regulation of the weed infestation at acceptable levels. Thus, classical biological control requires a time period of one to several years to achieve adequate control while the agent population builds up to levels to impact the weed population.

The inundative strategy attempts to overwhelm a weed infestation with massive numbers of a biotic agent in order to attain weed control in the year of release. In contrast to classical biological control, inundation involves timing of agent release to coincide with weed susceptibility to the agent and formulation of the agent to provide rapid attack of the weed host. A development of the inundative strategy is the *bioherbicide* approach, which involves application of weed pathogens in a manner similar to herbicide applications. Since most bioherbicides have been developed using selected plant pathogenic fungi that cause such diseases on weeds as anthracnose and rust, the term *mycoherbicides* is often used in reference to these fungal preparations.

The objectives of this chapter are to identify the place for bioherbicides in weed management including their integration into current systems, to develop an understanding of factors affecting their successful use in both conventional and alternative management systems, and to assess the prospects of developing strategies for using bioherbicides in biologically-based weed management.

Current status of bioherbicide development

Several microbial agents have been extensively evaluated and developed or are under development for commercial application (Table 9.1). The discovery, development, practical application and commercialization of early mycoherbicides (i.e. COLLEGO, DEVINE, BIOMAL) have been extensively described (Charudattan, 1991; TeBeest, 1996). The early mycoherbicides consisted of "classical" fungal plant pathogens that infected the aerial portions of weed host resulting in visible disease symptoms. These fungi could also be mass cultured in artificial media to produce large quantities of inocula needed for field application. Microbial agents comprising more recent bioherbicides can be broadly categorized to include obligate fungal parasites, soil-borne fungal pathogens, non-phytopathogenic fungi, bacteria and nematodes. Many of these organisms have different cultural and application requirements compared to the early mycoherbicides. This presents a curious dilemma in that even though the number of potential bioherbicides and target weeds in diverse habitats has expanded over the past 20 years, the production and formulation requirements and application methods have become more complex. This is a disadvantage for developing production facilities devoted to bioherbicides since standard culturing and processing techniques cannot be used in the production of all biocontrol agents necessary for a given set of weeds targeted by a bioherbicide market.

The bioherbicides listed in Table 9.1 include organisms that have been extensively evaluated for commercial development including several that have undergone field testing and evaluation as required by regulatory agencies. The original bioherbicide or mycoherbicide concept was based on mass artificial culture of organism to obtain large quantities of inoculum for inundative application to the weed host to achieve rapid epidemic buildup and high levels of disease (Charudattan, 1991). Since many of the recent bioherbicide candidates differ from the original definition in requirements for mass production and application, the bioherbicide concept has been redefined as living products that control specific weeds in agriculture similar to chemicals. Selected exam-

Table 9.1 Stage of development or commercialization of representative bioherbicides

Agent	*Target weed*	*Ecosystem*	*Status/Trade name*
Foliar/Stem Fungal Pathogens			
Colletotrichum gloeosporioides f. sp. *aeschynomene*	Northern jointvetch (*Aeschynomene virginica*)	Rice, soybean	Commercialized – COLLEGO Encore Technologies
Colletotrichum gloeosporioides f. sp. *malvae*	Round-leave mallow (*Malva pusilla*)	Flax, lentils, strawberries	Commercialized – BIOMAL Philom Bios (Canada)
Phytophthera palmivora	Strangler or milkweed vine (*Morrenia odorata*)	Citrus groves	Commercialized – DEVINE Abbott Laboratories
Alternaria cassiae	Sicklepod (*Cassia obtusifolia*)	Soybean	In process of commercial development – CASST – Mycogen Corp.
Colletotrichum truncatum	Hemp sesbania (*Sesbania exaltata*)	Soybean, cotton, rice	Under evaluation for formulations – COLTRU – United Agricultural Products (UAP)
Colletotrichum coccodes	Velvetleaf (*Abutilon theophrasti*)	Maize, soybean	Awaiting commercial development – VELGO
Puccinia canaliculata	Yellow nutsedge (*Cyperus esculentus*)	Horticultural crops	In process of commercial development – DR. BIOSEDGE
Colletotrichum gloeosporioides f. sp. *cuscutae*	Dodders (*Cuscuta* spp.)	Soybean	Product in use in China – LUBAO 2
Alternaria sp.	Dodders	Cranberry	Under development by UAP
Cephalosporium diospyri	Common persimmon (*Diospyros virginiana*)	Pastures, rangelands	Available for treatment of cut stumps and as "impregnated buckshot" – Noble Foundation
Chondrostereum purpureum	Black cherry (*Prunus serotina*) Red alder (*Alnus rubra*)	Forest	Commercialized in The Netherlands – BIOCHON Under commercialization in Canada – ECO-clear
Nectria ditissima	Red alder	Forest	Under evaluation; patented in Canada as PFC-ALDERKIL
Cylindrobasidium laeve	*Acacia* spp.	Forest, rangelands	Use registered by South Africa Dept of Agriculture
Cercospora rodmanii	Water hyacinth (*Eichornia crassipes*)	Aquatic sites	Under commercial development
Soilborne Fungal Pathogens			
Sclerotinia minor	Dandelion (*Taraxacum officinale*)	Turf	Under field testing and formulation evaluation
Rhizoctonia solani	Leafy spurge (*Euphorbia esula*)	Rangelands	Under field testing and formulation evaluation
Fusarium solani f. sp. *cucurbitae*	Texas gourd (*Cucurbita texana*)	Cotton, soybean	Under field testing and formulation evaluation
Fusarium oxysporum f. sp. *erythroxyli*	Coca (*Erythoxylum coca* var. *coca*)	Illicit narcotic crops	Under greenhouse testing and formulation evaluation

Table 9.1 (*Continued*)

Agent	*Target weed*	*Ecosystem*	*Status/Trade name*
Non-phytopathogenic Soilborne Fungi			
Gliocladium virens	Several	Row crops and horticultural crops	Under testing for optimal viridiol production, delivery and field application
Foliar Bacterial Pathogens			
Xanthomonas campestris pv. *poannua*	Annual bluegrass (*Poa annua*)	Turf, athletic fields	Marketed in Japan by Japan Tobacco – CAMPERICO Under development in U.S. by EcoSoil Systems
Pseudomonas syringae pv *tagetis*	Canada thistle, cocklebur (*Cirsium arvense, Xanthium strumarium*)	Maize, soybean	Under commercial development – Encore Technologies
P. syringae pv. *phaseolicola*	Kudzu (*Pueraria lobata*)	Non-cropland, pastures	Field testing
Non-pathogenic Bacteria			
Pseudomonas fluorescens D7	Downy brome (*Bromus tectorum*)	Cereal grains	Under development
Nematodes			
Subanguina picridis	Russian knapweed (*Acroptilon repens*)	Rangeland	Under field testing and formulation evaluation

Source: Julien and Griffiths, 1998; Jones and Hancock, 1990; Mortensen, 1998; Watson, 1993; personal communications, 1998, 1999

ples of production and application of more recent bioherbicides will be described below to illustrate the diversity of current bioherbicides.

The rust fungus, *Puccinia canaliculata*, an obligate parasite of yellow nutsedge (*Cyperus esculentus*), must be grown directly on host plants from which the uredospore inoculum must be harvested and stored in bulk prior to preparation of the product (Phatak *et al.*, 1983). The bioherbicide, *Chondrostereum purpureum*, must be applied to wounded stems or stumps of weedy tree species to inhibit re-sprouting by enhancing decay of the woody tissue (Prasad, 1996). Soilborne fungi have become important bioherbicide candidates since these fungi applied directly to soils can reduce weed populations through decay of seeds prior to emergence or kill seedlings shortly after emergence (Jones and Hancock, 1990). Plant pathogenic bacteria including *Xanthomonas campestris* pv *poannua* (Xcp) and *Pseudomonas syringae* pv *tagetis* (Pst), have been developed as bioherbicides for control of annual bluegrass (*Poa annua*) and Asteraceae (composite) weeds, respectively (Johnson *et al.*, 1996). The Xcp bioherbicide must be applied by spraying the bacterial suspension while mowing to allow bacterial cells to invade wounded tissue of the grass. The Pst bioherbicide is prepared with an organosilcone surfactant to enhance bacterial infection of leaf and stem tissue and onset of disease.

Another group of bacteria under intensive investigation for bioherbicidal potential are deleterious rhizobacteria (DRB), which differ from bacterial pathogens in that they are nonparasitic bacteria colonizing plant roots and able to suppress plant growth without invading the root tissues (Kremer and Kennedy, 1996). *Pseudomonas fluorescens* D7 is a soil applied DRB bioherbicide formulated as a liquid suspension or encapsulated in clay that effectively suppresses downy brome (*Bromus tectorum*) in cereal grain crops (Kennedy *et al.*, 1991). Progress has also been made in development and formulation of plant-specific nematodes for wide-scale management of rangeland weeds (Caesar-Thon That *et al.*, 1995).

The majority of bioherbicides presented in Table 9.1 typically has a narrow host range (one primary weed target) and is restricted to use in "minor crops" or other ecosystems with limited production areas. Many of the foliar/stem fungal pathogens require specific levels of humidity (dew period) and temperature for full effectiveness, which necessitates development of special formulations to assure effectiveness after delivery of the agent in the field. A recent review details the factors that must be addressed in bioherbicide formulation technology in order to maintain or enhance efficacy of the biocontrol agent as well as to be compatible with conventional field application systems (Boyetchko *et al.*, 1998). Factors of narrow host range, relative economic importance of target weed, use in minor crops and specific requirements for culturing and formulation to assure efficacy have limited commercial development of and interest in bioherbicides because of the likely low market potential for these products.

Bioherbicides in conventional cropping systems

Conventional cropping systems include large-scale production enterprises utilizing high-yielding crop varieties generally in monocultural systems, and large, costly inputs of chemical fertilizers and pesticides on the most fertile, productive soils available. Recently, certain crops (i.e. soybean, maize, cotton, etc.) have been developed transgenically for resistance to herbicides that allows producers to control a broad spectrum of weeds with a single herbicide. These transgenic crops have become very popular in conventional cropping systems. A notable feature of most conventional cropping systems is that the approach to controlling weeds involves using herbicides for reducing weed infestations to acceptable levels so that crops can be grown profitably with little regard given to a more long-term approach for weed management (Aldrich and Kremer, 1997; Zimadahl, 1993). It is not surprising that there would be little opportunity for replacement of broad-spectrum chemical herbicides with bioherbicides in a conventional management setting. Yet despite the advancements in herbicide technology and development of transgenic herbicide-resistant crops, a recent survey indicated that farmers perceive annual and perennial weed infestations as the most serious crop pest problems affecting their enterprises (Aref and Pike, 1998). This finding suggests that many farmers are open to alternative, nonchemical weed control methods.

Bioherbicides and selective weed control

Limited opportunities exist for potential practical use of bioherbicides in conventional systems. Charudattan (1990) proposed that bioherbicides most likely to be successfully adapted in conventional systems include those that effectively and consistently suppress weeds of economic importance on a very large scale ("economically attractive weeds") and those containing pathogens that may be manipulated to achieve broad- spectrum activity. An example of a bioherbicide with activity against an economically important weed is *Colletotrichum coccodes*, under commercial development as VELGO, a pathogen of velvetleaf, which is distributed throughout the continental United States and the southern region of the eastern Canadian provinces (Wymore *et al.*, 1988). Unfortunately, it is rare that weed infestations are limited to one predominant species, thus, it is questionable if use of a bioherbicide for one species in a mixture of weeds is justified if an expensive broad-spectrum herbicide is available to control multiple species. Fungal pathogens of the same species containing strains or subspecies with activity against several weeds might be developed through selective screening or through genetic recombination or hybridization into broad host-range pathotypes for use against multiple weed targets (Charudattan, 1990). This is a long- term tactic, however, as intense evaluation will be required to meet stringent regulations before approval is granted for release of genetically-altered organisms in the environment.

It is also possible to screen bioherbicide candidates for broad-spectrum activity during selection phases of development without resorting to genetic manipulation. An excellent example is Pst, a naturally-occurring bacterial phytopathogen that has been shown to be effective as a postemergence bioherbicide on several economically important weeds in the Asteraceae family including wild sunflower (*Helianthus* spp.), common cocklebur (*Xanthium strumarium*), common ragweed (*Ambrosia artimisiifolia*), and Canada thistle (*Cirsium arvense*) in soybean (*Glycine max*) (Johnson *et al.*, 1996).

Despite limitations of current bioherbicides, it is widely acknowledged that chemical herbicide use will continue to decrease in importance due to social and environmental concerns, development of herbicide-resistant weed biotypes and reduced availability of new, environmentally compatible herbicides while emphasis on "biologically-based pest management technology" will increase. Using these considerations Gressel *et al.* (1996) indicate several situations in which bioherbicides can be effective in conventional crop production including where: (i) weeds cannot be controlled by a herbicide because they are related to the crop; (ii) parasitic weeds that are without selective herbicides; (iii) individual weeds evolved resistance to a broad-spectrum herbicide; (iv) excessive rates of herbicides are required to control one weed species, thus bioherbicides would allow less herbicide use; (v) herbicides cannot be used due to cost or environmental limitations.

Bioherbicides and herbicide-resistant weeds

Bioherbicides based on DRB may soon be available for managing grass weeds such as downy brome (*Bromus tectorum*) and jointed goatgrass (*Aegilops cylindrica*), which are difficult to control with herbicides in cereal grain crops (Kremer and Kennedy, 1996). Although the problem of controlling weeds related to crops may be addressed in the short term through use of herbicide-resistant transgenic crops, herbicide-resistant weed biotypes will eventually develop after repeated applications of the same herbicide in a given field. A recent report of glyphosate resistance in rigid ryegrass (*Lolium rigidum*) occurred after repeated applications of glyphosate in an orchard to control grass weeds (Powles *et al.*, 1998). Even more serious is the development of weed biotypes with resistance to multiple herbicides that are widely used in row crop production (Foes *et al.*, 1998). As herbicide resistance becomes more problematic with many common weeds, strategies using bioherbicides will become more important in maintaining adequate weed control in conventional systems. The potential for successful use of bioherbicides in managing herbicide- resistant biotypes has been demonstrated where growth of an imazaquin-resistant common cocklebur biotype originating in soybean fields was suppressed with the mycoherbicide, *Alternaria helianthi* (Abbas and Barentine, 1995).

Bioherbicides and parasitic weeds

Many crop production regions throughout the world are infested with parasitic weeds that attack specific crop plants causing drastic yield reduction. There are no selective herbicides for satisfactorily controlling parasitic weeds, therefore, bioherbicides could be potentially highly effective on these weeds. Indeed, some of the most recent successful bioherbicides have been for control of dodders (*Cuscuta* spp.) in soybean and cranberry (Table 9.1). Also, a soil fungus under evaluation has suppressed germination and attachment of witchweed (*Striga* sp.) seedlings to grain sorghum roots and increased grain sorghum yield (Ciotola *et al.*, 1995). Development of this pathogen as a biotic agent could have a significant impact on food production in regions where *Striga* spp. are the dominant weed problem.

Bioherbicides and developing weed problem

Bioherbicides may also have a place in checking weed species that have not yet reached a competitive threshold level. An example might be their use against perennial weeds that are increasing under reduced-tillage farming systems. Shifts in weed composition in response to changes in crop practices, such as tillage, result in development of sub-competitive populations during the early years of the shift. The general tendency is for perennial species to increase as amount of tillage decreases. For example, common milkweed (*Asclepias syriaca*) infestations are increasing in the Midwestern United States due at least partially to reduced tillage. Although common milkweed occurs in wheat, corn and soybeans, it probably is not an important cause of reduced crop yields. Bioherbicides might provide a way of keeping the weed from becoming an economic weed problem. The discovery of a bacterial disease affecting common milkweed (Flynn and Vidaver, 1995) may lead to the development of the causal pathogen, *Xanthomonas campestris* pv. *asclepiadis*, as a biotic agent for maintenance of common milkweed stands below economic threshold levels. The disease is a systemic blight and reduces overall plant vigor and stand density. The report of milkweed bacterial blight is significant because an opportunity exists to test a "preventive biological control approach" and because of optimism for discovering similar agents on other perennial weeds for which no options for chemical control exist.

Bioherbicides in integrated weed management systems

An expanded and long-term approach to weed control is integrated weed management in which all available strategies including tillage, cultural practices, herbicides, allelopathy, and biological control are used to reduce the weed seedbank in soil, prevent weed emergence, and minimize competition from weeds growing with desired plants (Aldrich and Kremer, 1997). Like chemical herbicides, bioherbicides may be most effective as a component in an overall management program rather than as a single tactic approach. This may be the most promising situation for bioherbicides to be considered practical management options in cropping systems. When considered as a three-part system, weed management offers several opportunities for integration of bioherbicides at the critical stages during weed development: as seeds in soil, as growing and competitive plants, and during seed production (Aldrich and Kremer, 1997). Figure 9.1 illustrates how herbicides can be selected for attacking weeds at specific stages during the growing season in a total weed management system.

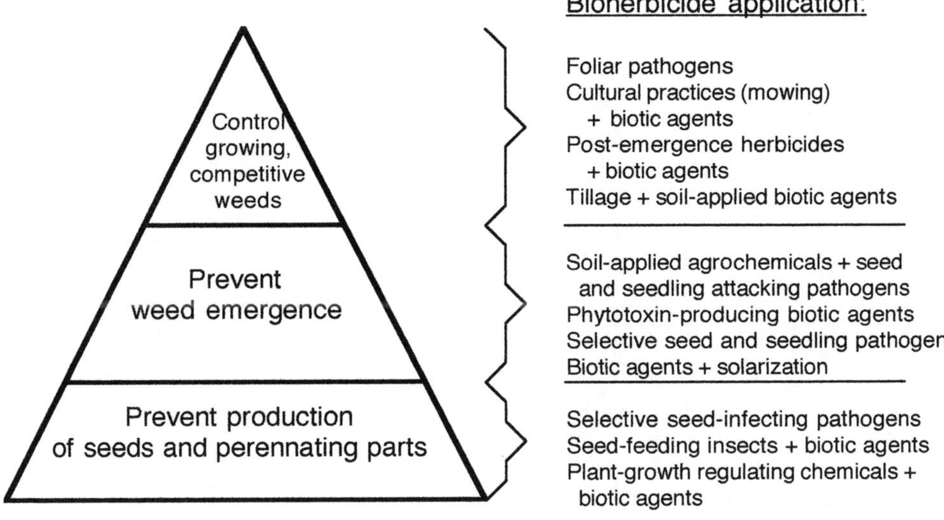

Figure 9.1 A three-part weed management system illustrating how bioherbicide components can be integrated in each management stage (Modified from Aldrich and Kremer, 1997).

Integrating bioherbicides with chemical herbicides

Several scenarios for integrating bioherbicides into weed management programs can be developed (Table 9.2). Since most biological control agents are specific toward one weed and most production fields contain several predominant weed species, the use of bioherbicides for control of a single species in conjunction with herbicides selected for control of other weeds present is a logical approach. Compatibility of the bioherbicide, *Fusarium solani* f. sp. *cucurbitae*, which controls Texas gourd (*Cucurbita texana*), a problem weed in soybean and cotton in the southern United States demonstrated that it could be integrated into a weed management strategy to broaden the spectrum of weed control within the crop (Weidermann and Templeton, 1988).

Integration with reduced rates of herbicides can successfully improve activity of mycoherbicides toward weeds. For example, *Phoma proboscis* was more effective in controlling field bindweed (*Convolvulus arvensis*) when combined with sub-lethal doses of 2,4-D than when applied alone (Heiny, 1994). The fungus *Colletotrichum gloeosporioides* f. sp. *malvae*, endemic on round-leaved mallow, provides adequate control (about 75 per cent kill) when applied alone as a bioherbicide (Grant *et al.*, 1990). Several chemical herbicides are only effective on round-leafed mallow in the early seedling stage. Combinations of the bioherbicide with several herbicides at recommended rates were evaluated for post-emergence control at the 4- to 5-leaf stage of growth. Tank mixes of the fungus with either metribuzin or imazethapyr greatly enhanced control and reduced biomass production over the fungus or the herbicide alone. These results clearly demonstrate that in some cases no single method is adequate for weed control and that combinations of methods are most effective.

Table 9.2 Scenarios for integrating bioherbicides into weed management strategies and the potential impact on weed management

Approach	*Agroecosystem*	*Impact on bioherbicide*
Combine with herbicides	Row crops; pastures; horticulture; IWMS[1]	Improve control of weeds difficult to control with herbicides
Apply to herbicide – resistant weed biotypes	Row crops; IWMS	Reduce potential increase in future resistant biotype populations
Combine with tillage	Row crops; IWMS, BWMS[2]	Reduce herbicide use; soil placement
Combine with cover crops, mulches	Row crops; Orchards; IWMS, BWMS	Increase weed control efficacy; reduce herbicide use
Apply to crop residues	Row, horticultural crops; IWMS, BWMS	Promote proliferation of biotic agent in soil, weed seed zone
Introduce on crop seeds	Row, horticultural crops; IWMS, BWMS	Optimize placement of agent for direct attack on weed seeds in soil
Apply soilborne pathogens, deleterious rhizobacteria	Row, horticultural, cereal crops; IWMS, BWMS	Attack weed seeds and seedlings; reduce seed bank in soil
Combine with crop varieties selected for competitiveness, allelopathy	Row, horticultural, cereal crops; IWMS, BWMS	Increase weed control efficacy; reduce herbicide use
Combine with plant-growth-regulating chemicals	Row, horticultural, cereal crops; IWMS	Reduce viable weed seed production, maintain seed banks at low densities
Combine soilborne pathogen with organic soil amendments	Row, horticultural crops; BWMS	Increase synthesis of phytotoxins, pre-emergence weed control
Combine with solarization	Horticultural crops; BWMS	Enhance weed seed and seedling elimination
Multiple agents (synergisms):		
Bioherbicides + classical agents (i.e. insects)	Pastures, rangeland IWMS, BWMS	Enhance rate of control by classical agents
Bioherbicides + seed-feeding insects	Row, horticultural crops; pastures, rangeland, forests; IWMS, BWMS	Enhance reduction of viable seeds; Manage weeds escaping early season control methods
Combine different bioherbicides	Row, horticultural crops; pastures, rangeland; BWMS	Enhance weed suppression and widen weed spectrum
Combine bioherbicide with natural pathogens	All systems; BWMS	Increase bioherbicide efficacy toward weed host

[1] IWMS: Integrated Weed Management Systems
[2] BWMS: Biological Weed Management Systems

Integrating bioherbicides with cultural practices

Cultural practices offer convenient application methods for integrating bioherbicides in cropping systems. Crop rotation is a practice that may also be manipulated to encourage development of specific inhibitory bacteria on weed rots. Tillage can influence the frequency of inhibitory bacteria occurring in soil and their growth-suppressive activity. Greater proportions of indigenous rhizobacteria inhibitory to downey brome and jointed goatgrass were detected under either conventional or reduced tillage compared to no-till. This finding suggests that application of selected deleterious rhizobacteria during tillage may be effective in integrated weed management (Kremer and Kennedy, 1996). Vegetative residues at or near the soil surface could serve as substrates for production of weed-suppressive chemicals by deleterious rhizobacteria applied as bioherbicides directly to the residues. Previous work reporting a rotation effect in corn was due partly to certain rhizobacteria specifically associated with corn roots illustrates the potential for using DRB to achieve suppression of weeds in crop rotation systems (Turco *et al.*, 1990). Increasing crop interference in the field by manipulating row spacing, seeding rates and other cultural practices to suppress early weed growth has been proposed as a viable component of integrated weed management (Jordan, 1993). Selection of highly competitive and allelopathic soybean varieties (Rose *et al.*, 1984) and matched with compatible bioherbicides may provide early-season weed suppression and require only minimal subsequent postemergence weed control.

Bioherbicide and management of seed banks and seedlings

Prevention of seed germination and seedling emergence is fundamental to maximally effective long-term weed management (Fig. 9.1). Thus, bioherbicides can play a significant role in reducing weed infestations by attacking seeds and seedlings before they become competitive with crop plants. Several approaches for managing the seed bank and seedling emergence have been described including direct application of biotic agents to soil or crop residues, or to crop seeds to prevent emergence of weeds in the crop seed-germinating zone, and in combination with solarization for enhancing seed deterioration in soil (Kremer, 1993). Also, certain agrochemicals known to stimulate seed imbibition or germination can be incorporated into soil combined with selected seed-attacking microorganisms in bioherbicidal preparations to kill germinating weed seeds (Kremer and Schulte, 1989).

Alternative agricultural systems

The current trends among enterprises that avoid the use of chemical herbicides and the restriction or banning of herbicides in areas considered environmentally sensitive favor the adoption of bioherbicides. For simplicity, the term alternative agriculture encompasses similar systems referred to as sustainable, natural, organic, biological, ecological and biodynamic farming systems. All of these involve a range of technological and management options to reduce costs, protect health and environmental quality, and enhance beneficial biological interactions and natural processes (National Research Council, 1989). In nearly all cases, little, if any, synthetic chemicals (including herbicides) are used. Alternative agriculture systems offer the greatest opportunities to study and refine non-chemical weed management (Liebman and Gallandt, 1997) yielding valuable information useful in developing improved herbicides and advancing their use in broader biologically based weed management systems. Because the demand for bioherbicides for alternative agriculture and ecosystem management is currently small relative to chemical herbicides, such products may be provided most efficiently through small-scale, specialized industries or even on-farm production facilities focused on "niche markets" (Auld and Morin, 1995; Charudattan, 1990).

Niche markets for bioherbicides

Charudattan (1990) indicated that the early successfully commercialized bioherbicides were indeed developed for special needs where chemical herbicides were unsatisfactory and suggests that considerable potential exists for other bioherbicide-weed combinations fitting the specialty designation. For cropping systems, a need exists for specialized bioherbicides effective on perennial and parasitic weeds for which there are no chemical herbicides. Likewise, bioherbicides for pasture weeds and poisonous plants such as the bracken fern (*Pteridium aquilinum*) (Womack *et al.*, 1996) would be of great potential use.The inevitable removal of methyl bromide from the pesticide arsenal will speed the development of bioherbicides for controlling weeds such as nutsedges in horticultural crops. Bioherbicides will be of significant value in managing weeds in areas where herbicides are not effective due to regulations that severely restrict or prohibit herbicide use and where preservation of the environment is the primary goal. These special situations include restoration of native ecosystems, wetlands, national parks, wildlife refuges and areas bordering waterways. For example, red alder (*Alnus rubra*) is a forest weed that interferes with timber production. Red alder infestations can be suppressed using a biocontrol fungus that is inoculated into the woody stems with a special injecting device (Dorworth, 1995). The fungus is useful for control of red alder along streams where herbicide application is prohibited and causes "slow killing" allowing slow release of nutrients from dying vegetation for use by desirable tree species as well as a gradual incursion of the crop trees.

Bioherbicides and biological weed management

Bioherbicides may be most effective in managing weeds as a component in a biological weed management system that is associated with alternative agriculture. Biological weed management involves the use of a diversity of biological agents including bioherbicides and biological approaches including allelopathy, crop competition, and other cultural practices to obtain similar dramatic reduction in weed densities as may be realized with chemical herbicides (Cardina, 1995). Examples of biological weed management approaches are listed in Table 9.2. Since alternative agriculture emphasizes pesticide-free crop production, the approach to weed control resembles a biological weed management system of which bioherbicides will likely be major components. Many of the approaches are similar to those for integrated weed management except that herbicides are not involved. Therefore, bioherbicides that would not be used in conventional integrated weed management because of unacceptable efficacy or too much time required for realizing an effect would be under practical use in biological weed management. For example, prevention of weed seed production and reduction of the seed bank could be attained using a bioherbicide consisting of seed pathogens applied to weeds occurring in a crop (Medd and Campbell, 1996). However, the impact of the bioherbicide would not be evident for one to two years until noticeable decreases in weed seedling densities occur due to the reduced seed bank size.

Cover crops and mulches as components of alternative management systems may be used for integrating bioherbicides by delivering the agents on seeds and promoting their establishment in soils for attack of weeds and seedlings prior to planting the main crop. Recent research demonstrated that several species inoculated with a DRB bioherbicide at planting maintained DRB populations on their roots and in soil and promoted colonization of giant foxtail (*Setaria faberi*) seedling roots in the early growing season of the main crop after the cover crop was terminated (Kremer, 1997). Combined effects of the DRB and allelopathic activity of the cover crop residues suppressed the growth of the weeds. Similarly, a "system management" approach where a crop is underseeded with a living green cover and treated with a post-emergence bioherbicide resulted in successful control of the target weed as well as the remaining weed flora by the cover crop (Pfirter *et al.*, 1997). Also, the agents in formulations applied at planting (Skipper *et al.*, 1996) can attack weed seeds and seedlings through delivery of bioherbicides to soil by either direct inoculation of crop seeds or by promoting colonization of crop roots. Crop roots not only may deliver microbial agents to adjacent roots of weeds but may also maintain or even enhance the agent's numbers for attack of seedlings emerging later in the season.

Alternative agricultural systems offer opportunities to explore the use of "synergisms" where the combined use of two or more methods enable bioherbicides to control weeds more effectively than when used alone (Gressel *et al.*, 1996; TeBeest, 1996). Bioherbicide efficacy on hemp sesbania (*Sesbania exaltata*) was increased by combining selected bacteria with the fungal pathogen, *Colletotrichum truncatum* (Schisler *et al.*, 1991). Combination of a *Colletotrichum* sp. bioherbicide with a naturally occurring rust fungus allows the bioherbicide to infect the weed host (*Xanthium* sp.) through rust lesions resulting in death of the plant (Morin *et al.*, 1993). A seed-feeding insect combined with seed-attacking fungi significantly decreased velvetleaf seed viability and seedling emergence and increased seed infection compared to either the insect or fungus alone (Kremer and Spencer, 1989). Pre-dispersal seed mortality of weeds escaping herbicide control may be effective in manipulating and reducing seed banks in soil. A practical application of soil-applied detrimental bacteria combined with insects would be in situations where the insect feeds on roots or crowns of target weeds (Kremer and Kennedy, 1996).

The very nature of high inputs of organic amendments and green manure in alternative agricultural systems promotes the ability of crops to compete more vigorously with weeds, which intuitively suggests that efficacy of bioherbicides would also be enhanced when used with these amendments (Gallandt *et al.*, 1998). Indeed the bioherbicidal fungus, *Gliocladium virens*, produces the phytotoxin viridol when grown in organic substrates such as peat and composts making this bioherbicide ideal for use in biological weed management (Jones and Hancock, 1990; Heraux and Weller, 1999). Furthermore, this fungus also produces fungicidal compounds effective against fungal plant pathogens suggesting the possibility of developing biotic agents with efficacy toward multiple pests.

Biorational approach for biological weed control

There is some question regarding whether compounds produced by microorganisms should be considered bioherbicides or not since the living organism is not applied to the target weeds. The objective of both biorational and biological weed control strategies is to suppress weed growth and adversely affect infestation by natural means. Natural compounds for use in weed management are of interest because they are highly effective against weeds, are not toxic against non-target organisms, cause no damage to the environment, and are readily biodegradable. Currently, no less than ten microbial herbicides have been discovered and developed for use in Japan (Okuda, 1992). Quimby and Birdsall (1995) argue that biorational compounds should indeed be referred to as bioherbicides since the herbicide "moiety" implies a chemical. They also propose current bioherbicides be redefined as "bioca", an acronym of biological control agent and to be used in a manner similar as pesticide is used for chemicals.

Conclusions

Despite apparent advances in biological control as a reliable strategy for weed management, little progress has been made in developing tactics for practical application in agro-ecosystems, especially those involving cropping systems. Efficacious strategies that target multiple weed species are needed. Best success in achieving this may well involve selection of several "core strains" of agents that are adapted to soils and climates in specific regions and are able to suppress growth of weeds comprising the dominant species at that site. To gain acceptance of these strategies, integration of biological control into current management systems is imperative so that the potential effectiveness of the agents can be demonstrated. Bioherbicides targeted for niche markets and for use in alternative agricultural systems will likely demonstrate the greatest effectiveness in biological weed management in the short term and generate impetus for continued discovery and development of bioherbicides for more widespread use. From a weed management standpoint, the integration of multiple tactics, including a diversity of potential bioherbicides and biologically-based approaches favors the effectiveness and stability required for long-term weed management (Cardina, 1995).

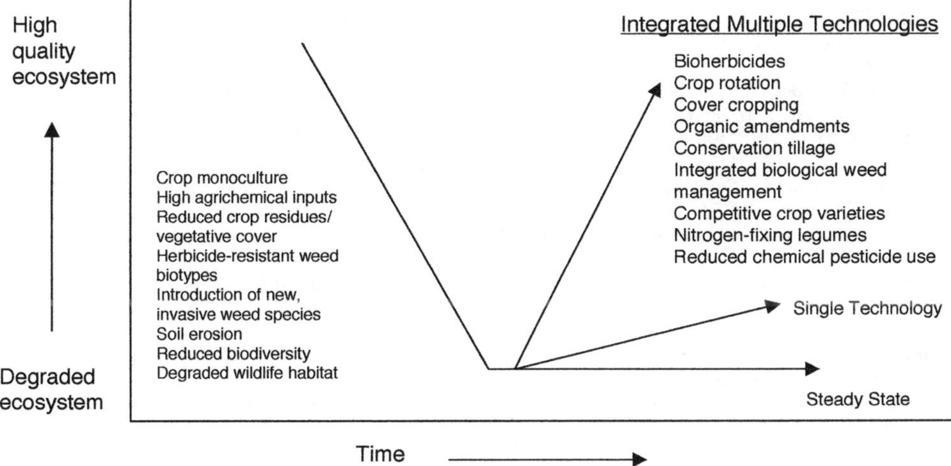

Figure 9.2 Successional process model for a generic ecosystem under agricultural use. Retrogression due to intensive conventional cropping results from combined effects of several factors including crop monoculture, continuous agrochemical inputs, development of herbicide-resistant weeds, and poor soil management. Retrogression leads to a steady state condition of low soil and organismal biodiversity and increased presence of herbicide-resistance and invasive weeds. Opportunities to remediate degraded ecosystems are determined by use of sustainable management technologies (bioherbicides, crop rotation, cover cropping, organic amendment, conservation tillage, etc.) in appropriate combinations and sequences to achieve acceptable weed management and profitable crop yields. Dependence on a single technology will likely result in a slow rate of recovery of ecosystem quality. Integration of different technologies including bioherbicides accelerates recovery toward a high quality ecosystem with undesirable weeds held in check (Modified from Masters *et al.*, 1996).

The integration of biological control into current systems also offers augmentative weed control options, as herbicide use becomes more restricted. It is well known that continued use of single herbicide control tactics favors resistance development in certain weed populations and conventional cropping systems. Bioherbicide technology used in appropriate integrated weed management in diversified cropping systems may aid in restoring fertility and productivity to degraded ecosystems and avoid the buildup of herbicide-resistant and invasive weeds (Fig. 9.2). Bioherbicides appropriately integrated in agricultural and environmental restoration systems can play a major role in reclaiming and restoring biodiversity to ecosystems degraded through continuous implementation of conventional cropping systems. Situations in which environmental quality can be restored in both ecologically sound farming systems and native habitats will benefit from use of effective bioherbicides.

References

Abbas, H.K., and Barrentine, W.L. (1995) *Alternaria helianthi* and imazaquin for control of imazaquin susceptible and resistant cocklebur (*Xanthium strumarium*) biotypes. *Weed Sci.*, **43**, 425–428.

Aldrich, R.J., and Kremer, R.J. (1997) *Principles in Weed Management*, Iowa State University Press, Ames, Iowa, USA.

Aref, S., and Pike, D.R. (1998) Midwest farmers' perception of crop pest infestation. *Agron. J.*, **90**, 819–925.

Auld, B.A., and Morin, L. (1995) Constraints in the development of bioherbicides. *Weed Technol.*, **9**, 638–652.

Boyetchko, S., Pedersen, E., Punja, Z., and Reddy, M. (1998) Formulations of biopesticides. In F.R. Hall and J.J. Menn (eds.), *Methods in Biotechnology, Vol. 5: Biopesticides: Use and Delivery*, Humana Press, Totowa, NJ, pp. 487–508.

Caesar-ThonThat, T.-C., Dyer, W.E., Quimby, P.C. Jr., and Rosenthal, S.S. (1995) Formulation of an endoarasitic nematode, *Subanguina picridis* Brzeski, a biological control agent for Russian knapweed, *Acroptilon repens* (l.) DC. *Biol. Contr.*, **5**, 262–266.

Cardina, J. (1995) Biological weed management. In A.E. Smith (ed.), *Handbook of Weed Management Systems*, Marcel Dekker Inc., New York, pp. 279–341.

Charudattan, R. (1990) Pathogens with potential for weed control. In R.E. Hoagland (ed.), *Microbes and Microbial Products as Microbial Herbicides*, American Chemical Society, Washington DC, pp. 133–154.

Charudattan, R. (1991) The mycoherbicide approach with plant pathogens. In D.O. TeBeest (ed.), *Microbial Control of Weeds*, Chapman & Hall, New York, pp. 24– 57.

Ciotola, M., Watson, A.K., and Hallet, S.G. (1995) Discovery of an isolate of *Fusarium oxysporum* with potential to control *Striga hermonthica* in Africa. *Weed Res.*, **35**, 303–309.

Dorworth, C.E. (1995) Biological control of red alder (*Alnus rubra*) with the fungus *Nectria ditissima. Weed Technol.*, **9**, 243–248.

Flynn, P., and Vidaver, A.K. (1995) *Xanthomonas campestris* pv. *asclepiadis*, pv. nov., causative agent of bacterial blight of milkweed (*Asclepias* spp.). *Plant Dis.*, **79**, 1176–1180.

Foes, M.J., Liiu, L., Tranel, P.J., Wax, L.M., and Stoller, E.W. (1998) A biotype of common waterhemp (*Amaranthus rudis*) resistant to triazine and ALS herbicides. *Weed Sci.*, **46**, 514–520.

Gallandt, E.R., Liebman, M., Corson, S., Porter, G.A., and Ullrich, S.D. (1998) Effects of pest and soil management systems on weed dynamics in potato. *Weed Sci.*, **46**, 238–248.

Grant, N.T., Prusinkiewicz, E., Mortensen, K., and Makowski, R.M.D. (1990) Herbicide interactions with *Colletotrichum gloeosporioides* f. sp. *malvae*, a bioherbicide for round leaved mallow (*Malwa pusilla*) control. *Weed Technol.*, **4**, 716–723.

Gressel, J., Amsellem, Z., Warshawsky, A., Kampel, V., and Michaeli, D. (1996) Biocontrol of weeds: Overcoming evolution for efficacy. *J. Environ. Sci. Health*, **B31**, 399–405.

Harley, K.L.S., and Forno, I.W. (1992) *Biological Control of Weeds: A Handbook for Practitioners and Students*, Inkata Press, Melbourne.

Heiny, D.K. (1994) Field survival of *Phoma proboscis* and synergism with herbicides for control of field bindweed. *Plant Dis.*, **78**, 1156–1164.

Heraux, F., and Weller, S.C. (1999) Development of a delivery system for field application of the biological weed control agent *Gilocladium virens. Proc. Weed Sci. Soc. Am.*, **39**, 73.

Johnson, D.R., Wyse, D.L., and Jones, K.L. (1996) Controlling weeds with phytopathogenic bacteria. *Weed Technol.*, **10**, 621–624.

Jones, R.W., and Hancock, J.G. (1990) Soilborne fungi for biological control of weeds. In R.E. Hoagland (ed.), *Microbes and Microbial Products as Microbial Herbicides*, Am. Chem. Soc., Washington DC, pp. 276–286.

Jordan, N. (1993) Prospects for weed control through crop interference. *Ecol Appl.*, **3**, 84–91.

Julien, M.H., and Griffiths, M.W. (1998) *Biological Control of Weeds. A World Catalogue of Agents and Their Target Weeds*, CAB International, Wallingford, UK.

Kennedy, A.C., Elliott, L.F., Young, F.L., and Douglas, C.L. (1991) Rhizobacteria suppressive to the weed downy brome. *Soil Sci. Soc. Am. J.*, **55**, 722–727.

Kremer, R.J. (1993) Management of weed seed banks with microorganisms. *Ecol Appl.*, **3**, 42–52.

Kremer, R.J. (1997) Weed distribution on a claypan soil affected by cover crops. *Proc. North Central Weed Sci. Soc. Am.*, **52**, 50–53.

Kremer, R.J., and Spencer, N.R. (1989) Interaction of insects, fungi, and burial on velvetleaf (*Abutilon theophrasti*) seed viability. *Weed Technol.*, **3**, 322–328.

Kremer, R.J., and Schulte, L.K. (1989) Influence of chemical treatment and *Fusarium oxysporum* on velvetleaf (*Abutilon theophrasti*) seed viability. *Weed Technol.*, **3**, 369–374.

Kremer, R.J., and Kennedy, A.C. (1996) Rhizobacteria as biological control agents of weeds. *Weed Technol.*, **10**, 601–609.

Liebman, M., and Gallandt, E.R. (1997) Many little hammers: Ecological management of crop-weed interactions. In L.E. Jackson (ed.), *Ecology in Agriculture*, Academic Press, San Diego, pp. 291–343.

Masters, R.A., Nissen, S.J., Gaussoin, R.E., Beran, D.D., and Stougaard, R.N. (1996) Imidazolinone herbicides improve restoration of Great Plains grasslands. *Weed Technol.*, **10**, 392–403.

Medd, R.W., and Campbell, M.A. (1996) A rationale for the use of a non-specific fungal seed pathogen to control annual grass-weeds of arable lands. In V.C. Moran and G.H. Hoffmann (eds.), *Proc. of IX Int. Symp. Biological Control of Weeds*, University of Capetown, Stellenbosch, South Africa, pp. 193–197.

Morin, L., Auld, B.A., and Brown, J.F. (1993) Synergy between *Puccinia xanthii* and *Colletotrichum orbiculare* on *Xanthium occidentale. Biol. Contr.*, **3**, 296–310.

Mortensen, K. (1998) Biological control of weeds using microorganisms. In G.J. Boland and L.D. Kuykendall (eds.), *Plant-Microbe Interactions and Biological Control*, Marcel Dekker Inc., New York, pp. 223–248.

National Research Council (1989) *Alternative Agriculture*, National Academy Press, Washington DC.

Okuda, S. (1992) Herbicides. In S. Omura (ed.), *The Search for Bioactive Compounds from Microorganisms*, Springer Verlag, New York.

Pfirter, H.A., Ammon, H.U., Guntli, D., Greaves, M. P., and Defago, G. (1997) Towards the management of field bindweed (*Convolvulus arvensis*) and hedge bindweed (*Calystegia sepium*) with fungal pathogens and cover crops. *Integrat. Pest Manage. Rev.*, **2**, 61–69.

Phatak, S.C., Summer, D.R., Wells, H.D., Bell, D.K., and Glaze, N.C. (1983) Biological control of yellow nutsedge with the indigenous rust fungus *Puccinia canaliculata. Science*, **219**, 1446–1447.

Powles, S.B., Lorraine-Colwill, D.F., Dellow, J.J., and Preston, C. (1998) Evolved resistance to glyphosate in rigid ryegrass (*Lolium rigidum*) in Australia. *Weed Sci.*, **46**, 604–607.

Prasad, R. (1996) Development of bioherbicides for integrated weed management in forestry. In H. Brown *et al.* (eds.), *Proc. 2nd Int. Weed Control Congress*, Department of Weed Control and Pesticide Ecology, Slagelse, Denmark, pp. 1197– 1203.

Quimby, P.C.Jr., and Birdsall, J.L. (1995) Fungal agents for biological control of weeds: Classical and augmentative approaches. In R. Reuveni (ed.), *Novel Approaches to Integrated Pest Management*, CRC Press Inc., Boca Raton, Florida, pp. 293–308.

Rose, S.J., Burnside, O.C., Specht, J.C., and Swisher, B.A. (1984) Competition and allelopathy between soybeans and weeds. *Agronomy J.*, **76**, 523–528.

Schisler, D.A., Howard, K.M., and Bothast, R.J. (1991) Enhancement of disease caused by *Colletotrichum truncatum* in *Sesbania exaltata* by coinoculating with epiphytic bacteria. *Biol. Contr.*, **1**, 261–268.

Skipper, H.D., Ogg, A.G.Jr., and Kennedy, A.C. (1996) Root biology of grasses and ecology of rhizobacteria for biological control. *Weed Technol*, **10**, 610–620.

TeBeest, D.O. (1996) Biological control of weeds with plant pathogens and microbial pesticides. *Adv. Agron.*, **56**, 105–113.

Turco, R.F., Bischoff, M., Breakwell, D.P., and Griffith, D.R. (1990) Contribution of soilborne bacteria to the rotation effect in corn. *Plant Soil*, **122**, 115–120.

Watson, A.K. (1993) *Handbook of Biological Control Agents for Weeds*, Weed Science Society of America, Champaign, IL.

Weidemann, G.J., and Templeton, G.E. (1988) Control of Texas gourd, *Cucurbita texana*, with *Fusarium solani* f. sp. *cucurbitae. Weed Technol.*, **2**, 271–274.

Womack, J.C., Burge, M.N., and Eccliston, G.M. (1996) Progress in formulation of a vegetable-oil-based invert emulsion for mycoherbicidal control of bracken, *Pteridium aquilinum.* In V.C. Moran and J.H. Hoffmann (eds.), *Proc. IX Int. Symp. Biological Control of Weeds*, University of Capetown, Stellenbosch, South Africa, pp. 535–539.

Wymore, L.A., Poirier, C., Watson, A.K., and Gotlieb, A.R. (1988) *Colletotrichum coccodes*, a potential bioherbicide for control of velvetleaf (*Abutilon theophrasti*). *Plant Dis.*, **72**, 534–538.

Zimdahl, R.L. (1993) *Fundamentals of Weed Science*, Academic Press, San Diego.

SPECIES INDEX

SUBJECT INDEX

CARDIFF UNIVERSITY
PRIFYSGOL CAERDYDD

MICROBIAL BIOPESTICIDES

Edited by

Opender Koul
Insect Biopesticide Research Centre, Jalandhar, India

and

G.S. Dhaliwal
Department of Entomology, Punjab Agricultural University, India

Over-dependence on chemical insecticides is a growing problem which could have devastating environmental and economic consequences. Insect pest management by the use of microbial pesticides is a widely accepted strategy to target this problem.

Biotechnological research has provided key developments in pest control agents, focusing on pathogens of insect pests as formulated biological pesticides. Emphasis has been placed on bacteria and viruses as they are well understood and easily manipulated.

This volume provides a comprehensive overview of the advances made in the use of bacteria, fungi and viruses, focusing on behavioural, chemical and molecular aspects. The authors discuss the potential of nematode-based biochemical agents and bioherbicides and explore the role of microbial biopesticides in integrated pest management and their prospects for commercial exploitation.

Microbial Biopesticides is an essential volume which provides up to date information to graduate students, research scientists and professionals in biological control, insect toxicology, biotechnology and microbial development.

About the Editors

Opender Koul, Fellow of the National Academy of Agricultural Sciences and the Indian Academy of Entomology, is an insect toxicologist/physiologist and currently the Director of the Insect Biopesticide Research Centre, Jalandhar, India. After obtaining his PhD in 1975 he joined the Regional Research Laboratory (CSIR), Jammu and then became Senior Group Leader of Entomology at Malti-Chem Research Centre, Vadodara, India (1980–1988). He has been a visiting scientist at the Universities of Kanazawa, Japan (1985–1986) and British Columbia, Canada (1988–1991). His extensive research experience concerns insect–plant interactions, spanning toxicological, physiological and agricultural aspects. Honoured with an Indian National Science Academy Medal and the Kothari Scientific Research Institute Award, he has authored over 100 research papers and articles, and is the editor of *Insecticides of Natural Origin* and *Phytochemical Biopesticides.* He has also been an informal consultant to BOSTID, NRC of USA and at ICIPE, Nairobi.

G.S. Dhaliwal, a fellow of the National Environmental Science Academy (NESA) and Entomological Society of India, is Professor of Ecology in the Department of Entomology at the Punjab Agricultural University, Ludhiana, India. Having completed his PhD in Entomology at the Indian Agricultural Research Institute, New Delhi in 1972, he was awarded the Gurprasad Pradhan Gold Medal and became a Post-doctoral Fellow at the International Rice Research Institute, Philippines, for two years. He has written extensively on different aspects of pest management, environment and sustainable agriculture. Honoured with best scientist award of NESA, he is the founding President of the Indian Society for the Advancement of Insect Science and the Society of Biopesticide Sciences, India; President of the Indian Ecological Society as well as Vice President of the Indian Society of Allelopathy and the Society of Pesticide Science, India.

Volume 2 of the book series **Advances in Biopesticide Research**